Advances in Environmental Sciences and Engineering

The Editors

Dr. Aditya Kishore Dash has M. Sc., M. Phil., M. Tech. (Environmental Sciences and Engineering) and Ph. D. (Environment), presently a faculty in Environmental Engineering at the Institute of Technical Education and Research (ITER) under S'O'A University, Bhubaneswar, Odisha. Before joining at ITER, Dr. Dash was the Professor and Head in Environmental Engineering at EAST Engineering College, Bhubaneswar. He has more than seventeen years of post graduate teaching and research experience in the area of Environmental Sciences and Engineering. He has more than eighteen research publications in the journals of both national and international repute and has attended more than fifteen national and international conferences. Dr. Dash has worked with a number of research projects including two major projects funded by the State Pollution Control Board, Odisha. His area of research interest includes biological treatment of wastewater and air pollution control study and its management. Dr. Dash has worked as the board of study member for different universities.

Dr. Mira Das is a professor in chemistry and presently the Associate Dean, School of Sciences, Humanities and Social Sciences, Institute of Technical Education and Research, Siksha 'O' Anusandhan University, Bhubaneswar. After completion of Master of Science in Chemistry from Ravenshaw College, she undertook research in a sensitive area of Environmental Chemistry. The focus of her research work was on heavy metal toxicity in rice and rice soils with special reference to chromium. On her successful work Utkal University conferred on her Ph. D. in 1990. In a teaching career spanning over 25 years she contributed to the curriculum of Chemistry and Environmental Engineering for engineering students under Utkal University, Biju Patnaik University of Technology and Siksha 'O' Anusandhan University. Her research focus has remained on Environmental Chemistry in which she has publications in journals of international and national repute. Her enthusiastic involvement in classroom teaching is quite motivating for the students.

Advances in Environmental Sciences and Engineering

— Editors —

Dr. Aditya Kishore Dash

Faculty in Environmental Engineering

Dr. Mira Das

Associate Dean, School of Sciences, Humanities and Social Sciences,

*Institute of Technical Education and Research (ITER),
Siksha 'O' Anusandhan University, Bhubaneswar, Odisha, India*

2015

Daya Publishing House®

A Division of

Astral International Pvt. Ltd.

New Delhi – 110 002

ISBN 978-93-5130-300-8

Published by   :   **Daya Publishing House®**
                          A Division of
                          Astral International Pvt. Ltd.
                          – ISO 9001:2008 Certified Company –
                          House No. 96, Gali No. 6,
                          Block-C, 30ft Road, Tomar Colony, Burari
                          New Delhi-110 084
                          E-mail: info@astralint.com
                          Website: www.astralint.com

Sales Office   :   4760-61/23, Ansari Road, Darya Ganj
                          New Delhi-110 002   Ph. 011-23245578, 23244987

Laser Typesetting   :   **Classic Computer Services**, Delhi - 110 035

Printed at   :   **Thomson Press India Limited**

PRINTED IN INDIA

राष्ट्रीय हरित अधिकरण
प्रधान न्यायपीठ, नई दिल्ली
**NATIONAL GREEN TRIBUNAL
PRINCIPAL BENCH**
Faridkot House, 1 Copernicus Marg,
New Delhi – 110 001

Prof. (Dr.) P.C.Mishra
Expert Member

Foreword

I am delighted to write this foreword to the present compilation of 30 research papers for the book "*Advances in Environmental Science and Engineering* " edited by Dr. Aditya Kishore Dash and Dr. Mira Das. The scope of Environmental Science and Engineering continues to expand in terms of new air and water quality problems encountered by the people and in terms of new pollutants which has both international and global impacts. Our ecological footprints weigh heavily on Earth's natural resources. Due to diligent of Environmental Engineers and Scientists, great progress has been made in our understanding of the fate and transport of substances that contaminate our air, surface water, soil and groundwater systems. Such progress has led to better technologies for controlling wastewater pollution, emissions and for cleaning up contaminated sites. Simultaneously, the continued rapid population growth and urbanization occurring in the developing countries of the world are causing unparalleled environmental health risk. Environmental Engineering and Technology play an increasingly important role in improving the plight of such people.

The present book contains quality research papers in the areas of the pollution status in air, water, soil and noise pollution and pollution control technologies with objectives of cost effectiveness and environmental compatibility. The papers embody materials which will be of immense use by the Scientists, Research students and Teachers engaged in research on the subject Environmental Science and Engineering.

I congratulate the Editors and the Publisher for bringing out such a book of present day relevance.

P.C. Mishra

Preface

In recent years, with growing concern for the environment and its management, the accumulated scientific and engineering concepts on environment have witnessed the emergence of the multidisciplinary subject *i.e.* Environmental Science and Engineering. Because of the interdisciplinary nature of the subject, sometimes, it poses a complex task for students, teachers, researches, scientists and engineers to keep track of the ongoing research in the field of Environmental Science and Engineering.

The book on "Advances in Environmental Sciences and Engineering" is a compilation of thirty insightful and exploratory research and review papers contributed by senior faculty members of some universities, scientific and engineering institutions in India and abroad. The sub themes are Environmental Pollution Monitoring and Control, Noise Pollution Modeling, Reclamation of Degraded and Contaminated Land, Environmental Toxicology, Modeling for Fate of Pesticides, Bioindicators and Biomarkers, Wastewater Treatment, Environmental Biotechnology, Climate Change, Application of Remote Sensing and GIS in Environmental Monitoring, Forest Ecology etc. All the papers are of relevance and have high application value in the context of contemporary industrial and mining activities as well as disposal of urban wastes without proper treatment. Presently, many researchers are working in the field of Environmental Science and Engineering. This book will be a handy reference book for all such researchers engaged in their respective fields.This book will also be useful for all researchrs in their pursuit of M. Phil, M. Tech. and Ph. D. degree in the field of Environmental Science and Engineering.

Dr. Aditya Kishore Dash
Dr. Mira Das

Contents

Chapter 1

Metals and Fluoride Dynamics in Fly Ash Mound under Plantation

P.C. Mishra[1]*, A.S.P. Mishra[2] and Suraj Tandon[3]

[1]*Professor Department of Environmental Sciences, Sambalpur University Jyoti Vihar, Sambalpur – 768 019, Odisha, India*
[2]*AGM, (Environment) Vedenta Aluminum, Jharsuguda, Odisha, India*
[3]*Suraj Tandon, Manager, Skipper Limited, Uluberia, Howrah – 711 303, West Bengal, India*

ABSTRACT

The present study was a part of a comprehensive research work on the vegetation succession and changes that are occurring in soil quality, litter decomposition, metal and fluoride dynamics etc. during the reclamation of ash mound of Hindalco Aluminium Industries, Hirakud under plantation strategy. Ash mound is located at a distance of 5 km from Hindalco Industries and it covers an area of around 103 acres. Three sampling sites were selected which includes 5year and 10year old plantation on ash mound and a natural site for comparison. The leachable metal and fluoride content in the top soil of the natural site was less as compared to 5 year and 10 year plantation sites on ash mound. The metal content (mg/kg) in natural site top soil was in the order of Fe (3104.89) > Mn (16.236) > Zn (14.21) > Ni (13.44) > Cd (8.120) > Cu (4.817) > and Cr (2.082). In 5 year plantation site the order was Fe (17215.08) > Zn (43.93) > Cu (36.72) > Mn (14.29) > Ni (14.09) > Cr (9.24) > Cd (8.81. In 10 year plantation site the metals were in the order of Fe (1334.59) > Cu (40.32) > Zn (40.03) > Mn (14.24) > Ni (14.04) > Cr (10.26) > Cd (4.34). Thus, the order of metal accumulation in top soil was more or less same in all the sites with presence of Fe in highest concentration in

* Present address: National Green Tribunal, Faridkot House, 1 Copernicus Marg, New Delhi – 110 001.

natural as well as ash mound sites. Metal content in top soil was relatively less in 10 year site compared to 5 year site. The fluoride content in the ash mound top soil were 52.0, 50.62 and 52.66 mg/kg respectively in natural, 5 year and 10 years plantation sites. Mobility of Cd has been recorded from soil to leaf of *Caesalpinia pulcherrima* with Biological Accumulation Coefficient (BAC) of 1.304. Rest of the plant species in 5 year site showed no bioaccumulation with BAC value less than one. Same is the case with 10 year and natural site, where no bioaccumulation was noted for different plant species. The BAC of Chromium in leaves of *Holarrhena antidysenterica* was 1.126, however, rest of the plant species in 5 year site have BAC less than one exhibiting no bioaccumulation of metal in leaves. Leaves of plants species from 10 year and natural site exhibited no bioaccumulation of heavy metal with all having BAC value less than one. Copper did not show a BAC of more than one in 5 year and 10 year plantation sites. However, tree vegetation from natural forest showed a remarkable increase in metal content, while moving from soil to leaf with the exception of *Butea monosperma*, the only plant species having BAC less than one. Fe also did not show any bioaccumulation in plant species from 5 year and 10 year plantation site. BAC value ranges from 0.009 to 0.028 and 0.017 to 0.024 in 5 year and 10 year site respectively. Same is also the case with natural forest site, where BAC ranged from 0.057 to 0.444. All the plants species sampled from 5 year site showed below detectable level of Mn in leaves, except *Acacia moniliform* is having Mn content of 7.15 with BAC of 0.244. *Delonix regia* from 10 year site showed a remarkable biomagnification in Mn content with BAC of 26.21 and the rest two species showed no such biomagnification. In natural forest site only *Mangifera indica* showed a slight biomagnification with BAC of 1.029 and rest of the species showed no bioconcentration of metals into leaves with BAC value less than one. Nickel behaved like manganese in the sense, it does not show any biomagnification as far as 5 year plantation is concerned. However, *Delonix regia* from 10 year site showed a noticeable increase in metal from top soil to leaf *i.e.* from 12.49 mg/kg in top soil to 332.48 mg/kg in leaf which clearly reveals a high mobility of Ni in case of *Delonix regia* as compared to other species grown in the ash mound. As far as natural site is concerned, except *Mangifera indica* with BAC of 1.245, no other plant species showed biomagnification of Ni. For Zinc, all the plant species from 5 year site have a BAC of less than one, indicative of no biomagnification of Zn from top soil to leaf except for *Holarrhena antidysenterica* whose BAC was 1.767. Plants species from 10 year site, did not show any such biomagnification with all the species having BAC of less than one. However, natural vegetation showed remarkable biomagnification in all the plant species with BAC value ranging from 1.527 in *Holarrhena antidysenterica* to 25.43 in *Polyalthia longifolia*. *Citrus medica* was the only species in natural site whose leaves showed no biomagnification (BAC = 0.626). BAC analysis of fluoride revealed that the plant species from 5 year site showed no biomagnification of F except in *Delonix regia*, where BAC was 2.372. Same was the case with 10 year plantation site where *Caesalpinia pulcherrima* and *Acacia moniliformis* showed no biomagnification with BAC value of 0.476 and 0.491 respectively but *Delonix regia* with BAC value of 1.442 showed biomagnification of F. However, natural site vegetation showed no bioaccumulation of F, with all species having BAC value less than one. Present study also indicates that among all the plant species analysed from natural and ash mound plantation sites, *Delonix regia and Polyalthia longifolia* showed remarkable biomagnification of metals such as Ni, Mn, Zn and Cu. This may be due nature of the plant species which make them resistance to metals.

Keywords: *Ash mound, Metals, Fluoride, Mobility, Biological accumulation coefficient.*

Introduction

Fly ash is that portion of the residue from the combustion of coal which enters in the flue gas stream and collected at precipitators (ESP). The other fraction contains coarser material called as bottom ash and is collected at the bottom of the furnace. About 75 per cent of the total ash production is fly ash. The constituents of fly ash are oxides of iron, silica, carbon, potassium, alumina, traces of sulphur, phosphorus, cadmium, chromium, lead, mercury, arsenic, zinc, boron, selenium, strontium etc. Indian ash contains calcium, sodium, magnesium and titanium along with the above said elements. Carbon and nitrogen are usually present in negligible quantities as they are likely to be oxidized into gaseous forms during combustion (Hodgson and Holiday, 1996). However, available phosphorus and potassium levels are generally higher as compared to the parent material *i.e.*, coal. Fly ash being an amorphous ferro-alumino silicate mineral contains all the naturally occurring elements which occur in coal (Klein *et al.*, 1975). The major matrix elements in fly ash are Si, Al and Fe (Natusch *et al.*, 1975). Generally fly ash contains higher concentration of all essential plant nutrients like sodium, potassium, calcium, magnesium, boron, sulphates and other nutrients except nitrogen in comparison to soil (Cope, 1962; Martens, 1970). The pH of fly ash varies from as low as 4.5 to as high as 12.0 (Plank and Martens, 1974). High sulphur content in coal produces ashes high in pH (Patel and Pandey, 1986). Mineralogical analysis of fly ash showed that between 70 to 90 per cent of the particles are glassy spheres, the remainder consists of quartz (SiO_2), mullite ($3Al_2O_3.2 SiO_2$), hematite (Fe_2O_3), magnetite (Fe_3O_4), a small portion of unburnt carbon (1-2 per cent) and gypsum ($CaSO_4.2H_2O$). Intensively investigated Indian fly ash reveals that 1000 tones of coal burnt in the city either in industries, boiler or domestic kitchen produces about 100 kg Pb, 40 kg Zn, 15 kg Cu, 8 kg Cd, 40 kg Ni, 50 kg Cr, 3 kg As etc. which ultimately found their way into the surrounding atmospheric and aquatic ecosystems (Sahu, 1990). Most of the toxic elements (*e.g.* As, Pb, Cd etc.) and hydrocarbons (like benzopyrine) present in the Indian coal and coal ash are well recognized carcinogens and potential risk of lungs and bone cancer of the population living in the adjoining affected areas (Prasad and Soman, 1990). Few additional toxic element (*e.g.* Co, Ni, Cu, Sb and Bi) other than indicated above, have been identified as very toxic and relatively accessible into the biological species (Wood, 1974) and thus causing a serious breakdown in the natural bio-geochemical cycles (Fulekar and Dave, 1986). A number of the trace elements are essential at low concentration to either plants or animals life but are toxic at higher concentration (Schwarz, 1974). Some of the trace elements like As, Cd, Pb and Se are more dangerous to man because they concentrate in particulate of respirable size range. Fly ash often contains high concentration of potentially toxic trace elements that damage the living and non living constituents in the environment (Block and Dams, 1975; Bose, 1982; Rainbow

1985, Patel and Pandey, 1987; Hopkin 1989). Some heavy metals such as Cd, Cr, Ni, Pb and Zn are enriched in fly ash while others lie Al, Mn, Fe, and Si have intermediate enrichment and few like Ca, Co, Cu, K etc. are present in low amount that is why, when waste products like fly ash containing toxic and hazardous trace elements are added to the soil it results in the solubilisation and leaching of heavy metals to surface and groundwater posing challenging problems in the form of usages, health hazards and environmental danger.

Study Site

Hirakud is small industrial town located at a distance of 16 km north of Sambalpur and at 21° 31′ north latitude and 82° 54′ east latitude. The climate of Hirakud is characterized by hot dry summer extending from March to mid-June followed by southwest monsoon rain from mid-June to September and the winter extends from October to February. The average annual rainfall varies from 1312-1662 mm. The air temperature varies from a minimum of 7-10°C during winter to a maximum of 45°C during summer. Hirakud power, operating within the Hirakud township, is a thermal power plant with present capacity of 367.5 MW. Electrostatic precipitator collects the fly ash in ground, which is disposed off through dry disposal method to ash mound. Bottom and fly ash generated by combustion of coal in the Hirakud Power Plant are conveyed to ash soil where it is temporarily stored. Fly ash is generated to the tune of 2000 tons per day and is transported to the ash mound through dumpers where disposal takes place adopting dry disposal method. After attaining the desired height plantations are taken up over the ash mound in order to prevent erosion and also to stabilize the system leading to the process of reclamation as tree plantation act as a tool for ash mound restoration by initiating the process of succession at a faster rate than natural one. Dumping of ash in the ash mound was started from November 1993 and has taken place phase-wise followed by plantation. Area of ash holding site is 6 acres with capacity of 47500 cubic meters.

Materials and Methods

Composite top soil samples (0-10 cm depth) were collected from each of the sites (Natural forest site, 5 year and 10 year old plantation ash mound sites). Similarly, leaf samples were also collected from selected tree species and the samples were processed for quantification of various metals *i.e.* Cd, Cr, Cu, Fe, Mn, Ni, Zn and Fluoride etc. Soil samples were collected within 10 feet radius of the tree from which leaf samples were collected for analysis. The soil and leaf sample were then analysed using Atomic Absorption Spectroscopy (AAS) for heavy metals and fluoride content was measured by fluoride analyzer after fusion.

Results

To assess the mobility of metals and fluoride from soil to leaf sample, selected plant species were analysed for heavy metals like Cd, Cr, Cu, Fe, Mn, Ni and Zn and fluoride. Analysis of ash mound soil indicated lead (Pb) below detectable limit.

Cadmium

Cadmium occurs in nature in association with zinc minerals. Growing plants acquire Zn and they also take up and concentrate Cd with the same biochemical set up. Cadmium is also released from particulate matter under the influence of increasing chloride concentration due to formation of chloride complexes (Elbay-Poulichet *et al.*, 1987). At high levels Cd causes kidney problems, anemia and bone marrow disorders. Figures 1.1a to 1.1f reveal the variation in cadmium content (mg/kg) in soil and leaf

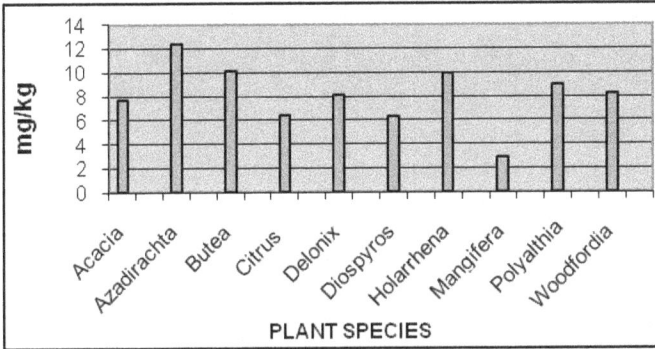

Figure 1.1a: Cd (mg/kg) in Top Soil of Natural Site

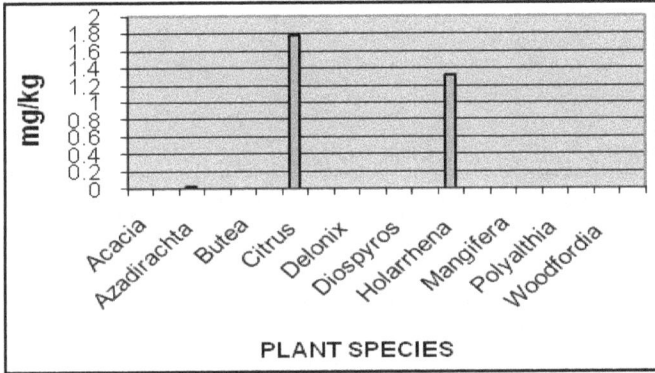

Figure 1.1b: Cd (mg/kg) in Leaves of Natural Site

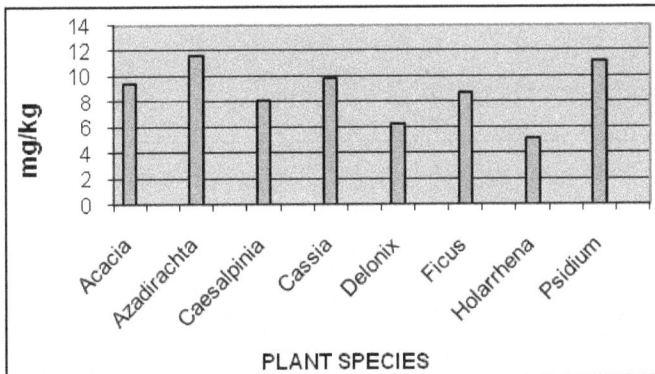

Figure 1.1c: Cd (mg/kg) in Top Soil of 5 Year Site

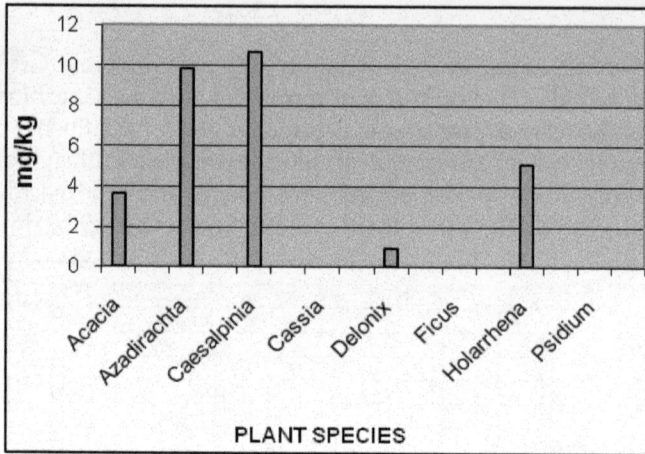

Figure 1.1d: Cd (mg/kg) in Leaves of 5 Year Site

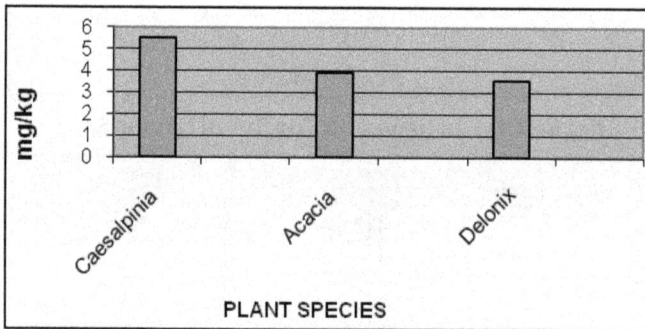

Figure 1.1e: Cd (mg/kg) in Top Soil of 10 Year Site

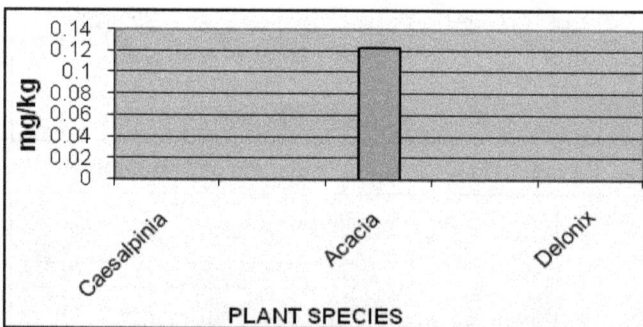

Figure 1.1f: Cd (mg/kg) in Leaves of 10 Year Site

of plant sp. in natural and ash mound sites. In natural site cadmium content (mg/Kg) in top soil and leaves of different plant species were found in the range of 2.92-12.37 and 0-1.78 respectively. Maximum was found in the top soil of *Azadirachta indica* (12.37) and minimum in *Mangifera indica* (2.92) where as Cd content in top soil and leaves of 5 year for different plant species site ranged from 5.10 to 11.658 and 0-10.658

respectively with maximum in soil near *Azadirachta indica* and minimum in *Holarrhena antidysenterica*. The Cd content in leaf was maximum in *Caesalpinia pulcherrima*. Cd concentration ranged from 3.56-5.53 in the top soil and 0-0.123 in the leaves of various plant species of 10 year site. Maximum Cd concentration in soil was found near *Caesalpinia pulcherrima* whereas for leaf it was in *Acacia moniliformis*.

Chromium

In soil, chromium (III) is relatively immobile due to its strong adsorption capacity on to soils. In contrast chromium (VI) is highly unstable and mobile, since it is poorly adsorbed on to soil under natural conditions. Figures 1.2a to 1.2f reveal the variation in chromium content in soil and leaves of plant species in natural and ash mound sites. Chromium content (mg/kg) ranged from 0.985-4.52 in the top soil of natural site with maximum near *Butea monosperma* (2.46) and minimum in *Azadirachta indica* (0.99). Leaves of plants species in natural site showed a range of 0.99-7.21 with maximum in *Polyalthia longifolia* and minimum in *Acacia moniliformis*. 5 year and 10 year plantation sites on ash mound showed a range of 1.2-12.35 and 6.5-14.5 of Cr content in top soil near different plant species present. Leaves of plant species in 5 year and 10 year site showed a range of 0.235-1.032 and 1.984-6.521 respectively with maximum Cr content in *Cassia angustifolia* and minimum in *Acacia moniliformis* for 5 year and for 10 year site. *Delonix regia* showed maximum Cr content while minimum was recorded in leaves of *Acacia moniliformis*.

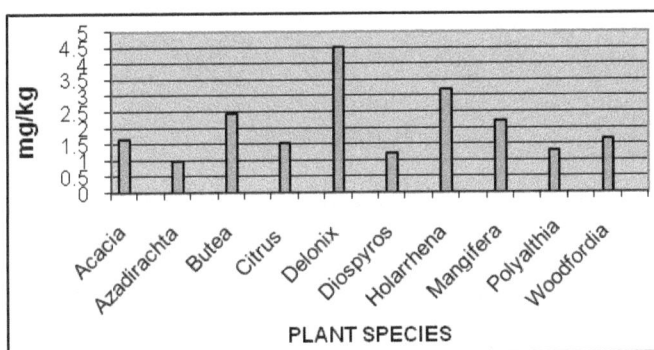

Figure 1.2a. Cr (mg/kg) in Top Soil of Natural Site.

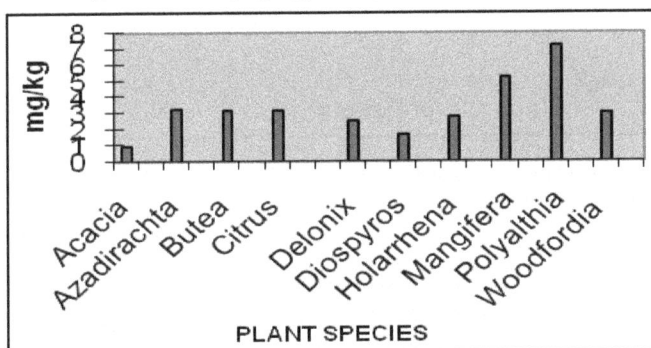

Figure 1.2b: Cr (mg/kg) in Leaves of Natural Site.

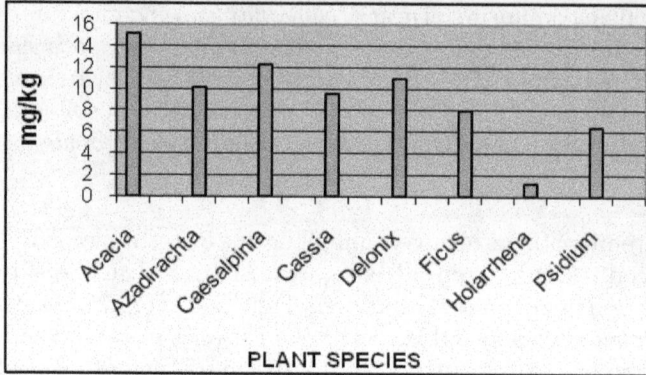

Figure 1.2c: Cr (mg/kg) in Top Soil of 5 Year Site.

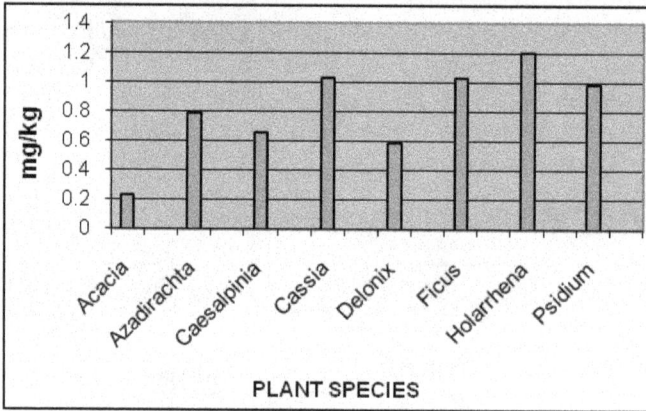

Figure 1.2d: Cr (mg/kg) in Leaves of 5 Year Site.

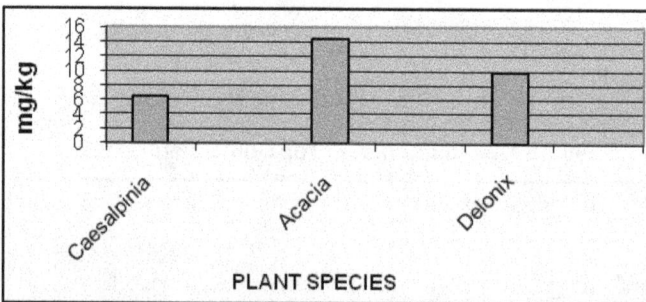

Figure 1.2e: Cr (mg/kg) in Top Soil of 10 Year Site.

Iron

Iron is abundantly present in the rocks in ferrous and ferric states. It is received both by plants and animal and has special role in the oxidation under anaerobic condition. Iron plays a very important role in human metabolism and its deficiency can lead to health disorders. The variation in iron content (mg/kg) in soil and leaves

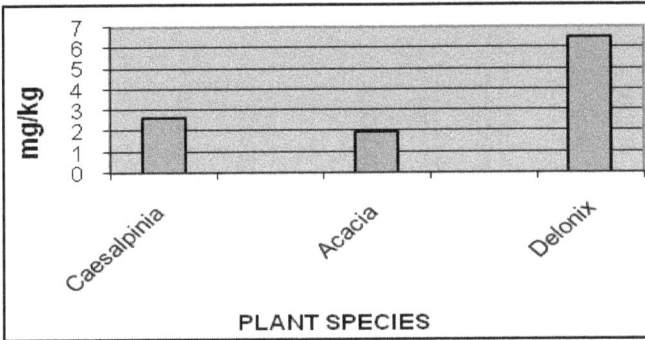

Figure 1.2f: Cr (mg/kg) in Leaves of 10 Year Site.

collected from both natural and ash mound site are presented in Figures 1.3a to 1.3f. The top soil collected from the base of different plant species at natural site showed Fe content in the range of 1079.61-4316.05. On the other hand leaves showed a range of 185.59-964.93 with maximum Fe content in the leaves of *Citrus medica* and minimum in *Mangifera indica*. In the same way 5 year ash mound site showed a range of 13633.98 - 22467.0 in top soil under tree vegetation and 176.24-14016.0 in the leaves of tree. Maximum Fe content was found in top soil of *Acacia moniliformis* (22467.46 mg/kg)

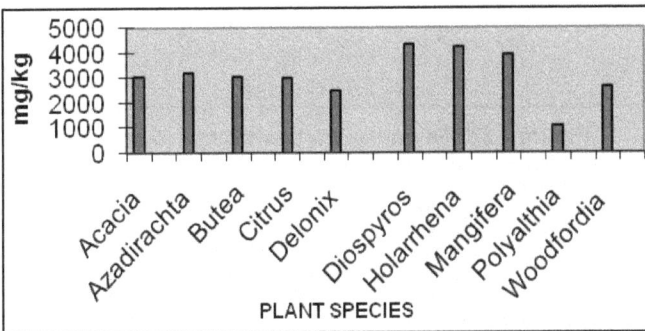

Figure 1.3a: Fe (mg/kg) in Top Soil of Natural Site.

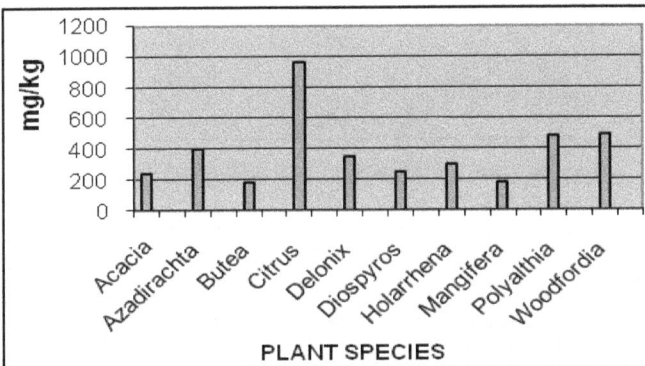

Figure 1.3b: Fe (mg/kg) in Leaves of Natural Site.

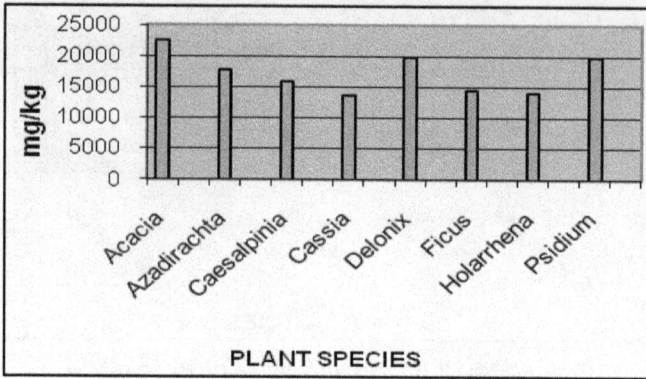

Figure 1.3c: Fe (mg/kg) in Top Soil of 5 Year Site.

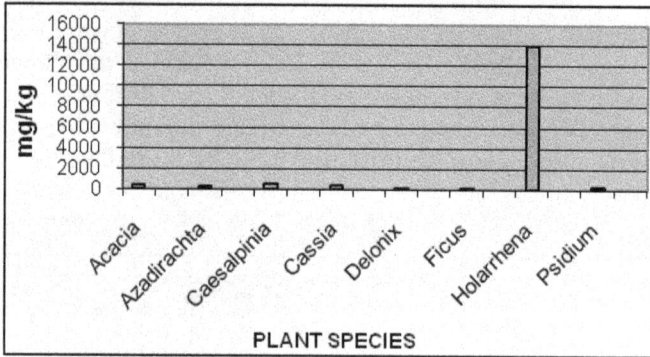

Figure 1.3d: Fe (mg/kg) in Leaves of 5 Year Site.

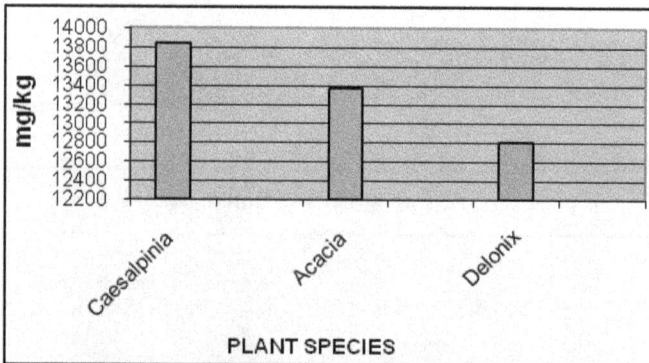

Figure 1.3e: Fe (mg/kg) in Top Soil of 10 Year Site.

and minimum in *Cassia angustifolia* (13633.99). 10 year ash mound site showed a variation in Fe content from 12801.3 to 13833 in top soil with maximum near *Caesalpinia pulcherrima* and minimum near *Delonix regia* and in case of leaves, Fe varied from 235.481-304.559 with maximum in *Delonix regia* where as minimum was observed in *Caesalpinia pulcherrima*.

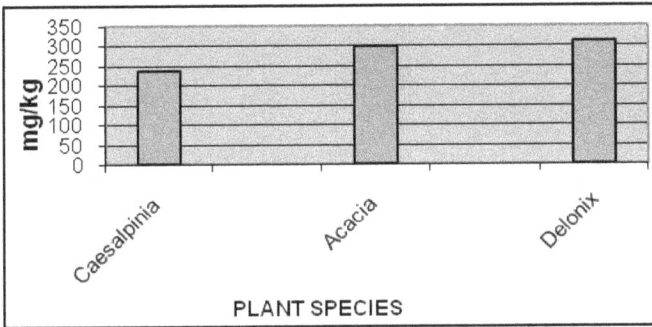

Figure 1.3f: Fe (mg/kg) in Leaves of 10 Year Site.

Zinc

Zinc in soil often remains strongly absorbed and in aquatic environments it will predominantly bind to suspended material before finally accumulating in the sediment. Variation in zinc content (mg/kg) in the natural and ash mound sites are presented in Figures 1.4a to 1.4f. Zn content varied from 1.24-40.97 in the top soil under tree vegetation in natural site with maximum value near *Citrus medica* and

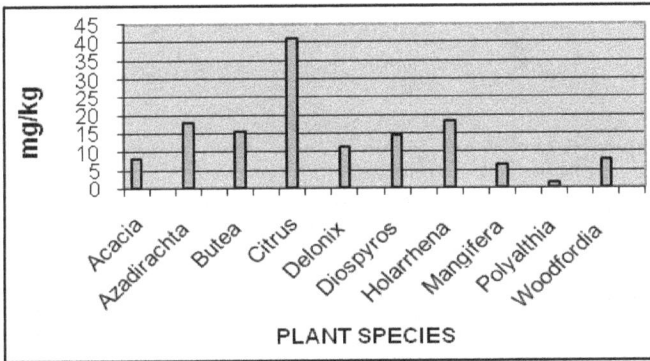

Figure 1.4a: Zn (mg/kg) in Top Soil of Natural Site.

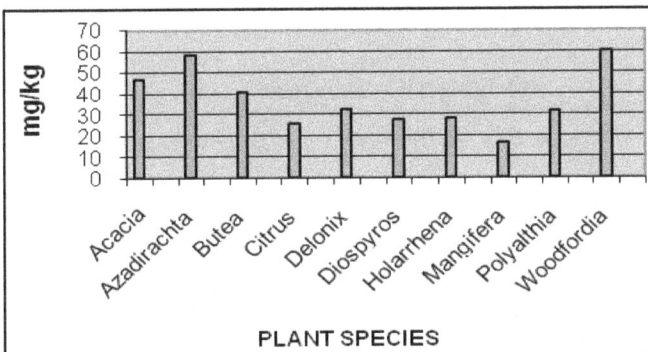

Figure 1.4b: Zn (mg/kg) in Leaves of Natural Site.

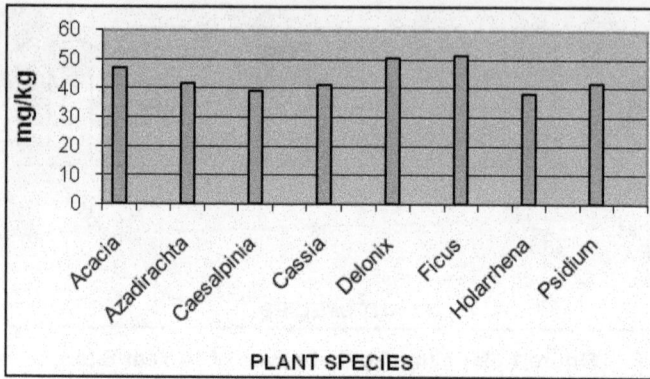

Figure 1.4c: Zn (mg/kg) in Top Soil of 5 Year Site.

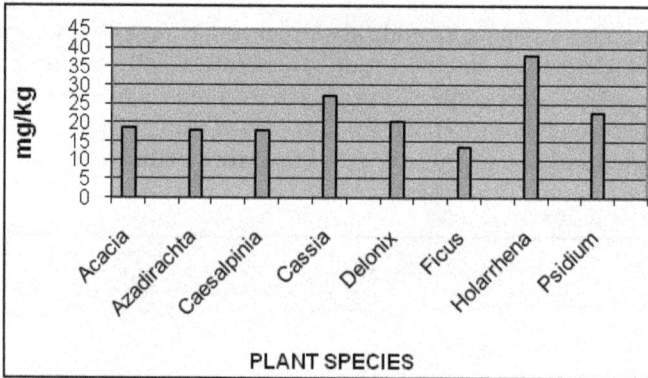

Figure 1.4d: Zn (mg/kg) in Leaves of 5 Year Site.

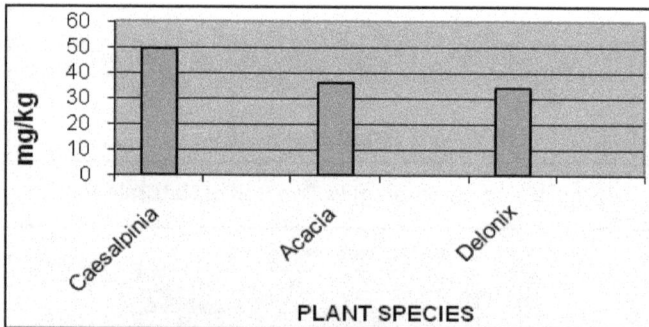

Figure 1.4e: Zn (mg/kg) in Top Soil of 10 Year Site.

minimum near *Polyalthia longifolia*. In leaf samples of natural site Zn showed a range of 16.28-60.408 mg/kg with maximum in *Woodfordia fruticosa* and minimum in *Mangifera indica*, whereas 5 year site showed a range of 38.29-51.36 in the top soil with maximum near *Ficus carica* and minimum near *Holarrhena antidysentrica*. On the other hand leaves of 5 year plants showed a range of 13.74-38.29 with maximum in *Holarrhena antidysentrica* and minimum in *Ficus carica*. 10 year old plantation on ash

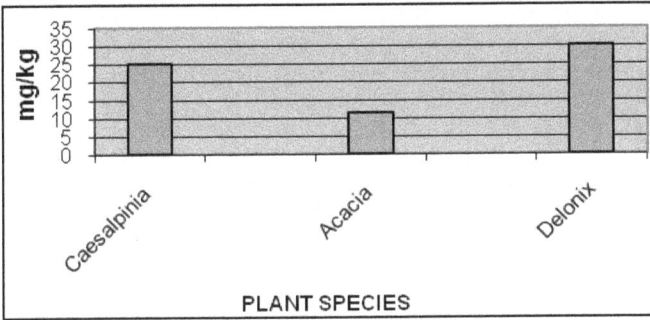

Figure 1.4f: Zn (mg/kg) in Leaves of 10 Year Site.

mound showed a range of 34.18 - 49.74 for top soil and 11.37 - 29.93 for leaves in tree vegetation. Zn content was maximum in the top soil near *Caesalpinia pulcherrima* (49.738) and minimum near *Delonix regia* (34.18).

Copper

Copper is very common substance that occurs naturally in the environment and spreads into the environment by natural phenomenon as well as human activities. Most copper compounds settles down and bounds to either water sediments or soil particles. When copper ends up in soil, it strongly attaches to organic matter and minerals as a result it does not travel very far after release and it hardly ever enters groundwater. Copper can interrupt the activity in soil, as it negatively influences the activity of microorganisms and earthworms. The variation in copper content (mg/kg) in soil and leafs collected from natural and ash mound sites are presented in Figures 1.5a to 1.5f. Copper content in top soil of natural site ranged from 0.675 -12.22 where as in leaves it varied from 4.74 -12.43. Maximum copper in top soil was found near *Citrus medica* and minimum near *Polyalthia longifolia*. For leaves maximum Cu content was shown by *Woodfordia fruticosa* and minimum by *Butea monosperma*. Likewise 5 years site showed a range of 21.24 -53.06 in the top soil collected from the base of different plants and for leafs, it varied from 3.93-21.00. Maximum Cu for top soil was found near *Acacia moniliformis* and for leaves it was *Psidium guaja*.10 years

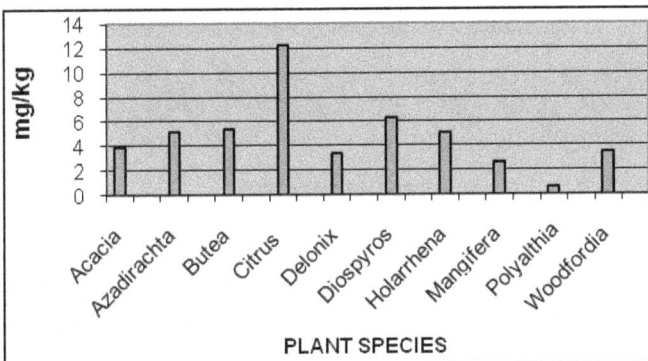

Figure 1.5a: Cu (mg/kg) in Top Soil of Natural Site.

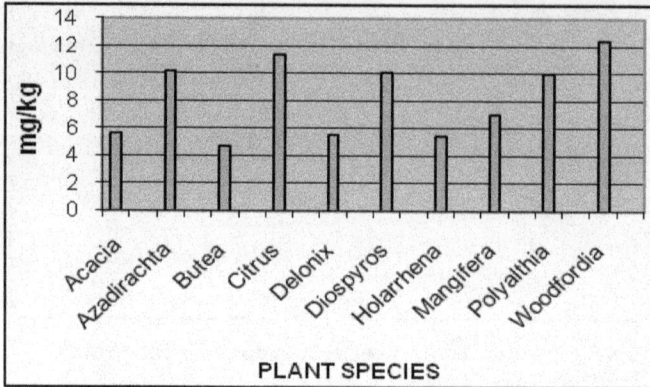

Figure 1.5b: Cu (mg/kg) in Leaves of Natural Site.

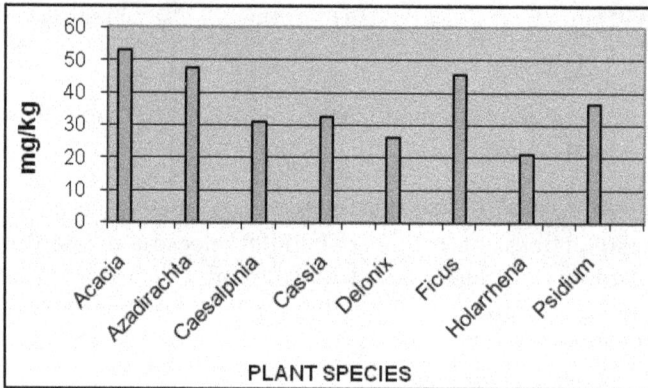

Figure 1.5c: Cu (mg/kg) in Top Soil of 5 Year Site.

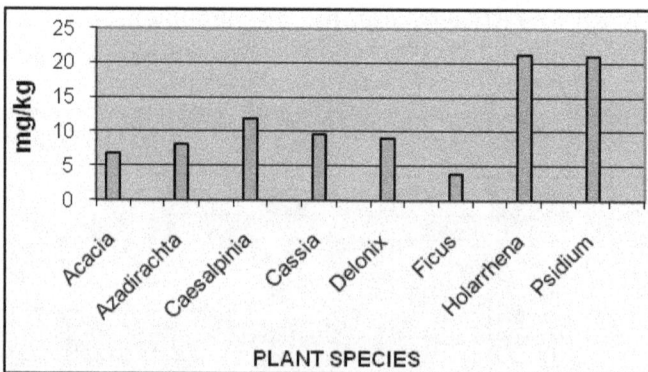

Figure 1.5d: Cu (mg/kg) in Leaves of 5 Year Site.

site showed a range of 33.13-53.4 for top soil with maximum near *Delonix regia* and minimum near *Caesalpinia pulcherrima*. Likewise for leaves copper content showed a variation of 5.49-12.52 with maximum in *Acacia moniliformis* and minimum in *Caesalpinia pulcherrima*.

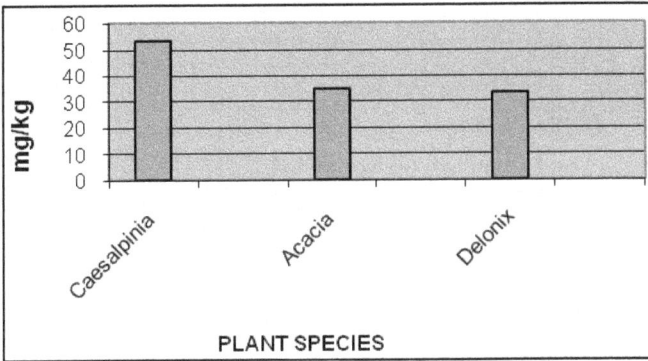

Figure 1.5e: Cu (mg/kg) in Top Soil of 10 Year Site.

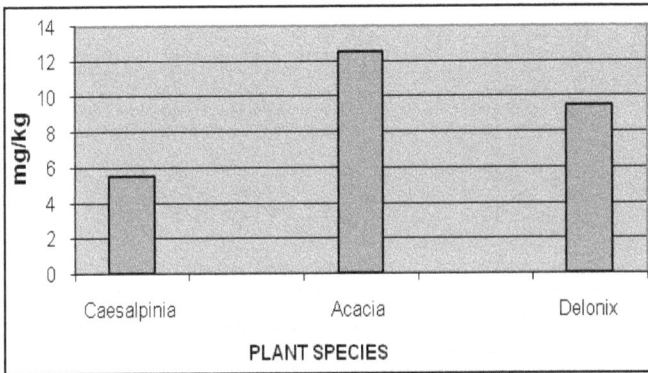

Figure 1.5f: Cu (mg/kg) in Leaves of 10 Year Site.

Manganese

Manganese is one of the most abundant metals in soil which occurs as oxides and hydroxides, and it cycles through its various oxidation states. Highly toxic concentrations of Mn in soil can cause swelling of cell walls, weathering of leaves and brown spots on leaves. The variation in manganese content (mg/kg) in soil and leaves collected from natural and ash mound sites are presented in Figures 1.6a to 1.6f. Top soil under different plants in natural site showed a variation of 1.0 to 37.77 in Mn content where as in leaves it varied from 0 to 15.21 with maximum near *Mangifera indica*. 5 year plantation site showed a variation of 3.32 -36.07 in top soil with maximum near *Psidium guaja* and minimum near *Holarrhena antidysentrica*. Mn content in leaves varied from 0 to 7.15 with maximum in *Acacia moniliformis*. 10 year plantation site exhibited a variation from 5.19 to 24.86 in top soil near different plant species with maximum near *Acacia moniliformis* and minimum near *Caesalpinia pulcherrima*. Mn content in leaves was found in the range of 0 - 332.012 mg/kg with maximum in *Delonix regia*.

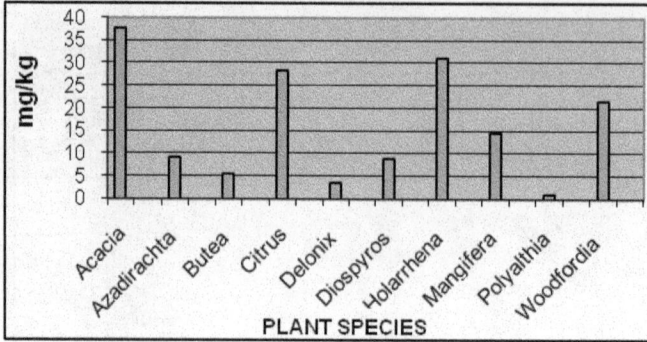

Figure 1.6a: Mn (mg/kg) in Top Soil of Natural Site.

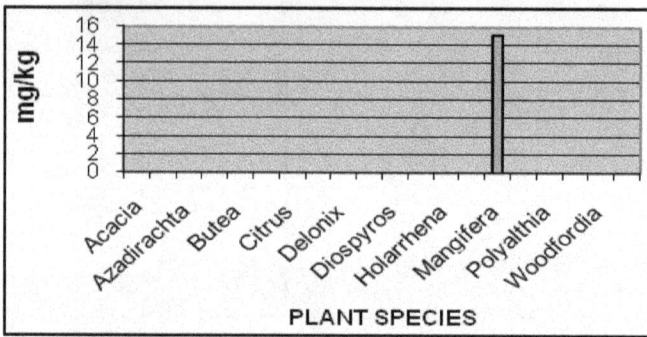

Figure 1.6b: Mn (mg/kg) in Leaves of Natural Site.

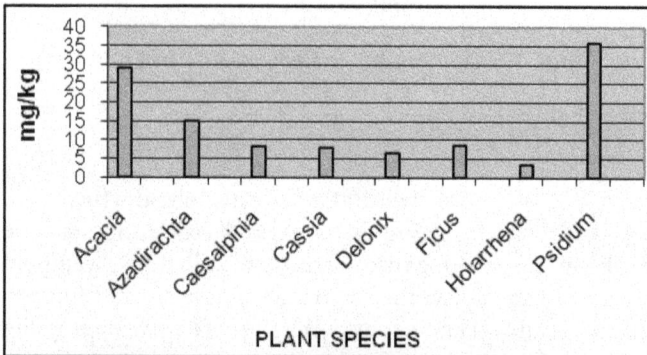

Figure 1.6c: Mn (mg/kg) in Top Soil of 5 Year Site.

Nickel

Variation in Nickel content (mg/kg) in soil and leaves collected from natural and ash mound sites are presented in Figures 1.7a to 1.7f. Nickel content in natural site top soil varied from 0.521 to 32.27 with maximum near *Acacia moniliformis* and minimum near *Butea monosperma* while for leaves it varied from 0 to 20.19 mg/kg with maximum in *Acacia moniliformis*. In 5 year site, ash mound top soil Ni content

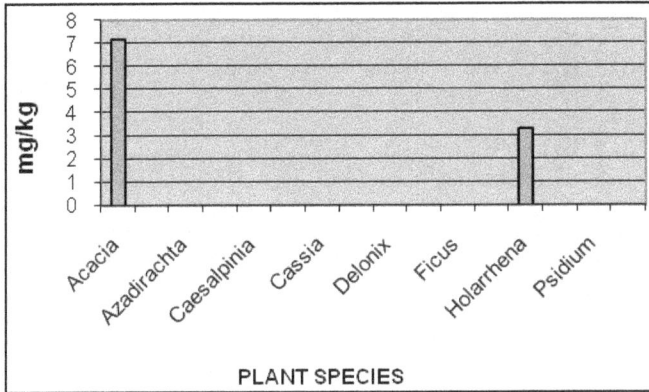

Figure 1.6d: Mn (mg/kg) in Leaves of 5 Year Site.

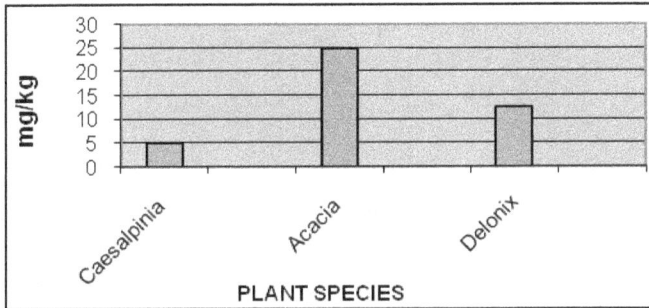

Figure 1.6e: Mn (mg/kg) in Top Soil of 10 Year Site.

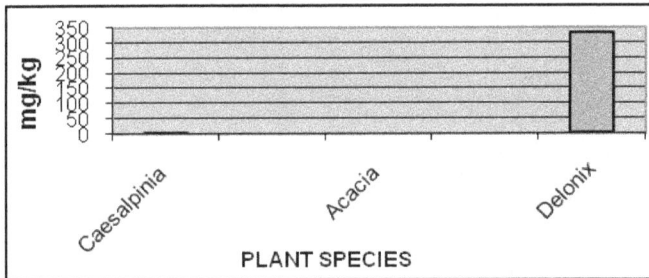

Figure 1.6f: Mn (mg/kg) in Leaves of 10 Year Site.

was found in the range of 3.27 - 35.57 near different plants where as in leaves it varied from 0 to 23.792 mg/kg. In the top soil as well as leaves of *Psidium guaja* Ni was found to be in maximum concentration. In the top soil of 10 year plantation site Ni content varied from 5.13 to 24.52 mg/kg with maximum near *Caesalpinia pulcherrima* where as for leaves it varied from 0 to 332.48 with maximum in *Delonix regia*.

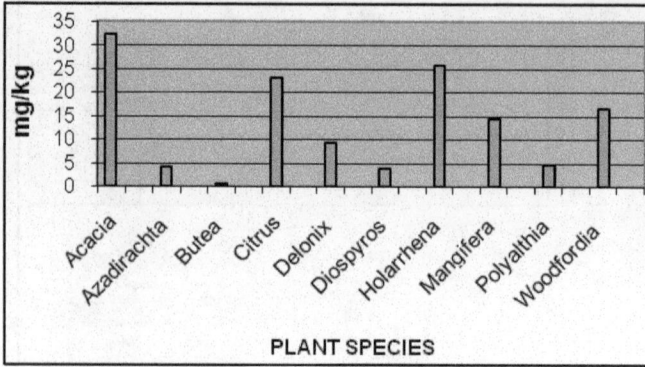

Figure 1.7a: Ni (mg/kg) in Top Soil of Natural Site.

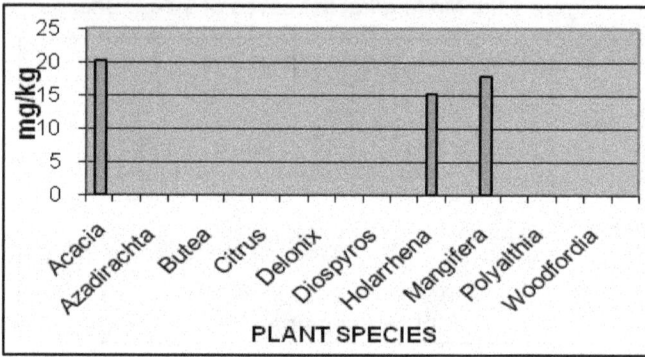

Figure 1.7b: Ni (mg/kg) in Leaves of Natural Site.

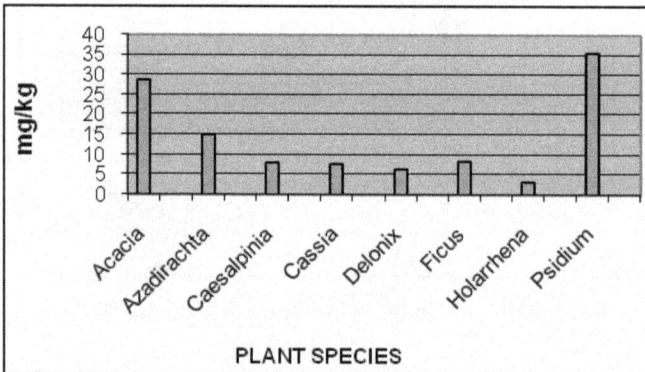

Figure 1.7c: Ni (mg/kg) in Top Soil of 5 Year Site.

Fluoride

Fly ash is a source of fluoride and soil receives fluoride when amended with flyash. Fluoride can cause many damaging effects not only to human beings but also to animals, aquatic creatures and vegetation. The variation in fluoride content (mg/kg) in soil collected from both top soil and leafs of natural and ash mound site are

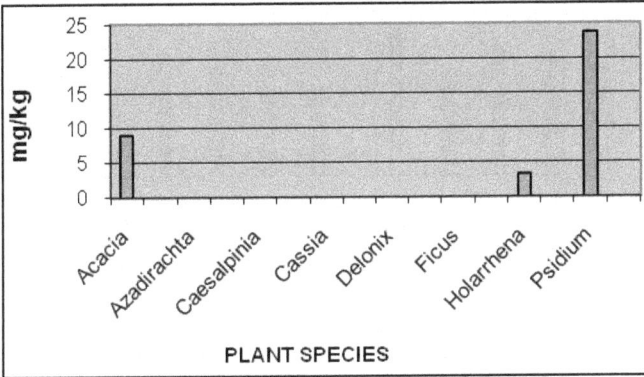

Figure 1.7d: Ni (mg/kg) in Leaves of 5 Year Site.

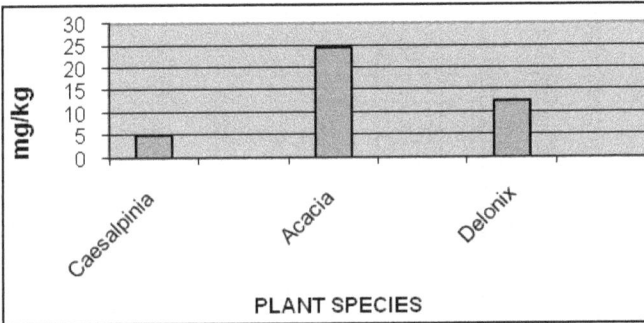

Figure 1.7e: Ni (mg/kg) in Top Soil of 10 Year Site.

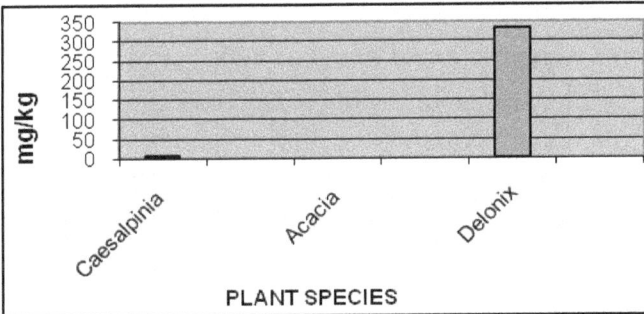

Figure 1.7f: Ni (mg/kg) in Leaves of 10 Year Site.

presented in Figures 1.8a to 1.8f. Top soil in natural site showed a variation of 40.0 to 65.0 F content with maximum near *Citrus medica* and minimum near *Woodfordia fruticosa*, whereas leaves showed a variation of 25.0-44.0 mg/kg with maximum in *Citrus medica* and minimum in *Woodfordia fruticosa*. 5 year plantation site on ash mound showed a range of 39.5 - 74.0 for top soil with maximum near *Psidium guaja* and minimum near *Azadirachta indica*. Leaves of tree vegetation in 5 year site depict a variation of 20.0 to 45.0 with maximum in *Acacia moniliformis* and minimum in *Ficus carica*. In case of 10 year old plantation F content in top soil near different plants

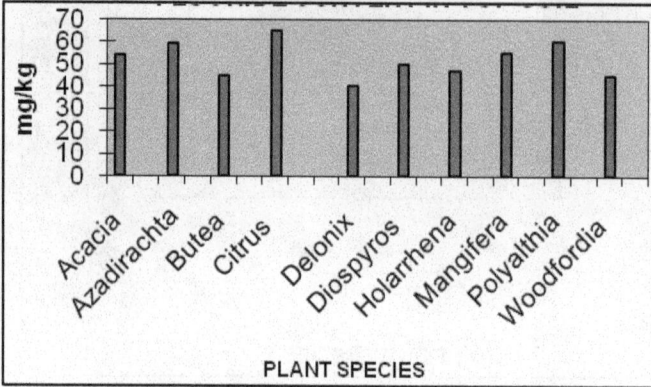

Figure 1.8a: F (mg/kg) in Top Soil of Natural Site.

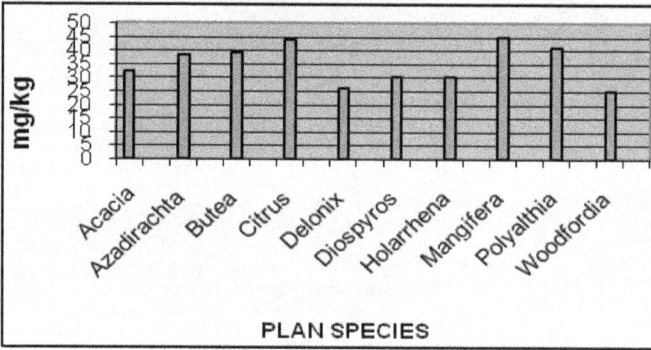

Figure 1.8b: F (mg/kg) in Leaves of Natural Site.

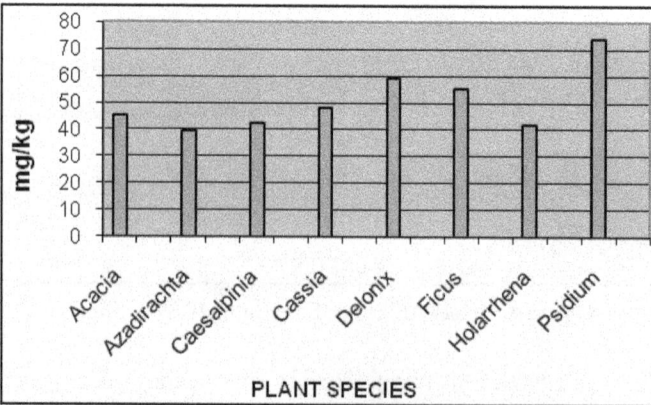

Figure 1.8c: F (mg/kg) in Top Soil of 5 Year Site.

varied from 42.0 to 64.0 with maximum near *Caesalpinia pulcherrima* and minimum near *Acacia moniliformis*. Leaves of tree vegetation showed a range of 20.0 - 75.0 mg/ kg with maximum in *Delonix regia* (75.0) and minimum in *Acacia moniliformis* (20.0).

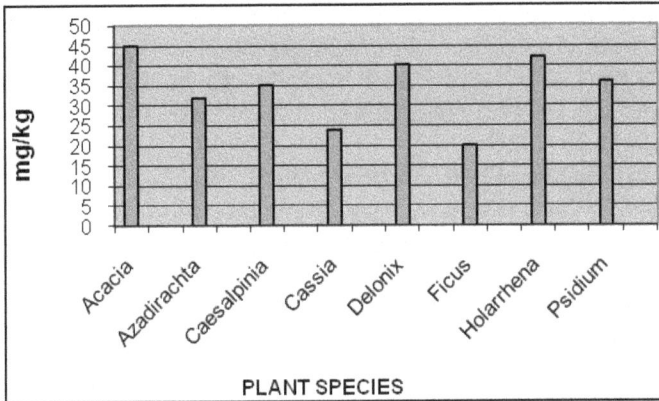

Figure 1.8d: F (mg/kg) in Leaves of 5 Year Site.

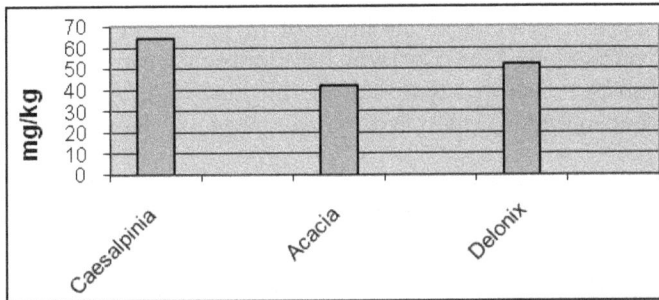

Figure 1.8e: F (mg/kg) in Top Soil of 10 Year Site.

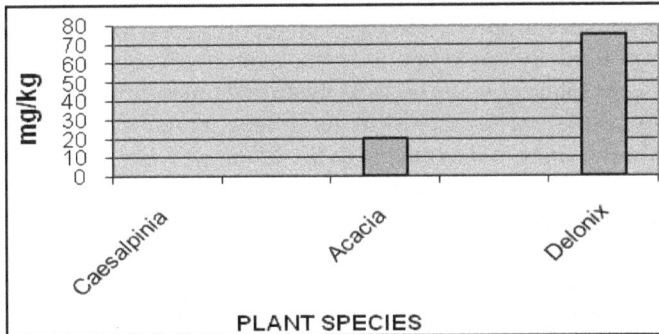

Figure 1.8f: F (mg/kg) in Leaves of 10 Year Site.

Discussions

The leachable heavy metals content in natural site top soil was estimated to be relatively less as compared to 5 year and 10 year plantation sites on ash mound which can be attributed to high organic matter content in natural site that chelatise the heavy metal and can also be due to enrichment of heavy metal during combustion of coal ultimately passing to fly ash (Anon, 2002).

As revealed from Tables 1.1–1.8, metal content (mg/kg) in natural site top soil occurs in the following order – Fe (3104.89±950.11) > Mn (16.24±12.80) > Zn (14.21±10.88) > Ni (13.44±10.80) > Cd (8.12±2.57) > Cu (4.82±3.06) > Cr (2.08±1.08). In 5 year site order is Fe (17215.08± 3258.58) > Zn (43.93±5.07) > Cu (36.72±11.05) > Mn (14.29±11.91) > Ni (14.09±11.75) > Cr (9.24±4.19) > Cd (8.81±2.27). In 10 year site the metals are in the order of Fe (1334.59±516.89) > Cu (40.32±13.34) > Zn (40.03±8.46) > Mn (14.24±9.92) > Ni (14.04±9.78) > Cr (10.26±4.02) > Cd (4.34±1.04).

As revealed from the above data the order of metal accumulation in top soil is more or less same in all the sites with Fe having maximum concentration in natural as well as fly ash amended soil. Metal content in top soil was noted to be relatively less in 10 year site as compared to 5 year site which is most probably due to high organic matter accumulation in 10 year site which after decomposition forms humus and Humic acid helps in chelating the metals as a result of which mobility metals in soil slows down with passage of time.

Fluoride content in soil is in the order of 10yr (52.66±11.01) > natural forest (52.0±7.93) > 5yr (50.62±11.57mg/kg) which is more or less same in all the three sites. Therefore, it can be concluded that with the passage of time as plantation grows older, the organic matter production can take care of problem of toxic metal in ash mound site.

Biological Accumulation Coefficient

It is a common conception that the total concentration of metals in soils are not a good indicator of phyto-availability, or a good tool for potential risk assessment due to different and complex distribution patterns of metals among various chemical species or solid phases (Wang and Sweigbarg, 1996). In most of the studies, the total content of the metal in soil and fly ash has been reported (Khan and Khan, 1996). The analysis of the potentially bio-available metal fraction is probably more significant than the analysis of total contents, because the former allows prediction of the risk of metal uptake by plants and its mobility in the system (Sims and Kline, 1991). The total metal present in the soil is not available to the plant grown therein and Diethylene Triamine Pentaacetic Acid (DTPA) extractable metals can be used as an indicator of bio-availability and toxicity of the metals.

Metal accumulation by plants can be evaluated using a simple index, termed as Biological Accumulation Coefficient (BAC). It is the ratio of leachable metal content in plant leaves to soil. Alloway (1990) has used the BAC ratio as total metal content in plants to total metal in soil to calculate the metal accumulation in plants.

BAC (Biological accumulation coefficient) =
Metal content in leaf/Leachable metal content in soil

Metals and Fluoride Accumulation in Plants and their Biological Accumulation Coefficient

Cadmium (Cd)

It is quite evident from Table 1.1 that biomagnification of Cd has taken place from soil to leaf of *Caesalpinia pulcherrima* with a BAC of 1.304. Rest of the plant species in 5 year site showed no biomagnification with BAC value less than one. Same is true with 10 year and natural site, where no biomagnification was noted for different plant species (Table 1.1).

Table 1.1: Cadmium (mg/kg) Dynamics in Ash Mound Plantation Sites and Natural Forest Site

Sl.No.	Tree Species	Top Soil	Leaf	BAC
	Five Year Plantation Site			
1.	Acacia moniliformis	9.455	3.64	0.384
2.	Delonix regia	9.86	0.915	0.092
3.	Caesalpinia pulcherrima	8.17	10.658	1.304
4.	Azadirachta indica	11.658	9.845	0.844
5.	Ficus carica	8.715	0	0
6.	Holarrhena antidysenterica	5.10	0	0
7.	Psidium guaja	11.270	0	0
8.	Cassia angustifolia	6.295	0	0
	Mean ± S.D	8.815 ± 2.273	3.132 ± 4.568	
	Ten Year Plantation Site			
1.	Caesalpinia pulcherrima	3.93	0.123	0.031
2.	Acacia moniliformis	5.53	0	0
3.	Delonix regia	3.565	0	0
	Mean ± S.D	4.342 ± 1.045	0.041 ± 0.071	
	Natural Forest Site			
1.	Mangifera indica	2.915	0	0
2.	Polyalthia longifolia	9.025	0	0
3.	Acacia moniliformis	7.66	0	0
4.	Diospyros melanoxylon	6.285	0	0
5.	Woodfordia fruticosa	8.254	0	0
6.	Delonix regia	8.123	0	0
7.	Azadirachta indica	12.37	0.023	0.001
8.	Butea monosperma	10.198	0	0
9.	Holarrhena antidysenterica	9.92	1.325	0.133
10.	Citrus medica	6.4587	1.778	0.276
	Mean ± S.D	8.120 ± 2.579	0.3126 ± 0.662	

Chromium (Cr)

Chromium showed biomagnification in leaves of *Holarrhena antidysenterica* with BAC of 1.126. However, rest of the plant species in 5 year site have BAC less than one exhibiting no bioconcentration of metals in leaves. Leaves of plants species from 10 year and natural site exhibited no bioconcentration of this toxic metal with all having BAC of less than one (Table 1.2).

Table 1.2: Chromium (mg/kg) Dynamics in Ash Mound Plantation Sites and Natural Forest Site

Sl.No.	Tree Species	Top Soil	Leaf	BAC
	Five Year Plantation Site			
1.	Acacia moniliformis	15.2	0.235	0.015
2.	Delonix regia	11.025	0.589	0.053
3.	Caesalpinia pulcherrima	12.352	0.658	0.053
4.	Azadirachta indica	10.2	0.789	0.077
5.	Ficus carica	7.98	1.025	0.128
6.	Holarrhena antidysenterica	1.2	1.352	1.126
7.	Psidium guaja	6.5	0.987	0.151
8.	Cassia angustifolia	9.5	1.032	0.108
	Mean ± S.D	9.244± 4.193	0.833± 0.342	
	Ten Year Plantation Site			
1.	Caesalpinia pulcherrima	14.5	1.984	0.136
2.	Acacia moniliformis	6.5	2.654	0.408
3.	Delonix regia	9.8	6.521	0.665
	Mean ± S.D	10.266±4.020	3.719±2.45	
	Natural Forest Site			
1.	Mangifera indica	2.21	5.231	2.366
2	Polyalthia longifolia	1.325	7.214	5.444
3.	Acacia moniliformis	1.65	0.987	0.598
4.	Diospyros melanoxylon	1.236	1.654	1.338
5.	Woodfordia fruticosa	1.659	2.987	1.8
6.	Delonix regia	4.52	2.546	0.563
7.	Azadirachta indica	0.985	3.214	3.262
8.	Butea monosperma	2.458	3.2	1.301
9.	Holarrhena antidysenterica	3.214	2.789	0.867
10.	Citrus medica	1.564	3.154	2.016
	Mean ± S.D	2.082±1.080	3.297±1.766	

Copper (Cu)

Plants species from 5 year and 10 year plantation site showed no biomagnification of the metal in leaves with all having BAC less than one. However, tree vegetation from natural forest showed a remarkable increase in metal content, with the exception of *Butea monosperma*, the only plant species having BAC less than one (Table 1.3).

Table 1.3: Copper (mg/kg) Dynamics in Ash Mound Plantation Sites and Natural Forest Site

Sl.No.	Tree Species	Top Soil	Leaf	BAC
	Five Year Plantation Site			
1.	Acacia moniliformis	53.06	6.838	0.128
2.	Delonix regia	26.465	9.23	0.348
3.	Caesalpinia pulcherrima	30.885	11.955	0.387
4.	Azadirachta indica	47.54	8.22	0.172
5.	Ficus carica	45.364	3.93	0.086
6.	Holarrhena antidysenterica	21.24	13.375	0.629
7.	Psidium guaja	36.656	21.005	0.573
8.	Cassia angustifolia	32.57	9.615	0.295
	Mean ± S.D	36.722±11.050	10.521±5.139	
	Ten Year Plantation Site			
1.	Caesalpinia pulcherrima	34.45	12.515	0.363
2.	Acacia moniliformis	53.4	5.495	0.102
3.	Delonix regia	33.13	9.495	0.286
	Mean ± S.D	40.326±11.341	9.168±3.521	
	Natural Forest Site			
1.	Mangifera indica	2.64	7.088	2.684
2.	Polyalthia longifolia	0.675	9.995	14.807
3.	Acacia moniliformis	3.925	5.64	1.436
4.	Diospyros melanoxylon	6.315	10.085	1.596
5.	Woodfordia fruticosa	3.455	12.425	3.596
6.	Delonix regia	3.375	5.6	1.659
7.	Azadirachta indica	5.12	10.24	2.0
8.	Butea monosperma	5.415	4.735	0.874
9.	Holarrhena antidysenterica	5.03	5.425	1.023
10.	Citrus medica	12.22	11.36	2.684
	Mean ± S.D	4.817±3.058	8.259±2.847	

Iron (Fe)

Iron also showed no biomagnification in plant species from 5 year and 10 year plantation site which is most probably due to very low mobility of iron from fly ash to leaves. BAC value ranged from 0.009 to 0.028 and 0.017 to 0.024 in 5 year and 10 year site respectively. Same is also the case with natural forest site, where BAC ranged from 0.057 to 0.444 (Table 1.4).

Table 1.4: Manganese (mg/kg) Dynamics in Ash Mound Plantation Sites and Natural Forest Site

Sl.No.	Tree Species	Top Soil	Leaf	BAC
	Five Year Plantation Site			
1.	Acacia moniliformis	22467.46	360.827	0.016
2.	Delonix regia	19775.1	185.403	0.009
3.	Caesalpinia pulcherrima	15807.039	470.652	0.029
4.	Azadirachta indica	17855.313	215.064	0.012
5.	Ficus carica	14371.453	176.236	0.012
6.	Holarrhena antidysenterica	14016.275	2610.55	0.186
7.	Psidium guaja	19794.02	230.863	0.011
8.	Cassia angustifolia	13633.985	385.364	0.028
	Mean ± S.D	17215.08±3258.588	579.369±827.631	
	Ten Year Plantation Site			
1.	Caesalpinia pulcherrima	13368.381	299.017	0.022
2.	Acacia moniliformis	13833.76	235.481	0.017
3.	Delonix regia	12801.63	309.559	0.024
	Mean ± S.D	1334.59±516.894	281.352±40.073	
	Natural Forest Site			
1.	Mangifera indica	3945.933	185.588	0.047
2.	Polyalthia longifolia	1079.613	480.010	0.444
3.	Acacia moniliformis	3070.373	241.284	0.078
4.	Diospyros melanoxylon	4316.047	250.275	0.057
5.	Woodfordia fruticosa	2639.252	495.297	0.187
6.	Delonix regia	2501.356	353.942	0.141
7.	Azadirachta indica	3189.336	396.286	0.124
8.	Butea monosperma	3050.088	186.246	0.061
9.	Holarrhena antidysenterica	4223.351	301.65	0.071
10.	Citrus medica	3033.631	964.933	0.318
	Mean ± S.D	3104.898±950.113	385.551±231.810	

Manganese (Mn)

All the plants species sampled from 5 year site showed no Mn in leaves, except *Acacia moniliformis* having Mn content of 7.145 with BAC value of 0.244. *Delonix regia* from 10 year site showed a remarkable biomagnification of Mn with BAC of 26.21 and the rest two species showed no such biomagnifications. In natural forest site biomagnifiction was only seen in *Mangifera indica* showed having BAC of 1.029 and rest of the species showed no bioaccumulation with BAC less than one (Table 1.5).

Table 1.5: Manganese (mg/kg) Dynamics in Ash Mound Plantation Sites and Natural Forest Site

Sl.No.	Tree Species	Top Soil	Leaf	BAC
	Five Year Plantation Site			
1.	Acacia moniliformis	29.186	7.145	0.244
2.	Delonix regia	6.449	0	0
3.	Caesalpinia pulcherrima	8.111	0	0
4.	Azadirachta indica	15.101	0	0
5.	Ficus carica	8.381	0	0
6.	Holarrhena antidysenterica	3.321	0	0
7.	Psidium guaja	36.065	0	0
8.	Cassia angustifolia	7.746	0	0
	Mean ±S.D	14.295±11.918	0.893±2.526	
	Ten Year Plantation Site			
1.	Caesalpinia pulcherrima	24.86	0	0
2.	Acacia moniliformis	5.197	0.548	0.105
3.	Delonix regia	12.667	332.012	26.21
	Mean ±S.D	14.241±9.925	110.853±191.529	
	Natural Forest Site			
1.	Mangifera indica	14.595	15.021	1.029
2.	Polyalthia longifolia	1.0	0	0
3.	Acacia moniliformis	37.772	0	0
4.	Diospyros melanoxylon	8.967	0	0
5.	Woodfordia fruticosa	21.745	0	0
6.	Delonix regia	3.745	0	0
7.	Azadirachta indica	9.252	0	0
8.	Butea monosperma	5.557	0	0
9.	Holarrhena antidysenterica	31.293	0	0
10.	Citrus medica	28.439	0	0
	Mean ±S.D	16.236±12.807	1.502±4.750	

Nickel (Ni)

Nickel behaved like manganese in the sense, it does not show any biomagnification as far as 5 year plantation is concerned. However, *Delonix regia* from 10 year site showed a noticeable increase in metal from top soil to leaf *i.e.* from 12.49 in top soil to 332.48 mg/kg in leaf which clearly reveals that mobility of Ni is more in case of *Delonix regia* as compared to other species available in the ash mound. As far as natural site vegetation is concerned, except *Mangifera indica* with BAC of 1.245, no other plant species showed biomagnification of Ni (Table 1.6).

Table 1.6: Nickel (mg/kg) Dynamics in Ash Mound Plantation Sites and Natural Forest Site

Sl.No.	Tree Species	Top Soil	Leaf	BAC
	Five Year Plantation Site			
1.	*Acacia moniliformis*	28.78	9.041	0.314
2.	*Delonix regia*	6.365	0	0
3.	*Caesalpinia pulcherrima*	8.0	0	0
4.	*Azadirachta indica*	14.894	0	0
5.	*Ficus carica*	8.266	0	0
6.	*Holarrhena antidysenterica*	3.274	0	0
7.	*Psidium guaja*	35.57	23.792	0.667
8.	*Cassia angustifolia*	7.64	0	0
	Mean ± S.D	14.098±11.754	4.104±8.561	
	Ten Year Plantation Site			
1.	*Caesalpinia pulcherrima*	24.519	0	0
2.	*Acacia moniliformis*	5.125	6.77	1.32
3.	*Delonix regia*	12.493	332.484	26.613
	Mean ± S.D	14.045±9.789	113.084±190.035	
	Natural Forest Site			
1.	*Caesalpinia pulcherrima*	14.353	17.881	1.245
2.	*Acacia moniliformis*	4.711	0	0
3.	*Delonix regia*	32.274	20.1876	0.625
4.	*Caesalpinia pulcherrima*	3.864	0	0
5.	*Acacia moniliformis*	16.467	0	0
6.	*Delonix regia*	9.179	0	0
7.	*Caesalpinia pulcherrima*	4.146	0	0
8.	*Acacia moniliformis*	0.521	0	0
9.	*Delonix regia*	25.884	15.111	0.583
10.	*Caesalpinia pulcherrima*	23.069	0	0
	Mean ± S.D	13.446±10.809	5.317±8.646	

Zinc (Zn)

Table 1.7 reveals that Zn in plants species from 5 year site has a BAC of less than one, which means no biomagnification of Zn from top soil to leaf except for *Holarrhena antidysenterica* whose BAC is 1.767. Plants species from 10 year site did not show any such biomagnification with all having BAC less than one. However, natural vegetation showed remarkable biomagnification of Zn in all the plant species with BAC ranging from 1.53 in *Holarrhena antidysenterica* to 25.43 in *Polyalthia longifolia*. *Citrus medica* was the only species in natural site whose leaves showed no biomagnification of Zn (BAC = 0.626).

Table 1.7: Zinc (mg/kg) Dynamics in Ash Mound Plantation Sites and Natural Forest Site

Sl.No.	Tree Species	Top Soil	Leaf	BAC
	Five Year Plantation Site			
1.	Acacia moniliformis	46.911	18.916	0.403
2.	Delonix regia	50.747	20.307	0.4
3.	Caesalpinia pulcherrima	39.254	17.912	0.456
4.	Azadirachta indica	41.836	18.072	0.431
5.	Ficus carica	51.357	13.743	0.267
6.	Holarrhena antidysenterica	38.288	67.655	1.767
7.	Psidium guaja	41.857	22.857	0.546
8.	Cassia angustifolia	41.24	27.244	0.66
	Mean ± S.D	43.936±5.072	25.838±17.351	
	Ten Year Plantation Site			
1.	Caesalpinia pulcherrima	36.184	11.365	0.314
2.	Acacia moniliformis	49.738	25.06	0.503
3.	Delonix regia	34.176	29.93	0.875
	Mean ± S.D	40.032±8.464	22.118±9.625	
	Natural Forest Site			
1.	Caesalpinia pulcherrima	6.36	16.284	2.56
2.	Acacia moniliformis	1.235	31.4	25.425
3.	Delonix regia	7.989	46.897	5.87
4.	Caesalpinia pulcherrima	14.557	27.48	1.887
5.	Acacia moniliformis	7.872	60.408	7.673
6.	Delonix regia	11.255	32.073	2.849
7.	Caesalpinia pulcherrima	18.072	58.149	3.217
8.	Acacia moniliformis	15.571	40.852	2.623
9.	Delonix regia	18.222	27.831	1.527
10.	Caesalpinia pulcherrima	40.968	25.654	0.626
	Mean ± S.D	14.21±10.886	36.702±14.511	

Fluoride (F)

BAC analysis of fluoride revealed no biomagnification of F in plant species from 5 year site except in *Delonix regia* where BAC was 2.37. Same is the case with 10 year plantation site where *Caesalpinia pulcherrima* and *Acacia moniliformis* showed no biomagnification with BAC of 0.48 and 0.49 respectively but *Delonix regia* with BAC of 1.44 showed biomagnification of F from top soil to leaf. However, natural site vegetation showed no biomagnification of F, with all species having BAC less than one (Table 1.8).

Table 1.8: Fluoride (mg/kg) Dynamics in Ash Mound Plantation Sites and Natural Forest Site

Sl.No.	Tree Species	Top Soil	Leaf	BAC
	Five Year Plantation Site			
1.	Acacia moniliformis	45.0	45.0	1
2.	Delonix regia	59.0	40.0	2.372
3.	Caesalpinia pulcherrima	42.5	35.0	0.823
4.	Azadirachta indica	39.5	32.0	0.81
5.	Ficus carica	55.0	20.0	0.363
6.	Holarrhena antidysenterica	42.0	35.0	0.833
7.	Psidium guaja	74.0	36.0	0.486
8.	Cassia angustifolia	48.0	24.0	0.5
	Mean ± S.D	50.625±11.578	33.375±8.105	
	Ten Year Plantation Site			
1.	Caesalpinia pulcherrima	42.0	20.0	0.476
2.	Acacia moniliformis	64.0	30.0	0.491
3.	Delonix regia	52.0	75.0	1.442
	Mean ± S.D	52.666±11.015	41.666±29.297	
	Natural Forest Site			
1.	Caesalpinia pulcherrima	55.0	45.0	0.818
2.	Acacia moniliformis	60.0	41.0	0.683
3.	Delonix regia	54.0	32.0	0.592
4.	Caesalpinia pulcherrima	50.0	30.0	0.6
5.	Acacia moniliformis	45.0	25.0	0.555
6.	Delonix regia	40.0	26.0	0.65
7.	Caesalpinia pulcherrima	59.0	38.0	0.644
8.	Acacia moniliformis	45.0	39.0	0.866
9.	Delonix regia	47.0	30.0	0.638
10.	Caesalpinia pulcherrima	65.0	44.0	0.676
	Mean ± S.D	52±7.93	35±7.318	

Among all the plants studied, highest BAC was recorded for Ni in *Denolix regia* (26.613) from ten year plantation site, which means mobility as well as biomagnification of Ni was maximum among all the heavy metal analysed, as pointed out by Alloway (1990). Ni is considered to be highly mobile element and it behaves in a similar fashion to Mn, Zn and Cu. Disposal of fly ash on land is the single largest input for both Cr and Ni to soil (Adriano *et al.*, 1980). A pilot scale study conducted by Maiti and Nandini (2005) reported Ni content in the naturally growing vegetation between 1.2 to 50 mg/kg which can also be established from present study where Ni concentration in natural site ranges from 0-20.18 mg/kg. The Ni accumulation in trees (*Cassia seamea*) growing on fly ash was reported between 60-120 mg/kg (Tripathi *et al.*, 2004). Alloway (1990) reported that critical concentration of Ni in plant ranges between 10-100 mg/kg above which toxicity effects are likely and if the concentration ranges from 8-220 mg/kg it is likely to cause 10 per cent depression in yield. General range of Ni in plants is 10-50 mg/kg (Alloway, 1990). This present study infer that Ni concentration in the vegetation from natural and ash mound sites are within safe limits except in *Delonix regia* from 10 year site, where Ni concentration goes up to 332.48 mg/kg which is very alarming.

Further study reveals that Mn, Zn and Cu also shows remarkable biomagnification with high BAC value of 26.21, 25.42 and 14.80 respectively in leaves of *Delonix regia* (10 year site), *Polyalthia longifolia* (Natural site) and again *Polyalthia longifolia* from natural site. The toxicity limit of Cu for the plants growing in the metalliferrous soil was reported between 5-25 mg/kg (Reeves, 2002). Alloway (1990) reported that the critical concentration of copper in plants ranges between 20-100 mg/kg, above which toxicity effects are likely and if the concentration ranges from 5-64 mg/kg, it is likely to cause 10 per cent depression in yield. General range of Cu in plants has been reported as 5-20 mg/kg (Alloway, 1990; Reeves, 2002). The present study indicates that Cu accumulation in leaves of different plant species from natural and ash mound sites are within the safe limits as proposed by Alloway (1990), Reeves (2002) and Kabata-Pendias and Pendias (1992).

Higher plants predominantly absorb Zn as a divalent cation (Zn^{2+}). Availability of Zn to the plants depends on total content, pH, organic matter, adsorption sites, microbial activity and moisture regime (Alloway, 1990). Johnson *et al.* (1979) reported that the highest concentration of Zn in plants occurred, when the maximum root penetration for a particular plant was equivalent to the depth of the contaminant. The normal range for plant Zn is between 10-300 mg/kg (Alloway, 1990) and Chaney (1993) found that plant Zn level of 500 mg/kg reduced plant yield. Maiti and Nandini (2006) reported Zn concentration in the natural growing vegetation between 15 and 61 mg/kg which can also be established from present study where Zn concentration in natural site ranged from 16.28 to 60.40 mg/kg. In fly ash amended soil, Zn concentration for *Cassia seamea* is between 22 and 67 mg/kg (Tripathi *et al.*, 2004). The toxicity level of Zn in the plant tissue reported by Kabata Pendias and Pendias (1992) was between 100-400 mg/kg. The present study infers that Zn concentration in vegetation from natural and ash mound site are within safe limit or normal as proposed by Alloway (1990) and Kabata-Pendias and Pendias (1992).

Uptake of Mn and Co by plants is a function of the concentration of these elements in ionic form in the soil solution and the concentration present in the exchange sites of the cation exchange complex *i.e.* the "available" on labile pool. The availability of Mn is largely governed by the supply of H^+ ions and electrons which reduces the higher valency states to Mn^{+2}. Maiti and Nandini (2005) reported Mn concentration in the naturally growing vegetation between 17-26 mg/kg. Reeves (2002) reported the range of 20-400 mg/kg as normal in plants growing in metalliferous soil. Alloway (1990) reported that the critical concentration of Mn in plants ranges between 300-500 mg/kg above which toxicity effects are likely and if the concentration ranges from 1000-7000 mg/kg it is likely to cause 10 per cent depression in yield. Generally

Mn in plants ranges from 20-1000 mg/kg (Alloway, 1990; Reeves, 2002). The present investigation infers that Mn concentration in vegetation of natural and ash mound sites are within safe limits as proposed by Alloway (1990) and Kabata-Pendias and Pendias (1992).

As far as fluoride is concerned, the present investigation reveals that F has biomagnified in leaves of *Delonix regia*, in both 5 year as well as 10 year site with BAC value of 2.372 and 1.442 respectively. The results clearly indicates that mobility of F from soil to leaf is very less and is not expected to harm the plant species as it accumulation rate is very slow.

Present study also indicates that among all the plant species analysed from natural and ash mound plantation sites, *Delonix regia* and *Polyalthia longifolia* showed remarkable biomagnification of heavy metal such as Ni, Mn, Zn and Cu. This may be due nature of the plant species which make them susceptible to metal accumulation.

Conclusion

Quantification of leachable heavy metals in top soil and tree leaves has been made. Lead considered to be a toxic metal is present in negligible amount (below detectable limit), which is a boon in disguise. Out of Fe, Mn, Zn, Ni, Cd, Cu and Cr, Iron is present in highest amount in all the three sites. It is interesting to note that metal contents decreased in 10yr site in comparison to 5yr site indicating the role of organic matter accumulation (humus) effecting chelation of metals.The Biological Accumulation Coefficient (BAC), a ratio of metal in leaf to soil, indicates biomagnification of the metal. The BAC of more than one indicates some level of bioaccumulation and biomagnification. The following metals showed biomagnification in the leaves of trees as indicated below in ash mound sites.

☆ Cadmium - *Caeseliania pulcherrima*

☆ Manganese - *Acacia moniliformis*

☆ Nickel - *Delonix regia*

☆ Zinc - *Holarrhena antidysenterica*

In both the plantation sites over ash mound Fluoride showed BAC of more than one in *Delonix regia*.

Acknowledgement

The authors are thankful to Hindalco Industries Limited, Hirakud for extending financial support in the form of a Research Project to carry out the work. Laboratory facilities provided by the Head, P.G. Department of Environmental Sciences, Sambalpur University is gratefully acknowledged.

References

Alloway, B.J. (1990): Heavy metals in soils. Glasgow, UK : Blackie. 339p.

Anon, J.K. (2002): The potential for beneficial reuse of coal fly ash in Southwest Virginia mining environments. Virginia cooperative extension, Pub. No. 460-134, January 2002.

Block, C.R. and Dams, R. (1975): Inorganic composition of Belgian coal and coal ashes. Environ. Sci Technol., 9(2): 148.

Bose, D. (1982): Leaching of heavy metals in Calcutta's sewage sludge Proc. Status and impact of heavy metal Pollution in India.

Cheney, R.L. (1993): Zinc phytotoxicity. In S.D. Robson (Ed), Proc. of the Int. Symp. Zinc in Soils and Plants, The Univ. of Western Australia, 27-28 Sept. 1993, (pp. 135-150). Dordrecht, The Netherlands : Kluwer.

Cope, F. (1962): The development of a soil from an industrial waste ash. Int. Soc. Soil Sci. Trans. Comm. IV, Palmerston, New Zealand, 859-863.

Elbay-Poulichet, F., Martin, J.M., Huang, W.W. and Zhu, J.X. (1987): Dissolved Cd behavior in some selected French and Chinese estuaries Consequences on Cd supply to the ocean. Mar. Chem., 322: 125-136.

Fulekar, M.H. and Dave, J. M. (1986): Disposal of fly ash- An environmental problem. Int. J. Environ. Stud., 26: 191-215.

Hodgson, D.R. and Holiday, R. (1996): The agronomic properties of pulverized fuel ash. Chem. India, 785-790

Hopkin, S.P. (1989): Ecophysiology of metals in terrestrial invertebrates. Elsevier Applied Science Publishers, London.

Johnson, M.S., McNeilly, T. and Putwain, P.D. (1977): Revegetation of metalliferrous mine soil contaminated by Lead and Zinc. Environmental Pollution, 12: 261-277.

Kabata-Pendias, A. and Pendias, H. (1992): Trace elements in soils and plants (2nd ed.) (p. 365). Boca Raton, FL : CRC Press.

Khan, M.R. and Khan, M.R. (1996): The effect of fly ash on plant growth and yields of tomato. Environmental Pollution, 92 (2): 105-111.

Klein, D.H., Andren, A.W., Carter, J.A. and Emery. (1975): Pathways of thirty seven trace elements through coal-fired power plants. Environ. Sci. technol., 9 : 973-979.

Maiti, S.K. and Nandhini, S. (2005): Heavy metal distribution pattern in pattern fly ash in CTTP (Jharkhand) and in spontaneously occurring vegetation. In D.P. Tripathy and B.K. Pal (Eds.), Proceedings of technological development and environmental challenges in mining and allied industries in the 21st century (TECMAC, 2005), NIT Rourkela, February 05-06. 477-486.

Maiti, S.K. and Nandhini, S. (2006): Bioavailability of metals in fly ash and their bioaccumulation in naturally occurring vegetation. Environmental Monitoring and Assessment, 116: 263-273.

Martens, D.C. (1970): Availiability of plants nutrients in flyash. Bums. Rep. inv. M. Nov-Dec. 15- 18.

Natusch, D.F.S., Bauer, C.F., Matusiewicz, H. and Evans (1975): Characterization of trace elements in flyash. In: Heavy metal in the Environment. Int. Conf., Toronto, Ontario,Canada, II(2): 553-575.

Patel, C.B. and Pandey, G.S. (1986) Alkalinization of soil through thermal power plant fly ash fallout. Sci. Tot. Environ., 57: 67-72.

Plank, C.O and Martens, D.C. (1974): Boron availiability as influenced by application of fly ash to soil, Soil Sci. Amer. Proc., 38 (6): 974-976.

Prasad, A.N. and Soman, S.D. (1990): Nuclear waste: How much of a hazard. 10 Apr. Hindustan Times, 6p.

Rainbow, P.S. (1985): The Biology of heavy metals in the sea. Int. J. Environ. Studies., 25: 195- 211.

Reeves, R.D. (2002): Metal tolerance and metal accumulating plant exploration and exploitation. In 9th New Phytologist symp. On Heavy metals and plants. Philadelphia.

Sahu, K.C. (1990): Heavy Metal dispertion around aquatic regime of Bombay. In: Protection of Environment of City Water Fronts. Natl. Symp., Central Water Commission, New Delhi, 13-23.

Schwarz, K.(1974): Recent dietary trace element research exemplified by tin, fluorine and silicon. Fed. Proc., 33: 1748-1757.

Sims, J.T. and Kline, J.S. (1991): Chemical fractionation and plant uptake of heavy metals in soil amended with compost sewage sludge. Journal of Environmental Quality, 20: 387-395.

Tripathi, R.D., Vajpayee, P., Singh, N., Rai, U.N., Kumar, A. and Ali, M.B. (2004): Efficacy of various amendments for amelioration of fly ash toxicity : Growth performance and metal composition of *Cassia seamea* lamk. Chemosphere, 54: 1581-1588.

Wang, D. and Sweigard, R. (1996): Characterization of fly ash and bottom ash from a coal fired power plant. International Journal of Surface Mining, Reclamation and Environment, 10: 181-186.

Wood, J.M. (1974): Biological cycles for toxic elements in the environment. Sci., 183: 1049-1052.

Chapter 2

Predicting the Fate of Pesticides in Multimedia Environmental using Level IV Fugacity Model: A Case Study of Kelowna (BC)

Sarah Moffat, Amin Zargar and Rehan Sadiq

School of Engineering, University of British Columbia (Okanagan Campus)
Kelowna, British Columbia, Canada, V1V 1V7

ABSTRACT

The Okanagan Basin is a semi-arid watershed in British Columbia, Canada. Kelowna is the largest city within this Basin, where agriculture plays a major role in the economy of the city as well as for the province. Agricultural land is dispersed throughout the city, and pesticides have been used to protect crops from insects and weeds. Recently, the City of Kelowna banned the use of pesticides for cosmetic and non-essential use on residential properties. However, the pesticides have still been sprayed on the orchards as well as oncity property dedicated to public use. A Level IV fugacity model has been developed to determine the multimedia environmental fate of three pesticides commonly used in city orchards: 2,4 *Dichlorophenoxyacetic* acid, carbaryl, and glyphosate. The City of Kelowna and a portion of Okanagan Lake are used as the area of interest. The Level IV fugacity model has been developed using four standard compartments, which includes air, soil, water, and sediment. Numeric simulations were run over one year period and maximum concentrations in the Kelowna area were used as exposure concentrations for adults. The results show that the current pesticide control measures for the City of Kelowna maintain a very low hazard for the general population.

Keywords: *Level IV fugacity model, Pesticides, Exposure concentration, Coupled differential equations.*

Introduction

The Okanagan Basin is a semi-arid watershed in British Columbia (BC), Canada. This basin is a major producer of fruits, vegetables and forage crops (accountingfor 25 per cent of total value of province's agriculture (NRC, 2008). Kelowna is the largest city in the Basin, and its agriculture sector is an integral part of the city and province's economy. The agricultural land in Kelowna is dispersed throughout the city where pesticides have been used extensively to protect crops from potential damage caused by insects and weeds. The use of these pesticides has however been a concern for the residents of Kelowna. A pesticide is a chemical product used to control or destroy pests, and includes herbicides, insecticides, fungicides and rodenticides.

Effective January 1, 2009, the City of Kelowna banned the use of pesticides for cosmetic and non-essential use on residential properties. The aim of this ban was to protect the environment and human health by reducing the use of products containing carbaryl, 2,4-*dichlorophenoxyacetic acid*, (2,4-D), glyphosate, mecoprop, dicambia, and diazinon. There are several exceptions to the ban, including farmers that use the pesticides to protect their crops, any property owned by the City of Kelowna that is for public use and certain excluded pesticides.

The health risk from these chemicals remains to be verified using environmental fate and transport models. Modelling the environmental fate of these chemicals can provide an indication as to whether the current restrictions are sufficient. Comparable models have been used in other regions around the world. For example, Di Guardo *et al.* (1994) analyzed the fate of pesticide runoff in surface waters in two Italian water basins using an advanced fugacity model what they referred as Level V. Similarly, Paraíba *et al.* (2007) used a Level IV fugacity model to estimate the concentration of carbofuran in rice fields and found that the model can reasonably predict concentrations in the rice growing environment. Some recent applications of Level IV fugacity model have been summarized in Table 2.1.

This paper models the environmental fate of three chemicals of concern in the Kelowna area using a four-compartment Level IV Fugacity model. Level IV fugacity models deals with non-steady state and non-equilibrium conditions. This is ideal for modeling pesticides, since most are sprayed only one or two times per growing season and then degrade over time in the environment. Level IV fugacity models will give both the maximum concentration in each compartment, as well as how long (time) after spraying it will take to reach this concentration. These four compartments include air, soil, water, and sediment. Chemical movement is modelled using diffusion between compartments, advection (bulk movement) and first order degradation in various compartments.

Table 2.1: Some Recent Applications of Level IV Fugacity Model (L4FM)

Reference	Application	Notes
Sweetman *et al.* (2002)	Fate of PCB	Using a four-compartment L4FM the fate of three PCB since their introduction into commerce until the present in the UK is simulated.
Li *et al.* (2006)	Fate of DDT	A four-compartment L4FM system estimates risk of DDT to human health in Tianji, China from the fate and transfer of *p,p2*-DDT before and after its ban in 1983.
Tao *et al.* (2006)	Concentration of g-hexachloro-cyclohexane	L4FM was used to simulate the dynamic changes of g-hexachlorocyclohexane concentrations in four environmental media in Tianjin, China.
Huang *et al.* (2007)	Distribution and fate of nonylphenol in aquatic environment	Four-compartment L3FM and L4FM systems were developed to investigate the distribution and fate of nonylphenol in Tianjin, China.
Lang *et al.* (2007)	Seasonal variations of polycyclic aromatic hydrocarbons	L4FM was applied on four bulk compartments comprised of nine sub-compartments and the seasonal variation of polycyclic aromatic hydrocarbons in bulk media in Pearl River Delta, China were simulated.
Contreras *et al.* (2008)	Fate of pesticides in rice fields	Numerical simulation of L4FM is coupled with a dispersion-advection equation to simulate the fate of Carbofuran (2,3-*dihydro*-2,2-*dimethyl*-7-*benzofuranolmethylcarbamate*).
Nazir *et al.* (2008)	Fate of spilled oil in the marine environment	Oil weathering processes were coupled with L4FM in a two-compartment (water and sediment) system to determine the fate of crude oil spill of Statfjord.

Apples and other tree fruits (peaches, cherries, etc.) make up the majority of the crop farmland in Kelowna. Orchards are susceptible to a variety of pests, particularly to the coddling moth, leafroller caterpillar, bud moth, and weeds. Spray records obtained from a local apple orchardist (Harvie, 2010) indicated that 2, 4-D, carbaryl, and glyphosate were used to control pests. These three pesticides were modelled in this study as the chemicals of concern. The data from the spray records were assumed to be an average emission rate for the agricultural area within the multimedia boundaries, and were extrapolated to estimate the total chemical emission to the environment and the concentrations in each compartment.

Evaluative Environment

For the proposed model an evaluative environment was established. The multimedia boundary for the evaluative environment includes the city limits of Kelowna (BC) (Figure 2.1) and also a portion Okanagan Lake proportional to the ratio of Kelowna shoreline to the total Okanagan Lake shoreline. The environment is

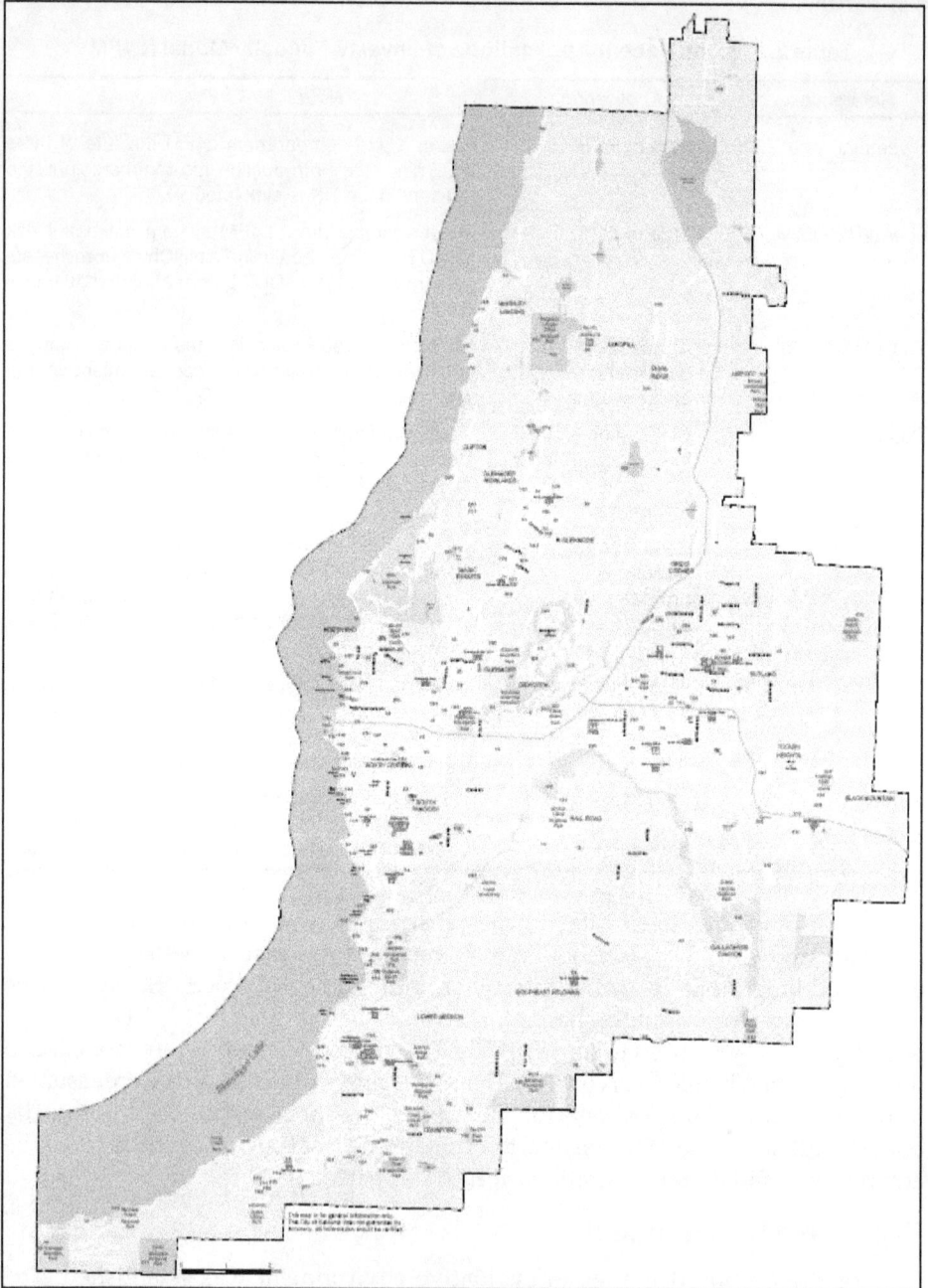

**Figure 2.1: Map of Model Boundaries
(City of Kelowna, 2010).**

divided into four main compartments: air, soil, water, and sediment and their parameters are denoted by respective subscripts: A, S, W, and D.

Material Volumes

The volumes for the four multimedia compartments may be calculated in the first step. For soil, Menon and Gopal (2003) studied carbaryl however did not detect its presence in soil beyond the depth of 15 cm. Similarly, Zablotowicz*et al.* (2009) studied glyphosate and found detectable concentration to a depth of 10 cm. Veeh *et al.* (1996) studied 2,4-D degradation in soil up to depth of 120 cm and found that most of it degrades in the top 30cm. In this study we assumed soil depth of 30 cm to determine the volume of soil. The land area of city of Kelowna is approximately 21388 ha (BC Ministry of Agriculture, Food and Fisheries, 2001), this yields a total soil volume of V_S = 6.42×10^7 m^3.

The Kelowna shoreline on Okanagan Lake is approximately one-eighth of the total shoreline. Therefore we used one-eighth of the volume of the Okanagan Lake in the model system boundary. Moreover the volumes of other small lakes within the Kelowna limits are also considered in water volume calculations. Okanagan Lake has a total surface area of 351 km^2 and a mean depth of 76 m (WLD, 1999). Other small lakes in Kelowna cover an area of 1452 ha (BCMAFF, 2001). For simplification, we assumed a mean depth of 1 m, since most of the surface waters in Kelowna are shallow streams or creeks. Therefore the total volume of water amounts toV_W = 3.35×10^9 m^3.

Relevant air mixing height was assumed as 6000 m, therefore the total air volume amounts to V_A = 1.63×10^{12} m^3. Finally, we assumed a 3 cm layer of sediment over the area covered by water, yielding a total sediment volume of V_D = 1.75×10^6 m^3.

Fugacity Capacities

The fugacity capacity is the ability of a medium to sorb and retain a specific chemical of interest. The fugacitycapacities of the four multimedia compartments dependon the physio-chemical properties of the pesticide (*cf.* Table 2.2 for complete multimedia properties).

The fugacity capacity of air is

$$Z_A = \frac{1}{RT} \tag{1}$$

where, R is the ideal gas constant and T is the temperature of the air. All three pesticides are assumed to be sprayed in mid-May to early June; therefore an average air temperature is assumed to be 20°C or 293.15 °K.

The fugacity capacity of water is given as

$$Z_W = \frac{1}{H_p} = \frac{S_A}{m_p P_V} \tag{2}$$

where, H is the Henry's constant for the desired chemical, calculated using $H_p = \dfrac{m_p P_V}{S_A}$

where m_p, P_V, and S_A are the molar mass, vapour pressure, and aqueous solubility of the pesticides, respectively.

The fugacity capacity of the soil is given as

$$Z_S = oc_S K_{OC} \rho_S Z_W \tag{3}$$

where, oc_S is the fraction of organic carbon in the soil, assumed to be 0.02 (MacKay, 2001), K_{OC} is the organic carbon partitioning coefficient that equals to $0.41 K_{OW}$ (MacKay *et al.*, 1985), where K_{OW} is the octanol water partitioning coefficient, ρ_S is the soil density and Z_W is the fugacity capacity of water as defined above.

Table 2.2: Environmental Multimedia Properties

Temperature (K)	293.15[1]
R (m³Pa/mol K)	8.314[2]
Soil density, ρ_S(g/cm³)	1.5[3]
Soil organic carbon content, oc_S	2 per cent[3]
Soil porosity, ϕ_S	0.36[1]
Soil water content (kg/kg)	0.32[1]
Sediment density, ρ_D(g/cm³)	2.65[4]
Sediment organic carbon content, oc_D	4 per cent[3]
Sediment clay content, cl_D	10 per cent[1]
Sediment silt content, st_D	40 per cent[1]
Sediment sand content, sd_D	46 per cent[1]
Sediment porosity, ϕ_D	0.43[1]
Molar volume of air, v_A (cm³/mol)	20.1[2]
Molar mass of air, m_A (g/mol)	28.9[2]
Molar mass of water, m_W (g/mol)	18[2]
Viscosity of water, μ_W (cp)	0.89[2]
A_{WA} (m²)	5.84E+07[5,6]
A_{AS} (m²)	2.14E+08[5]
A_{WS} (m²)	2.57E+06[1]
δ_{AW} (m)	0.001[7]
δ_{AS} (m)	0.001[7]
δ_{WD} (m)	0.001[7]
δ_{SW} (m)	0.001[7]
Airshed residence time, τ_A (hr)	120[1]
Okanagan Lake residence time, τ_W (hr)	462528[6]

1: Estimated value; 2: Contreras *et al.*, 2008; 3: MacKay, 1985; 4: Morris and Fan, 1997; 5: BCMAFF, 2001; 6: WLD, 1999; 7: Paraíba *et al.*, 2002.

The fugacity capacity of the sediment is calculated in the same way as for soils. However a higher fraction of organic carbon, $oc_D = 0.04$ (MacKay, 2001) and higher density ρ_D is assumed.

$$Z_D = oc_D K_{OC} \rho_D Z_W \tag{4}$$

Area between Compartments

The area between air and soil is the surface area used in the volume calculations, $A_{AS} = 2.14 \times 10^8 \, m^2$. Similarly, the area between the air and water is $A_{AW} = 5.87 \times 10^7 \, m^2$. The area between the water and sediment is determined empirically by the following formula:

$$A_{WD} = \rho_D S_{SA} V_D \tag{5}$$

where, S_{SA} is the specific superficial area of the sediment, estimated by

$$S_{SA} = (1313.78 oc_D + 117.00 cl_D + 116.90 st_D + 5.15 sd_D) \times 10^3 \tag{6}$$

where, oc_D, cl_D, st_D, and sd_D are the fractions of organic carbon, clay, silt, and sand in the sediment, respectively (Paraiba *et al.*, 2004). Based on the results of Shaw (1977), which showed high levels of sand and silt in samples taken from East Okanagan Lake, a sediment composition of 4 per cent organic carbon, 10 per cent clay, 40 per cent sand and 46 per cent silt was estimated, which yields an area of $A_{WD} = 5.26 \times 10^{11} \, m^2$.

The area between the water and soil is estimated using one eighth on the Okanagan Lake shoreline times the average lake depth. This gives an area of $A_{WS} = 2.57 \times 10^6 \, m^2$.

Diffusion

Diffusion parameters in equations (7) – (11) were derived using empirical formulas proposed by Contreras *et al.* (2008). The diffusivity of pesticide in air is estimated by

$$D_{PA} = \frac{3.4 \times 10^{-6} T^{1.75} \sqrt{M_{PA}}}{\left(\sqrt[3]{v_P} + \sqrt[3]{v_A}^{-2} \right)^2} \tag{7}$$

where, v_P and v_A are the molar volumes (cm^3/mol) of the pesticide and air, respectively. M_{PA} is defined by

$$M_{PA} = \frac{m_P + m_A}{m_P m_A} \tag{8}$$

where, m_P and m_A are the molar mass of the pesticide and air, respectively. The diffusivity of the pesticide in water can be estimated using the following formula

$$D_{PW} = \frac{2.664 \times 10^{-8} T \sqrt{2.6 m_W}}{\mu_W v_P^{0.6}} \tag{9}$$

where, m_w is the molar mass of water, μ_w is the viscosity of water (0.89 cp), and v_p is the molar volume of the pesticide. The diffusivity of the pesticide in soil is calculated using following formula

$$D_{PS} = \frac{\sqrt[3]{\theta^{10}}}{\phi_S^2} D_{PW} \tag{10}$$

where, θ is the soil water volumetric fraction, assumed to be 0.32 for Kelowna, and ϕ_s is the soil porosity, assumed as 0.36. Finally, the diffusivity in the sediment is approximated as

$$D_{PD} = \frac{D_{PW}\phi_D^2}{(1-\phi_D)\rho_D oc_D K_{oc} + \phi_D} \tag{11}$$

where, ϕ_D, ρ_D, and oc_D are the porosity, density, and organic carbon content of the sediment, respectively.

Transfer Coefficients

A graphical representation of the compartments used can be seen in figure 2. The transfer coefficients between compartments (d_{ij}) were computed using the following relationship:

$$d_{ij} = \frac{A_{ij} D_{pi} D_{pj} Z_i Z_j}{\delta_{ij}(D_{pi}Z_i + D_{pj}Z_j)} \tag{12}$$

where, A_{ij} is the area between compartments i and j, D_{pi} is the diffusivity of the pesticide in compartment i, Z_i is the fugacity capacity of compartment i, and δ_{ij} is the thickness of the diffusion layer between compartments i and j. The values for δ_{ij} were estimated as proposed by Paraíba et al. (2002). The transfer coefficients $d_{ij} = d_{ji}$ represents a transfer between air and soil, air and water, and sediment and water. Transfer is assumed to take place in only one direction from soil to water.

Pesticide loss is also accounted for through advection in the air and water. The volumetric flow rates of air and water, G_A and G_W, were calculated using the Kelowna airshed and Okanagan Lake residence times. The residence time of the airshed is approximately $\tau_A = 5$ days and for Okanagan Lake is approximately $\tau_W = 52.8$ years (WLD, 1999). The flow rate values were then calculated using

$$G_i = \frac{V_i}{\tau_i} \tag{13}$$

Fugacities

Based on collected and calculated input data, four coupled differential equations

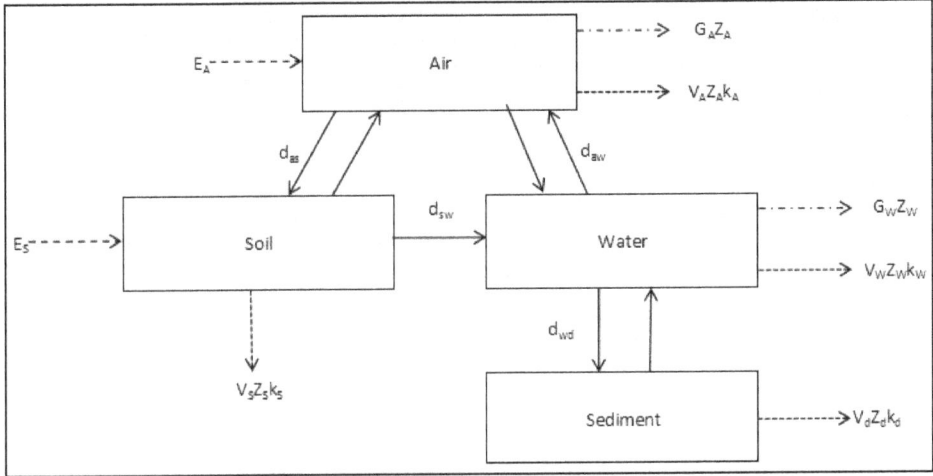

**Figure 2.2: Graphical Representation of Compartments Used in
Level IV Fugacity Model.**

are developed that describe the rate of change of the fugacity of each compartment
with respect to time. The rate of change of the amount of the chemical in each
compartment at a given time is the sum of the amount coming in minus the amount
leaving at that point in time. This can be explained mathematically with the following
equation:

$$V_i Z_i \frac{df_i(t)}{dt} = \sum_j d_{ji} f_j(t) - \sum_j d_{ij} f_i(t) - G_i Z_i f_i(t) - V_i Z_i k_i f_i(t) \tag{14}$$

This leads us to following coupled differential equations

$$\frac{df_A(t)}{dt} = \frac{d_{SA} f_S(t)}{V_A Z_A} + \frac{d_{WA} f_W(t)}{V_A Z_A} - \frac{d_{AS} f_A(t)}{V_A Z_A} - \frac{d_{AW} f_A(t)}{V_A Z_A} - \frac{G_A f_A(t)}{V_A} - k_A f_A(t) \tag{15}$$

$$\frac{df_S(t)}{dt} = \frac{d_{AS} f_A(t)}{V_S Z_S} - \frac{d_{SA} f_S(t)}{V_S Z_S} - \frac{d_{SW} f_S(t)}{V_S Z_S} - k_s f_S(t) \tag{16}$$

$$\frac{df_W(t)}{dt} = \frac{d_{SW} f_S(t)}{V_W Z_W} + \frac{d_{AW} f_A(t)}{V_W Z_W} - \frac{d_{WA} f_W(t)}{V_W Z_W} - \frac{d_{WD} f_D(t)}{V_W Z_W} - \frac{G_W f_W(t)}{V_W} - k_W f_W(t) \tag{17}$$

$$\frac{df_D(t)}{dt} = \frac{d_{WD} f_W(t)}{V_D Z_D} - \frac{d_{DW} f_D(t)}{V_D Z_D} - k_D f_D(t) \tag{18}$$

where, $f_A(t), f_S(t), f_W(t)$, and $f_D(t)$ are the fugacities of air, soil, water, and sediment.

Physical Properties

Physical properties for 2,4-D and carbaryl were obtained from MacKay(2001),
while properties for glyphosate were taken from the World Health Organization data

sheets on pesticides (1996). These data are summarized in Table 2.3. From this data, the molar volume of the pesticide was calculated as following:

$$v_p = \frac{m_p}{\rho_p} \tag{19}$$

where v_p is the molar volume, m_p is the molar mass, and ρ_p is the density of the pesticide. In each compartment, the chemical is assumed to naturally degrade with a first order rate constant, k_i. The values for k_i are calculated by the following equation

$$k_i = \frac{\ln(2)}{\tau_{1/2i}} \tag{20}$$

where $\tau_{1/2i}$ is the half-life of the chemical in compartment i.

Table 2.3: Calculated Multimedia Environmental Properties

Specific superficial area, S_{SA}	1.13E+05
Area between water and sediment, A_{WD} (m²)	5.26E+11
V_A (m³)	1.63E+12
V_W (m³)	3.35E+09
V_S (m³)	6.42E+07
V_D (m³)	1.75E+06
G_A (m³/hr)	1.36E+10
G_W (m³/hr)	7.24E+03

Pesticide Application

Spray records were obtained for 2008 and 2009 for local apple orchards (Harvie, 2010). The data was recorded as the volume of pesticide added to each tank of spray. Each tank can spray roughly two acres of the orchard. The spray records were assumed as an average application rate and extrapolated for the 4307 hectares of total crop farmland in Kelowna (BCMAFF, 2001).

According to the spray records, 3.8 litres/tank of the pesticide Sevin, with an active ingredient carbaryl, were applied in the first week of June. Using data from (Bayer, 2008) this results in carbaryl application rate of 0.228 g/m². Extrapolating this application rate for all crop land in Kelowna yields a total emission of 4.87×10^4 moles of carbaryl. From this, the initial fugacity of each compartment can be calculated, by assuming that 90 per cent of the pesticide is applied to the soil and 10 per cent goes into the air. We also assumed that all the pesticide is emitted at the same time, since the crops need to be sprayed at the beginning of the growing season when the weather is good. This gives the initial values for carbaryl: $f_A(0) = 7.27 \times 10^{-6}$ Pa, $f_S(0) = 1.09 \times 10^{-5}$ Pa, $f_W(0) = 0$ Pa, and $f_D(0) = 0$ Pa (*cf.* Table 2.4 for properties of selected pesticides).

Table 2.4: Properties of Selected Pesticides

Properties	2,4-D	Carbaryl	Glyphosate
Molar Mass (g/mol)	2.21E+02[1]	2.01E+02[1]	1.69E+02[2]
Vapour Pressure (Pa)	8.00E-05[1]	2.67E-05[1]	1.31E-05[2]
Aqueous Solubility (g/m³)	4.00E+02[1]	1.20E+02[1]	1.05E+04[2]
Density (g/cm³)	1.10E+00[1]	1.20E+00[1]	1.43E+00[2]
Log K_{ow}	2.81E+00[1]	2.36E+00[1]	−3.70E+00[2]
$\tau_{1/2\,A}$ (hr)	1.70E+01[1]	5.50E+01[1]	0.00E+00[3]
$\tau_{1/2\,W}$ (hr)	5.50E+01[1]	1.70E+02[1]	1.85E+03[3]
$\tau_{1/2\,S}$ (hr)	5.50E+02[1]	5.50E+02[1]	3.12E+03[3]
$\tau_{1/2\,D}$ (hr)	1.70E+03[1]	1.70E+03[1]	1.44E+03[4]

1: MacKay, 2001; 2: Pesticide Manual, 2010; 3: Schuette, 1998; 4: Zaranyika and Nyandoro, 1993.

For 2,4-D, the spray records indicate that 3 litres/tank of 2,4-D were applied in the second week of May. Based on information collected from UAP (2008), the application rate is estimated at 7.04×10^{-2} g/m². Extrapolating in the same manner, this gives a total emission of 1.37×10^4 moles of 2,4-D. Based on similar assumptions, the initial fugacites for 2,4-D are obtained as: $f_A(0) = 2.04 \times 10^{-6}$ Pa, $f_S(0) = 3.05 \times 10^{-6}$ Pa, $f_W(0) = 0$ Pa, and $f_D(0) = 0$ Pa.

Glyphosate spray records indicate that 3.0 litres/tank of the pesticide was sprayed in the second week of May. Based on information collected from Cheminova (1999), the application rates are estimated at 1.78×10^{-1} g/m². Extrapolating this data we get a total glyphosate emission of 4.12×10^4 moles. Therefore, the initial fugacities for glyphosate are: $f_A(0) = 6.15 \times 10^{-6}$ Pa, $f_S(0) = 9.18 \times 10^{-6}$ Pa, $f_W(0) = 0$ Pa, and $f_D(0) = 0$ Pa.

Model Development and Application

Based on the information provided in last two sections, the behaviour of each pesticide can be predicted over time in four compartments. First, we simplify equations (15) – (18) into the following form:

$$\frac{df_A(t)}{dt} = P_{AA}f_A(t) + P_{AS}f_S(t) + P_{AW}f_W(t) + P_{AD}f_D(t) \tag{21}$$

$$\frac{df_S(t)}{dt} = P_{SA}f_A(t) + P_{SS}f_S(t) + P_{SW}f_W(t) + P_{SD}f_D(t) \tag{22}$$

$$\frac{df_W(t)}{dt} = P_{WA}f_A(t) + P_{WS}f_S(t) + P_{WW}f_W(t) + P_{WD}f_D(t) \tag{23}$$

$$\frac{df_D(t)}{dt} = P_{DA}f_A(t) + P_{DS}f_S(t) + P_{DW}f_W(t) + P_{DD}f_D(t) \tag{24}$$

where, the constants P_{ij} are the coefficients of the fugacities, as defined in equations (15) – (18). Then we used the backwards Euler method to numerically approximate

the solution to the differential equations using the following matrix equation and the
initial fugacities calculated earlier.

$$
\begin{bmatrix} f_A \\ f_S \\ f_W \\ f_D \end{bmatrix}_n =
\begin{bmatrix}
(1-\Delta t P_{AA}) & -\Delta t P_{AS} & -\Delta t P_{AW} & -\Delta t P_{AD} \\
-\Delta t P_{SA} & (1-\Delta t P_{SS}) & -\Delta t P_{SW} & -\Delta t P_{SD} \\
-\Delta t P_{WD} & -\Delta t P_{WS} & (1-\Delta t P_{WW}) & -\Delta t P_{WD} \\
-\Delta t P_{DA} & -\Delta t P_{DS} & -\Delta t P_{DW} & (1-\Delta t P_{DD})
\end{bmatrix}
\begin{bmatrix} f_A \\ f_S \\ f_W \\ f_D \end{bmatrix}_{n+1} \quad (25)
$$

A program was written using Microsoft Excel and Maple 12 software to
numerically approximate a solution for the model. Data for the system and each of
the chemicals was entered into Excel and the coefficients and initial fugacity values
were calculated. Then the data was entered into the program which ran 8766 iterations
(365×24) with a step size of $\Delta t = 1$ hour, which corresponds to one year. The resulting
data was multiplied by the fugacity capacity for each compartment and molar mass
for each pesticide to calculate the concentration in grams per meter cubed for each
time step. For each compartment, the concentrations with respect to time are plotted
in Figure 2.3–2.6. Calculated values of parameters for three pesticides are shown in
Table 2.5.

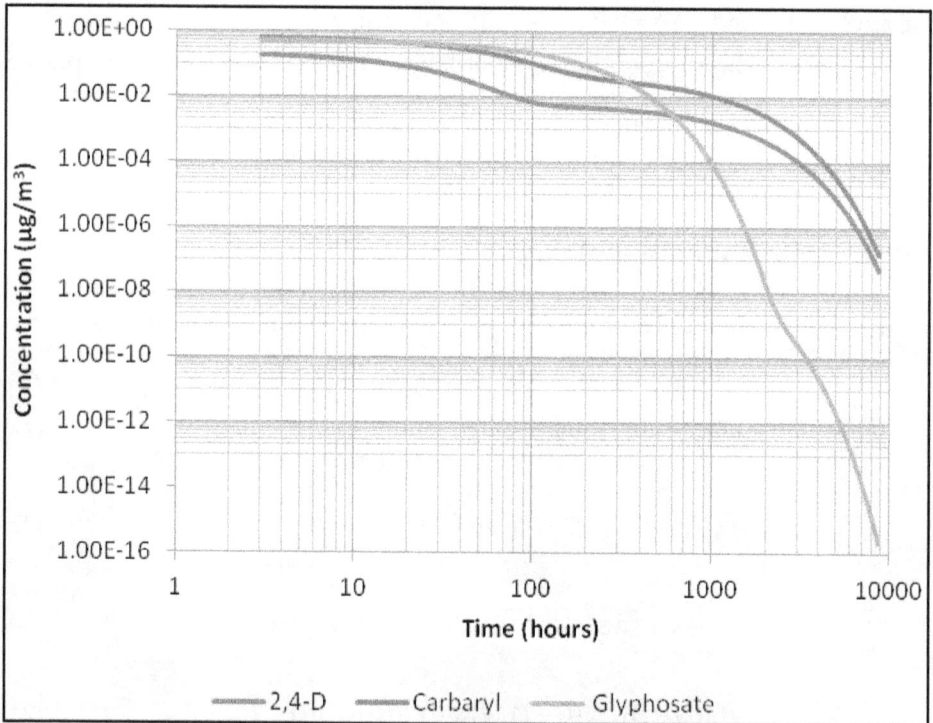

Figure 2.3: Concentration of Pesticide in Air with Respect to Time.

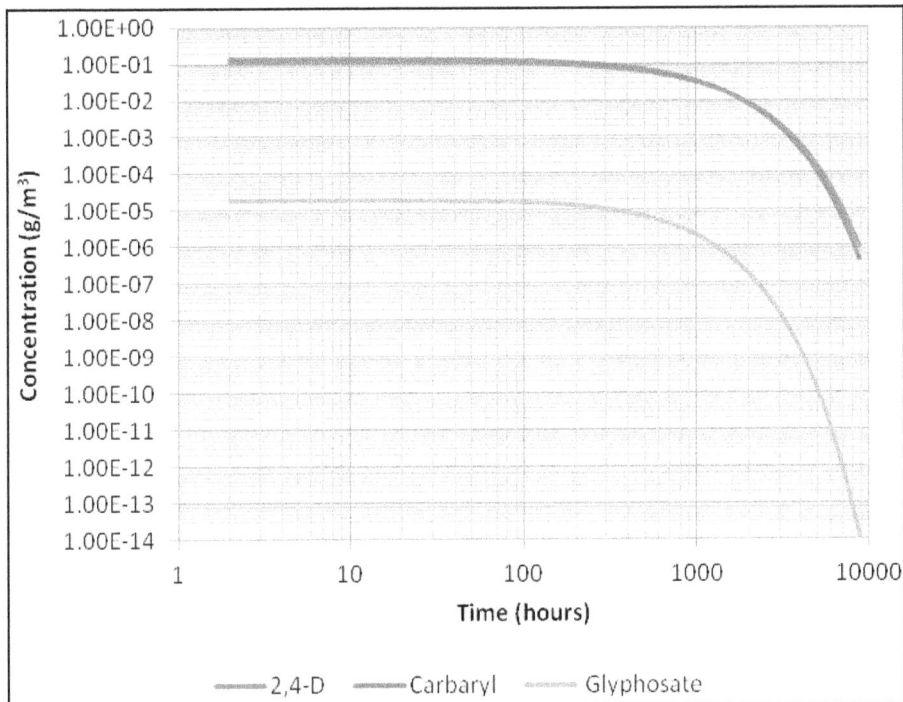

Figure 2.4: Concentration of Pesticides in Soil with Respect to Time.

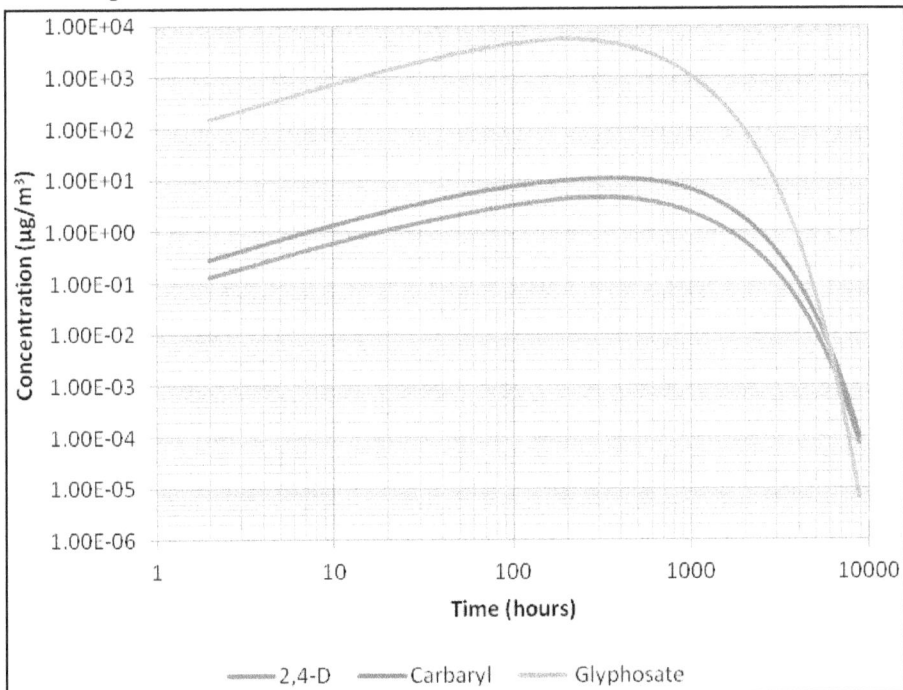

Figure 2.5: Concentration of Pesticides in Water with Respect to Time.

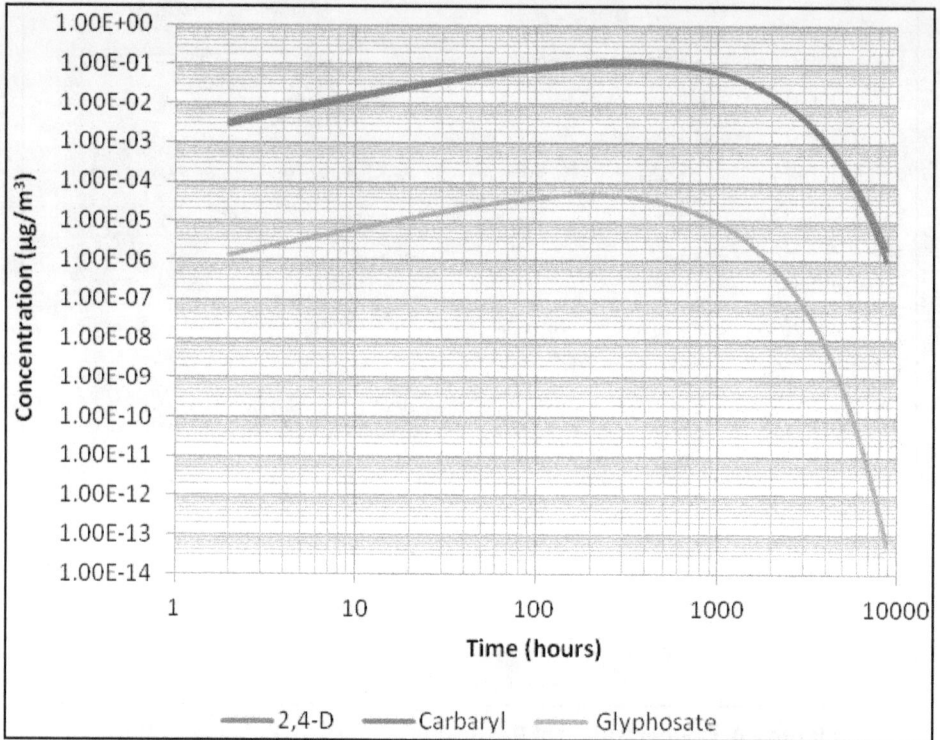

Figure 2.6: Concentration of Pesticides in Sediment with Respect to Time.

Note on Uncertainty Analysis

The analysis performed in this work is in a deterministic setting, *i.e.*, variables and parameters were represented by single values and uncertainty surrounding variables and model was ignored. There is however increased appreciation of the role of the uncertainty in environmental risk assessment. Uncertainty in input variables can propagate and undermine subsequent analyses including decision-making. In addition to input variables, uncertainty also arises from parameters and the model itself. Uncertainty model for steady-state Level III have been developed in Kühne *et al.* (1997) and Citra (2004). For non-steady-state Level IV fugacity models uncertainty analysis has however largely been missing (contrary cases include Tao *et al.*, 2006; Lang *et al.*, 2007). In addition, uncertainty models for Level III framework are mostly relevant for handling certain uncertainties of probabilistic nature, including, error and variability. Probabilistic risk assessment is based on the probability theory (either classic or Bayesian) where the mapping of likely values to a *feature* (*e.g.*, a variable or a parameter) follows probability theory principles. However, in addition to variability (also termed *stochasticity* and *aleatory uncertainty*),uncertainty may also result from *epistemic uncertainty*. Such human-induced uncertainty includes vagueness resulting from *e.g.*, linguistic expressions, and incompleteness from incomplete information. Such uncertainties are considered best handled using generalized theories of information such as fuzzy sets theory (Zadeh, 1965) and

Table 2.5: Calculated Values of Parameters for Three Pesticides

Parameters	2,4-D	Carbaryl	Glyphosate
K_{ow}	6.46E+02	2.29E+02	2.00E-04
K_{oc} (L/kg)	2.65E-01	9.39E-02	8.18E-08
K_A (hr^{-1})	4.08E-02	1.26E-02	0.00E+00
K_W (hr^{-1})	1.26E-02	4.08E-03	4.58E-04
K_S (hr^{-1})	1.26E-03	1.26E-03	2.22E-04
K_D (hr^{-1})	4.08E-04	4.08E-04	4.81E-04
H_P (m^{3}Pa/mol)	4.42E-05	4.48E-05	2.11E-04
Z_A (mol/m^{3}Pa)	4.10E-04	4.10E-04	4.10E-04
Z_W (mol/m^{3}Pa)	2.26E+04	2.23E+04	4.74E+03
Z_S (mol/m^{3}Pa)	1.80E+02	6.29E+01	1.16E-05
Z_D (mol/m^{3}Pa)	6.35E+02	2.22E+02	4.11E-05
M_{pa} (g/mol)$^{-1}$	3.91E-02	3.96E-02	4.05E-02
D_{pa} (m^2/hr)	8.42E-03	8.93E-03	1.04E-02
D_{pw} (m^2/hr)	2.49E-06	2.78E-06	3.81E-06
D_{ps} (m^2/hr)	4.31E-07	4.80E-07	6.58E-07
D_{pd} (m^2/hr)	1.04E-06	1.19E-06	1.65E-06
d_{aw} (mol/hrPa)	2.02E+05	2.14E+05	2.49E+05
d_{as} (mol/hrPa)	7.08E+05	6.99E+05	1.64E+00
d_{ws} (mol/hrPa)	1.98E+05	7.75E+04	1.97E-02
d_{wd} (mol/hrPa)	3.44E+11	1.39E+11	3.58E+04
$f_A(0)$ (Pa)	2.04E-06	7.27E-06	6.15E-06
$f_S(0)$ (Pa)	3.05E-06	1.09E-05	9.18E-06
$f_W(0)$ (Pa)	0.00E+00	0.00E+00	0.00E+00
$f_D(0)$ (Pa)	0.00E+00	0.00E+00	0.00E+00

Dempster-Shafer theory (DST) (Dempster, 1967; Shafer, 1976) that allow for more precise mapping of uncertainty to features. *For example*, fuzzy sets theory provides a *possibilistic* framework where data are mapped to a range of possible values with a variable *membership* (instead of probability). DST generalizes traditional Bayesian theory allowing probability from incomplete information to be assigned to multiple hypotheses or ignorance (*i.e.*, all hypotheses).

Results and Discussions

The results from the one-year deterministic simulation show that the concentrations are very low in each compartment. It can safely be assumed that there is very little *likelihood* of any accumulation of pesticides in the environment over years. The maximum concentrations calculated in each compartment are reported in

Table 2.6. To determine if there are any human health concerns, average daily potential doses were calculated based on maximum estimated concentrations

$$ADD = \frac{C \cdot IR}{BW} \tag{26}$$

where, ADD is the average daily dose, *C* is the maximum concentration, *IR* is the intake rate (see values in Table 2.7) and *BW* is body weight. The total daily dose was calculated taking into account inhalation, soil ingestion, and water ingestion using the maximum concentrations in air, soil, and water. For 2,4-D the expected dose is 4.1 × 10^{-5} mg/kg/day, for carbaryl it is 1.3 × 10^{-4} mg/kg/day and for glyphosate it is 0.11 mg/kg/day. Oral reference doses gathered from (USEPA IRIS, 2010) indicate the safe level of consumption for 2,4-D to be 0.01 mg/kg/day, for carbaryl 0.1 mg/kg/day, and for glyphosate 0.1 mg/kg/day. This results in hazard quotients of 0.0041 for 2,4-D, 0.0013 for carbaryl, and 1.1 for glyphosate. The hazard index for glyphosate reveals that the population could be at risk of increased incidence of renal tubular dilation in offspring due to consumption of contaminated water if exposed to maximum concentration (USEPA IRIS, 2010). However, this value is calculated using population averages and a detailed Monte Carlo analysis will be helpful to determine variability in this hazard index value by taking into account various sensitive subpopulations (*e.g.*, children).

Table 2.6: Maximum Concentrations in Multimedia Environment

Concentration	2,4-D	Carbaryl	Glyphosate
$C_A(t)$ (μg/m³)	0.185	0.600	0.469
$C_S(t)$ (g/m³)	0.121	0.138	1.81E-05
$C_W(t)$ (μg/m³)	4.57	10.5	5.40E3
$C_D(t)$ (g/m³)	0.128	0.104	4.69E-05

Table 2.7: Parameters Used for Exposure Assessment

Body weight (Adults ≥ 20 years)	70.7 kg[1]
Soil ingestion rate	0.02 g/day[2]
Inhalation rate	15.8 m³/day[1]
Water ingestion rate	1.5 L/day[1]

1: Richardson, 1997; 2: MADEP, 2002.

The maximum water concentrations for 2,4-D, carbaryl, and glyphosate were found at 311, 363, and 204 hours after spraying occurred, which is about 13, 15, and 8.5 days. This implies that the best time to take samples from the lake is in early to mid-June. The City of Kelowna currently monitors the lake for pesticides in July and has never found sample that tested positive for pesticides (City of Kelowna, 2009-2). However, the modelled water concentration does not vary greatly between June and July, and as the calculated water concentrations were quite low. It is likely that concentrations are below the detectable limit and this implies that the model predictions are in reasonable agreement with the city's monitoring data.

Conclusions

The City of Kelowna and a portion of Okanagan Lake was modelled using a deterministic Level IV fugacity model to determine the multimedia fate of 2,4-D, carbaryl and glyphosate in the environment. A one-year simulation was run using spray records from local orchards and the concentrations in air, soil, water, and sediment were calculated at each time step. Concentrations at the end of the one year simulation were very low and will not result in accumulation in the environment. As concentrations were found to be very low, a detailed probabilistic analysis was not performed, however in case of availability of more data; Monte Carlo analysis will be helpful to determine variability in this hazard index value by considering various sensitive subpopulations (*e.g.*, children).

The single-value maximum concentrations were used to calculate human exposure through inhalation, soil ingestion, and water ingestion and hazard quotients were calculated for each pesticide. The hazard quotients found for 2,4-D, carbaryl, and glyphosate indicate that the current pesticide control measures should result in very low hazard for the population of Kelowna (BC Canada) and therefore the current pesticide control measures are likely sufficient.

The analysis presented in this work follows the principles of deterministic analysis. There is however increased realization of the role of uncertainty on analysis results. Future analysis can be strengthened by enhancing this deterministic setting to probabilistic setting, *e.g.*, by Monte-Carlo, but also beyond probabilistic analysis to ones that model epistemic uncertainties.

Acknowledgements

This chapter presents the results of an ongoing research funded under Canada NSERC Discovery Grant program of the last author.

References

Bayer Crop Science (2008): Material Safety Data Sheet Sevin XLR Carbaryl Insecticide Liquid Suspension. Available online: http://www.bayercropscience.ca/English/LabelMSDS/221/File.ashx.

BC Ministry of Agriculture, Food, and Fisheries (2001): City of Kelowna Agriculture in Brief. Available online: http://www.agf.gov.bc.ca/resmgmt/sf/agbriefs/Kelowna.pdf.

Cheminova (1999): Material Safety Data Sheet: Glyfos "Available online: http://www.sentinelpestcontrol.com/msds/GlyfosMSDS.pdfCity of Kelowna (2010)". Parks and Schools Map. Available online: http://www.kelowna.ca/CityPage/Docs/PDFs//Maps/Parks per cent 20and per cent 20Schools.pdf.

Citra, M. J. (2004): Incorporating montecarlo analysis into multimedia environmental fate models. Environmental Toxicology and Chemistry, 23(7): 1629–1633.

City of Kelowna (2009-2): Water Quality Frequently Asked Questions. Available online: http://www.kelowna.ca/CM/Page 402.aspx.

Dempster, A. P. (1967): Upper and Lower Probabilities Induced by a Multivalued Mapping. The Annals of Mathematical Statistics, 38(2): 325–339.

Di Guardo, A., Calamari, D., Zanin, G., Consalter, A., and Mackay, D. (1994): A fugacity model of pesticide runoff to surface water: Development and validation. Chemosphere, 28(3): 511-531.

Harvie, K. (2010): Personal communication.

Kühne, R., Breitkopf, C., and Schüürmann, G. (1997): Error propagation in fugacity level-III models in the case of uncertain physicochemical compound properties. Environmental Toxicology and Chemistry, 16(10): 2067–2069.

Lang, C., Tao, Shu., Wang, X., Zhang, G., Li, J., and Fu, J. (2007): Seasonal variation of polycyclic aromatic hydrocarbons (PAHs) in Pearl River Delta region, China. Atmospheric Environment, 41(37): 8370–8379.

MacKay, D. (2001): Multimedia Environmental Models.CRC Press.

Mackay, D., Paterson, S., Cheung, B., and Neely, W.B. (1985): Evaluating the environmental behaviour of chemicals with a level III fugacity model.Chemosphere, 14(3/4): 335-374.

Menon, P. and Gopal, M. (2003): Dissipation of ^{14}C carbaryl and quinalphos in soil under a groundnut crop (*Arachis hypogaea* L.) in semi-arid India.Chemosphere, 53(8): 1023-1031.

NRC (2008):Natural Resources Canada. Okanagan Basin Waterscape. Available online at: http://www.nrcan.gc.ca/earth-sciences/products-services/mapping product/geoscape/waterscape/okanagan-basin/6271. Accessed 20/10/2012.

Paraíba, L.C., Bru, R., Carrasco, J.M. (2002): Level IV fugacity model depending on temperature by a periodic control system. Ecological Modelling, 147: 221–232.

Paraíba, L.C., de M. Plese, L. P., Foloni, L. L., Carrasco, J. M. (2007): Simulation of the fate of the insecticide carbofuran in a rice field using a level IV fugacity model. Spanish Journal of Agricultural Research, 5(1): 43-50.

Paraíba, L.C., Luiz, A.J.B., Pérez, D.V. (2004): Estimativa do coeficiente de sorç o no solo de pesticidas.Arq. Inst. Biol, 71 (supl): 701-704.

Shafer, G. (1976): A mathematical theory of evidence. Princeton University Press, Princeton, NJ, 297 pp.

Shaw, J. (1977): Sedimentation in an Alpine Lake during Deglaciation, Okanagan Valley, British Columbia, Canada. *Geografiska Annaler. Series A*, Physical Geography, 59(3): 221-240.

Tao, S., Yang, Y., Cao, H. Y., Liu, W. X., Coveney, R. M., Xu, F. L., Cao, J., Li, B. G., Wang, X. J., Hu, J. Y., and Fang, J. Y. (2006): Modeling the dynamic changes in concentrations of gamma-hexachlorocyclohexane (gamma-HCH) in Tianjin region from 1953 to 2020. Environmental pollution, 139(1): 183–93.

UAP Canada (2008): Material Safety Data Sheet PAR III Turf Herbicide. Available online: http://www.uap.ca/products/documents/2008-ParIIITurfHerbicide PCP27884revised.pdf.

United States Environmental Protection Agency Integrated Risk Information System (USEPA IRIS). (2010): Available online: http://www.epa.gov/iris.

Veeh, R. H., Inskeep, W. P., and Camper, A. K. (1996): Soil Depth and Temperature Effects on Microbial Degradation of 2,4-D. Journal of Environmental Quality, 25: 5-12.

World Health Organization. (1996): WHO/FAO Data Sheets on Pesticides No. 91 Glyphosate. Available online: http://www.inchem.org/documents/pds/pds/pest91_e.htm.

World Lake Database. (1999): Okanagan Lake. Available online: http://wldb.ilec.or.jp.

Zablotowicz, R. M., Accinelli, C., Krutz, L. J., and Reddy, K. N. (2009): Soil Depth and Tillage Effects on Glyphosate Degradation. Journal of Agricultural Food Chemistry, 57: 4867–4871.

Zadeh, L. (1965): Fuzzy sets. Information and Control, 8(3): 338–353.

Symbols

δ_{ij}: thickness of the diffusion layer between compartments i and j

θ: soil water volumetric fraction

μ_W: viscosity of water

ρ_i: density of compartment i

$\tau_{1/2i}$: half-life of the chemical in compartment i

τ_A: residence time of Kelowna airshed

τ_W: residence time of Okanagan Lake

v_i: molar volume of chemical i

ϕ_i: porosity of compartment i

A_{ij}: area between compartments i and j

cl_D: fraction of clay in sediment

d_{ij}: transfer coefficient from compartment i to compartment j

D_{pi}: diffusivity of the pesticide in compartment i

G_i: volumetric flow rate of compartment i

H_P: Henry's constant for pesticide P

K_{OC}: organic carbon partitioning coefficient

K_{OW}: octanol-water partitioning coefficient

m_i: molar mass of chemical i

oc_i: percentage of organic carbon in compartment i

P_V: vapour pressure

R: ideal gas constant

S_A: aqueous solubility

S_{SA}: specific superficial area of the sediment

sd_D: fraction of sand in sediment

st_D: fraction of silt in sediment

T: average air temperature

V_i: volume of compartment i

Z_i: fugacity capacity of compartment i

Chapter 3

Effectiveness of Simultaneous Adsorption and Biodegradation over Indivisual Processes for Reduction of Cyanide from Effluents

Rajesh Roshan Dash[1], Chandrajit Balomajumder[2] and Rakesh Roshan Dash[3]

[1]School of Infrastructure, IIT Bhubaneswar – 751 013, Odisha, India
[2]Department of Chemical Engineering, IIT Roorkee – 247 667, Uttarakhand, India
[3]VSS University of Technology, Burla – 768 018, Odisha, India

ABSTRACT

The present investigation focused on a comparative study for removal of cyanide from separate solutions of sodium, zinc and iron cyanide compounds by three different processes such as; adsorption, biodegradation and simultaneous adsorption and biodegradation (SAB) processes. Adsorption studies were carried out on commercial granular activated carbon (GAC). Biodegradation and SAB studies were conducted with suspended and immobilised cultures of *Pseudomonas fluorescens* respectively. Effect of pH, temperature, agitation time on percentage removal of cyanide was measured at an initial cyanide concentration of 100 mg/L and it was found that these parameters significantly affect the three process performances. The SAB process was found to have better removal efficiency as compared to adsorption and biodegradation processes. SAB process could achieve more than 99 per cent removal efficiency for cyanide concentrations up to 100 mg/L in sodium, zinc and iron cyanide solutions. Higher percentage of cyanide removal and specific uptake was achieved in case of zinc cyanide complexes as compared to other cyanide complexes. Adsorption isotherms evaluated the uptake and degradation on cyanide in SAB process.

Keywords: Pseudomonas fluorescens, Zinc cyanide, Iron cyanide, Sodium cyanide, SAB.

Introduction

The cyanide compounds are the strictly regulated compounds world-wide because of their extreme toxicity (Ebbs, 2004). Although cyanides are present in small concentrations (1-4000 mg HCN/kg of plant weight) in these plants and microorganisms, their large-scale presence (10-10000 mg CN^-/L) in the environment is attributed to the human activities as cyanide compounds are extensively used in industrial applications (Patil and Paknikar, 2000; Dzombak *et al.*, 2006; Zhang *et al.*, 2010). Large amounts of cyanides are released into solid wastes and wastewaters due to different industrial activities; for example, metal plating, aluminum electrolysis, coal gasification, coal coking, ore leaching, pharmaceuticals, synthetic fibers, and plastics (Chen *et al.*, 2009). Since cyanide is a metabolic inhibitor, particularly in terminal cytochromes of electron transport chain, cyanide pollution causes great damage to ecosystems. Moreover, these waste products generally contain other contaminants, including heavy metals such as nickel, copper, zinc, and iron (Chen *et al.*, 2009; Gupta *et al.*, 2012). Cyanide, in some forms, is a very powerful and fast acting toxin. To protect the environment and water bodies, wastewater containing cyanide must be treated before discharging into the environment. Typical cyanide consent to discharge standards to sewers is range from 0.5 mg/L to 1.0 mg/L. U.S. EPA standards for drinking and aquatic-biota waters regarding total cyanide are 200 and 50 ppb, respectively. The German and Swiss regulations have set limit of 0.01 mg/L for cyanide for surface water and 0.5 mg/L for sewers. The Mexican ministry of the environment and natural resources (SEMARNAT) and Central Pollution Control Board (CPCB), India have set a limit for cyanide in effluent as 0.2 mg/L (Dash *et al.*, 2009a,b, 2013). Due to their toxic effects, cyanide-containing effluents cannot be discharged without detoxification to the environment. Cyanide compounds present in environmental matrices and waste streams as free, simple and complex cyanides, cyanates and nitriles (Ebbs, 2004). The metal-cyano complexes are the major forms in metal containing waste due to the quick binding of cyanide with metals (Chen *et al.*, 2009). These cyano complexes are assumed to constitute a significant fraction of cyanide-related waste. Such metal-cyano complexes are highly stable and more resistant to biological attack compared with free cyanide. Cyanide and its related compounds can be treated and removed by several processes (Dash *et al.*, 2009a,b; Dwivedi *et al.*, 2011; Dash *et al.*, 2013).

Adsorption and bio-degradation are two significant methods for treatment of wastewater bearing cyanide compounds. Biological treatment is a cost-effective and environmentally acceptable method for cyanide removal compared with the other techniques currently in use (Chen *et al.*, 2008). Most reports demonstrated the ability of microorganisms to utilize cyanide as a source of nitrogen or both carbon and nitrogen by specific enzymes and pathways (Chapatwala *et al.*, 1998; Ebbs, 2004). Metal cyanide compounds could serve as a source of nitrogen for *Pseudomonas fluorescens* NCIMB 11764, originally isolated on cyanide (Harris and Knowles, 1983; Chen *et al.*, 2009). Metabolism of cyanide by various strains of *Pseudomonas, Acinetobacter, Bacillus,* and *Alcaligenes* has been reported (Dursun *et al.*, 1999; Dash *et al.*, 2009a,b; Dash *et al.*, 2013). Degradation of cyanide compounds by fungi such as *Fusarium solani, F. oxysporum, Gloeocerocospora sorghi, Fusarium lateritum, Stemphylium loti* (Campos *et al.*, 2006; Dash *et al.*, 2006, 2008, 2010) has also been reported. Bio degradation is performed in presence of microbes either in mobilized or immobilized phase. The immobilization of living microbial cells on a suitable adsorbent improves the removal efficiency (Oh *et al.*, 2011, Dash *et al.*, 2013). This improvement is due to the monolayer/bio-layer formation on the adsorbent bed where adsorption and biodegradation occurs simultaneously (known as simultaneous adsorption and biodegradation process) (Dash *et al.*, 2008; Depci, 2012). It is also pointed out that attached growth processes and combined processes such as oxic/anoxic processes may prove advantageous for the cyanide detoxification (Patil and Paknikar, 2000; Dash *et al.*, 2008). Recently developed methods dealing with the removal of toxic compounds by adsorption and biological treatment, either operated separately or simultaneously in one unit (Yazici *et al.*, 2009; Oh *et al.*, 2012). In most cases, the presence of both processes in one unit results in a better removal and process performance (Oh *et al.*, 2011; Dash *et al.*, 2013). Microbial mass can, in some extent, adsorb the substances, but at the same time it also degrades them. On the other hand, adsorption of the substances onto adsorbent reduces the inhibitory effect of the substances for microbial mass. Accordingly, the process is expected to be more stable and the toxic compounds may be converted into less harmful substances. Activated carbon is able to enrich dissolved oxygen; probably this oxygen can also be utilized by microorganisms adsorbed on activated carbon (Mordocco *et al.*, 1999; Dash *et al.*, 2008; Oh *et al.*, 2012).

In the present study, cyanide was removed from sodium, zinc and iron cyanide contaminated synthetic solutions by adsorption, biodegradation and SAB in separate batch experiments. Granular Activated Carbon (GAC) was used as adsorbent. A variety of enzymatic pathways for cyanide degradation has been reported. These hydrolytic pathways lead to the production of formamide or formate plus ammonia or the direct formation of bicarbonate plus ammonia *via* cyanide oxidase or by a dioxygenase (Harris and Knowles, 1983). Cell-free extracts from *Pseudomonas fluorescens* NCIMB 11764 catalysed the degradation of cyanide into products that included CO_2, formic acid, formamide and ammonia. Cyanide-degrading activity (CDA) was localized to cytosolic cell fractions and was observed at substrate concentrations as high as 100 mM (2600 mg CN^-/L) (Kunz *et al.*, 1992). There are various reports which describe the cyanide degradation ability of *P. fluorescens* for various cyanide compounds (Kunz *et al.*, 1992; Dash *et al.*, 2008). Suspended and immobilised cultures of *Pseudomonas fluorescens* (MTCC Code 103) was used to evaluate the performance of the culture for removal of the three different cyanide compounds (sodium, zinc and iron cyanide) by biodegradation and SAB. The effect of pH, temperature, contact time and initial concentration of cyanide on removal of cyanide from the cyanide contaminated synthetic solutions were studied separately in batch experiments.

Materials and Methods

All the chemicals were of analytical grade and solutions were prepared by Milli-Q water. The synthetic solutions of sodium cyanide (NaCN), Ferro cyanide/iron cyanide (FeCN) were prepared by dissolving NaCN (1.88 g/l), $K_4Fe(CN)_6.3H_2O$ (2.7 g/l) in Milli-Q water respectively, where as for zinc cyanide (ZnCN) solution sulfates of zinc (3.14 g/l) was added to KCN (3.25 g/l). The three cyanide compounds in the initial concentration ranges of 50 to 400 mg CN^-/L were used in the present study separately. The commercial GAC obtained from M/s. S. d. fine-chem, Ltd, India (bulk density 400 g/L, particle size of 2-4 mm, BET Surface area 583.35 m ^{2}/g and micropore (<2nm) volume 0.2112 cm^3/g) was used in the study after purification with Milli-Q water and dried at 110 °C for 24 h. Adsorption studies were conducted in 250 ml conical flasks containing 100 ml of synthetic cyanide solution with 20 g/L of GAC.

Freeze-dried culture of *Pseudomonas fluorescens* (*P. fluorescens*) (MTCC Code 103) species was obtained from Institute of Microbial Technology (IMTECH), Chandigarh, India, was revived in the nutrient broth at 26 °C and pH 7.5. The sterile buffer medium (pH 7.5) used for growth study of *P. fluorescens* contained: glucose (5.0 g/L), peptone (1.0 g/L), yeast extract (1.0 g/L), K_2HPO_4 (0.5 g/L), KH_2PO_4 (0.5 g/L), $MgSO_4$ $7H_2O$ (0.05 g/L), and NH_4SO_4 (0.5 mg/L). After the culture was inoculated into 100 ml of growth medium (in 1:100 ratio) in a 250 ml conical flask, it was incubated at 26 °C in an agitated shaker (120 rpm) for 24 h (Dash *et al.*, 2008). Microorganisms were transferred (in 1:100 ratio) into the growth medium containing cyanide ions as only source of carbon and nitrogen replacing glucose, peptone yeast extract and NH_4SO_4 from the enrichment medium. Sodium cyanide, Zinc cyanide and Ferrocyanide was first used as both carbon and nitrogen source with the mineral salts. But no growth was observed in media containing cyanide compound as the sole source of carbon and nitrogen. Addition of glucose into the medium had major influence on degradation of cyanide compounds and led to bacterial growth and biodegradation of cyanide. Hence, glucose was used as the carbon and energy source in the biodegradation medium. Again, addition of NH_4SO_4 to the medium reduced the utilization of cyanide compounds as nitrogen source. The biodegradation medium was prepared by mixing the cyanide solution (as the only source of nitrogen) autoclaved separately and the sterilized solution containing other ingredients such as glucose (5 g/L), KH_2PO_4 (0.5 g/L), K_2HPO_4 (0.5 g/L), $MgSO_4$ (0.05 g/L). The microorganisms were adapted to cyanide in the medium starting from a lower concentration of 5-50 mg CN^-/L. The initial pH of the solution was adjusted to the desired value by using sterile dilute and concentrated H_2SO_4 or NaOH. Sterilization of the GAC, glassware and medium was performed in an autoclave at 121 °C for at least 20 min. All biodegradation studies were performed separately in 250 ml conical flasks containing *P. fluorescens* culture (1:100) in 100 ml of the biodegradation medium.

For immobilization of microbes on GAC, a known volume of thick slurry of *P. fluorescens* cells, whose concentration was determined by drying a sample to a constant weight, was added to purified and sterilized GAC. The GAC was shaken in an incubator shaker for 6 h at 120 rpm and left for 1 day. The supernatant liquor was taken out, and the biologically activated GAC on which the cells of *P. fluorescens* were immobilized were used for SAB study. *P. fluorescens* immobilized GAC was added to the biodegradation medium (containing 50-400 mg CN^-/L) to maintain 20 g/L biological activated carbon (BAC) concentration for SAB study (Dash *et al.*, 2013). Effect of initial pH, temperature, cyanide concentration and agitation time on adsorption, biodegradation and SAB for removal of cyanide was studied. pH and temperature was optimized for the cyanide compounds with initial concentration of 100 mg/L from a range of pH 4-11 and temperature 20-45 °C keeping one parameter constant at a time. All studies were conducted in 250 ml conical flasks in rotary incubator shaker at 150 rpm for an agitation period of 120 h (Dash *et al.*, 2008, 2009). Figure 3.1 presents the photomicrographs of GAC and BAC by scanning electron microscope (SEM) (LEO® 435 VP, UK). Parameters such as pH, total cyanide content, optical density (550 nm) were checked periodically. Total cyanide was determined by pyridine–barbituric acid colourimetric method (578 nm) after distillation as described in Standard Methods with a precision up to 0.001 mg/L. pH was measured using pH meter as specified by standard methods (APHA, 2001).

Figure 3.1: Photomicrogrph of BAC Immobilised with *P. fluorescens*.

Results and Discussions

Effect of pH

pH plays an important role for the adsorption of metal cyanides on GAC (Dash *et al.*, 2009b). Effect of pH on removal of cyanide from aqueous solutions by adsorption, biodegradation and SAB processes was investigated. The effect of initial pH of the sample solution on cyanide removal has been given in Table 3.1. The significant increase in adsorption of cyanide was above pH 7 in NaCN solution. Maximum adsorption occurred in pH 9-10. Again there was decrease in adsorption above pH 10. In case of ZnCN solutions the adsorption was optimum in the pH range of 6-7. At higher or lower pH values there was decrease in adsorption. The FeCN complexes showed adsorption was greater at lower pH values and decreased as pH increased. Above pH 10, ferrocyanide adsorption was negligible. From the results it was observed that, percentage removal of metal cyanide complexes were maximum at neutral and slight acidic pH. Whereas there was increase in percentage removal for sodium cyanide in alkaline conditions.

Table 3.1: Effect of pH on Removal of Cyanides by Adsorption, Biodegradation and SAB

pH	Per cent Removal of Cyanide								
	Adsorption			Biodegradation			SAB		
	NaCN	ZnCN	FeCN	NaCN	ZnCN	FeCN	NaCN	ZnCN	FeCN
4	45.2	71.6	81.7	—	—	84.8	46.2	73.9	91.8
5	50.2	80.2	83.3	36.4	53.8	94.1	72.1	89.8	98.7
6	52.4	84.2	82.6	73.0	88.4	94.4	92.6	98.5	99.9
7	53.7	84.4	79.6	89.1	93.4	93.8	99.9	99.9	99.2
8	59.9	82.5	72.8	89.0	90.8	82.6	99.9	98.1	87.6
9	63.2	79.4	64.2	85.3	84.9	69.4	99.5	87.6	68.9
10	63.2	66.0	43.3	75.4	67.4	43.6	88.3	74.5	47.7
11	61.7	42.4	31.2	44.0	23.2	32.8	84.6	56.3	33.8

At pK_a 9.39 value of cyanide, pH had a marked effect on the stability of sodium cyanide (Davidson and Veronese, 1979; Dash *et al.*, 2009b). ZnCN is a weak-acid dissociable (WAD) cyanide complex as it is easily dissolved under mildly acidic conditions (pH = 4-6) (Dzombak *et al.*, 2006). Hence in lower pH ranges (pH<6) there is a possibility of dissociation of ZnCN, which reduced the adsorption efficiency. Hence the optimum removal was found at a pH above 6. The adsorption of ferrocyanide on activated carbon is reported as a function of pH and found greater adsorption at low pH, but below pH 3 there was possibility of volatilization of ferrocyanide (Dzombak *et al.*, 2006; Dash *et al.*, 2013). At these pH values cyanide ion exists as HCN which was a weak acid and is highly soluble in water (Davidson and Veronese, 1979; Van der Merwe, 1991). Davidson and Veronese (1979) related the effect of pH on the adsorption of metal cyanide complexes to the strong adsorption of both hydroxide and hydronium ions, whereas Adams *et al*. (1987) and Adams and Fleming (1989) favoured a mechanism in which OH^- reacts with the functional groups on the surface of the carbon. Huang and Ostovic (1978) proposed that hydration of the activated carbon may result in the formation of reactive surface functional groups. The relative proportion of the surface functional groups may vary with the method of preparation of the activated carbon, thereby causing a difference in the value of the surface acidity constants (equilibrium constants). The cyanide ion is a nucleophile, and in contact with the surface of activated carbon, could replace the OH^- present in various surface functional groups. Both the processes would be equivalent to CN^- uptake by ion exchange. In acidic medium (pH < 3), complexes such as hexacyano-iron (II) or (III) probably absorbs as the protonated anion (Dash *et al.*, 2007, 2013).

The optimum pH for maximum percentage of removal of cyanide in the presence of suspended cultures and immobilized cultures of *P. fluorescens* from NaCN, ZnCN and FeCN solutions was found to be 7, 7 and 6 respectively (Table 3.1). At pH above 8 there was a sudden decrease of cyanide removal for all the three compounds. However, below pH 5, no biological activity was found for NaCN and ZnCN in the presence of suspended cultures, but there was possibility of degradation of FeCN at pH less than 5. Also, the percentage cyanide removal from FeCN solution decreased significantly above pH 7. At acidic pH conditions the percentage removal of cyanide was higher in FeCN solution as compared to NaCN and ZnCN solutions. There was decrease in CN^- removal rate above and below pH 6-7 for all three processes. At higher pH conditions the efficiency of adsorption decreased, and below pH 6, there was possibility of volatilization of zinc cyanide (Dzombak *et al.*, 2006). In WAD complexes, cyanide is readily released from the complexes when the pH is lowered to 4.5 to 6. Weakly complexed metal-cyanides decompose at pH values lower than 4 and evolve HCN (Pohlandt *et al.*, 1983; Dash *et al.*, 2013). Therefore, WAD refers to any free cyanide already present and cyanide released from nickel, zinc, copper, and cadmium complexes (but not from iron or cobalt complexes) (Luque-Almagro *et al.*, 2005). The iron complexes are very stable and for this reason less toxic and more recalcitrant. The stability of the complexes depends on the pH; they are, in general, more stable as the pH increases (Dash *et al.*, 2008). The zinc-cyanide containing wastewaters emanating from plating industries have pH values in the alkaline range (7.5–10) (Patil and Paknikar, 2000). These results are conducive for the practical use of the process, as very little pH adjustment of the effluents may be required.

It is known that the relationship between simple and complex cyanides in water is dependent on factors such as the pH and the heavy metal concentrations capable of forming metal-cyanide complexes (Dash *et al.*, 2013). Under alkaline conditions, free cyanide (HCN, CN^-) is completely ionised and stable metal complexes are formed. In neutral and acidic conditions, the free cyanide is weakly ionised and the formation and partial liberation of HCN is favoured. The growth and biological activity of the microbes depend on the pH conditions of the medium. Probably, the optimum pH for degrading cyanide complexes is the result of a balance between the stability of the complex and the optimum pH for growth. Patil and Paknikar (2000) obtained an optimum pH 7 for cyanide biodegradation by *Pseudomonas* sp. Degradation of cyanide compounds by bacterial and fungal species was commonly observed at neutral and alkaline pH values. In any case, any bacterium is able to use both free cyanide and its metal complexes at alkaline pH, thus providing a clear advantage since both cyanide volatilization and precipitation of its metal complexes are prevented. In the present study, pH 7 was taken as the optimum pH value for the three processes for the three cyanide solutions.

Effect of Temperature

The influence of temperature on removal cyanide compounds by the three processes has been presented in Table 3.2. It was observed that the difference was not great at different temperatures, but adsorption increased slightly with rise in temperature in FeCN solution. Increased adsorption at higher temperature is difficult to explain as it is against the general adsorption behaviour. However, such adsorption nature may be explained carefully by examining the mode and type of the adsorption process. Diffusion of adsorbate species from the bulk phase into pores of adsorbent have been observed in some of the adsorption processes of endothermic nature (Dash *et al.*, 2009b). Here the rise of temperature favours the adsorbate transport within the pores of the adsorbent. The increase in adsorption with temperature was mainly due to an increase in number of adsorption sites caused by breaking of some of the internal bonds near the edge of the active surface sites of the adsorbent (Dash *et al.*, 2009b, 2013). With increasing temperature chemisorption predominate physical sorption of cyanide onto activated carbon. Hydrolytic decomposition reaction was more important than oxidation to cyanate at high temperatures in presence of activated carbon (Adams, 1990). Increase in adsorption behaviour at higher temperatures for FeCN may be due to the dissociation of iron cyanide complexes.

Table 3.2: Effect of Temperature on Removal of Cyanides by Adsorption, Biodegradation and SAB

pH	Per cent Removal of Cyanide								
	Adsorption			Biodegradation			SAB		
	NaCN	ZnCN	FeCN	NaCN	ZnCN	FeCN	NaCN	ZnCN	FeCN
20	62.6	81.6	78.9	89.6	91.4	92.3	95.3	93.7	96.4
25	65.8	83.7	80.8	96.4	93.4	94.3	99.9	99.9	99.9
30	67.0	83.9	82.6	96.4	93.4	94.4	99.2	99.5	99.7
35	67.2	83.9	82.6	94.4	91.3	93.6	95.7	97.1	98.0
40	67.0	84.1	83.5	80.2	85.7	90.4	93.7	95.7	94.9
45	67.3	84.2	83.5	76.2	79.2	70.9	89.7	90.7	92.2

In case of biodegradation and SAB maximum growth and removal of CN⁻ was observed at temperature 25-30 °C and growth of *P. fluorescens* ceased above 40 °C. The percentage removal of cyanide by biodegradation in the presence of the microbes mainly depends on the optimum growth condition of the microbes. The optimum temperature for removal of cyanide complexes was the result of a balance between the stability of the complex and the optimum temperature for growth. The dissociation of cyanide from metal complexes increases with the increase in temperature. The dissociation rate varies depending on the stability of the complexes. The dissociation constant of ZnCN complexes are higher than FeCN complexes. 30 °C was taken as the optimal temperature for all the studies.

Effect of Contact/Agitation Time

Figures 3.2(a), (b) and (c) represent the percentage removal of 100 mg/L of sodium, zinc and iron cyanide complexes respectively with increase in agitation time by adsorption, biodegradation and SAB at 30 °C and pH 7. The rate of increase in the percentage removal of cyanide species with the increase in agitation time is appreciably fast at the initial stage. However, after a period of ~ 42 h in case of NaCN solutions, ~ 30 h in case of ZnCN solutions and ~ 36 h in case of FeCN solutions, rate of increase in percentage removal of cyanide species by adsorption with increase in agitation is less. It was observed that after 84 h, 72 h and 78 h in case of NaCN, ZnCN and FeCN solutions respectively there was no further increase in percentage removal of cyanide by GAC and a steady state had arrived. The highest percentage of cyanide removal was achieved in case of ZnCN solutions as compared to NaCN and FeCN

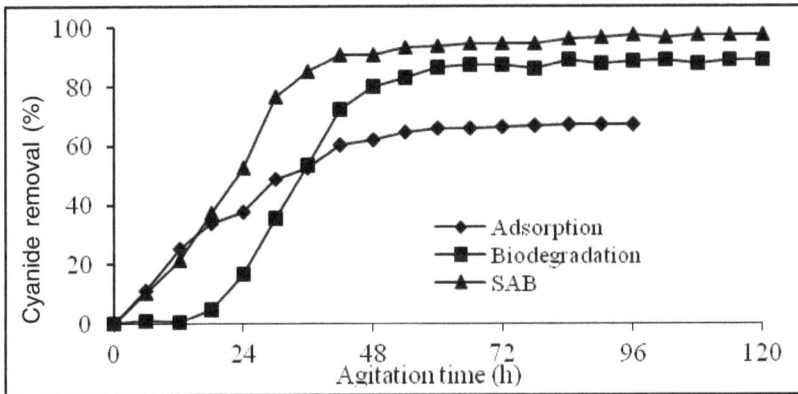

Figure 3.2(a): Effect of Agitation Time on Removal of NaCN.

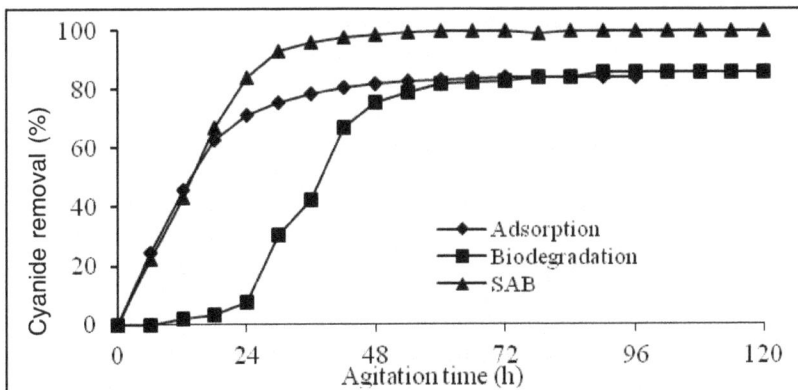

Figure 3.2(b): Effect of Agitation Time on Removal of ZnCN.

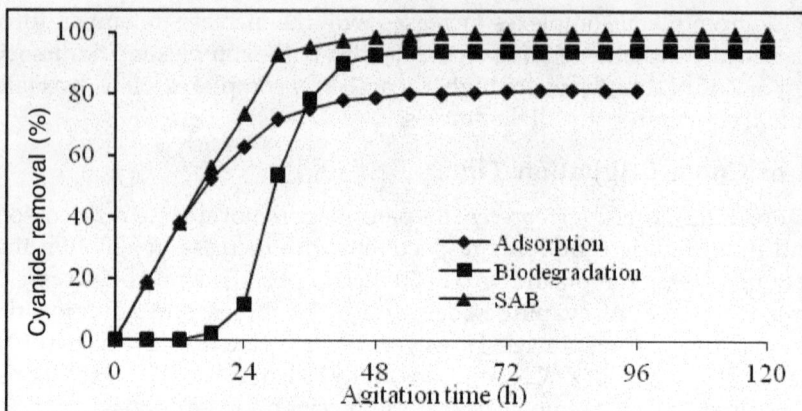

Figure 3.2(c): Effect of Agitation Time on Removal of FeCN.

solutions. Further, it was also observed that, the steady state and equilibrium condition arrived earlier in case of cyanide adsorption from ZnCN solutions as compared to NaCN and FeCN solutions. Cyanide removal from NaCN solution was found to be the lowest and it takes more time to reach the equilibrium condition. The slow step in the NaCN solution was considered to be the diffusion of CN^- from the bulk solution to the active surface sites. This process would be influenced by the concentration gradient between those two points and the thickness of the diffusion layer which was a function of agitation process. Available adsorption results revealed that the uptake of cyanide species were fast at the initial stage of contact period, and thereafter, it becomes slower near the equilibrium. During the initial stage of the experiment a large number of active surface sites of GAC are available for adsorption (Deveci *et al.*, 2006). Thus, the concentration of cyanide in the solutions as well as the driving force for adsorption of cyanide on the GAC surface is maximum. Further, the resistance to mass transfer between bulk phase and adsorbent is overcome by the energy provided by agitation to bring the cyanide species from bulk of the solutions to the active sites of the adsorbent. In physical adsorption most of the adsorbate species are adsorbed within a short interval of contact time (Dash *et al.*, 2013). GAC contains both positive as well as negative sites on its surface, with negative charges predominating over positive charges (Mondal *et al.*, 2008). The presence of positive charge of the metal ions get attracted towards the negatively charged GAC surfaces sites (Adhoum and Monser, 2002) neutralizing some negative charges of GAC surface after their adsorption on the negative sites of GAC and will create some additional positive sites on the GAC surface. The positive sites created on the GAC surface make the cyanide ion adsorbed on the surface of the GAC. Here chemisorption predominates over physical adsorption for adsorption of cyanides on GAC. Hence, strong chemical binding of adsorbates with adsorbent required longer contact time for the attainment of equilibrium.

From the figures it was evident that no biodegradation or growth of microbe was found in the staring 12-18 h of agitation. This may be due the lag phase of microbe.

Degradation started at 24 h and after 72 h of agitation there was no significant increase in the percentage removal of cyanide. This represents the removal of cyanide from the medium occurred during the initial stages of growth. This suggested that the complex was taken up by the biomass prior to utilization. The maximum percentage removal was found in the log phase of growth of microbes. Although it was evident that biodegradation delayed for 12-18 h due to delayed growth of microbes in the presence of cyanide ions, but in SAB process the percentage removal of cyanide was started earlier. This may be due to adsorption occurred in the first phase followed by biodegradation. With the increase in agitation period, greater numbers of cyanide ions were moved to the surface of BAC; as a result the percentage removal of cyanide increased with the increase in agitation time. The maximum percentage of removal of cyanides from the three cyanide solutions was faster in SAB process as compared to its corresponding adsorption and biodegradation processes. The combined process was more effective and less time consuming. The cyanide adsorbed on the biologically active GAC surface could be easily biodegraded by microbes as GAC acted as an enrichment surface and attached growth gave better efficiency (Yazici, 2009). Moreover during SAB there was the possibility of bio-regeneration of BAC, which increased the adsorption capacity and prolonged the time of adsorption process. The continuous bio-regeneration of GAC and biodegradation of cyanide adsorbed on GAC surface increase the possibility of higher efficiency and the increase in percentage removal (Dash *et al.*, 2008, 2009a). SAB process took a long duration to reach the exhaustion condition only after 108-120 h of agitation for all the three cyanide solutions. Hence, SAB is more efficient and prolonged process for the removal of cyanide as compared to adsorption and biodegradation alone.

Effectiveness of Adsorption, Biodegradation and SAB Process at Various Initial Cyanide Concentrations

Figures 3.3(a), (b) and (c) represent the percentage removal of cyanide from of sodium, zinc and iron cyanides solutions respectively at various initial cyanide concentrations by adsorption, biodegradation and SAB process. From the experimental results it was observed that, the removal efficiency decreased with increase in initial cyanide concentration for all processes. It can be noted that the increase in percentage removal of cyanide compounds with decrease in initial cyanide concentration was less for higher value of initial concentration of cyanide than that for lower value. Removal of higher concentrations of cyanide could be achieved with immobilized cultures in SAB process as compared to adsorption and biodegradation. Better removal efficiency was found for iron cyanide complex with higher concentrations by biodegradation and combined process of adsorption and biodegradation. Biodegradation of zinc cyanide was less as at higher concentrations, but in the presence of BAC, there was possibility of biological growth and degradation at high concentrated cyanide contaminated solutions. Biodegradation of sodium cyanide was more as compared zinc cyanide may be due to its readily availability to microbes, where as removal of iron cyanide in higher percentage may be due to the affinity of the microbe to iron. For zinc and iron cyanide complexes, removal due to adsorption was more as compared to biodegradation at higher concentrations of cyanide.

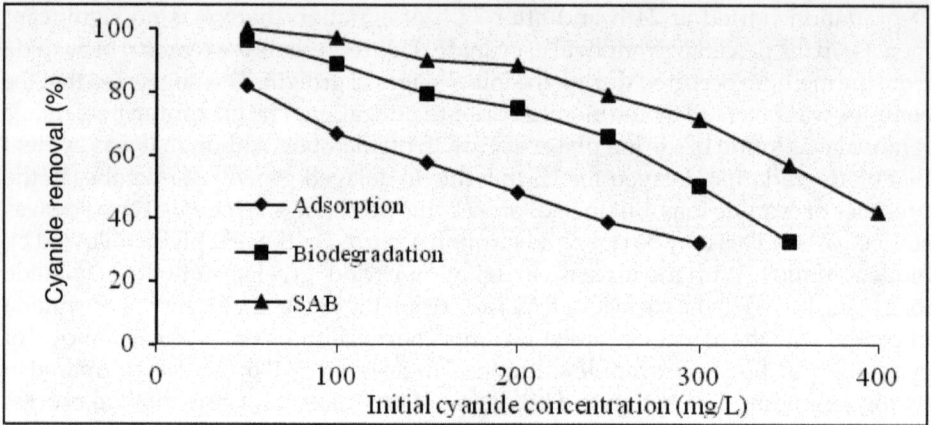

Figure 3.3(a): Effect of Initial Cyanide Concentration on Removal of NaCN.

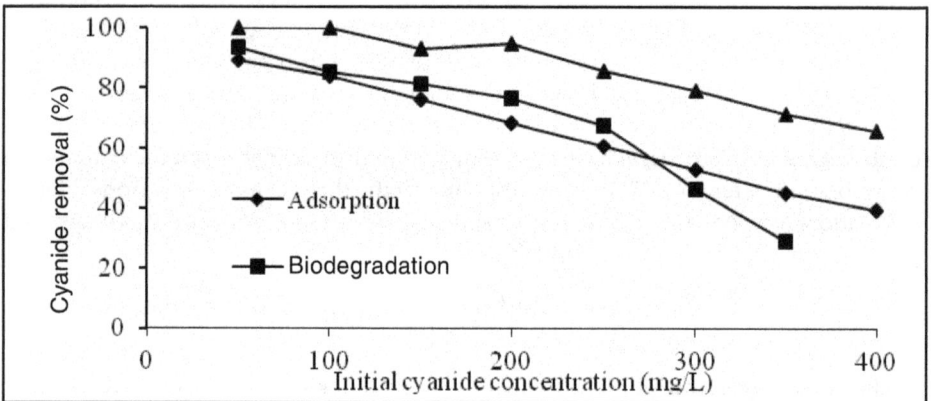

Figure 3.3(b): Effect of Initial Cyanide Concentration on Removal of ZnCN.

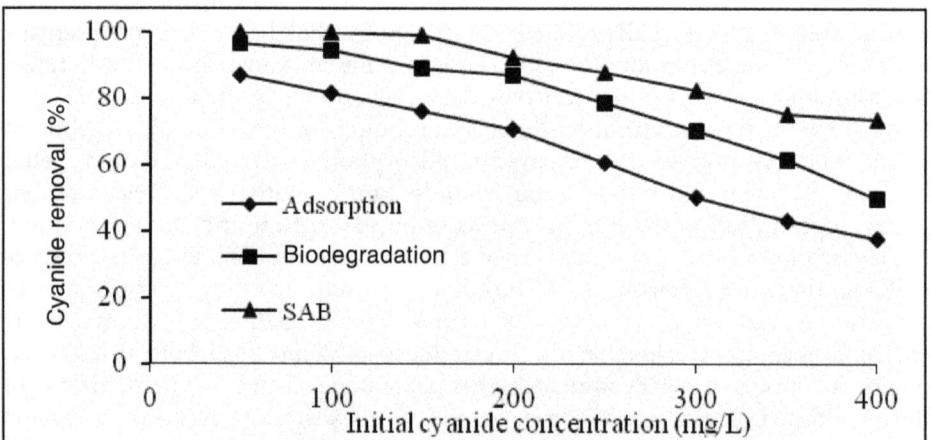

Figure 3.3(c): Effect of Initial Cyanide Concentration on Removal of FeCN.

It is a well-known fact of an adsorption process that at a particular environment the percentage removal depends upon the ratio of the number of adsorbate moiety to the available active sites of adsorbent. This ratio is also related to the surface coverage of the adsorbent (number of active sites occupied/number of active sites available) that increases with increase in the number of adsorbate moiety per unit volume of solution at a fixed dose of adsorbent. Less is the value of this ratio more is the percentage removal. At higher cyanide concentration this ratio was high and decreased gradually with the decrease in cyanide concentration as a result the percentage removal was increased with decrease in cyanide concentration (Dash *et al.*, 2009b, 2013). In biodegradation process, the toxicity of cyanide compounds exerts difficulties in bacteria capable of using these as a carbon source for growth (Dursun *et al.*, 1999). The growth of *P. fluorescens* was found in the absence of any external carbon source. However, the removal efficiency was increased with the addition of external glucose source. It could be easier to utilize cyanide as a source of nitrogen in the presence of another source of carbon and energy, as the amount of nitrogen needed for the growth is less than the requirement for carbon. The decrease in removal of cyanide with increase in initial concentration may be due to the toxicity of cyanide compounds to *P. fluorescens* at higher concentration. No significant biological activity was found in the medium above cyanide concentrations of 300 mg/L in the solutions. The initial concentration up to which biodegradation was possible is less in zinc cyanide as compared to other compounds may be due to the presence of free cyanide ions and zinc ions. Although metal-cyanide complexes by themselves are much less toxic than free cyanide, their dissociation releases free cyanide as well as the metal cation, which can also be toxic (Dash *et al.*, 2013). It was also evident from the Figure 3.3 that in SAB process, due to attach growth and combined process performance, resistance to cyanide toxicity by *P. fluorescens* was more and could achieved good efficiency even at higher concentrations. SAB process has been used successfully for degradation of various compounds such as phenol, toxic metals, dye etc., however its fate was not known for removal of cyanide compounds.

In the present case removal of cyanide was due to physical and chemical adsorption on GAC surface as well as due to the bioadsorption and biodegradation. The adsorption of cyanide ions was mostly due to monolayer adsorption. Hence in case of SAB process, due to the monolayer adsorption of cyanide on the GAC, there was a possibility of the reduction in toxicity of cyanide and metal ions to microbes, which increase the possibility of biodegradation rather than bioadsorption. It took some time for the immobilised biomass to be active in the biodegradation medium with cyanide solution. Hence, at the initial stage of agitation, the adsorption on GAC surface predominates over the biodegradation. Due to this no lag phase was found in case of the SAB process as compared to biodegradation. The initial rapid increase in the percentage removal of cyanide may be attributed to the dominating role of only adsorption process at the initial stage of SAB process. The increase in percentage removal in adsorption process was due to the presence of active sites on the GAC surface. However, in case of SAB process the adsorption decreased up to some extent due to the formation of biofilms on the active sites of adsorbent. But in that case biodegradation predominated over adsorption process for removal of cyanide from various cyanide solutions. The toxicity of the cyanide in the solution was reduced due to adsorption on monolayer and cyanide was easily available for degradation. This made the combined process more effective than the single process and gave better efficiency. Although, the process was continuous theoretically, due to the formation of biofilm on the active sites of GAC, equilibrium condition arrived. However, it had taken a long duration to reach the equilibrium condition as compared to adsorption process.

Isotherms for Adsorption and SAB

The efficiency of an adsorptive removal process is described either by specific uptake or percentage removal. Specific uptake is used to describe the total removal of adsorbate per unit mass of adsorbent, whereas percentage removal gives a comparison between the concentrations of adsorbate in the solution before and after treatment. Table 3.3 represents the specific uptake of cyanide from the three cyanide (NaCN, ZnCN and FeCN) solutions by GAC and BAC. The specific uptake of cyanide was found to be more by BAC, as compared to GAC for all the three cyanide solutions. It was also observed that, specific uptake of cyanide from ZnCN solution was found more in both adsorption and SAB process, may be due its easy dissociation as compared to cyanide in FeCN solutions. From Table 3.3 it was observed that, for adsorption process the equilibrium condition was arrived at an initial cyanide concentration of 200-250 mg/L, however in case of SAB process specific uptake increased upto higher concentrations. However, there is a decrease in specific uptake in SAB process at cyanide concentrations of 400 mg/L due the toxicity of cyanide to immobilised microbes, and adsorption predominated biodegradation at such a high concentration.

Table 3.3: Specific Uptake of Cyanide by GAC and BAC

Cyanide Concentration	Specific Uptake (mg CN-/g GAC)			Specific Uptake (mg CN-/g BAC)		
	NaCN	ZnCN	FeCN	NaCN	ZnCN	FeCN
50 mg/L	2.052	2.227	2.170	2.497	2.497	2.497
100 mg/L	3.350	4.210	4.075	4.995	4.995	4.995
150 mg/L	4.115	5.722	5.572	6.757	6.990	6.682
200 mg/L	4.850	6.840	7.070	9.740	9.510	9.240
250 mg/L	4.837	7.575	7.530	11.325	10.780	10.487
300 mg/L	4.845	7.920	7.560	13.290	12.825	12.330
350 mg/L		7.930	7.580	13.415	13.155	12.152
400 mg/L		7.940		13.180	13.200	13.060

The experimental adsorption and bio-adsorption equilibrium data for removal of cyanide from solutions of NaCN, ZnCN and FeCN on GAC and BAC were fitted with three adsorption isotherm models. Large numbers of researchers in the field of environmental engineering have used Freundlich and Langmuir isotherm equations to represent equilibrium adsorption data using plain and biological activated carbon for adsorption and SAB processes (Aksu and Gönen, 2004; Dash *et al.*, 2009a; Dai *et al.*, 2010). In the present study along with the two isotherm models, Redlich and Peterson (R-P) (Dash *et al.*, 2013) isotherm model has also been used to evaluate the adsorption and SAB isotherms. Table 4 represents the Freundlich, Langmuir and R-P isotherm constants and MPSD for GAC and BAC.

The Freundlich isotherm model is given in Eq. 2

$$q_e = X/M = q_e = K_F C_e^{\frac{1}{n}} \qquad \text{Eq. (2)}$$

where, K_F is the Freundlich constant $(mg/g)/(mg/L)^{1/n}$, the heterogeneity factor is $1/n$,

Langmuir isotherm model, given in Eq. 3

$$q_e = \frac{q_m K_L C_e}{1 + K_L C_e} \qquad \text{Eq. (3)}$$

where, K_L is the Langmuir adsorption constant $(1/mg)$ related to the energy of adsorption and q_m signifies adsorption capacity (mg/g). The essential characteristics of a Langmuir isotherm could be expressed in terms of a dimensionless separation factor, R_L which describes the type of isotherm and is defined by Eq. 4 (Dash *et al.*, 2009b, 2013):

$$R_L = 1/(1 + K_L C_e) \qquad \text{Eq. (4)}$$

If, $R_L > 1$, unfavourable;

$R_L = 1$, linear;

$0 < R_L < 1$, favourable;

$R_L = 0$, irreversible.

The values of R_L for $C_e = 100$ mg/L are given in Table 3.4.

R-P equation incorporates three parameters into an empirical isotherm, and therefore, can be applied either in homogenous or heterogeneous systems due to the high versatility of the equation. It can be described as follows in Eq. 5:

$$q_e = \frac{K_R C_e}{1 + a_R C_e^{\beta}} \qquad \text{Eq. (5)}$$

where, K_R is R-P isotherm constant (L/g), a_R is R-P isotherm constant $(L/mg)^{1/\beta}$ and β is the exponent which lies between 0 and 1, C_e is the equilibrium liquid phase concentration (mg/L).

Due to the inherent bias resulting from linearization, Marquardt's per cent standard deviation (MPSD) error function was used from a number of different error functions of non-linear regression basin to find out the best-fit isotherm model to the experimental equilibrium data. To determine the best-fit isotherm, Marquardt's per cent standard deviation (MPSD) error function may be generated as follows (Dash *et al.*, 2013):

$$\text{MPSD} = 100 \sqrt{\frac{1}{n_m - n_p} \sum_{i=1}^{n} \left[\frac{q_{e,cal} - q_{e,exp}}{q_{e,exp}} \right]_i^2}$$ Eq. (1)

Table 3.4: Isotherm Constants for GAC and BAC

Freundlich Isotherm Model Constants

Cyanide Compounds	n	K_F	R^2	MPSD
Adsorption on GAC				
NaCN	3.48432	1.15718	0.9544	9.2913
ZnCN	3.06185	1.55668	0.9132	14.7587
FeCN	2.99133	1.38388	0.8920	18.3558
SAB on P. fluorescens immobilised BAC				
NaCN	5.841121	4.39036	0.9129	18.20581
ZnCN	5.780347	5.67414	0.9152	19.74848
FeCN	5.506608	5.724	0.9309	17.35424

Langmuir Isotherm Model Constants

Cyanide Compounds	q_m	K_L	R_L	R^2	MPSD
Adsorption on GAC					
NaCN	5.15730	0.07262	0.12103	0.9867	6.29131
ZnCN	8.56898	0.06410	0.13495	0.9993	2.52841
FeCN	8.53242	0.05158	0.16238	0.9967	5.43892
SAB on P. fluorescens immobilised BAC					
NaCN	8.285004	8.56028	0.00117	0.8951	31.8121
ZnCN	12.04819	5.53333	0.0018	0.9766	14.3665
FeCN	13.6612	4.69231	0.00213	0.9742	20.1547

R-P Isotherm Model Constants

Cyanide Compounds	K_r	a_r	β	R^2	MPSD
Adsorption on GAC					
NaCN	0.47492	0.14946	0.9030	0.9982	6.17291
ZnCN	0.57474	0.07933	0.9667	0.9996	2.45786
FeCN	0.42054	0.03988	1.0418	0.9982	5.22377
SAB on P. fluorescens immobilised BAC					
NaCN	166.0019	32.0053	0.8686	0.9966	19.2319
ZnCN	128.0644	16.9421	0.8938	0.9965	13.4399
FeCN	191.2297	27.7879	0.8562	0.9965	16.7139

Here, $q_{e,cal}$ and $q_{e,exp}$ are the calculated and the experimental value of the equilibrium adsorbate solid concentration in the solid phase (mg/g) and n_m is the number of data points, where n_p is the number of parameters in the isotherm equation.

From Table 3.4, it was found that, similar to the adsorptive removal of cyanide from various cyanide solutions, various isotherm models were fitted to the SAB removal of cyanide. The correlation coefficient for R-P isotherm model was found closer to unity as compared to other isotherm models. The MPSD values for R-P isotherm models were found to be lower comparative to the other isotherm models for the three cyanide compounds for SAB studies immobilized BAC. Hence R-P isotherm model was found to be best fitted to the experimental results.

Conclusion

It was found that pH and agitation time have significant effect on the removal of cyanide. It was evident from the results that SAB process is more effective and less time consuming. The cyanide removal is much higher in SAB process as compared to biodegradation. It was observed from the biodegradation process that biodegradation delayed due to increase in lag phase of *P. fluorescens* in higher concentration of cyanide. In case of the SAB process due to adsorption a certain amount of cyanide ions was removed, and then the biodegradation process was started along with the adsorption. The toxicity of the cyanide in the solution was reduced due to adsorption on monolayer and cyanide was easily available for degradation. The cyanide adsorbed on the biologically active GAC surface could be easily biodegraded by microbes as GAC acts as enrichment surface and attached growth on it has been given better efficiency. This made the attach growth process more effective than the suspended growth process and better reduction of cyanide was observed.

References

Adams, M.D. (1990): The chemical behaviour of cyanide in the extraction of gold. 1. Kinetics of cyanide loss in the presence and absence of activated carbon. J. S. Afr. Inst. Min. Metall., 90: 37-44.

Adams, M.D. and Fleming, C.A. (1989): Mechanism of adsorption of aurocyanide onto activated carbon. Metallurgical Transactions B (Process Metallurgy), 20: 315-325.

Adams, M.D., McDougall, G.J. and Hancock, R.D. (1987): Models for the adsorption of aurocyanide onto activated carbon. Part III: Comparison between the extraction of aurocyanide by activated carbon, polymeric adsorbents and 1-pentanol, Hydrometall., 19: 95-115.

Adhoum, N. and Monser, L. (2002): Removal of cyanide from aqueous solution using impregnated activated carbon. Chem. Eng. Processing 41: 17-21.

Aksu, Z. and Gonen, F. (2004): Biosorption of phenol by immobilized activated sludge in a continuous packed bed: prediction of breakthrough curves. Process Biochem., 39: 599-613.

APHA, (2001): Standards Methods for the Examination of Water and Wastewater, American Public Health Association, Washington, DC.

Campos, M.G., Pereira, P. and Roseiro, J.C. (2006): Packed-bed reactor for the integrated biodegradation of cyanide and formamide by immobilised *Fusarium oxysporum* CCMI 876 and *Methylobacterium* sp. RXM CCMI 908. Enzyme. Microb. Tech., 38: 848-854.

Chapatwala, K.D., Babu, G.R.V., Vijaya, O.K., Kumar, K.P. and Wolfram, J.H. (1998): Biodegradation of cyanides, cyanates and thiocyanates to ammonia and carbon dioxide by immobilized cells of *Pseudomonas putida*. J. Ind. Microbiol., 20: 28-33.

Chen, C.Y., Kao, C.M. and Chen, S.C. (2008): Application of *Klebsiella oxytoca* immobilized cells on the treatment of cyanide wastewater, Chemosphere, 71: 133-139.

Chen, C.Y., Kao, C.M., Chen, S.C. and Chen, T.Y. (2009): Biodegradation of tetracyanonickelate by *Klebsiella oxytoca* under anaerobic conditions. Desalination, 249: 1212-1216.

Dai, X., Jeffrey, M.I. and Breuer, P.L. (2010): A mechanistic model of the equilibrium adsorption of copper cyanide species onto activated carbon. Hydrometallurgy, 101: 99-107.

Dash, R.R., Balomajumder, C. and Kumar, A. (2008): Treatment of metal cyanide bearing wastewater by Simultaneous Adsorption Biodegradation (SAB). J. Hazard. Mater. 152: 387-396.

Dash, R.R., Dash, R.R. and Majumdar, C.B. (2013): Treatment of cyanide bearing effluents by adsorption, biodegradation and combined processes: effect of process parameters. Desalination and Water Treatment, doi: 10.1080/ 19443994.2013.800330.

Dash, R.R., Majumdar, C.B. and Kumar, A. (2006): Cyanide Removal by Combined Adsorption and Biodegradation Process. Iranian J. Environ. Health Sci. Eng., 3: 91-96.

Dash, R.R., Majumdar, C.B. and Kumar, A. (2007): Removal of metal cyanide complexes from wastewaters by adsorption on granular activated carbon, Chem. Eng. World, 42: 89-93.

Dash, R.R., Majumdar, C.B. and Kumar, A. (2009a): Removal of metal cyanides from aqueous solutions by suspended and immobilized cells of *Rhizopus oryzae* (MTCC 2541). Eng. Life Sci. J., 9: 53-59.

Dash, R.R., Majumdar, C.B. and Kumar, A. (2009b): Removal of cyanide from water and wastewater using granular activated carbon. Chem. Eng. J., 146: 408-413.

Dash, R.R., Majumdar, C.B. and Kumar, A. (2010): Biodegradation of metal cyanides from cyanide bearing simulated wastewaters by suspended cultures of *Stemphylium loti* (MTCC 2542). Int. J. Geotech. Environ., 2: 45-54.

Davidson, R.J. and Veronese, V. (1979): Further studies on the elution of gold from activated carbon using water as the eluant. J. S. Afr. Inst. Min. Metall., 79: 437-469.

Depci, T. (2012): Comparison of activated carbon and iron impregnated activated carbon derived from Gölbaþý lignite to remove cyanide from water. Chem. Eng. J., 181-182: 467-478.

Deveci, H., Yazici, E.Y., Alp, I. and Uslu, T. (2006): Removal of cyanide from aqueous solutions by plain and metal-impregnated granular activated carbons, Int. J. Mineral Processing, 79: 198-208.

Dursun, A.Y., Calik, A. and Aksu, Z. (1999): Degradation of ferrous(II) cyanide complex ion by *Pseudomonas fluorescens*. Process Biochem., 34: 901-908.

Dwivedi, N. Majumder, C.B., Mondal, P. and Dwivedi, S. (2011) Biological Treatment of Cyanide Containing Wastewater. Res. J. Chem. Sci., 1: 15-21.

Dzombak, D.A., Ghosh, R.S. and Wong-Chong, G.M. (2006): Cyanide in water and soil Chemistry, risk and management. Taylor and Francis Group, CRC Press, NW.

Ebbs, S. (2004): Biological degradation of cyanide compounds. Curr. Opin. Biotech., 15: 1-6.

Gupta, N., Balomajumder C. and Agarwal, V. K. (2012): Adsorption of cyanide ion on pressmud surface: A modeling approach. Chem. Eng. J., 191: 548- 556.

Harris, R. and Knowles, C. J. (1983): Isolation and growth of a *Pseudomonas* species that utilizes cyanide as a source of nitrogen. J. Gen. Microbiol., 129: 1005-1011.

Huang, C.P. and Ostovic, F.B. (1978): Removal of Cadmium (II) by activated carbon adsorption. J. Env. Engg. Div. (ASCE), 104: 863-878.

Kunz, D.A., Nagappan, O., Silva, A. J. and deLong, G.T. (1992): Utilization of cyanide as a nitrogenous substrate by *Pseudomonas fluorescens* NCIMB 11764: Evidence for multiple pathways of metabolic conversion. Appl. Environ. Microbiol., 58: 2022-2029.

Luque-Almagro, V.M., Huertas, M.J., Martý´nez-Luque, M., Vivia´n, C.M., Dolores Rolda´n, M., Garcý´a-Gil, L.J., Castillo, F. and Blasco, R. (2005): Bacterial degradation of cyanide and its metal complexes under alkaline conditions. American Society for Microbiology, 940-947.

Mondal, P., Majumder, C.B. and Mohanty, B. (2008): Effects of adsorbent dose, its particle size and initial arsenic concentration on the removal of arsenic, iron and manganese from simulated groundwater by Fe^{3+} impregnated activated carbon. J. Hazard. Mater., 150: 695-702.

Mordocco, A., Kuek, C. and Jenkins, R. (1999): Continuous degradation of phenol at low concentration using immobilized *Pseudomonas putida*. Enzyme Microb. Tech., 25: 530-536.

Oh, W., Lim, P., Seng, C., Ngilmi, A. and Sujari, A. (2011): Bioregeneration of granular activated carbon in simultaneous adsorption and biodegradation of chlorophenols. Bioresource Tech., 102: 9497-9502.

Oh, W., Lim, P., Seng, C., Ngilmi, A. and Sujari, A. (2012): Kinetic modeling of bioregeneration of chlorophenol-loaded granular activated carbon in simultaneous adsorption and biodegradation processes. J. Bioresource Tech., 114: 179-187.

Patil, Y.B. and Paknikar, K.M. (2000): Development of a process for biodetoxification of metal cyanides from wastewater. Process Biochem., 35: 1139-1151.

Pohlandt, C., Jones, E.A and Lee, A.F. (1983): A critical evaluation of methods applicable to the determination of cyanides. J. S. Afr. Inst. Min. Metall., 83: 11-19.

Yazici,E.Y., Deveci, H. and Alp, I. (2009): Treatment of cyanide effluents by oxidation and adsorption in batch and column studies. J. Hazard. Mater. 166: 1362-1366.

Zhang, W., Liu, W., Lv, Y., Li, B. and Ying, W. (2010): Enhanced carbon adsorption treatment for removing cyanide from coking plant effluent. J. Hazard. Mater., 184: 135-140.

Chapter 4

Water Pollution, Wastewater Management Practices and Regulatory Mechanism in India

Akhila Kumar Swar

Senior Environmental Engineer,
State Pollution Control Board, Odisha, Bhubaneswar, India

ABSTRACT

Rapid increase in population, industrialization and urbanization has increased the demand of freshwater proportionately and at the same time, huge quantities of wastewater being discharged from all these point and non-point sources are also increasing day by day. In view of this, there is an urgent need to restrict water pollution and process the wastewater for reuse. The origin of wastewater from various sources like domestic, manufacturing processes from utilities, agricultural and the regulatory frame work to prevent and control water pollution has been described in this chapter. Suggestions have been made regarding waste minimization through reducing water use, reducing contamination of water and enhance water reuse. The different treatment technologies of wastewater from various sources and the issues on water pollution control have been highlighted.

Keywords: *Water quality, Water pollution, Wastewater generation, Regulatory frame work, Waste minimization, Effluent treatment, Pollution control, GPRS based real time monitoring.*

Introduction

Increased urbanization and industrialization leads to improved socio-economic conditions of people and better standard of living. This creates conditions for people to expect and demand not only sufficient quantity of water but of good quality too. On the other side the large-scale industrialization, agricultural activities and huge wastewater discharges from industries and urban area pollutes the major water bodies (State of Environment report, Orissa-2006). Therefore, it is the need of the day to minimize industrial water consumption, prevent and control water pollution to the maximum possible extent so that sufficient quantity of good quality water is made available for human use. The desirable and permissible limit for different parameters drinking water quality and undesirable effect outside the desirable limit as per IS 10500 (Revised) has been summarized in Table 4.1. Various laws have been promulgated to prevent and control water pollution and improve quality of the water bodies. A lot of emphasis has been given by the Government bodies and voluntary actions are being taken by the industries, NGOs to conserve water resource by adopting reduction of water consumption, proper treatment of wastewater, reuse and recycle the treated wastewater in order to maintain the wholesomeness of water bodies. By doing so more volume of good quality water can be made available in water bodies. But there remains a lot more to be done in this field.

Origin of Wastewater

Wastewater from Domestic Sources

Major towns are located by riverside. For domestic consumption water is drawn from the rivers or from groundwater source and after use, wastes and sewage water is discharged to the nearby rivers. With the increase of population, water consumption has increased so also the discharge quantity and the pollution load. Earlier, the low lying areas, ponds and ditches inside the towns/cities were used to absorb large quantity of wastewater/rain water but now all those voids have disappeared and most of the drains which were kutchha type are improved and converted to pucca drains. All these have increased the domestic pollution load on the rivers. In our state, full-fledged sewerage scheme with treatment plant have not been executed in most of the ULBs. With the abolition of dry latrines, all towns have changed over to septic tank and soak pits. At many places sewage is directly discharged to drains leading to the nearby nallah or river. Although all the major industrial townships are having effluent treatment plants, most of the Urban Local Bodies have not installed sewage treatment plants. Thus, the major part of pollution load on rivers is contributed by discharges of untreated domestic effluent.

Wastewater from Manufacturing Processes

Ultimately every process produces a waste stream and, although it may be possible to reduce its volume or change its phase from liquid to solid or gas, it is not possible to eliminate it completely. It is possible to operate industrial processes with zero liquid discharge but there is invariably a solid residue left which has to be disposed of or a gaseous emission to atmosphere. Industrial wastewaters arise from three principal sources.

Manufacturing processes, as has already been stated, vary considerably and no two factories, even those operating the same manufacturing processes, produce identical wastewaters. We can generally categories wastewaters according to the nature of their major contaminants.

Table 4.1: Test Characteristics for Drinking Water (IS 10500, Revised)

Sl.No.	Substance or Characteristic	Requirement (Desirable Limit)	Undesirable Effect Outside the Desirable Limit	Permissible Limit in the Absence of Alternate Source	Remarks
			ORGANOLEPTIC AND PHYSICAL PARAMETERS		
1.	Colour, Hazen units, Max	5	Above 5, consumer acceptance decreases	25	Extended to 25 only if toxic substances are not suspected, in absence of alternate sources
2.	Odour	Agreeable	—	Agreeable	—
3.	Taste	Agreeable	—	Agreeable	—
4.	Turbidity, NTU, Max	5	Above 5, consumer acceptance decreases	10	Turbidity level of 1 NTU is achievable with the available treatment technologies
5.	Dissolved solids mg/l, Max	500	Beyond this palatability decreases and may cause gastro intestinal irritation	2000	—
6.	pH value	6.5 to 8.5	Beyond this range the water will affect the mucous membrane and/or water supply system	No relaxation	—
7.	Total Hardness(as $CaCO_3$) mg/l, Max	300	Encrustation in water supply structure and adverse effects on domestic use	600	—
			GENERAL PARAMETERS CONCERNING SUBSTANCES UNDESIRABLE IN EXCESSIVE AMOUNT		
8.	Copper (as Cu) mg/l, Max	0.05	Astringent taste, discolouration and corrosion of pipes, fitting and utensils will be caused beyond this limit	1.5	—
9.	Iron (as Fe) mg/l, Max	0.3	Beyond this limit taste/appearance are affected, has adverse effect on domestic uses and water supply structures, and promotes iron bacteria	1.0	—
10.	Manganese (as Mn) mg/l, Max	0.1	Beyond this limit taste/appearance are affected, has adverse effect on domestic uses and water supply structures	0.3	WHO guideline specify 0.4 mg/l and EU specify 0.05 mg/l

Contd...

Table 4.1–Contd...

Sl.No.	Substance or Characteristic	Requirement (Desirable Limit)	Undesirable Effect Outside the Desirable Limit	Permissible Limit in the Absence of Alternate Source	Remarks
11.	Nitrate (as NO_3) mg/l, Max	45	Beyond this methaemoglobinemia takes place	No relaxation	—
12.	Fluoride (as F) mg/l, Max	1.0	Fluoride may be kept as low as possible. High fluoride may cause fluorosis	1.5	—
13.	Zinc (as Zn) mg/l, Max	5	Beyond this limit it can cause astringent taste and an opalescence in water	15	—
14.	Aluminium (as Al), mg/l, Max	0.03	Cumulative effect is reported to cause dementia	0.2	—
15.	Chlorides (as Cl) mg/l, Max	250	Beyond this limit, taste, corrosion and palatability are affected	1000	—
16.	Selenium (as Se), mg/l, Max	0.01	Beyond this, the water becomes toxic	No relaxation	To be tested when pollution is suspected
17.	Sulphate (as SO_4) mg/l, Max	200	Beyond this causes gastro intestinal irritation when magnesium or sodium are present	400	May be extended upto 400 provided (as Mg) does not exceed 30
18.	Alkalinity as Calcium Carbonate, mg/l, Max	200	Beyond this limit taste becomes unpleasant	600	—
19.	Calcium (as Ca) mg/l, Max	75	Encrustation in water supply structure and adverse effects on domestic use	200	—
20.	Magnesium (as Mg) mg/l, Max	30	Encrustation to water supply structure and adverse effects on domestic use	100	—
21.	Residual Free chlorine, mg/l, Max. Applicable only when water is chlorinated. Tested at consumer end.	0.2	—	1	To be applicable only when water is chlorinated. Tested at consumer end. When protection against viral infection is required, it should be Min 0.5 mg/l

Contd...

Table 4.1–Contd...

Sl.No.	Substance or Characteristic	Requirement (Desirable Limit)	Undesirable Effect Outside the Desirable Limit	Permissible Limit in the Absence of Alternate Source	Remarks
22.	Phenolic compounds (as C_6H_5OH) mg/l, Max	0.001	Beyond this, it may cause objectionable taste and odour	0.002	—
23.	Mineral oil mg/l, Max	0.01	Beyond this limit undesirable taste and odour after chlorination take place	0.03	To be tested when pollution is suspected
24.	Anionic detergents (as MBAS) mg/l, Max	0.2	Beyond this limit it can cause a light forth in water	1.0	To be tested when pollution is suspected
25.	Boron (as B), mg/l, Max	0.3	Boron has serious ill effects on central nervous system	1.5	WHO specify 0.5, but permissible limit should be 1 mg/l
PARAMETERS CONCERTING TOXIC SUBSTANCES					
26.	Barium (as Ba) mg/l, Max	0.7	—	No relaxation	—
27.	Molybdenum (as Mo) mg/l, Max	0.07	—	No relaxation	—
28.	Sulphide (as H_2S), mg/l, Max	0.05	—	No relaxation	—
29.	Mercury (as Hg) mg/l Max	0.001	Beyond this, the water becomes toxic	No relaxation	To be tested when pollution is suspected
30.	Cadmium (as Cd), mg/l Max	0.003	Beyond this the water becomes toxic	No relaxation	To be tested when pollution is suspected
31.	Total Arsenic (as As), mg/l, Max	0.01	Beyond this, the water becomes toxic	0.05	To be tested when pollution is suspected
32.	Cyanide (as CN), mg/l, Max	0.05	Beyond this, the water becomes toxic	No relaxation	To be tested when pollution is suspected
33.	Lead (as Pb), mg/l, Max	0.01	Beyond this, the water becomes toxic	No relaxation	To be tested when pollution/plumbo-solvency is suspected
34.	Chromium (as Cr^{+6}) mg/l, Max	0.05	May be carcinogenic above this limit	No relaxation	USEPA, EU and WHO specify 0.02, 0.05 and 0.07 respectively
35.	Polynuclear aromatic hydro-carbons (as PAH) g/l, Max	0.0001	May be carcinogenic	No relaxation	—

Contd...

Table 4.1–*Contd...*

Sl.No.	Substance or Characteristic	Requirement (Desirable Limit)	Undesirable Effect Outside the Desirable Limit	Permissible Limit in the Absence of Alternate Source	Remarks
36.	Pesticides:	Toxic			
	a) DDT (o,p and p,p-Isomers of DDT, DDE and DDD)	1 µg/l Max		No relaxation	—
	b) Gamma – HCH (Lindane)	2 µg/l Max		No relaxation	—
	c) 2,4-D	3 µg/l Max		No relaxation	—
	d) Isoproturon	9 µg/l Max		No relaxation	—
	e) Alachor	20 µg/l Max		No relaxation	—
	f) Atrazine	2 µg/l Max		No relaxation	—
	g) Aldrin/Dieldrin	0.03 µg/l Max		No relaxation	—
	h) Alpha HCH	0.01 µg/l Max		No relaxation	—
	i) Beta HCH	0.04 µg/l Max		No relaxation	—
	j) Delta HCH	0.04 µg/l Max		No relaxation	—
	k) Endosulfan (alpha, beta, and sulphate)	0.4 µg/l Max		No relaxation	—
	l) Monocrotophos	1 µg/l Max		No relaxation	—
	m) Ethion	3 µg/l Max		No relaxation	—
	n) Chlopyriphos	30 µg/l Max		No relaxation	—
	o) Phorate	2 µg/l Max		No relaxation	—
	p) Butachlor	125 µg/l Max		No relaxation	—
	q) Methyl Parathion	0.3 µg/l Max		No relaxation	—
	r) Malathion	190 µg/l Max		No relaxation	—

Contd...

Table 4.1–Contd...

Sl.No.	Substance or Characteristic	Requirement (Desirable Limit)	Undesirable Effect Outside the Desirable Limit	Permissible Limit in the Absence of Alternate Source	Remarks
37.	Nickel (as Ni), mg/l, Max	0.02		No relaxation	—
38.	Polychlorinated Biphenyls, mg/l, Max	0.0005		No relaxation	—
39.	Trihalomethanes:				
	a) Boromoform mg/l, Max	0.1	May be carcinogenic above this limit	No relaxation	—
	b) Dibromochloromethane mg/l, Max	0.1	May be carcinogenic above this limit	No relaxation	—
	c) Boromodichloromethane mg/l, Max	0.06	May be carcinogenic above this limit	No relaxation	—
	d) Chloroform mg/l, Max	0.2	May be carcinogenic above this limit	No relaxation	—
PARAMETERS CONCERNING RADIOACTIVE SUBSTANCES					
40.	Radioactive materials				—
	a) Alpha emitters Bq/l, Max	0.1	May be carcinogenic above 0.1	0.1	—
	b) Beta emitters Bq/l, Max	1.0	May be carcinogenic above 1.0	1	—

In the first category are those wastewaters which contain mostly inorganic contaminants. Typical examples include metal finishing processes, which may contain acids, alkalis, heavy metals, cyanides and chromates; tanning effluents which often contain chromates and sulphides; textile effluents which may contain heavy metals, salt, surfactants, sulphide, detergents, solvents, colour acids, alkalis; and inorganic chemical industry wastes.

The second broad category of wastewaters is those containing easily biodegradable. Food industry wastewaters fall into this category. Since most of its wastes are biological in origin they would be expected to be readily biodegradable either aerobically or anaerobically. Meat processing to cooked products, soft drinks, sugar processing and brewing wastes is obvious examples, but tanning, textile industry, paper-making, tobacco processing and oil refining also produce easily biodegradable wastewaters.

The third category of manufacturing wastewaters is those containing "hard" COD which is not capable of being removed by biological treatment. It includes wastewaters from a wide variety of industries such as pharmaceuticals and organic chemical manufacture, oil and petrochemicals, timber treatment and pesticide and insecticide manufacture. These wastes are generally difficult to treat and may require advanced and expensive technologies and are not easily biodegradable. Many of them have high toxicity and treatment to reduce COD may not always reduce toxicity.

Wastewater from Utilities

Evaporation in steam boilers and cooling towers removes water as steam and leaves behind the dissolved salts which become concentrated. The concentration of salts in the boiler or cooling tower is controlled by discharging some of the water as blow down. Boiler blow down water is hot and contains "conditioning" chemicals which have been added to control scale and corrosion. These include oxygen scavengers, like sulphites, which remove oxygen from water and prevent aqueous corrosion taking place, film forming amines, which form a protective coating on metal surfaces, and phosphates and acrylates for scale prevention and sludge conditioning. It is, therefore, potentially harmful to the environment or sewage treatment plant.

A similar situation exists in evaporative cooling towers where, once again, water is lost as vapour leaving behind a concentrated solution of salts. The blow down water contains a variety of conditioning chemicals including:

☆ Scale inhibiting chemicals, such as polyphosphonates and polyacrylates which prevent scale formation by interfering with crystal growth and promoting the precipitation of hardness salts as sludge.

☆ Of corrosion inhibitors such as zinc salts, molybdates and nitrites which inhibit corrosion reactions.

☆ Biocides to control bacterial and algal proliferation.

As in the case of boiler systems, these chemicals are expensive as well as being potentially harmful to the environment, minimizing their consumption by minimizing blow down will be beneficial to both the cooling tower operator and the environment. Cooling water is often close to saturation or even super-saturated with calcium carbonate so there is the possibility of precipitation in drains causing blocking.

One of the main sources of wastewater under the general heading of utilities is water and wastewater treatment plant. Filter backwash water contains high levels of suspended solids (typically about 1000 mg/l) and oil separator wastes contain very high levels of free and emulsified oil. Clearly wastewater streams of this type cannot be safely discharged to the environment and therefore an alternative disposal route has to be found.

Agriculture Run-off

The run-off from agriculture field is contaminated with pesticides and insecticide chemicals. Agriculture Department should take up awareness campaign among farmers to use environment friendly chemicals and restrict their use to required limit. Excess use of chemicals should be checked and controlled. Organic farming should be encouraged.

Regulatory Frame Work to Prevent and Control Water Pollution

Provisions of the Pollution Control Acts

As per the provision of the Water (Prevention and Control of Pollution) Act, 1974 all the industries, industrial townships and urban local bodies are required to obtain consent of the State Pollution Control Board for establishment and operation of their industries or any treatment and disposal system, which is likely to discharge sewage and industrial effluent in to streams, well, sewer or water bodies (CPCB, 2010). Under this act, the quality and quantity of discharge is regulated by the State Pollution Control Board (SPCB) and in case of non-compliances like inadequate treatment of effluent or discharge of untreated effluent to any water body in excess of the stipulated standard, the Board can initiate legal action including issue of direction of closure to the defaulting unit.

As per the provision of the above Act, industrial/domestic effluent should be treated in Effluent Treatment Plants before discharged to outside water bodies/on land for plantation. The effluent standards (industry wise) and the general standards of discharge has been prescribed by the Ministry of Environment and Forest (MoEF), Govt. of India, under the provision of Environment Protection Act, 1986 and Rules framed thereunder, presented in Table 4.2 and monitored periodically by SPCB and Central Pollution Control Board (CPCB). Similarly water quality of all the major wastewater recipient bodies are also monitored at different locations by SPCB and CPCB. The standards and classification for river water quality has been prescribed in IS handbooks.

As per the provisions of Water (Prevention and Control of Pollution) Cess Act, 1977 and Amendments made during 2003, the industries are required to pay water cess to SPCB on the quantity of water consumed by them every month for different purposes like cooling, spraying, boiler feed, domestic purposes and manufacturing processes. Different rates have been fixed under this Act for different purposes. This

Table 4.2: General Standards for Discharge of Environmental Pollutants through Effluent

Sl.No.	Parameter	Inland Surface Water	Public Sewers	Land for Irrigation	Marine/Coastal Areas
		(a)	(b)	(c)	(d)
1.	Suspended solids mg/l, max.	100	600	200	(a) For process wastewater (b) For cooling water effluent 10 per cent above total suspended matter of influent.
2.	Particle size of suspended solids	Shall pass 850 micron IS Sieve	—	—	(a) Floatable solids, solids max. 3 mm (b) Settleable solids, max 856 microns
3.	pH value	5.5 to 9.0	5.5 to 9.0	5.5 to 9.0	5.5 to 9.0
4.	Temperature	Shall not exceed 5°C above the receiving water temperature			Shall not exceed 5°C above the receiving water temperature
5.	Oil and grease, mg/l max.	10	20	10	20
6.	Total residual chlorine, mg/l max	1.0	—	—	1.0
7.	Ammonical nitrogen (as N),mg/l, max.	50	50	—	50
8.	Total kjeldahl nitrogen (as N); mg/l, max. mg/l, max.	100	—	—	100
9.	Free ammonia (as NH_3), mg/l, max.	5.0	—	—	5.0
10.	Biochemical oxygen demand (3 days at 27°C), mg/l, max.	30	350	100	100
11.	Chemical oxygen demand, mg/l, max.	250	—	—	250
12.	Arsenic(as As).	0.2	0.2	0.2	0.2
13.	Mercury (As Hg), mg/l, max.	0.01	0.01	—	0.01
14.	Lead (as Pb) mg/l, max.	0.1	1.0	—	2.0
15.	Cadmium (as Cd) mg/l, max.	2.0	1.0	—	2.0

Contd...

Table 4.2–Contd...

Sl.No.	Parameter	Inland Surface Water (a)	Public Sewers (b)	Land for Irrigation (c)	Marine/Coastal Areas (d)
16.	Hexavalent chromium (as Cr + 6), mg/l, max.	0.1	2.0	–	1.0
17.	Total chromium (as Cr) mg/l, max.	2.0	2.0	–	2.0
18.	Copper (as Cu) mg/l, max.	3.0	3.0	–	3.0
19.	Zinc (as Zn) mg/l, max.	5.0	15	–	15
20.	Selenium (as Se)	0.05	0.05	–	0.05
21.	Nickel (as Ni) mg/l, max.	3.0	3.0	–	5.0
22.	Cyanide (as CN) mg/l, max.	0.2	2.0	0.2	0.2
23.	Fluoride (as F) mg/l, max.	2.0	15	–	15
24.	Dissolved phosphates (as P), mg/l, max.	5.0	–	–	–
25.	Sulphide (as S) mg/l, max.	2.0	–	–	5.0
26.	Phenolic compounds (as C_6H_5OH) mg/l, max.	1.0	5.0	–	5.0
22.	Radioactive materials: (a) Alpha emitters micro curie mg/l, max. (b) Beta emitters micro curie mg/l	10^{-7} 10^{-6}	10^{-7} 10^{-6}	10^{-8} 10^{-7}	10^{-7} 10^{-6}
23.	Bio-assay test	90 per cent survival of fish after 96 hours in 100 per cent effluent	90 per cent survival of fish after 96 hours in 100 per cent effluent	90 per cent survival of fish after 96 hours in 100 per cent effluent	90 per cent survival of fish after 96 hours in 100 per cent effluent
24.	Manganese	2 mg/l	2 mg/l	–	2 mg/l
25.	Iron (as Fe)	3 mg/l	3 mg/l	–	3 mg/l
26.	Vanadium (as V)	0.2 mg/l	0.2 mg/l	–	0.2 mg/l
27.	Nitrate Nitrogen	10 mg/l	–	–	20 mg/l

* These standards shall be applicable for industries, operations or processes other than those industries, operations or process for which standards have been specified in Schedule of the Environment Protection Rules, 1989.

Act serves as an economic instrument in controlling water pollution and regulates the quantity/quality of discharge to the water bodies by encouraging reduction of water consumption, recycling and reuse of treated wastewater.

The State Pollution Control Board, impose bank guarantee on the defaulting industries and mines to compel them to install adequate effluent treatment plants and take up proper water pollution control measures in a time bound manner to ensure that the treated effluent meets the norms before discharged to water bodies of land. In case the unit fails to comply within the stipulated time and continues to pollute water, legal action is taken or direction of closure as per the provision of Water (PCP) Act, 1974 is issued by the Board against the defaulting industry or mine to prevent further deterioration of water quality of the water bodies in the surrounding area.

Guidelines Under Charter on Corporate Responsibility for Environmental Protection (CREP), CPCB

It is a statutory requirement to comply with the regulatory norms for prevention and control of pollution. Besides, it is imperative to go beyond compliance level through adoption of clean technologies and improvement in management practices. Commitment and voluntary initiatives of the industry for responsible care of the environment will help in building partnership for pollution control. After a number of interactions meet among the industries, State Pollution Control Boards and Central Pollution Control Boards, action points and target to achieve effluent, emission standards and wastewater recycling are fixed up (CPCB, 2003). The targets set for some of the 17 category of highly polluting industries for reduction of water consumption and generation of wastewater is summarized in Table 4.3.

Waste Minimization Options

Water recycling and compliance with discharge conditions both require the introduction of treatment either at source or at end of pipe. We should look at the ways of reducing water consumption by reducing the quantity of water used in processes and re-using water/recycling water directly reduces the amount of contamination entering the water stream.

Reducing Water Use

Large volume of water is often wasted and it needs only common sense to achieve significant cost savings. We can have savings that can be made by identifying and repairing leaks in pipe work, but there is often unnecessary/excess use of water. Poor maintenance and poor housekeeping lead to water losses. Many factories have very obvious steam leaks to the atmosphere and water leaking from valves which have not been properly closed or are leaking due to inadequate maintenance. The cost of maintenance will normally be rapidly repaid by savings in water.

Uncontrolled blow down from boilers and cooling towers is another potential source of waste. The purpose of the blow down is to control the water quality but, all too often, blow down valves are left open at a fixed flow rate irrespective of changes in operating conditions. A simple monitor and automatic valve is a very cost effective

Table 4.3: Guidelines of CREP for Reduction of Water Consumption and Wastewater Discharge

Sl.No.	Industry	Restriction on Water Consumption/Wastewater Discharge
1.	Copper industry	100 per cent recycling of treated wastewater and zero discharge.
2.	Distilleries	100 per cent utilization of spent wash. Restricted discharge can be allowed during rainy season to river.
3.	Fertilizer industry	Specific water consumption shall not exceed 10 m^3 per ton of product. Even storm water before discharge shall be treated.
4.	Integrated iron and steel plant	Specific water consumption shall not exceed 5 m^3 per ton of long product and 8 m^3 per ton of flat product.
5.	Oil refineries	Maximise reuse and recycling of treated effluent. The discharge of wastewater shall not exceed 0.4 m^3 per ton of product.
6.	Pesticides industry	Treated effluent to be taken to evaporation pond without any discharge to outside.
7.	Pulp and Paper industry	For large industries: Wastewater discharge shall be restricted to (i) 100 m^3/tonne of paper for units installed after 1992, (ii) 120 m^3/tonne of paper for units installed before 1992. For small industries: Wastewater discharge shall be restricted to 150 m^3/tonne of paper. In both cases, the treated water should be utilized for irrigation wherever possible.
8.	Sugar industry	Wastewater generation shall be restricted to 100 litres per tonne of cane crushed and there should not be any discharge into inland surface water.
9.	Tanneries	Water consumption rate shall be brought down to 28 m^3/tonne of hides by taking water consumption measures. The treated wastewater should be mixed with sewage and further treated. The combined treated effluent should be used on land for irrigation.
10.	Thermal power plants	Hundred per cent ash water recirculation systems should be provided.
11.	Zinc industry	Zero discharge through 100 per cent recycle/reuse of treated wastewater.

way of controlling blow down. Installing an automatic conductivity controlled blow down valve saves water and reduces costs.

Improved process control such as optimization of reaction temperatures and pressures together with equipment modifications such as changing from water lubricated seals to dry ones all have a part to play in waste minimization. In addition they may improve product quality.

Reducing Contamination of Water

Preventing a contaminant from entering the wastewater means that it will not need to be removed from the wastewater prior to discharge and a reduced level of contamination may allow the water to be re-used. Often contamination is the result of poor housekeeping and can be prevented both easily and cheaply.

Water Reuse

One can always consider whether the quality of water used at each stage of manufacture is correct. Quite often water of a quality higher than the requirement is supplied to processes because it is convenient. Similarly relatively good quality wastewater is discharged to drain. It may be possible to use this type of wastewater from one unit operation as feed to another. The treated sewage water can be used for gardening/flushing of the toilets, the industrial effluent after treatment can be reused and the ash pond overflow water can be used again for making ash slurry. This has been adopted now at many places. Besides the environmental issues of water conservation and waste minimization, all the cases of reuse and recycling gives financial saving. Good engineering design should give, at worst, a break even and at best a cost saving.

Wastewater Treatment Options

When all waste minimization options have been implemented, the quality of the wastewater streams should be reviewed for treatment either at source or at "end of pipe". There will usually be several different options for treatment, each of which will have a capital cost and each an operating cost. Only careful evaluation will identify which is the optimum solution for a particular factory. Often previous experience in the same or a similar industry will help but every industrial wastewater is unique and every solution is different. Before a sensible approach to wastewater treatment can be made, it is necessary to know what waste streams are to be treated. Characterization of the wastes means, for each stream, identifying the source, the quality and the quantity. In order to achieve this it may be necessary to undertake a programme of proper sampling, analysis and flow monitoring. The first decision, which must be made, is whether to treat "at source" or at "end of pipe". The degree of treatment of wastewater depends on the prescribed standards of the regulatory authority and to maintain condition of the recipient water body (WHO/UNEP, 1997).

At source treatment allows waste streams to be segregated so that each can be treated optimally with the possibility of local recycling of water or recovery of raw materials. Treatment at source will usually involve dealing with small volumes of high concentration — often a better option for chemical treatment. However, it clearly does not make sense to neutralize individual streams of high and low pH wastes when mixing them together may give neutralization at reduced costs, unless mixing them together forms complex ions which are difficult to remove.

We cannot destroy contaminants in wastewater; but we can only move them about. We can concentrate them into small volumes or move them from one phase to another to form sludge or a gaseous discharge but, ultimately, they will still be present in the environment as greenhouse gases, metal-bearing sludge, solid mineral salts and so on. We can only try to minimize the damage, which they might cause to the environment. We can dilute wastewaters to meet prescribed limits but, even if this is an economic solution, it is not environmentally acceptable. This means that effluents, which do not meet the prescribed standard of regulatory authority, must be treated adequately.

Treatment processes fall into two broad types: destruction processes and concentration processes. Destruction processes convert the influent organic matter into simpler, less toxic molecules that can be safely discharged to either the environment or to sewer. Ultimately all organic matter can be degraded to carbon dioxide, water together with nitrogen gas, NO_x and SO_x, however such degradation is likely to be uneconomic. Concentration processes remove the organic matter from the process stream into a more concentrated stream but do not actually destroy the organic matter.

Each of the wastewater flows should have been characterized in detail during the water audit under the headings for determining their treatment methods:

☆ Wastewater quality

☆ Wastewater flow

☆ Wastewater volume

☆ Pattern of flow

The quality of each waste stream is of major importance. Obviously it is necessary to know the composition — what type of contaminants are present and in what concentrations — because the nature of the contaminants is fundamental to the choice of technology. Waste streams containing high levels of inorganic salts require a different treatment process from wastes containing COD. However, it is also important to understand how the quality of the stream varies diurnally and seasonally.

It is also necessary to know how much wastewater will need to be treated and, for this, we need to know both the average daily or weekly volumetric flow, because this will control the treatment capacity of the treatment plant, but we also need to know the maximum instantaneous flow because this will control the hydraulic capacity of the plant and the requirements for flow and load balancing. Some industries, particularly batch processing industries like brewing, discharge wastewater intermittently, but even those which appear to be continuous, like plating shops, often have intermittent discharges which have to be handled. So the pattern of flow and its variability is an important design parameter. Different treatment options are described below and the summary of industrial waste, its origin, character and treatment is given in Table 4.4.

Table 4.4: Summary of Industrial Waste: Its Origin, Character and Treatment

Industries Producing Wastes	Sources of Generation of Wastewater	Major Characteristics of the Wastewater	Major Treatment and Disposal Methods
Apparel			
Textiles	Cooking of fibers : desizing of fabric	Highly alkaline, coloured, high BOD & temperature, high suspended solids	Neutralization, Chemical precipitation, biological treatment, aeration and/or trickling filtration
Leather goods	Un-hairing, soaking, de-liming, and bating of hides	High total solids, hardness, salt, sulfides, chromium, pH	Screening, chemical precipitation, flotation, and adsorption.
Food and Drugs			
Canned goods	Trimming, culling, juicing and blanching of fruits and vegetables	High in suspended solids, colloidal and dissolved organic matter	Screening, lagooning, soil absorption or spray irrigation.
Diary products	Dilutions of whole milk, separated milk, buttermilk, and whey	High in dissolved organic matter, mainly protein, fat, and lactose	Skimming, Biological treatment, aeration, trickling filtration, activated sludge process (ASP)
Brewed and distilled beverages	Steeping and pressing of grain; residue from distillation of alcohol; condensate from stillage evaporation.	High in dissolved organic solids, containing nitrogen and fermented starches or their products	Recovery, concentration by centrifugation and evaporation, trickling filtration; activated sludge process
Meat and poultry products	Stockyards; slaughtering of animals; rendering of bones and fats; residues in condensates; grease and wash water; pickling of chickens	High in dissolved and suspended organic matter, blood, other proteins, and fats	Screening, settling and/or flotation, activated sludge process (ASP)
Animal feedlots	Excreta for animals	High in organic suspended solids and BOD	Land disposal and anaerobic lagoons.
Beet sugar	Transfer, screening, and juicing water; draining from lime sludge; condensate after evaporator; juice and extracted sugar	High in dissolved and suspended organic matter, containing sugar and protein	Reuse of wastes, coagulation, and lagooning

Contd...

Table 4.4—Contd...

Industries Producing Wastes	Sources of Generation of Wastewater	Major Characteristics of the Wastewater	Major Treatment and Disposal Methods
Pharmaceutical products	Mycelium, spent filtrate, and wash water	High in suspended and dissolved organic matter, including vitamins	Evaporation and drying
Yeast	Residue from yeast filtration	High in solids (mainly organic) and BOD	Anaerobic digestion, trickling filtration
Pickles	Lime water; brine, alum and turmeric, syrup, seeds and pieces of cucumber	Variable pH, high suspended solids, colour, and organic matter	Good housekeeping, screening equalization, clarifier.
Coffee	Pulping and fermenting of coffee bean	High BOD and suspended solids	Screening, settling and trickling filtration.
Fish	Rejects from centrifuge; pressed fish; evaporator and other wash water wastes	Very high BOD, total organic solids, and odour	Screening, ASP, filtration, chlorination
Par-boiled Rice	Soaking, cooking and washing of rice	High BOD, total and suspended solids (mainly starch)	Equalization, UASBR followed by aerobic process like activated sludge process (ASP) followed by activated carbon filtration
Soft drinks	Bottle washing; floor and equipment cleaning; syrup storage – tank drains	High pH, suspended solids, and BOD	Screening, equalization, ASP and tertiary treatment
Bakeries	Washing and greasing of pans; brine; alum sludge	High BOD, grease, floor washings, sugars, flour, detergents	Biological oxidation
Drinking Water	Filter backwash; lime – soda sludge; brine; alum sludge	Minerals and suspended solids	Settling tanks/holding lagoons
Cane sugar	Spillage from extraction, clarification, etc. Evaporator entrainment in cooling and condenser waters	Variable pH, soluble organic matter with relatively high BOD of carbonaceous nature	Neutralization, recirculation, chemical treatment, some selected aerobic oxidation.
Agriculture	Variable origin depending upon exact source; agriculture chemicals, irrigation return flows, crop residues and liquid and solids animal wastes	Highly organic and BOD detergent cleaning solutions	Biological oxidation basins; composting and anaerobic digestion; land application.

Contd...

Table 4.4–Contd...

Industries Producing Wastes	Sources of Generation of Wastewater	Major Characteristics of the Wastewater	Major Treatment and Disposal Methods
		Materials	
Pulp and paper	Cooking, refining, washing of fibers, screening of paper pulp	High or low pH, colour, high suspended, colloidal and dissolved solids, inorganic fibers	Settling, lagooning, biological treatment, aeration, Clarifier, recovery of by-products.
Photographic products	Spent solutions of developer and fixer	Alkaline, containing various organic and inorganic reducing agents	Neutralisation, Recovery of silver; discharge of wastes into municipal sewer
Steel	Coking of coal, washing of blast – furnace flue gases, and pickling of steel	Low pH, acids, cyanogens, phenol, ore, coke limestone, alkali, oils, mill scale, and fine suspended solids	Neutralization, oil separators, recovery and reuse, chemical coagulation, clarifier, biological treatment, BOD plant.
Metal – plated products	Stripping of oxides, cleaning and plating of metals	Acid, metals, toxic, low volume, mainly mineral matter	Alkaline chlorination of cyanide; reduction and precipitation of chromium; lime precipitation of other metals
Iron – foundry products	Wasting of used sand by hydraulic discharge	High suspended solids, mainly sand; some clay and coal	Selective screening, drying of reclaimed sand
Oil fields and refineries	Drilling mud, salt, oil and some natural gas; acid sludge and miscellaneous oils from refining	High dissolved salts from field; high BOD, odour, phenol, and sulfur compounds from refinery	Recovery injection of salts; acidification and burning of alkaline sludge
Fuel oil use	Spills from fuel – tank filling waste; auto crankcase oils	High in emulsified and dissolved oils	Leak and spill prevention, flotation
Rubber	Washing of latex, coagulated rubber, exuded impurities from crude rubber	High BOD and odor, high suspended soils, variable pH, high Chlorides	Aeration, chlorination, sulfonation, biological treatment
Glass	Polishing and cleaning of glass	Red colour, alkaline non settle able suspended solids	Calcium chloride precipitation
Naval stores	Washing of stumps, drop solution, solvent recovery and oil recovery water	Acid, high BOD	By product recovery, equalization, recirculation and reuse trickling filtration.

Contd...

Table 4.4–Contd...

Industries Producing Wastes	Sources of Generation of Wastewater	Major Characteristics of the Wastewater	Major Treatment and Disposal Methods
Glue manufacturing	Lime wash, acid washes, extraction of nonspecific proteins	High COD, BOD, pH, chromium, periodic strong mineral acids	Amenable to aerobic biological treatment, flotation, chemical precipitation.
Wood preserving	Steam condensates	High in COD, BOD, solids phenols	Chemicals coagulation; oxidation pond and other aerobic biological treatment
Candle manufacturing	Was spills, stearic acid condensates	Organic (fatty) acids	Anaerobic digestion
Plywood manufacturing	Glue washings	High BOD, pH, phenols potential toxicity	Settling ponds, incineration
Metal container	Cutting and lubricating metals, cleaning can surface	Metal fines, lube, oils, variable pH, surfactants, dissolved metals	Oil separation, chemical precipitation, collection and reuse, lagoon storage. Final carbon absorption.
Petrochemicals	Contaminated water from chemical production and transportation of second generation oil compounds	High COD, T.D.S., metals, COD/BOD ratio, and cpds. Inhibitory to boil action.	Recovery and reuse, equalization and neutralization, chemical coagulation, settling or flotation, biological oxidation.
Cement	Fine and finish grinding of cement, dust leaching collection, dust control	Heated cooling water, suspended solids, some inorganic salts	Segregation of dust – contact streams, neutralization and sedimentation.
Wood furniture	Wet spray booths and laundries	Organics from staining and sealing wood products	Evaporation or burning.
Asbestos	Cleaning and crushing ore	Suspended asbestos and mineral solids	Detention in ponds, neutralization and land filling.
Plant and inks	Solvent – based rejected materials scrubbers from paint vapors; refining and/or removing inks	Contain organic solids from dyes, resins, oils, solvents, etc.	Settling ponds for detention of paints, lime coagulation of printing inks.
Chemicals			
Acids	Dilute wash waters; many varied dilute acids	Low pH, low organic content	Up-flow or straight neutralization, burning when some organic matter is present
Detergents	Washing and purifying soaps and detergents	High in BOD and saponified soaps	Flotation and skimming, precipitation with $CaCl_2$

Contd...

Table 4.4–Contd...

Industries Producing Wastes	Sources of Generation of Wastewater	Major Characteristics of the Wastewater	Major Treatment and Disposal Methods
Corn starch	Evaporator condensate or bottoms when not reused or recovered, syrup from final washes, wastes from "bottling up" process	High BOD and dissolved organic matter; mainly starch and related material	Equalization, biological filtration, anaerobic digestion.
Explosives	Washing TNT and guncotton for purification, washing and pickling of cartridges	TNT, coloured, acid, odorous; and contains organic acids and alcohol from powder and cotton, metals, acid, oils, and soaps	Flotation, chemical precipitation, biological treatment, aeration, chlorination, neutralization, adsorption
Pesticides	Washing and purification products such as 2,4-D and DDT	High organic matter, benzene-ring structure, toxic to bacteria and fish, acid	Dilution, storage, activated carbon adsorption, alkaline chlorination
Phosphate and phosphorus	Washing, screening, floating rock, condenser bleed off from phosphate reduction plant	Clays, slimes and tall oils, low pH, high suspended solids, phosphorus, silica and fluoride	Lagooning, mechanical clarification, coagulation and settling of refined waste
Formaldehyde	Residues from manufacturing synthetic resins and from dyeing synthetic fibers	Normally high BOD and HCHO, toxic to bacteria in high concentrations.	Trickling filtration, adsorption on activated charcoal.
Plastics and resins	Unit operations from polymer preparation and use; spills and equipment wash downs	Acids, caustic, dissolved organic matter such as phenols, formaldehyde, etc.	Neutralization, sedimentation, carbon adsorption
Fertilizer	Chemical reactions of basic elements, spills, cooling waters, washing of products, boiler blow downs	Sulfuric, phosphorus, and nitric acids; mineral elements, P, S, N, K, Al, NH_3, NO_3, etc, F, suspended solids	Neutralization, detain for reuse, sedimentation, air stripping of NH_3, lime precipitation etc.
Aluminium smelters	Rectifier cooling, spillage to drains, floor washings, toilets	Suspended solids, fluorides	Equalization, chemical dosing, clarification, ultrafiltration, reverse osmosis, reject containment and forced evaporation system

Contd...

Table 4.4—Contd...

Industries Producing Wastes	Sources of Generation of Wastewater	Major Characteristics of the Wastewater	Major Treatment and Disposal Methods
Toxic chemicals	Leaks, accidental spills, and refining of chemicals	Various toxic dissolved elements and compounds such as Hg and PCBs	Retention/containment and reuse
Mortuary	Body fluids, wash waters, spills	Blood salt, formaldehydes, high BOD, infectious diseases	Holding and chlorination before discharge to municipal sewer
Hospital-Res. Labs.	Washing, sterilizing of facilities, used solutions, spills	Bacteria, various chemicals radioactive materials	Holding and boil/chlorination. Aeration in large facilities, STP consisting of ASP.
Energy			
Thermal power	Cooling water, boiler blow down, coal drainage	Hot, high volume, high inorganic and dissolved solids	Cooling by aeration, neutralization of excess acid wastes, coal settling pits
	Ash pond overflow	Total suspended solids	Clarifloculation, collection tank and recycling
Scrubber power plant wastes	Scrubbing of gaseous combustion by liquid water	Particulates, SO_2, impure absorbents or NH_3, NaOH, etc.	Solids removal usually by settling, pH adjustment and reuse
Coal processing	Cleaning and classification of coal, leaching of sulfur strata with water	High suspended solids, mainly coal; low pH, high H_2SO_4 and $FeSO_4$	Settling, froth flotation, drainage control, recycling and sealing of mines
Nuclear power and radioactive materials			
	Processing ores; laundering of contaminated clothes; research-lab wastes	Radioactive elements can be very acid and hot	Concentration/containment or dilution as per atomic energy regulation
Non-point sources	Dirt, dust, combustion product run-off, salt run-off, organic matter run-off	Largely mineral and organic matter	Sealing sources, holding by settling basins/ponds, natural evaporation etc.
Mining			
Coal Mines	Discharge of Mine drainage water	pH, Total suspended solids	Equalization cum neutralization, Settling tanks in series or Clarifloculation and pressure sand filtration depending on use.

Contd...

Table 4.4–Contd...

Industries Producing Wastes	Sources of Generation of Wastewater	Major Characteristics of the Wastewater	Major Treatment and Disposal Methods
Chromite Mines	Discharge of Mine drainage water and surface run-off from overburden dumps	pH, Total suspended solids and Hexavalent chromium	Equalization cum neutralization tank, Chemical Precipitation and settling tank or clarifier and filtration.
Ash mound/Solid waste dump	Surface run-off	Total suspended solids	Garland drain followed by settling ponds.

At Source Treatment

If possible, individual waste streams from different parts of the manufacturing process should be segregated to allow the streams to be treated separately at source. Whilst stream segregation is easy to implement in new factories it is not easy to retro-fit in existing factories where the locations and routes of drains may not be clearly identifiable. Also the segregation of drains has less relevance in the case of batch manufacturing industries where flows are intermittent and effluent composition may vary from one product campaign to another. The advantages of at source treatment include:

☆ Lower flows which means that treatment plant is smaller

☆ The range of contaminants to be removed is greatly reduced which simplifies the treatment process

☆ Contaminant concentrations are higher so that treatment processes operate more efficiently

☆ There may be an opportunity for recovery of water, raw materials or energy

End of Pipe Treatment

Treatment at end of pipe means that all the wastewater from a factory is mixed together and treated as a single stream. This used to be the most common choice but economics suggest that it is not always the right one. Large balancing tanks are usually required to even out diurnal variations in flow and quality. The volumes to be treated are large and this may be further increased by storm water and this means a high capital cost for the plant. The wastewater at the end of pipe contains all the discharged contaminants and all the discharged water, which means that the concentrations of contaminants may be low and this can make treatment inefficient. The treatment process has to be a "broad spectrum" process to deal with a range of contaminants and will usually involve a number of unit operations in sequence. The order in which they proceed will depend on the nature of the waste and the limit to be achieved. For example, biological processes are often used in refinery effluent treatment to "polish" the effluent from preceding physical or chemical processes. In the food industry, physical filtration may be used as tertiary treatment after biological oxidation.

Common Effluent Treatment

The concept of effluent treatment, which is "by means of a collective effort", has assumed reasonable gravity by being especially purposeful for cluster of small scale industrial units. Common Effluent Treatment Plant (CETP) not only help the industries in easier control of pollution, but also act as a step towards cleaner environment and service to the society at large. Small scale industries, by their very nature of job cannot benefit much from economies of scale and therefore the burden of installing pollution control equipment, falls heavy on them. Realizing this practical problem, under the policy statement for abatement of pollution the Government felt to extend the scheme for promoting combined facilities for treatment of effluent and management of solid waste for clusters of small scale industrial units and also to provide technical support to them. Accordingly, Ministry of Environment and Forests, Govt. of India, had instructed various State Pollution Control Boards, to examine the possibilities of establishing CETPs in various industrial estates in the respective states.

The approach of joint or common effluent treatment provisions has many advantages. Wastewater of individual industries often contain significant concentration of pollutants; and to reduce them by individual treatment up to the desired concentration, become techno-economically difficult. The combined treatment provides a better and economical option because of the equalization and neutralization taking place in the CETP.

Other important issues for the merit of common treatment include scarcity of land at the industry's level and a comparatively easier availability of professional and trained staff for the operation of CETP, which can otherwise be difficult, at the individual industry level. For the regulatory authorities also, common treatment facility offers a comparatively easier means of ensuring compliance of stipulated norms. The handling and disposal of solid waste also becomes increasingly easier as the infrastructure is created in the project itself. The concept of common treatment, based on feasibility, should be part of the new industrial estates as essential component of infrastructure. In fact, the location of industries should always be such that units with compatible nature of activity are located in a cluster which in-turn can facilitate in providing common treatment. CETP can be explained as follows for better understanding.

C   COMFORTABLE FINANCE

Government incentives through its subsidiary institutions provide much needed financial inputs, a significant boost, especially for small scale sector.

E   EASIER OPERATION

Establishment of individual treatment plants becomes unacceptable by small scale industries (SSIs) due to extreme characteristics of untreated effluent and high qualitative variance therein. A skilled operational staff becomes an affordable option in CETP, wherein all the member units can identify a management group from within, for its effective operation.

T   TREATMENT AT LOW COST

Source specific effluent quality (extreme pH, BOD, colour, nutritive value and metals etc.) which would otherwise render their treatment as a costly preposition, becomes homogenized in CETP and significantly reduce the treatment cost.

P   PRAGMATIC REGULATION

Performance evaluation and its surveillance for implementation of qualitative measures in large numbers of individual plants, at times become a difficult task for regulatory authorities. All this become easily manageable in case of CETP.

The design criteria for CETP consist of the following:

☆ Inventory of Industries

☆ Qualitative and quantitative characterization of wastewater from industries

☆ Classification of industries based on wastewater generation

☆ Classification of wastewater based on bio-degradability

☆ Site-specific, effective and easy-to-maintain design of conveyance system

☆ Bench scale and pilot scale treatability study

☆ Assessment of appropriate treatment technology

☆ Segregation of wastewater

☆ Pre-treatment of wastewater

☆ Waste minimization and resource recovery

☆ Disposal mechanism of treated effluent and sludge

☆ Estimation of treatment cost

☆ Cost benefit analysis

☆ Selection of best suited cost sharing pattern

☆ Stress on cleaner technologies

Advantages of Common Treatment

☆ Saving in Capital and operating cost of treatment plant. Combined treatment is always cheaper than small scattered treatment units.

☆ Availability of land which is difficult to be ensured by all individual units in the event they go for individual treatment plants. This is particularly important in case of existing old industries which simply do not have any space.

☆ Contribution of nutrient and diluting potential, making the complex industrial waste more amenable to degradation.

☆ The neutralization and equalization of heterogeneous waste makes its treatment techno-economically viable.

☆ Professional and trained staff can be made available for operation of CETP which is not possible in case of individual plants.

☆ Disposal of treated wastewater and sludge becomes more organized.

☆ Reduced burden of various regulatory authorities in ensuring pollution control requirement.

Common effluent treatment plants have been successfully operating in the states of Gujarat, Maharastra, Andhra Pradesh and few other states. The would be adopted by other states.

Wastewater Treatment Processes

We may consider the range of treatment processes available by the type of technology: physical, chemical or biological and a new category of "advanced oxidation processes". The Table 4 provides a brief summary of the major liquid wastes, their origin, characteristics, and current methods of treatment for a ready reference (Kiely, 2007; Metcalf and Eddy, 2003; Peavy *et al.*, 1985; Tilche and Orhon, 2002).

Physical Processes

Physical separation processes are those, which do not involve a chemical change and are concentration processes. The processes includes the following:

☆ Clarification processes such as sedimentation and dissolved air flotation which produce sludge for disposal typically containing 1 – 5 per cent dry solids.

☆ Adsorption of dissolved organic molecules onto activated carbon, silica and a range of synthetic adsorbents which may be re-generable but, more frequently, have to be disposed of on exhaustion.

☆ Membrane micro filtration which gives removal of suspended particles upto 0.1 μm into a concentrated slurry typically 2 – 10 per cent of the influent volume.

☆ Ultrafiltration which will remove colloidal matter including emulsified oils and large organic molecules around 10,000 Dalton molecular weight into a concentrated liquid stream typically 10 – 20 per cent of the influent volume.

☆ Reverse osmosis which removes organic molecules of molecular weigh 200 Dalton and above as well as inorganic ions into a concentrated liquid stream typically 20 – 30 per cent of the influent volume.

☆ Stripping which removes dissolved gases and volatile organic compounds into an air stream which is discharged to atmosphere.

☆ Evaporation which removes water from dissolved solids allowing the water to be recovered and leaving a residue which can be either a highly concentrated liquid or a completely dry crystalline or amorphous solid.

☆ Incineration converts organic matter into carbon dioxide which is discharged to atmosphere and may liberate other gases. It is usually uneconomic for streams containing high levels water.

☆ Pyrolysis or gasification converts the waste material into a fuel gas.

Chemical Processes

Chemical processes include the following:

☆ Neutralization and precipitation processes which may produce a sludge containing potentially toxic materials.

☆ Coagulation to remove colloidal matter and break emulsions.

☆ Reduction reactions involving the addition of sodium bisulphite to convert chromates to chromium ions.

☆ Oxidation of cyanides, organic materials and so on involving the addition of oxidizing agents such as chlorine, sodium hypochlorite (often in combination with a catalysts like nickel), potassium permanganate, ozone, hydrogen peroxide, Fenton's Reagent (hydrogen peroxide catalysed with ferrous ions) and peroxydisulphate. Chlorine and similar oxidising agents can give rise to toxic chlorinated by-products.

☆ Ion exchange to remove ionic species.

☆ Electro-dialysis to remove ionic species.

Biological Processes

Biological processes for industrial wastewater are either aerobic or anaerobic. Aerobic processes, based on activated sludge, include:

☆ Conventional high rate percolating filters which offer low operating costs for easily degradable wastes.

☆ Completely mixed activated sludge systems using suspended growth or fixed film technologies which are more resistant to changes in influent composition than plug flow reactors.

☆ Nitrogen and phosphorus removal processes to meet current EU Directive requirements.

☆ Sequencing Batch Reactors which allow flexibility of operation including optional anoxic stages and variation in aeration time to treat variable feeds.

☆ PACT which adds powdered activated carbon to activated sludge to allow the treatment of potentially toxic wastes.

☆ The use of fungi, higher animals and specially selected and acclimated bacteria to treat difficult wastewater streams like pharmaceutical wastes.

☆ Anaerobic processes, using suspended growth or fixed film technologies have proved useful in treating high BOD wastes because they are compact, produce low sludge yields and generate methane which, if there is sufficient quantity, can be used as a fuel gas.

Online Monitoring of Water Quality and Real Time Data Transmission through GPRS Network

Online and continuous water quality anlysers can monitor the parameters like water temperature, conductivity, dissolved oxygen, pH, turbidity, CODMn, Ammoniacal Nitrogen, Total Phophorous, Total Nitrogen, TOC (Total Organic Carbon), Fluoride etc. The anlyser system mainly consists of water sampler with pump and pipe, automatic cleaning filtration sysem, water distribution control device, pretreatment, automatic online water quality analyzers, cleaning device, pure water device, PLC control unit, online display of measured parameters remote terminal units, remote GPRS transmission/communication devices, power supply etc as shown in Figures 4.1–4.4, which is found to be practised in P. R. China.

The water quality data management system consists of automatic storage, PC communications, data logger, GPRS (General Packet Radio Service) device with GSM SIM for wireless data transmission, antena and data acquisition channel as shown in Figure 4.4.

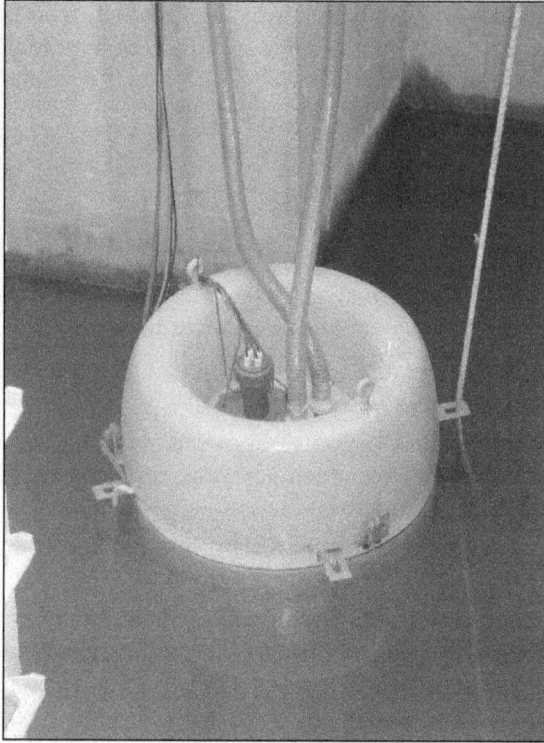

Figure 4.1: Sampling Float in Water Body.

Figure 4.2: Water Distributor and Water Quality Analysers at
Online Water Quality Monitoring Station.

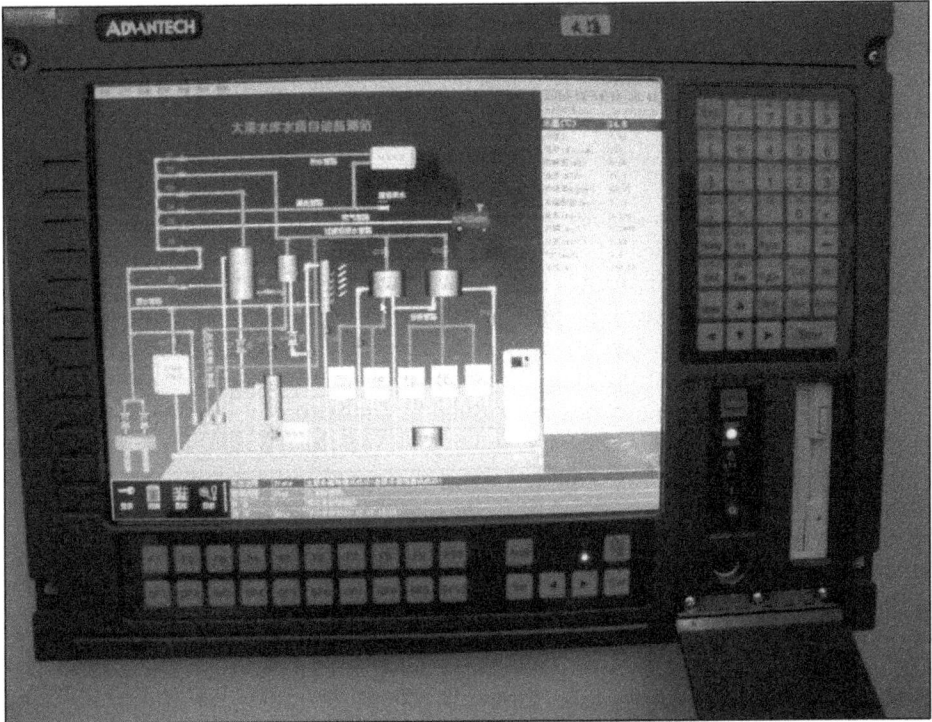

Figure 4.3: Online Display Monitor at the Station.

Figure 4.4: GPRS Device and Antenna for Data Transmission.

The technology on online water flow measurement and water quality monitoring with GPRS based real time data transmission system has been implemented in various parts of the world now-a-days. This has been developed and adopted now by the State Pollution Control Board, Odisha.

The real time water quality data from the online analysers to SPCB server is transferred through an advanced communication system like GPRS network consisting of GPRS device enabled with GSM SIM Card, transmitted through existing mobile phone network to SPCB server and also display the real-time data through an electronic display board in front of factory gate for public information. Mobile phone towers are located in most of the places including remote locations and suitable for data transmission with less recurring cost. A new innovative concept called "Y Cable to capture the actual data from analyzer prior to going to the plant's computer" along with GPRS based technology has been given by the author of this chapter to check manipulation and it has been introduced in RT-DAS (Real time Data Transmission system) adopted by SPCB, Odisha for the first time in India to capture the remote online environmental monitoring data on a real time basis. Online data prior to arrival at the plant's local computer will be transmitted through GPRS device to the server of the Board stationed in its Head office at Bhubaneswar directly without any chance of manipulation and delay. The GPRS network can simultaneously send real-time data to SPCB server, Server of the industry and Electronic display board installed in front of the factory gate. This can maintain more transparency with public and stakeholders. Nanjing Automation Institute of Water Conservancy and Hydrology, Ministry of Water Resources, Nanjing P.R. China is one of the expert organisation in Asia for implementation GPRS Network for Real Time Water Quality Data transmission from online Monitoring Stations from hundreds of stations of stations.

Real time data transmission keeps the regulator, industry alert and continuous information to public in a more transparent way. In event of high emission, the plant and their Head will receive SMS alert, so that they will immediately attend to control pollution. The system will also keep the process head of the plant responsible apart from the environmental management head. This will prompt the process engineers to take up preventive maintenance to avoid higher emission which is likely to be visible to all through RT-DAS so that they can be afraid of public disgrace and criticism. Automatically they have to maintain the air pollution control measures properly and continuously operate the same even in odd hours/night time. In case continuous violation is observe, the SPCB can issue show cause notice and closure direction to the defaulting industry to promptly curb water pollution.

Issues on Water Pollution Control

Socio Political Aspects

Often the phrase "appropriate technology" is interpreted in the developed world to mean that technology to be supplied to developing countries should be old technology that is technology which was used in developed countries may be twenty years ago. International companies often see, in developing countries, an opportunity to sell old technology to make a quick profit.

But, in correct sense the appropriate technology means, the technology that will achieve the desired output quality and quantity from the available input in the light of the available local resources. It is a concept equally applicable to the developed world. Let us consider a couple of examples.

In Northern Europe some treatment processes are very expensive because of the high power required for the UV irradiation system. In tropical countries this can be provided by sunlight, making this new technology much more appropriate to the less sunny and developed world. Activated sludge process is more appropriate for India than cold countries.

Economics

Economics is the language of industry and industry must be able to see an economic benefit in the implementation of water management. Clearly taxation can be an effective tool provided that it is not so restrictive as to force industry to close down factories and relocate to cheaper areas, for this would be detrimental both to the local community and, possibly, the national economy. The increasing cost of mains water supplies and sewer discharge is done much to focus attention on water management and the technologies employed. Recycling and reuse of wastewater in an industry saves lot of money.

The economic climate changes continuously. New processes become cheaper as their use becomes more widespread. Reverse osmosis membranes are now very cheaper (and more efficient) that they were even ten years ago as a result of the increased demand and competition amongst manufacturers.

Appropriate technology matches wastewater characteristics to discharge criteria and recycle quality taking account of resources, utilities and chemicals availability and the standards of operating and maintenance personnel and spares availability. It should be properly implemented, economically viable and meets the socio-political aspirations of industry, government and the community.

Conclusion

Water pollution can be controlled and the precious water resources can be conserved by various means as described in the chapter and by adopting following measures.

1. Process modernization and improvement reducing water consumption.
2. The existing industries should upgrade their ETP to meet stricter standards.
3. Water and wastewater management should be clearly measured in terms of cost reduction taking in account of reduction in water consumption, re-use water where possible, recycle water where practicable, reduce waste of all types.
4. Restricting the use of river water and enforcing stricter environment standard for discharge of polluted/treated wastewater in to the river or any water body.
5. Adopting optimum recirculation for reusing treated effluent.

6. Utilizing maximum quantity of treated effluent for plantation and agriculture.

7. Locating big industries consuming high quantity of water such as large Pulp and Paper Mill, Integrated Steel, Fertilizer etc. away from the river bank. This will help to check the instant water pollution by the industries at odd hours. Industries should store their treated water in large reservoirs meeting standards. They may reuse the water or partly discharge the stored water to river/drains in a controlled manner.

8. All ULBs and industrial townships should install STP to treat the drainage/sullage water and save degradation of river/surface water quality.

9. The treated wastewater may be permitted for discharge into the river meeting standards without altering the classification of the river (based on use). During summer when the dry water flow is minimum or even nil the discharge of wastewater in to the river should be stopped.

10. Improvement of Rural Sanitation by providing individual and community latrine (WHO/UNICEF, 2000).

11. Awareness campaign to stop open defecation on the banks of river, pond, lakes and nearby places which directly affects the water quality of these water resources during rainy season.

12. Other traditional, religious and cultural use of river/lake/pond water by which the water quality degrades should be checked. Organic colour should be used in place of chemical compounds. For protection these water resources, traditional practices must be changed. Since it is an age-old practice continuous awareness campaign in this field is necessary to change the mindset of the people.

13. Excess drawl of water for agricultural/industrial use by which the minimum dry weather flow is affected should not be permitted. In such condition the downstream users are affected directly.

14. Surface run-off study of the industrial premises should be conducted and provisions should be made to treat the surface run-off adequately to remove the pollutants before discharged to outside water bodies during monsoon.

15. Rain water harvesting structures should be developed by all the large industries to reduce water drawl from the rivers to maintain environmental flow even during lean season.

16. Much trust has to be given to contaminated agriculture run-off to the water bodies in terms of ban of use of pesticides/insecticides, promoting organic farming and monitoring toxicity of the portable water. Water pollution should the co-related to health effects through in-depth study by reputed research institutes and the Government.

17. National drinking water quality standards should be reviewed from time to time through experts and strictly implemented throughout the country. Facility for measuring drinking water quality should be made available for public at least in all district headquarters and review of policy to preserve water quality should be done on a regular interval (WHO/UNICEF, 1997).

18. Advanced technologies like GPRS based online and continuous monitoring of water quality in rivers, reservoirs, lakes should be implemented with close circuit camera and real time data transmission to headquarters for review and ensure to maintain proper water quality before it is consumed, giving much value to human and animal life.

References

Administrative Staff College of India, Hyderabad and State Pollution Control Board, Orissa, State of Environment, Orissa-2006 report.

Central Pollution Control Board, Pollution Control Acts and Rules and Notifications issued there under, Pollution Control Law Series, June, 2010, New Delhi.

Central Pollution Control Board, Charter on corporate responsibility for environmental protection (CREP), New Delhi, 2003.

Kiely, G. (2007): Environmental Engineering, The McGraw-Hill Co. Ltd., New Delhi.

Metcalf and Eddy, (2003): Wastewater Engineering – Treatment and reuse, Tata-McGraw-Hill Publishing Co. Ltd., New Delhi.

Peavy, H. S., Rowe, D. R. and Tchobanoglous, G. (1985): Environmental Engineering, McGraw-Hill, Inc, New York.

Tilche, A. and Orhon, D. (2002): Appropriate basis of effluent standards for industrial wastewaters, Water Science and Technology, 12: 1-11.

WHO/UNEP (1997): Water Pollution Control – A guide to the use of water quality management principles.

WHO/UNICEF (2000): Global Water Supply and Sanitation Assessment Report.

Chapter 5

Sequencing Batch Reactor (SBR) Technology for Treatment of Slaughterhouse Wastewater

Pradyut Kundu[1] and Somnath Mukherjee[2]

[1]Ph. D, Research Scholar, [2]Professor,
Civil Engineering Deparment (Environmental Engineering Division),
Jadavpur University, Kolkata – 32, West Bengal, India

ABSTRACT

Sequencing batch reactor (SBR) is a modification of activated sludge process which has been successfully used to treat municipal and industrial wastewater. In this system, wastewater is fed to a single reactor which operates in a fill-and-draw batch treatment mode repeating a cycle (sequence) continuously. All the operations such as fill, react, settle, draw, idle are achieved in a single batch reactor. The process could be applied for nutrients removal, high chemical oxygen demand containing industrial wastewater such as food processing industries effluents, landfill leachates, tannery wastewater etc. Slaughter house wastewater is mainly composed of diluted blood, protein, fat and suspended solids. The organic matter concentration in this wastewater is medium to high and the residues are partially solubilized, leading to a highly contaminating effect in riverbeds or sewer systems if the same is not properly treated. One of the serious problems that encounters is eutrophication. Laboratory was conducted by author to explore simultaneous removal of organic carbon and nitrogen from slaughterhouse wastewater using a SBR technology. The reactor was operated under three different variations of aerobic-anoxic sequence, *viz.* 4+4, 5+3 and 3+5 hours of total react period with influent soluble COD (SCOD) and ammonia nitrogen level 2000 ± 50mg/L, 90 ± 10 mg/L, 2000 ± 50 mg/L and 140 ± 10 mg/L, respectively. It has been observed that 88 to 90 per cent of SCOD removal was possible at the end of 8.0 hour of overall reaction period, irrespective of the length of the aerobic react period. In case of 4+4 aerobic-anoxic operating cycle, reasonable degree of nitrification 92.55 and 80.75 per cent corresponding to initial NH_4^+-N value of 94.58 and 142.52 mg/L respectively, along with 92.55 and 89.45 per cent of organic carbon removal corresponding to initial SCOD value of 1984.34 and 1975.35 mg/ L respectively, have been achieved for treatment of slaughter house wastewater in sequencing batch reactor.

Keywords: *Slaughterhouse wastewater, Sequencing batch reactor, Carbon oxidation, Ammonia oxidation.*

Introduction

With the advent of economic growth and increase in population, pollution becomes a significant feature all over the world. Treatment of pollution in an economic way is becoming more complex. Environmental engineers and scientists are endeavoring for finding an economic way to minimize the menace of pollution. Traditionally, wastewater treatment is looked in view of the removal of gross organic and inorganic constituents and pathogens from wastewater that primarily includes carbonaceous COD, suspended solids removal and disinfection processes. In addition to this nutrients *viz.* nitrogen and phosphorus are considered to be significant contributors of pollution problem. Industries like meat processing, slaughterhouse, dairy, fermentation product, fertilizer etc, emanate wastewater containing concentrated organic nitrogenous matter along with carbonaceous waste have given rise to high level of BOD and COD. The slaughterhouse and meat processing industry generates large volumes of wastewater, which requires considerable degree of treatment before getting released to the water environment. Slaughterhouse wastewater primarily contains diluted blood, protein, fat and suspended solids. In aerobic stabilization of such wastewater organic/inorganic nitrogen is ultimately converted into nitrate as end product of metabolism. However, high level of nitrates in treated effluent to be discharged into the surface wastes is not desirable since they are potentially hazardous both for aquatic and human life. Formations of nitrosoamine, methanomoglobinemia are the toxic effects to human being where as eutrophication lakes are caused by nitrogen discharge as supply of stimulant of algae and other aquatic plants. Due to the combined effect of carbonaceous and ammoniacal matters, dissolved oxygen (DO) level in aquatic body drastically depletes due to the consumption of DO. Hence, a treatment is very much necessary which would be combined in nature preferably by biological method. The real benefit of application of the process lies in its capability to remove both inorganic and organic type of nitrogen in a single stage to yield an effluent of high quality at reasonable cost.

In recent year, sequencing batch reactor (SBR) has been employed as an efficient technology for wastewater treatment, especially for domestic and industrial wastewaters because of its simple configuration (all necessary processes are taking place time- sequenced in a single basin) and high efficiency in COD and suspended solids removal. SBRs could achieve nutrient removal using alternation of anoxic and aerobic periods. The basic process involved in an effluent treatment scheme *viz.*, Fill,

React, Settle and Draw are carried out in a single reactor. An SBR operates in a true batch mode with aeration and sludge settlement occurring in the same tank. The major differences between SBR and conventional continuous flow activated sludge system is that the SBR tank carries out the functions of equalization, aeration and sedimentation in a time sequence rather that in the conventional space sequence of continuous flow systems. Thus, there is a degree of flexibility associated with working in a time rather than in a space sequence. It is reported that treatment in SBR is economical than conventional continuous mode activated sludge process in the tune of 20 per cent as well as it is equally efficient for organic removal.

SBR Operation

Irvine and Ketchum (1989) describe the SBR and its periods in detail. During the "fill" period, influent wastewater is added to the biomass that was left in the tank from the previous cycle. The length of the "fill" period depends on the number of SBRs, the volume of the SBRs, and the nature of the flow of the wastewater source, which can be intermittent or continuous. The reactor may or may not be mixed during this period. Filling ends when the wastewater has reached the maximum water level or at some fraction of that if multiple fill periods are used during a cycle. The "react" period occurs after a fill period. In most cases the reactor is mixed during this period. Aeration may or may not be used depending on the reactor's objective and operation. In addition, the react period may be interrupted with fill periods and sludge wastage. During the react period many reactions can take place such as nitrification, denitrification, COD removal, phosphorous removal, and many others. After the react period, a "settle" period takes place. During the settle period, the SBR acts as a clarifier. The solids, including biomass and particulate substrate, settle and leave the relatively clear effluent on top. The settle period normally lasts between 0.5 and 1.5 hours so that the solids blanket does not begin to float due to gas buildup. "Decant" occurs at the end of the settle period and is the time when the effluent is drawn off. This may take place with a pipe or weir. The decant level should be adjustable so as to make the SBR more adaptable to changes. Finally, some systems include an "idle period." This period is most necessary when several SBRs are being used with a continuous wastewater source. This allows a small amount of leeway when trying to match the cycles of the SBRs so that one SBR is always on the fill period. Figure 5.1 shows successive operational phase in one cycle period of SBR.

Advantages of SBR

SBR possess following advantages over conventional biological treatment units.

1. It serves as an equalizing basin during fill-period and can tolerate waste shock loads under extreme conditions.
2. Recycling of sludge is not necessary.
3. Solid-liquid separation occurs under nearly ideal quiescent conditions. Hence short circuiting is non-existing during settling.
4. Changing the fill level easily controls filamentous growth.
5. SBR can be operated to achieve nitrification/denitrification or phosphorous removal without chemical addition in the same single reactor.

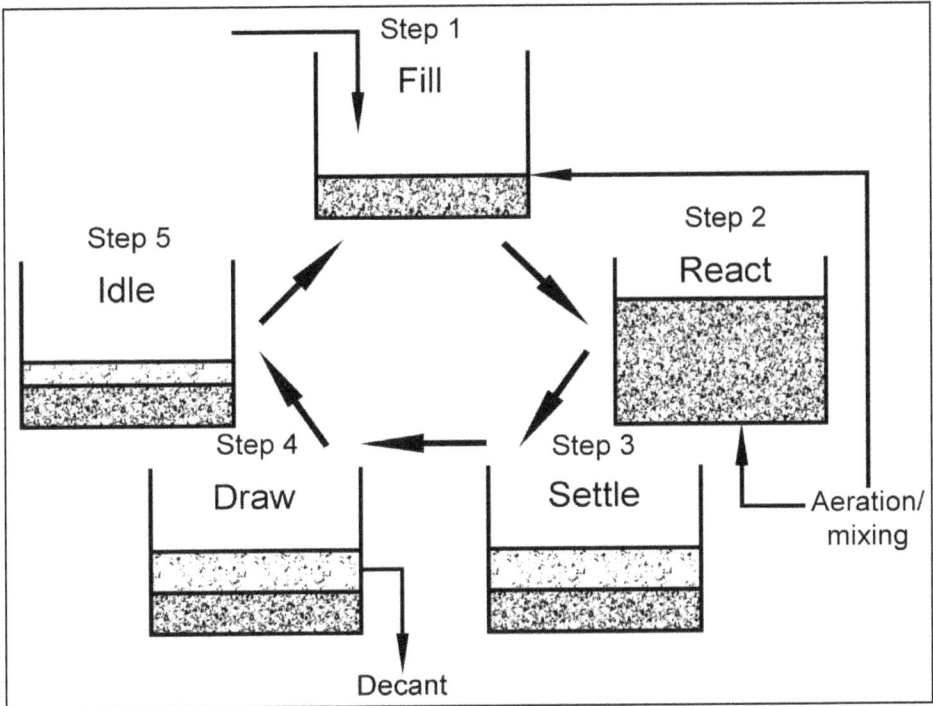

Figure 5.1: SBR Operation Cycle Showing Five Discrete Time Periods as Fill, React, Settle, Draw and Idle.

Process Limitation

The major limitations, which became apparent during the development stages of SBR are stated below:

1. It is a non-continuous flow system.
2. It was perceived to have value only in small system.

To overcome the situation, parallel SBR tanks may be provided.

Comparison of SBR with other Processes

Sequencing Batch Reactor (SBR), a promising scheme for effluent treatment, can be most reasonably compared to continuous flow systems. It falls in a broad category of unsteady activated sludge systems where the effluent flows from a particular process to the next sequentially in a single reactor. The basic processes involved in an effluent treatment scheme of Fill, React, Settle and Draw are carried out in a single reactor. The main difference between SBR and the conventional continuous flow systems is that the former provides in time where the latter provides in space. The major advantages in utilizing SBR are the flexibility of operation and processes. The effluent treatment could be adjusted as per their quantity and quality. In fact, a properly designed semi-batch reactor should achieve a higher effluent quality than Continuous Flow System (CFS).

In contrast to continuous flow system, both biological reactions and biomass separation take place in the same tank. This is a single tank batch system, sequencing on individual cycle and thus provides both low capital and operating costs. Conventional Waste Management technologies commonly adopted in tropical climates are not only expensive but also warrant exacting operation and maintenance requirements.

"Flexibility of Operation" is another important parameter associated with SBR and not found in CFS. The time of all the phases (fill, react, aeration etc.) can be easily changed which facilitates the SBR to tolerate variations in organic and hydraulic loading and to sustain the shock loads. SBR also has the advantage of a more flexible operation with true Plug Flow characteristics. Since the treatment process proceeds over time in one reactor, the process control can be conventionally executed by means of auto-analyzers (*e.g.* for DO, pH, nitrate) and computers.

On the other hand, continuous flow systems are more flexible with respect to recycle of sludge and wastewater. Recycle flows can be directed to any reactor in the system whereas the comparable operation is not possible in the SBR. SBR treatment profile functions similarly as in a Continuous Flow Stirred Tank Reactor (CFSTR) scheme, where the concentration of substrate reduces in similar pattern. Table 5.1 shows the comparison between SBR and continuous flow reactors following Figure 5.2 show the concentration profile of organic substrate in a Continuous Flow Stirred Tank Reactor and SBR.

Basic Theory of Biological Treatment

Organic Carbon Oxidation

Biological treatment of wastewater is carried out with the objective to coagulate and remove non-settleable colloidal solids and to stabilize the organic matter. The removal of carbonaceous matter, the coagulation of non-settleable colloidal solids and stabilization of organic matter are accomplished by micro-organisms, principally

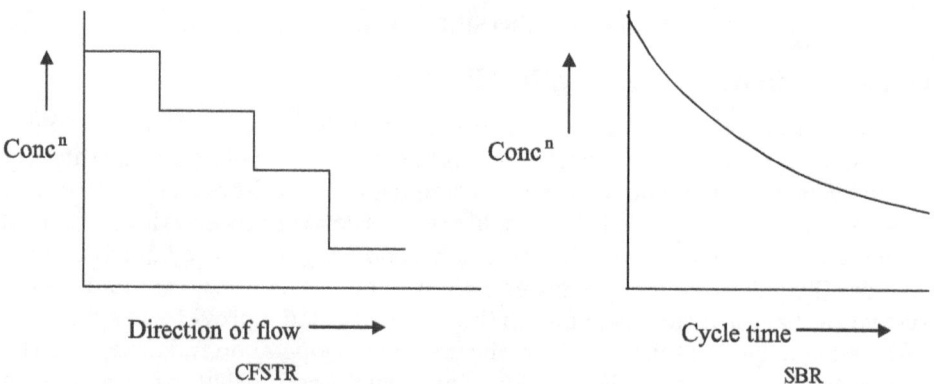

Figure 5.2: Concentration Profile of Organic Substrate in a Continuous Flow Stirred Tank Reactor and SBR.

Table 5.1: Comparison between SBR and Continuous Flow Reactors

Parameter	Batch Process	Continuous Process
Concept	Time Sequence	Spatial Sequence
Inflow to Reactor	Periodic	Continuous
Discharge from Reactor	Sequenced	Continuous
Aeration	Sequenced	Continuous
Mixed Liquor	Always in reactor and no recycle	Recycle through reactor and clarifier
Clarification	Quiescent hydraulics	Hydraulic motion
Flow Pattern	Perfect plug	Complete mix or approaching plug
Equalization	Inherent	None
Flexibility	Considerable	Limited
Hydraulic Sizing	Variable	Uniform
Capital Cost	Low	Medium
Physical structure	Compact	Larger than SBR
Sludge Yield	Low	Moderate
Effluent Quality	Good	Variable

bacteria. The microorganisms are used to convert the colloidal and dissolved carbonaceous organic matter into simple end products and additional biomass, as represented by the following equation (eqn. 1) for the aerobic biological oxidation of organic matter. In accomplishing this type of treatment, the chemoheterotrophic organisms are of primary importance because of their requirement for organic compounds as both carbon and energy source. In aerobic respiration process molecular oxygen is used as electron acceptor in respiratory metabolism. A definite stoichiometric relationship exists between the substrate removed, the amount of oxygen consumed during aerobic heterotrophic biodegradation and the observed biomass yield.

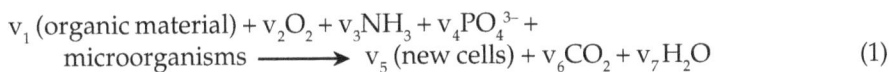

$$v_1 \text{ (organic material)} + v_2O_2 + v_3NH_3 + v_4PO_4^{3-} +$$
$$\text{microorganisms} \longrightarrow v_5 \text{ (new cells)} + v_6CO_2 + v_7H_2O \tag{1}$$

where, v_i = stoichiometric coefficient

Biological Transformations of Nitrogen

Removal of nitrogen from wastewater is usually achieved by physical and biological processes. The screening and settling processes are physical ways to remove organic nitrogen bound in suspended solids but it cannot remove most of the nutrients including the large fraction of nitrogen that is soluble. This leaves biological treatment as the next choice in nitrogen removal. Three major biological processes directly involved with biological nitrogen removal in wastewater treatment are ammonification, nitrification, and denitrification.

Ammonification

Ammonification occurs when organic nitrogen is converted to ammonia. It is an important mechanism that ultimately allows organic nitrogen to be removed from wastewaters through hydrolysis to amino acids, which are broken down to produce ammonium or directly incorporated into biosynthetic pathways in support of bacterial growth. Nitrogen as ammonia or nitrate can be assimilated by bacteria to form cellular mass.

Nitrification

Nitrification is the two-step process for the conversion of nitrogen in the form of ammonia/ammonium to the form of nitrate or nitrite which is shown in Figure 5.3. Ammonium nitrogen is oxidized to nitrite by ammonia oxidizing bacteria (AOB) and then to nitrate by nitrite oxidizing bacteria (NOB). Many AOBs and NOBs are autotrophic, although heterotrophic bacteria are known to function as nitrifiers (Painter, 1977).

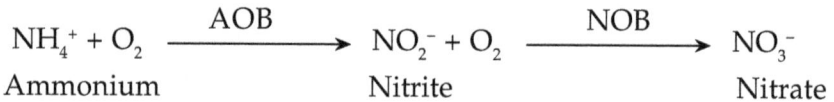

$$NH_4^+ + O_2 \xrightarrow{\text{AOB}} NO_2^- + O_2 \xrightarrow{\text{NOB}} NO_3^-$$

Ammonium   Nitrite   Nitrate

Figure 5.3: Biological Transformations of Ammonium Nitrogen.

The total ammoniacal nitrogen (TAN) in the wastewater originates from the breakdown of urea by the enzyme urease, which is present in fecal matter, and the breakdown of proteins in organic matter, which contain amine groups. The combination of the urine and feces releases a large amount of ammonia. Genera of microorganisms that scientists believe play roles in nitrification are *Nitrosomonas, Nitrobacter, Nitrosospira, Nitrosolobus, Nitrosovibrio,* and *Nitrosococcus* (Madigan *et al.,* 2000). These genera of organisms are autotrophic, so their carbon source is carbon dioxide (CO_2). Ammonia oxidizing bacteria, such as *Nitrosomonas,* utilize the reduced nitrogen in ammonia as the electron donor, or energy source. They oxidize it to form nitrite (NO_2^-), using oxygen (O_2) as the terminal electron acceptor (TEA). Nitrite-oxidizing bacteria, such as *Nitrobacter,* then use the nitrite as their energy source with oxygen as the TEA to form nitrate (NO_3^-) (Madigan *et al.,* 2000). Approximate equations for nitrification process are:

For Nitrosomonas

$$55NH_4^+ + 76O_2 + 109HCO_3^- \rightarrow C_5H_7O_2N + 54NO_2^- + 57H_2O + 104H_2CO_3 \qquad (2)$$

For Nitrobacter

$$400NO_2^- + NH_4^+ + 4H_2CO_3 + HCO_3^- + 195O_2 \rightarrow C_5H_7O_2N + 3H_2O + 400NO_3^- \qquad (3)$$

Denitrification

Denitrification involves conversion of nitrate or nitrite into various reduced nitrogen compounds. Microorganisms utilize the nitrate as an electron acceptor in the absence of oxygen. When microorganisms deplete the dissolved oxygen in a system, the system is often described as anaerobic or anoxic. In anaerobic conditions, fermentation takes place by those cells that are capable of adaptation to the lack of oxygen. However, if there is nitrate or nitrite present after the oxygen is depleted, some organisms will make metabolic adjustments in order to respire using the nitrate or nitrite instead of being forced to use fermentation, which yields much less energy, or to go dormant. Scientists refer to this environmental state as anoxic. The energy yield for nitrate respiration is not as high as that for aerobic respiration, but nitrate is still an effective terminal electron acceptor (Madigan *et al.*, 2000). The metabolism for this process generally is heterotrophic. Organic compounds act as both the carbon and energy sources. The first step of denitrification is the reduction of nitrate to nitrite. Then, depending on each species, the organisms reduce the nitrite further to various nitrogen oxides and nitrogen gas. Many different species of bacteria have the capability of denitrification, including those from *Pseudomonas, Achromobacter, Bacillus, Rhizobium, Aquaspirillum, Flavobacterium, Aeromonas, Moraxell.*

The reactions for denitrifying bacteria with glucose as carbon source are given by:

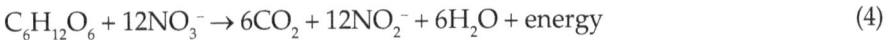

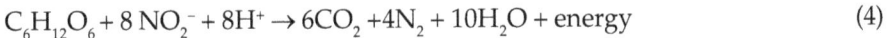

$$C_6H_{12}O_6 + 12NO_3^- \rightarrow 6CO_2 + 12NO_2^- + 6H_2O + energy \tag{4}$$

$$C_6H_{12}O_6 + 8\,NO_2^- + 8H^+ \rightarrow 6CO_2 + 4N_2 + 10H_2O + energy \tag{4}$$

General Description of Slaughterhouse Wastewater

Characterization of wastewater is an important part of the initial work of in the design treatment process. There are various industrial wastewaters which contain high organic and nitrogen and slaughterhouse is one of them. The organic matter concentration is this wastewater is medium to high and the residue are partially solubilized, leading to a highly contaminating effect in river beds and sewer systems if the wastewater is not previously treated.

Types of Slaughterhouses

There are different types of slaughter house depending on types of animals slaughter, variation in quality and quantity of wastewater and solid waste generation. They are as follows:

1. Large animal or bovine slaughterhouse (Cattle, Buffalo)
2. Goat and sheep slaughterhouse
3. Pig slaughterhouse
4. Poultry slaughterhouse

Slaughterhouse unit is water intensive, critically water polluting industry which generates large quantities of obnoxious liquid effluent with high organic load. As reported by the Ministry of Food Processing, Government of India, in 1989, a total of 3616 recognized slaughter houses slaughter over 2.0 million cattle and buffaloes. In recent times this number has only increased. In West Bengal, small-scale bovine slaughtering industry has thrived in places like Magrahat in South 24 Parganas. The main environmental issue associated with this industry is the discharge of high strength effluent, which is being dumped mostly into the local water bodies without appropriate degree of treatment. This waste stream is characterized by high concentration of fats, oil and grease, COD and nitrogenous substances. Water is consumed in all stages of the operation ranging from the entry of the animal in the lairage to the requirement for personal hygiene of the workers. The average wastewater consumption was estimated at about 0.25 m³ per head slaughtered. On an average, daily 8 to 10 bovines are being slaughtered per day in small-scale rural slaughterhouses. The solids present in the effluent do not breakdown readily in subsurface soil absorption systems. The effluent, being discharged mostly in the nearby water body, causes severe water pollution and spreads foul odour in the locality.

Description of Slaughtering Process

The operations involved during slaughtering of large animals are described underneath in brief and also shown in a flow chart (Figure 5.4).

Lairage

After ante mortem health inspection, the animals are given enough quantity of water but no fodder, for 12 hrs prior to slaughtering, in order to flush out the pathogenic microorganisms. However, it was observed that only very few slaughter houses (less than 1 per cent) have lairage facilities.

Slaughtering

In the *Halal* method large animals are slaughtered as per the Islamic Rites. The animal is pushed on the floor and the jugular vein is out manually by the butcher to drain blood. In majority of the slaughterhouses the blood is allowed to spill on the floor and join the wastewater drain. Only in a few large slaughterhouses, part of the blood is collected by some agencies for manufacture of machines/tonics.

Bleeding

In the *Halal* method of slaughtering, blood collection is not done immediately on slaughtering and most of the blood goes down into municipal drains causing pollution. Blood of the animals, which can be collected for making use in pharmaceutical industry, is thus by and large lost. Due to inadequate facilities at the slaughter houses and scattered illegal slaughtering of animals, a very few slaughter house collect blood.

Dressing

Due to lack of means and tools, dehiding of the carcasses is done on the floor itself, which causes contamination of the meat. The hides and skins are spread on the floor of the slaughtering area. Similarly legs, bones, hooves etc. are not removed immediately from the slaughtering area. The dressing operation consists of:

☆ Sticking of heart to ensure complete bleeding

☆ Removal of horns, hind legs, head trimming and demasking

☆ Flying of abdomen and chest

☆ Removal of hide

Cattle

Water

Dung (For its use as manure or cow dung cake)

Lairage

Slaughtering

Blood (For preparation of blood meal or manure)

Horn, Hind leg, Head, Trimming etc

Hide (Exported)

Bleeding

Offals (mostly edible and hence sold out)

Dressing

Evisceration

Raw Effluent to ETP

Carcass Splitting

Carcass Washing

Transport to Meat shop

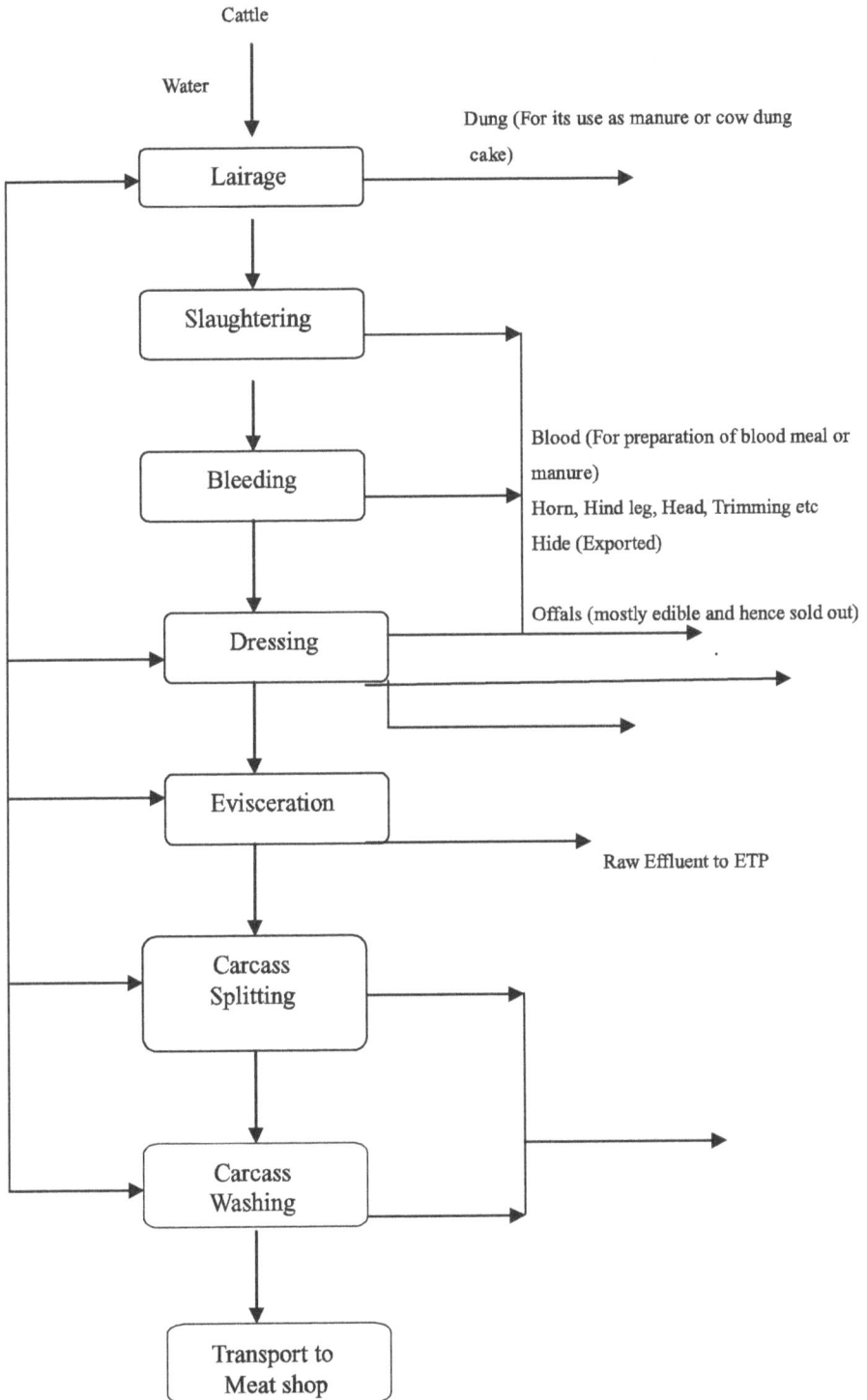

Figure 5.4: Process Flow Chart for Bovine Slaughtering in the Slaughterhouse.

Evisceration

This particular process during slaughtering generates maximum amount of waste. The butchers who carry out illegal slaughtering of animals generally throw visceral material at the community bins and wash the small intestines at their shops itself and thus create pollution problem.

Relevant Literature on Sequencing Batch Reactor

Research on SBR reactors began in the 1970s. SBR received considerable attention since Irvine and Davis described its application (Irvine and Davis 1971). Studies of SBR process were originally conducted at the University of Notre Dame Indiana (Irvine and Busch 1979). Irvine and Busch (1979) reported on bench scale studies of fill and draw of SBR and discussed control strategies for both single and multiple tank systems. Carbonaceous matter removal, nitrification, denitrification can be achieved in these systems.

Kargi and Uygur (2002) studied the Nutrient removal from synthetic wastewater by sequencing batch operation at different specific nutrient loading rates (SNLR) obtaining the highest COD (99 per cent), NH_4-N (99 per cent) and PO_4-P (97 per cent) removal efficiencies with the initial COD concentration of 600 mg/l, at COD loading rate of nearly 40 mg COD/(g biomass)/h.

A performance evolution study was conducted by Surampalli *et al.*, 1997 on the basis of data, collected from nineteen municipal and private SBR wastewater treatment plant in the United State. The average design flow for these plants ranged from 0.028 to 3.0 MGD. The average effluent BOD concentration ranged from 3.0 to 14.0 mg/l with removals ranging from 88.9 per cent to 98.1 per cent. Effluent NH_3-N concentrations ranged from 0.029 to 1.68 mg/l and ammonia removals were between 90.8 to 96.8 per cent. The average effluent phosphorous concentration were between 0.53 to 4.27 mg/l.

A study was conducted by Li *et al.* (2008) to examine the nitrogen removal from slaughter house wastewater treatment in laboratory scale sequential batch reactor. Operated at low dissolved oxygen level under two aeration strategies 1) Intermittent aeration, 2) Continuous aeration. It was proposed that *in situ* measurement of oxygen utilization rates used to control the SBR for nitrogen removal.

Nutrients in piggery wastewater with high organic matter, nitrogen and phosphorous content were biological removed by Obaja *et al.*, 2005 in a sequencing batch reactor (SBR) with anaerobic, aerobic and anoxic stages. The SBR was operated with three cycles per day, temperature 30c, SRT 1 day 7 HRT 11 days, with wastewater containing 1500 mg/l ammonium and 144 mg/l phosphate, a removal efficiency of 99.7 per cent for nitrogen and 97.3 per cent for phosphate was obtained.

Biological treatment of a piggery wastewater for organic carbon and nitrogen removal; in a combined anaerobic-aerobic system was investigated by Bernet *et al.*, 2000 using two laboratory scale sequencing batch reactors. The cycle length was 24 hr. In the anaerobic reactor, fed with raw waste matter recycling from the aerobic reactor, denitrification followed by anaerobic digestion of organic carbon was observed. In the aerobic reactor, more organic carbon removal and ammonia oxidation to mainly nitrite occurred. Denitrification was also observed in the aerobic reactor during the filling period, when mixed liquor dissolved oxygen concentration was very low.

Objective of the Present Work

The objective of the present work is to conduct the performance study of laboratory – scale sequencing batch reactor (SBR) for combined removal of soluble organics (COD) and nitrogen from slaughter house wastewater by changing operational condition like MLVSS concentration, initial substrate concentration cycle time, DO, anoxic/aerobic sequence, anoxic/aerobic time phase, etc.

Materials and Methods

Seed Acclimatization for Combined Carbon Oxidation and Nitrification

Seed acclimatization study was conducted in a measuring cylinder of 2.0 L volume. Heterotrophic microbes were actively acclimatized in laboratory environment by inoculating sludge collected from an aeration pond of M/S Mokami small-scale slaughterhouse located in the village Nazira, South 24 Parganas district (West Bengal) India, to a growth propagating media composed of 250 mL of nutrient solution. Finally 750 mL volume of distilled water was added to liquid mixture to made a volume of 1L and the mixture was continuously aerated with intermittent feeding with dextrose solution having concentrations of 1000 mg/L and ammonium chloride (NH_4Cl) having concentration of 500 mg/L as a carbon and nitrogen source respectively. The acclimatization process was continued for an overall period of 30 days. The biomass growth was indicated by the magnitude of sludge volume index (SVI) and MLVSS concentration in the reactor. pH in the reactor was maintained in the range 6.8-7.5 by adding required amount of sodium carbonate (Na_2CO_3) and phosphate buffer. The seed acclimatization phase was considered to be over when a steady state condition were observed in terms of equilibrium COD and NH_4^+-N reduction with respect to a steady level of MLVSS concentration in the reactor.

Experimental Set-up

The experimental work was carried out in a laboratory scale SBR, made of Perspex sheet of 6 mm thickness, having 20.0 L of effective volume. In order to assess the treatability of slaughterhouse wastewater in a SBR, the real life wastewater was taken as grab sample from outlet of the same slaughterhouse unit as described earlier. The wastewater was screened with a cotton cloth and then settled further for 3hr. The settled effluent was examined for determining its initial characteristics which are exhibited in Table 5.2. The settled effluent was diluted in different proportion and poured in reactor of 20.0 L capacity to carry out necessary experiments to explore potential of the biological removal efficiency. 750 mL of pre-acclimatized seed of mixed nature was added to the reactor containing 20.0L of pre-treated slaughterhouse wastewater to perform the necessary experiments. Oxygen was supplied through belt driven small air compressor. A stirrer of 0.3 KW capacity was installed at the center of the vessel for mixing the content of the reactor. Air supply was provided during aerobic phase of react period. However, during the anoxic phase the stirrer was allowed to operate for mixing purpose. A timer was also connected to compressor for controlling the sequence of different react period (aerobic and anoxic). Oxygen was supplied in the reactor through strainer type diffuser, placed at the bottom of the reactor. A schematic diagram of the experimental set-up is shown in the Figure 5.5.

The cycle period for the operation of SBR was taken as 10 hour, with a fill period of 0.5 hour, overall react period of 8.0 hours, settle period of 1.0 hour and idle/decant period 0.5 hour. The overall react period was divided into aerobic and anoxic react period in the following sequences.

☆ **Combination – 1 :** 4 hour aerobic react period and 4 hour anoxic react period.

☆ **Combination – 2 :** 5 hour aerobic react period and 3 hour anoxic react period.

☆ **Combination – 3 :** 3 hour aerobic react period and 5 hour anoxic react period.

During the time course of the study, 100mL of sample was collected from the outlet of the reactor at every 1.0 hour interval, on completion of the fill period. The samples were analyzed for the following parameters, *viz.* pH, DO, MLSS, MLVSS, COD, NH_4^+-N, NO_2^--N and NO_3^--N as per the methods described in Standard Methods (APHA, AWWA and WPCF, 1995).

Results and Discussions

The slaughter house wastewater possesses a complex composition as it contains dilute blood, fat, suspended solids, manure etc. The wastewater containing concentrated organic nutrient-laden matter along with carbonaceous waste giving rise to high level of COD and TKN. The observed typical characteristics of raw and pre-treated slaughterhouse wastewater used for kinetic study are shown in Table 5.2.

Table 5.2: Characteristics and Composition of Slaughterhouse Wastewater

Parameters	Raw Wastewater	Pre-treated Wastewater
	Range	Range
pH	8.0-8.5	7.5-8.5
Total Suspended Solids	10120-14225	1250-1680
Total Dissolved Solids	6345-7840	2800-4230
COD	6185-6840	2040-2235
BOD_5 at 20°C	3000-3500	1030-1204
Total Kjeldahl Nitrogen (TKN)	350-400	220-300
NH_4^+ N (as N)	145-200	85-157

All units are in mg/L except pH.

The performance of the present study was carried out taking wastewater having an average initial soluble chemical oxygen demand (SCOD) of 2000 ± 50 mg/L and ammonia nitrogen (NH_4^+-N) concentration of 90 ± 10 and 140 ± 10 mg/L as N.

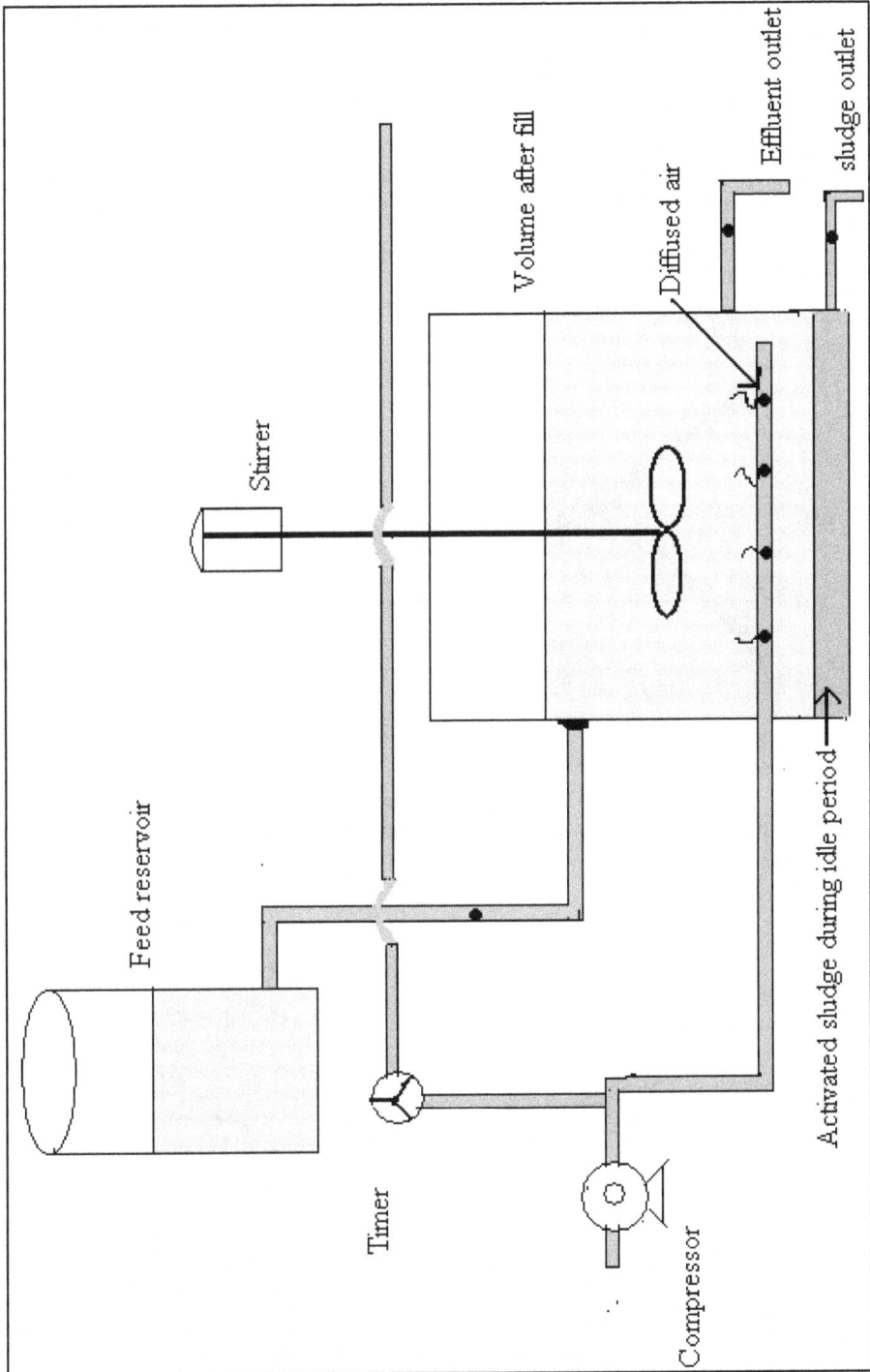

Figure 5.5: A Schematic Diagram of the Experimental Set-up.

Removal of Carbonaceous Matter

Organic carbon, which is the source of energy for microbial metabolism, has been estimated in terms of chemical oxygen demand (COD). In the case of a particular experiment, carried out with an initial SCOD of 1984.34 mg/L and initial NH_4^+-N concentration of 94.58 mg/L in 4 hour aerobic and 4 hour anoxic cycle period, it has been observed that the major fraction of SCOD removal (72.52 per cent) took place during aerobic react period. In the anoxic phase, further SCOD removal has been noticed to the extent of 90.14 per cent (Figure 5.6). A similar trend of removal pattern

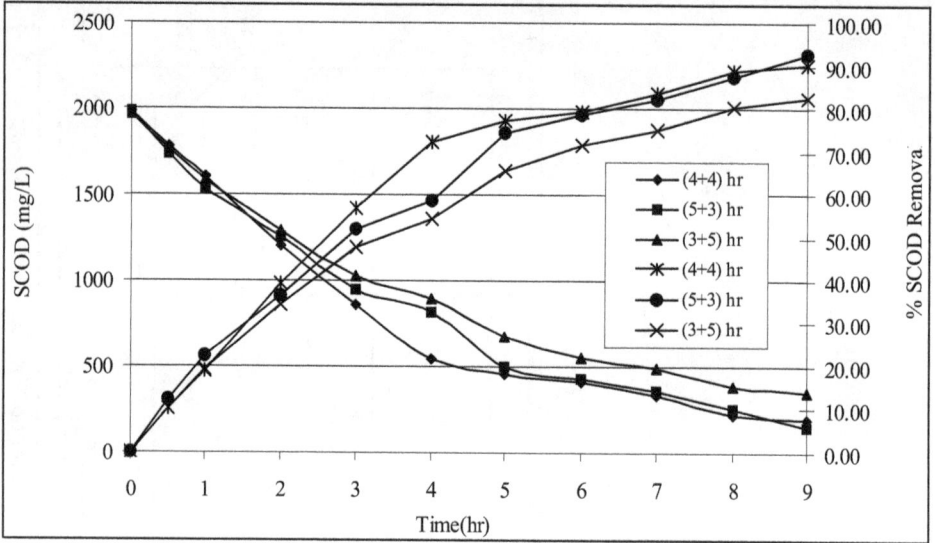

Figure 5.6: Carbon Oxidation Profile
[Initial SCOD = 2000 ± 50 mg/L; Initial NH_4^+-N = 90 ± 10 mg/L as N].

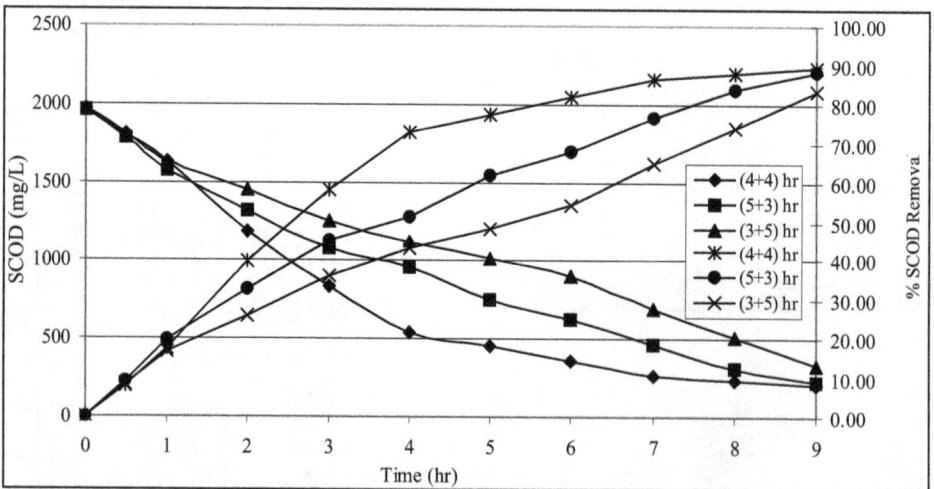

Figure 5.7: Carbon Oxidation Profile
[Initial SCOD = 2000 ± 10 mg/L; Initial NH_4^+-N = 140 ± 10 mg/L as N].

was also observed when initial NH_4^+-N concentration was increased to 142.52 mg/L as N with initial SCOD of 1975.35 mg/L in a separate set of experiment. It is revealed from Figures 5.6 and 5.7 that SCOD level has decreased rapidly during aerobic react period as compared to its rate of removal during anoxic condition due to dominance of organic heterotrophs. When the react period was changed into 5 hour aerobic followed by a reduced 3.0 hour anoxic, a marginal improvement of SCOD removal in aerobic phase and anoxic phase was observed due to enhanced aeration time. On the other hand, when the react period was subsequently changed to 3.0 hour aerobic period followed by 5.0 hour anoxic period, a marginal decrease of SCOD removal in aerobic phase and anoxic phase was obtained due to lag of aeration time.

Ammonia Oxidation

Ammonia oxidation took place due to the presence of previously acclimatized nitrifying organisms within the reactor. Considering the cycle period of 4 hr (aerobic) + 4 hr (anoxic), it has been observed that at the end of 8hr react period of reaction, ammonia oxidation was 92.55 per cent, when initial NH_4^+-N was approximately 94.58 mg/L as N. When the reactor system was operated in 5 hr (aerobic) + 3 hr (anoxic) mode of react cycle, an overall improvement of ammonia oxidation from 92.55 to 96.19 per cent and 92.45 per cent for initial NH_4^+-N of 93.24 mg/L and 143.78 mg/L as N respectively, was observed. The results reveal the fact that the extension of aeration period helped to enhance the oxidation efficiency for the present system. It was also observed that when aerobic period was reduced to 3.0 hr, ammonia oxidation reduced to 78.88 per cent and 73.48 per cent corresponding to initial NH_4^+-N value of 95.24 mg/L and 145.35 mg/L respectively, at the end of 8hr react period. The NH_4^+-N level at different time periods for different combinations of react periods and initial NH_4^+-N concentration are exhibited in Figures 5.8 and 5.9.

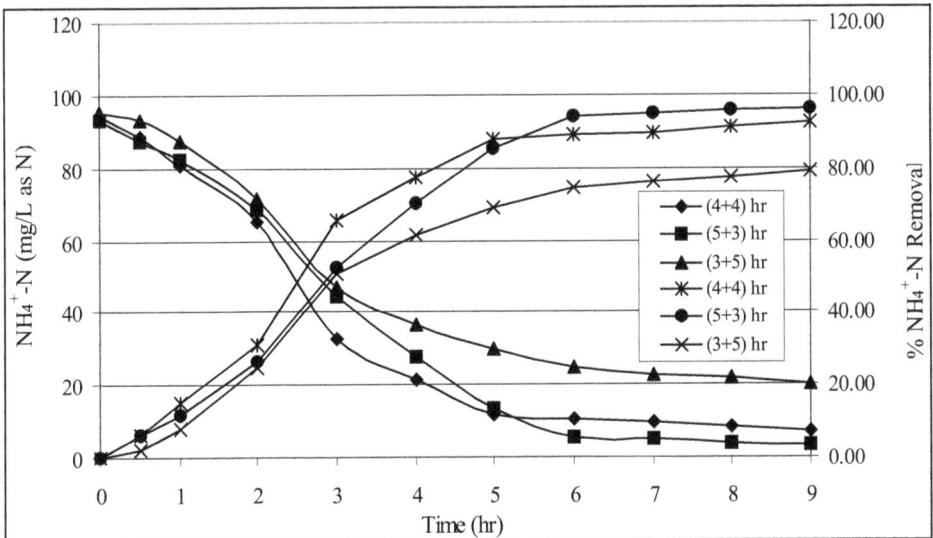

Figure 5.8: Ammonia Oxidation Profile
[Initial SCOD = 2000 ± 50 mg/L; Initial NH_4^+-N = 90 ± 10 mg/L as N).

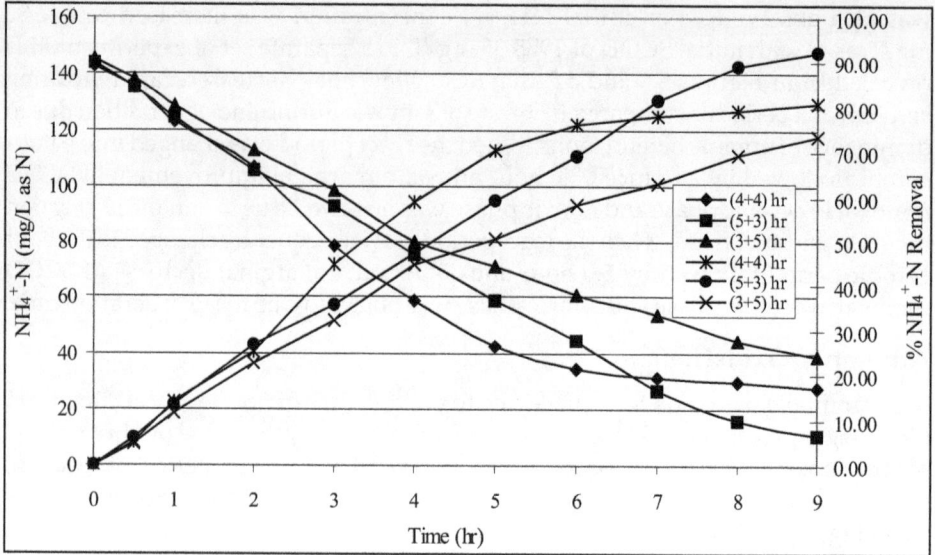

Figure 5.9: Ammonia Oxidation Profile
[Initial SCOD = 2000 ± 50 mg/L; Initial NH$_4^+$-N = 140 ± 10 mg/L as N).

Conclusion

The following conclusions are drawn based on the present work on simultaneous removal of organic carbon and nitrogen from slaughterhouse wastewater in a laboratory scale SBR.

1. The SCOD removal during aerobic react period was achieved due to utilization of substrate as required of aerobic microorganisms in the mixed culture. In the anoxic phase, the residual SCOD was utilized by the facultative microbes.

2. It is observed that the SBR can perform efficiently in achieving nitrification and denitrification sequentially along with oxidation of organic carbon. The combination of 4.0 hour aerobic react period and 4.0 hour anoxic react period has been found to be optimum from the view point of both nitrification and denitrification.

3. Length of aeration time in 5 hr (aerobic) + 3 hr (anoxic) mode react phase of the operation cycle, did not have any significant impact on SCOD reduction.

4. Longer aeration period (5 hour) has been found to be effective in achieving higher degree of nitrification. However, it affected the percent removal of nitrate due to the prevalence of shorter anoxic period essential for denitrification.

References

APHA, AWWA and WPCF, (1995): Standard Methods for the Examination of Water and Wastewater, 20th Edition, New York.

Bernet, N., Delgenes, N., Akunna, J.C., Delgenes,J.P. and Moletta, R. (2000) : Combined anaerobic-aerobic SBR for the treatment of piggery wastewater, Wat. Res. 34(2): 611- 619.

Irvine, R.L. and Busch, A.W. (1979): Sequencing Batch Biological Reactors: An Overview, Journal of Water Pollution Control Federation. 51 (2): 235-243.

Irvine,R.L. and Davis, W.B. (1971): Use of Sequencing Batch Reactors for Waste Treatment – CPC International, Corpus Christi, Texas, In Proceedings of the 26th Annual Purdue Industrial Waste Conference, Purdue University: West Lafeyette, IN, P450.

Irvine, R.L. and Ketchum, L.H. (1989): Sequencing Batch Reactors for Biological Wastewater Treatment, CRC Critical Reviews in Environmental Control. 18: pp. 255-294.

Kargi, F. and Uygur, A. (2002): Nutrient Removal Performance of a Sequencing Batch Reactor as a Function of the Sludge Age. Enzyme Microb. Technol. 31: 842–847.

Li, J.P., M.G, Zhan, X.M. and Rodgers,M. (2008): Nutrients removal from slaughter house wastewater in an intermittently aerated sequencing batch reactor, Bioresource Technol. 52: 163-167.

Madigan, M.T, Martinko, J.M, Parker, J. (2000): Brock Biology of Microorganisms, 8th ed. Englewood Cliffs, NJ: Prentice-Hall.

Obaja, D., Mac, S. and Mata-Alveraz, J. (2005): Biologicalnutrient removal by a sequencing batch reactor (SBR) using an internal organic carbon source in digested piggery wastewater, Bioresource Tech., 96 : 7 – 14.

Painter, H.A. (1977): "Microbial Transformations of Inorganic Nitrogen", Progress in Water Technology. 8: pp.-3-29.

Surampalli, R.Y., Tyagi, R.D., Scheible, O.K. and Heidman, J.A. (1997): Nitrification, Dinitrification and Phosphorous removal in a sequencing batch reactors, Bioresource Tech., 61: 151-157.

Chapter 6

Algae of Chilika Lagoon: A Documentation after Opening of the New Mouth

D. Mohanty, S.K. Das and S.P. Adhikary

Department of Biotechnology, Institute of Science, Visva-Bharati, Santineketan – 731 235, West Bengal, India

ABSTRACT

Eighty one algal taxa comprising of 24 species of Cyanophyta (Cyanobacteria), 2 Rhodophyta, 19 Chlorophyta, 6 Euglenophyta and 30 Bacillariophyceae under Heterokontophyta were recorded from four different sectors of Chilika lake during May 2011 to April 2012. Of these 19 species were common to those reported in the lake before opening of new mouth to the Bay of Bengal in 2000. Sixty two algal taxa comprised of 20 Cyanophytes, 17 Chlorophytes, 6 Euglenophytes and 19 Bacillariophyta members appeared in the lagoon after opening of the new mouth. Maximum algal diversity was observed in the Northern sector followed by Central, Southern and Outer channel sectors. The agarophyte *Gracilaria verrucosa* recorded in the Outer channel sector of the lagoon for the first time showed its extended distribution in the lake due to influx of saline water coupled with consistent wave action through the new opening to the sea.

Keywords: Algal diversity, Chilika lake, New mouth, Bay of Bengal, Salinity.

Introduction

Chilika lagoon is situated in the east coast of India between 19°28′ and 19°54′ N latitude and 85°5′ and 85°38′E longitude covering Puri, Khurdha and Ganjam district of Odisha state. The lake occupies an average area of 1165 to 906 km^2 during rainy and summer seasons respectively with an average depth of approximately 2m (Rath and Adhikary, 2005 a) and has been connected to Bay of Bengal through an outer channel running parallel to the sea. The lagoon receives freshwater from deltic branches of Daya and Bhargavi in the Northern part. As a result distinct salinity gradient has been found in the lagoon ranging from almost zero to that of sea level. Basing on salinity gradient and depth the lagoon is divided in to four sectors *i.e.* Southern, Central, Northern and Outer channel sector. Prior to 2000 in the Southern sector the salinity level recorded was moderate (8-20ppt), Central sector showed a seasonal variation in salinity (5-30ppt) whereas due to opening of several river mouths in the Northern sector the salinity level was very low leading to almost a freshwater habitat (0.1-3ppt). Due to tidal impact the salinity level in the Outer channel sector was nearly to the sea level (20-33 ppt) during most times of the year except the rainy season (Rath and Adhikary 2006). This spatial salinity regime gave rise to an unique character to the estuarine ecosystem in Chilika encouraging several species to occur, hence regarded as one of the hotspot of biodiversity.

Prior to 2000 due to chocking and degradation of drainage basin in the Outer channel sector the lagoon encountered increased siltation which adversely affected the ecology of the lake. Odisha government recognized that siltation was the main cause of hindrance to the tidal exchange and consequent decrease in the salinity level. To restore the lagoon quality appropriate flux of seawater into the lake, a new mouth was dredged in September 2000 near Sipakuda of the Outer channel sector. Subsequently after six years, in March 2008 a natural mouth was opened near Gabakunda close to earlier mouth (Patra *et al.* 2010). Both these openings facilitated mixing of salt water with freshwater resulting in a wide variation in the hydrology which possibly leads to a change in the biological diversity of the lagoon. Nayak and Behera (2004), Mohanty *et al.* (2009), Panigrahi *et al.* (2009) and Patra *et al.* (2010) have reported the changes in the physico-chemical parameters of the lagoon after opening of these new mouths. However, documentation of algal diversity after opening of these mouths had not been made so far. Algal forms occurring in different sectors of the lagoon after opening of these mouths was studied and their diversity was re-assessed upon comparing with the earlier reports (Biswas 1932; Roy 1954; Ahmed 1966; Patnaik 1973 and 1978; Raman 1990; Adhikary and Sahu 1992; Sahu and Adhikary 1999; Rath and Adhikary 2005b).

Materials and Methods

Samples were collected from several locations under four different sectors (Mangaljodi, Northern sector; Barakul, Kalijai, Pathara, Central sector; Rambha, Ghantashila, Southern sector; and Sipakuda at Outer Channel sector) of the lagoon at regular intervals during May 2011 to April 2012 (Figure 6.1). The samples were kept in Tarson made specimen tubes, fixed on the spot with 4 per cent formalin and brought to the laboratory for analysis. The planktonic samples were collected by using 45 μm pore size plankton net. Attached algae, *e.g.* epilithic, epiphytic samples were collected using forcep and scalpel. Each sample was assigned a voucher number and deposited at the Department of Biotechnology, Visva Bharati. The voucher is

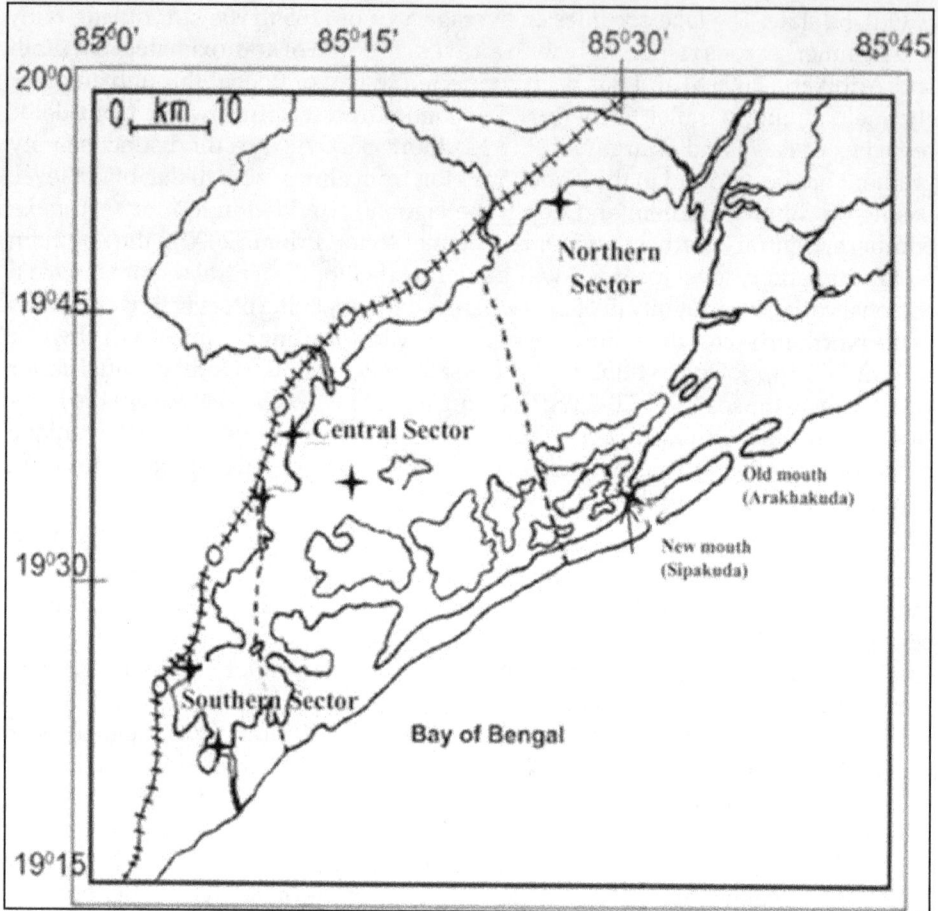

**Figure 6.1: Map Showing 7 Different Sampling Locactions
(Mangalajodi, Barakul, Pathara, Rambha, Kalijai, Ghantashila, Sipakuda).**

given in parenthesis after the sector it was collected against each taxa in the systematic enumeration below. Microscopic observation and microphotography was carried out using Olympus B47 trinocular fluorescence microscope fitted with Nikon 4500 coolpix digital camera. Measurement of length and breadth of microscopic forms was recorded with standard Erma micrometers. The algal species were identified following Kützing (1865), Huber-pestalozzi (1942), Desikachary (1959, 1987, 1988, 1989), Ramanathan (1964), Phillipose (1967), Komárek and Fott (1983), Ettl and Gärtner (1995), Komárek and Anagnostidis (1998, 2005), Wo³owski and Hindák (2005) and various research publications like Gandhi (1970), Hegewald *et al.* (1990), Pal and Santra (1990, 1993), Prasad and Misra (1992), Prasad and Srivastava (1992), Komárek and Jankovska (2001), Misra and Srivastava (2003), Rath and Adhikary (2005 a) and Tabassum and Saifullah (2010). The classification of Cyanophyta was as per Komárek and Anagnostidis (1989, 1998, 2005) and as per Lee (1999) for other divisions of algae.

Results

Systematic Enumeration of Algal Taxa Recored from Chilika Lake

Division – Cyanobacteria (Cyanophyta/Cyanoprokaryota)

Order – Chroococcales

Family – Synechococcaceae

Genus – *Cyanobacterium* Rippka *et* Cohen - Bazire

1. *Cyanobacterium diachloros* (Skuja) Komárek (Plate 6.1, Figure 1)

Komárek and Anagnostidis 1998, p. 48, Figure 25.

Cells solitary, pale green in colour, broadly oval or cylindrical in shape, 38–40 µm long and 21 – 25 µm broad, a thin transparent mucilage layer is found outside the cell.

Planktonic in Northern sector (2920) and Southern sector (4045), (4735).

Family – Merismopediaceae

Genus – *Aphanocapsa* Nägeli

2. *Aphanocapsa marina* Hansgirg (Plate 6.1, Figure 2)

Komárek and Anagnostidis 1998, p. 148, Figure 170.

Almost spherical colonies with thick yellowish mucilaginous envelope, cells spherical, loosely arranged 1.6 – 2.2 µm in diameter.

Epilithic in Northern sector (4039) and Central sector (4744) and Planktonic in Outer channel sector (4715).

Genus – *Merismopedia* Meyen

3. *Merismopedia glauca* (Ehrenberg) Kützing (Plate 6.1, Figure 3)

[Synonym - *Gonium glaucum* Ehrenberg, *Merismopedia aeruginea* Brébisson, *Merismopedia nova* Wood]

Komárek and Anagnostidis 1998, p. 177, Figure 225.

Colonies light blue green, almost rectangular with slightly sinuate margin, hemispherical cells, 1 – 2 µm in diameter; colony is 64 celled, colony is 14 – 15 µm long and 16 – 17 µm broad.

Planktonic in Central sector (4728) and Southern sector (4064).

4. *Merismopedia punctata* Meyen (Plate) (Plate 6.1, Figure 4)

Komárek and Anagnostidis 1998, p. 175, Figure 222.

Colonies 4 – 8 celled, plate like quadrangular, pale green in colour, cells spherical 1.6 – 2.6 µm in diameter, mucilage is not clear.

Planktonic in Central sector (4040, 4716, 4727) and Northern sector (4041, 4759).

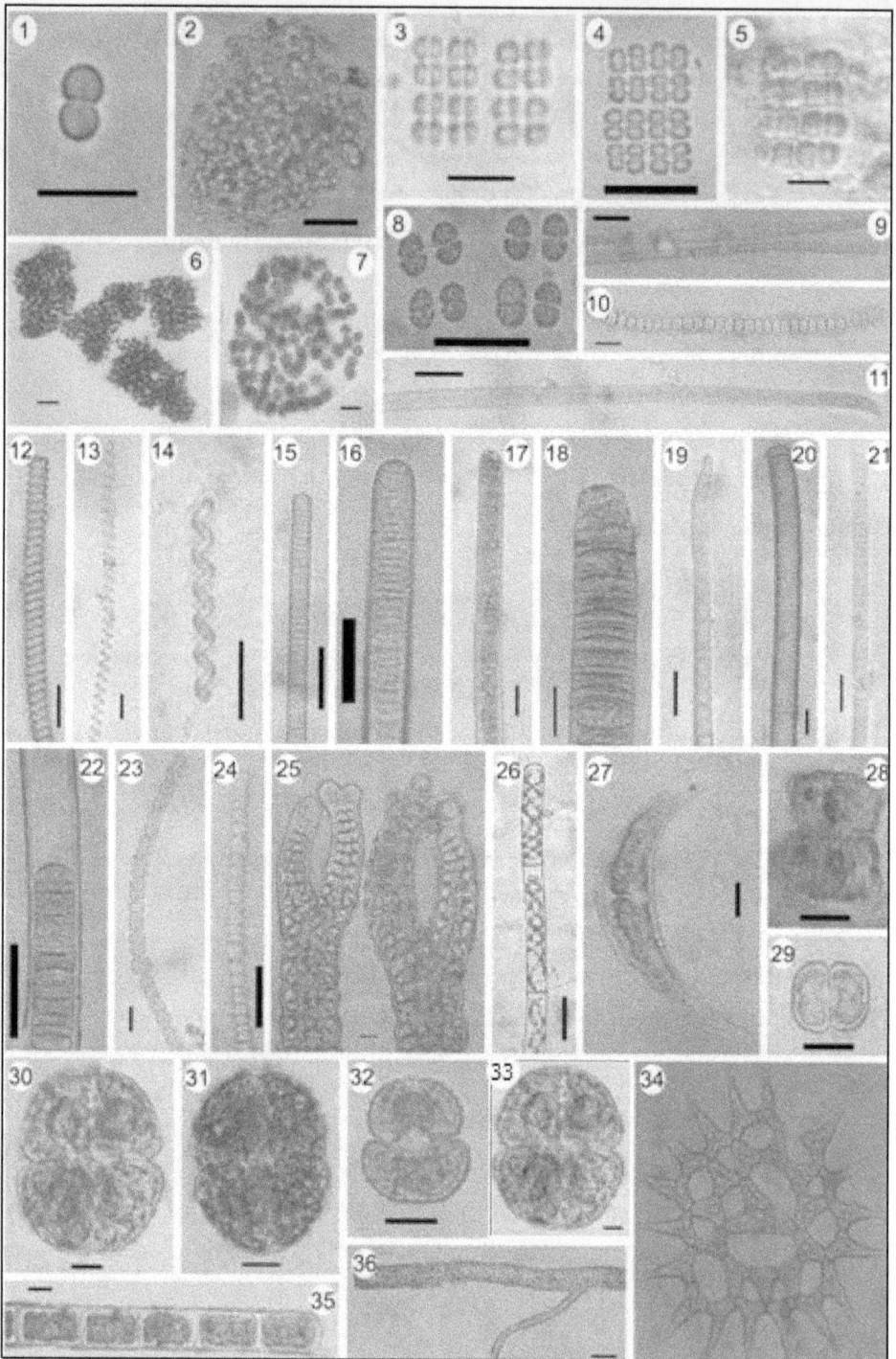

Plate 6.1

1. *Cyanobacterium diachloros*, 2. *Aphanocapsa marina*, 3. *Merismopedia glauca*, 4. *Merismopedia punctata*, 5. *Merismopedia warmingiana*, 6. *Microcystis aeruginosa*, 7. *Microcystis wesenbergii*, 8. *Chroococcus limneticus*, 9. *Pseudanabaena limnetica*, 10. *Pseudanabaena minima*, 11. *Geitlerinema earlei*, 12. *Spirulina labyrinthiformis*, 13. *Spirulina major*, 14. *Spirulina subtilissima*, 15. *Phormidium ambiguum*, 16. *Oscillatoria limosa*, 17. *Oscillatoria perornata*, 18. *Oscillatoria princeps*, 19. *Oscillatoria proteus*, 20. *Oscillatoria sancta*, 21. *Oscillatoria simplicissima*, 22. *Lyngbya aestuarii*, 23. *Anabaena oscillarioides*, 24. *Anabaena variabilis*, 25. *Ceramium diaphanum*, 26. *Spirogyra* sp., 27. *Closterium venus*, 28. *Euastrum dubium*, 29. *Cosmarium awadhense*, 30. *Cosmarium decoratum*, 31. *Cosmarium lundellii* var. *ellipticum*, 32. *Cosmarium miscellum*, 33. *Cosmarium punctulatum*, 34. *Pediastrum simplex* var. *simplex*, 35. *Microspora willeana*, 36. *Enteromorpha usneoides*.

(Scale bar: figure 26, 36 = 30 *μm*; figure 6, 31 = 20 *μm*; figure 1 – 5, 7 – 25, 27 – 30, 32 – 35 = 10 *μm*).

5. *Merismopedia warmingiana* Lagerheim (Plate 6.1, Figure 5)

Komárek and Anagnostidis 1998, p. 174, Figure 218.

Colonies of 4 – 16 – 32 celled, rectangular, microscopic, blue green in colour, cells hemispherical to almost spherical, cells 2.2 – 2.5 µm long and 1.5 – 1.6 µm broad, thin mucilage layer is found.

Planktonic in Central sector (4068).

Family - Microcystaceae

Genus – *Microcystis* Kützing ex Lemmermann

6. *Microcystis aeruginosa* (Kützing) Kützing (Plate 6.1, Figure 6)

[Synonym – *Clathrocystia aeruginosa* var. *major*, *Microcystis aeruginosa* f. *aeruginosa Kützing*, *Micraloa aeruginosa* Kützing, *Diplocystis aeruginosa* (Kützing) Trevisan, *Clathrocystis aeruginosa* (Kützing) Henfrey]

Komárek and Anagnostidis 1998, p. 232, Figure 304.

Colonies mucilaginous, microscopic, irregular, distinctly elongate, mucilage colourless, irregular structure, diffluent, cells spherical, pale blue green to brown in colour, with numerous aerotopes, 2.5 – 3.5 µm in diameter.

Planktonic in Northern sector (2066).

7. *Microcystis wesenbergii* (Komárek) Komárek in Kondrateva (Plate 6.1, Figure 7)

[Synonym - *Diplocystis wesenbergii* Komárek]

Komárek and Anagnostidis 1998, p. 232, Figure 305.

Colonies spherical when young, cells arranged randomly, densely in young colonies rather near the colonial surface, but also over the whole colonial content, mucilage colourless, structure irregular, cells spherical, with distinct aerotopes, 4.5 µm in diameter.

Planktonic in Northern sector (2066).

Family - Chroococcaceae

Genus – *Chroococcus* Nägeli

8. *Chroococcus limneticus* Lemmermann (Plate 6.1, Figure 8)

[Synonym – *Chroococcus limneticus* var. *carneus* (Chodat) Lemmermann, *Anacystis thermalis* f. *major* (Lagerheim) Drouet

Komárek and Anagnostidis, 1998, p. 290, Figure 382.

Free floating, irregular colonies, 16 celled, with 4 celled groups, mucilage colourless, irregular in structure, cell content homogenous, cells spherical, bright blue-green, 2.5 – 3.8 µm in diameter.

Planktonic in Southern sector (4065) and Nothern sector (4042, 4757).

Order – Oscillatoriales

Family – Pseudanabaenaceae

Genus – *Pseudanabaena* Lauterborn

9. *Pseudanabaena limnetica* (Lemmermann) Komárek (Plate 6.1, Figure 9)

[Synonym - *Oscillatoria limnetica* Lemmermann]

Komárek and Anagnostidis 2005, p. 84, Figure 60.

Trichome bluish green, solitary, straight, cross walls hyaline, translucent, slightly constricted, without aerotopes, not attenuated at the ends; cells long, cylindrical, 2.2 – 3.5 µm broad, 5 – 6.2 µm long, cell contents homogenous, apical cells rounded.

Planktonic in Nothern sector (2066), Central sector (4421) and Southern sector (4419).

10. *Pseudanabaena minima* (G.S. An) Anagnostidis (Plate 6.1, Figure 10)

Komárek and Anagnostidis 2005, p. 81, Figure 57.

Filaments solitary, almost straight, pale blue green in colour, cross walls deeply constricted and thickened, 3.2 – 4 µm broad, thin layer of sheath present around the filament, cells shorter than broad, 2.5 – 3.1 µm long, apical cell with obtusely rounded end.

Planktonic in Nothern sector (4044), Central sector (4067) and Southern sector (4045).

Genus – *Geitlerinema* (Anagnostidis *et* Komárek) Anagnostidis

11. *Geitlerinema earlei* (Gardner) Anagnostidis (Plate 6.1, Figure 11)

Komárek and Anagnostidis 2005, p. 134, Figure 147.

Filaments thin, straight, blue green, with no constriction at the cross walls, 2.6 – 3 µm broad, cells isodiametric or slightly longer than broad, 2.7 – 3.5 µm long, apical cell pointed.

Planktonic in Central sector (4066) and Epilithic in Southern sector (4045).

Genus - *Spirulina* Turpin ex Gomont

12. *Spirulina labyrinthiformis* Kützing ex Gomont (Plate 6.1, Figure 12)

Komárek and Anagnostidis 2005, p. 146, Figure 171.

Filaments solitary, pale blue green in colour, densely coiled in clock wise manner, cells shorter than broad, 1.8 – 2.2 µm broad and 1.5 – 2 µm long, apical cell rounded.

Epilithic in Southern sector (4430) and Central sector (4046, 4716, 4423).

13. *Spirulina major* Kützing ex Gomont (Plate 6.1, Figure 13)

[Synonym – *Spirulina oscillariorides* Turpin, *Arthrospira major* (Kützing) Crow]

Komárek and Anagnostidis 2005, p. 148, Figure 173.

Trichomes pale to bright blue green, 1.2 µm wide, regularly screw like coiled, coils left handed, and distance between spirals is 4 – 4.5 µm.

Epilithic in Northern sector (2060, 4043, 4019) and Central sector (4321).

14. *Spirulina subtilissima* Kützing ex Gomont (Plate 6.1, Figure 14)

Komárek and Anagnostidis 2005, p. 144, Figure 168.

Trichomes short, solitary free floating, bright blue green in colour, coiling is regular, clock wise, cells 0.8 – 1.4 µm broad and 1 – 1.5 µm long.

Epilithic in Central sector (4413).

Family – Phormidiaceae

Genus – *Phormidium* Kützing ex Gomont

15. *Phormidium ambiguum* Gomont ex Gomont (Plate 6.1, Figure 15)

[Synonym - *Amphithrix amoena* Kützing, *Lyngbya bourrellyana* Compére]

Komárek and Anagnostidis 2005, p. 479, Figure 718.

Filaments elongate, straight, trichomes bright blue green to olive green, 3 – 6.6 µm broad, slightly constricted at the granulated cross walls, not attenuated at the ends, cells shorter than wide, 0.8 – 2.3 µm long, cell content frequently with dispersed large granules, apical cell rounded.

Epilithic in Central sector (4718), Southern sector (4047) and Epiphytic in Nothern sector (4756, 4018).

Family- Oscillatoriaceae

Genus - *Oscillatoria* Vaucher ex Gomont

16. *Oscillatoria limosa* Agardh ex Gomont (Plate 6.1, Figure 16)

Komárek and Anagnostidis 2005, p. 593, Figure 886.

Thallus blackish blue green, trichomes bright green, straight, slightly bent, sheath absent, numerous constricted at the crosswalls, 4.4 – 5.6 µm wide, cells shorter than broad, 0.5 – 1 µm long, cell content is granular, apical cell flatly rounded, without calyptra.

Epiphytic in Central sector (4720, 4057, 4421, 4047, 4721) and Southern sector (2967).

17. *Oscillatoria perornata* Skuja (Plate 6.1, Figure 17)

Komárek and Anagnostidis, 2005, p. 586, Figure 873.

Solitary trichome, pale blue green, distinctly constricted at the cross walls, cells finely granulated, with aerotopes, 2.6 – 3.1 µm long and 7.4 µm broad, apical cell rounded- conical, without calyptra.

Planktonic in Northern sector (2066).

18. *Oscillatoria princeps* Vaucher ex Gomont (Plate 6.1, Figure 18)

[Synonym- *Oscillatoria princeps* Vaucher, *Trichophorus princeps* (Vaucher) Desvaux, *Oscillatoria princeps* (Vaucher) Gaillon, *Lyngbya princeps* (Vaucher) Hansgirg]

Komárek and Anagnostidis 2005, p. 590, Figure 883.

Trichomes blue green, mostly forming a thallus, mostly straight, not constricted at the cross walls, 13.7 – 16.5 µm broad, slightly or briefly attenuated at the apices and bent, cells 2.3 – 5 µm long, end cells flatly rounded, slightly capitate.

Floating in Central sector (4721), Epiphytic in Central sector (4723) and Southern sector (2941).

19. *Oscillatoria proteus* Skuja (Plate 6.1, Figure 19)

Komárek and Anagnostidis 2005, p.464, Figure 682.

Filaments solitary, blue green, 5 µm wide, constricted at the crosswalls, briefly attenuated and bent at the end, cells 2.8 – 5 µm long, apical cell conical- rounded.

Floating in Northern sector (2060).

20. *Oscillatoria sancta* Kützing ex Gomont (Plate 6.1, Figure 20)

Komárek and Anagnostidis 2005, p. 594, Figure 890.

Filaments straight, olive green in colour, slightly constricted at the granulated cross walls, 8.8 – 9.3 µm broad, cellular content granular, cells shorter than broad, 3.2 – 4.5 µm long, apical cell flatly rounded, capitate.

Epiphytic in Northern sector (4073) and Epilithic in Central sector (4056, 4737).

21. *Oscillatoria simplicissima* Gomont (Plate 6.1, Figure 21)

Komárek and Anagnostidis 2005, p. 586, Figure 876.

Trichome straight, blackish blue green, not constricted at the cross walls, 6.6 µm wide, end not attenuated, terminal cells rounded, without calyptra, cells 2.6 µm long.

Floating in Northern sector (2060).

Genus – *Lyngbya* Agardh ex Gomont

22. *Lyngbya aestuarii* Liebman ex Gomont (Plate 6.1, Figure 22)

Komárek and Anagnostidis 2005, p. 621, Figure 947, 948.

Trichomes straight, sometimes slightly curved, blue green, not constricted at the granulated crosswalls, 6.6 – 7.7 µm broad, sheath thin in young filaments but thickened in mature ones, cells plate like, much broader than long, 4.6 – 5 µm broad and 0.5 – 1 µm long.

Epilithic and also floating in Central sector (2946, 4721), Epiphytic and floating in Southern sector (4049, 4429, 4749).

Order – Nostocales

Family – Nostocaceae

Genus – *Anabaena* Bory

23. *Anabaena oscillatrioides* var. *angustus* Bharadwaj (Plate 6.1, Figure 23)

Desikachary 1959, p. 418, pl. 78, Figure 1.

Trichomes straight, single, end cells barrel shaped, 3.3 – 5 μm long and 4.2 μm broad, heterocysts intercalary, ellipsoidal, 6.6 μm long and 5.3 μm broad, yellowish brown epispores present on both the sides of the heterocyst, 11.7 μm long and 5 μm broad.

Planktonic in Northern sector (2060).

24. *Anabaena variabilis* Kützing ex Bornet et Flahault (Plate 6.1, Figure 24)

Desikachary 1959, p. 410, pl. 71, Figure 5

Filaments short, straight, pale blue green in colour, cells barrel shaped, iso-diametric, 2 – 3 μm long and 2 – 2.5 μm broad, heterocysts broadly elliptical, intercalary, 3 – 3.3 μm long and 2 – 2.5 μm broad.

Plaktonic in Northern sector (4409, 4757) and Central sector (4076), and Epiphytic in Central sector (4056).

Phylum – Rhodophyta

Class – Rhodophyceae

Order – Ceramiales

Family – Ceramiaceae

Genus – *Ceramium* Roth

25. *Ceramium diaphanum* var. *elegans* (Roth) Roth (Plate 6.1, Figure 25)

Rath and Adhikary 2005, p. 54, Figure 13.

Thallus hair like, forming reddish mats on rock surfaces, dichotomously branched, branches gradually attenuated, slightly bent inwards, terminal pairs forceps shaped, branches 25 – 50 μm wide, size of the cells gradually lowers towards the tip, at the apices cells become iso-diametric.

Epilithic in Central sector (4048, 4749) and floating in Central sector (4076, 4720).

Phylum – Chlorophyta

Class – Charophyceae

Order – Zygnematales

Family – Zygnemataceae

Genus – *Spirogyra* Link

26. *Spirogyra* sp. (Plate 6.1, Figure 26)

Filaments straight, floating, cells long, cylindrical, 48 – 56 μm long and 7 – 8 μm broad, chloroplasts 2, spirally coiled, making 4.5 – 5 turns.

Floating in Northern sector (4760, 4047), Central sector (4074, 4419, 4720, 4055, 4423, 4741) and Southern sector (4049, 4744).

Family – Desmidaceae

Genus – *Closterium* **Nitzsch ex Ralfs**

27. *Closterium venus* **Kützing (Plate 6.1, Figure 27)**

Brock 2002, p. 527, pl. 129, Figure O.

Cells small, much longer than broad, deeply curved, cells attenuated towards the apices to form acute apices, cell wall smooth, chloroplasts with 2 – 3 pyrenoids, cell length 184 – 198 µm, breadth 18 – 22 µm at the centre and 4 – 6 µm at the tip.

Epilithic in Northern sector (2939), Central sector (4740) and Southern sector (4746).

Genus – *Euastrum* **Ehrenberg ex Ralfs**

28. *Euastrum dubium* **Nägeli (Plate 6.1, Figure 28)**

[Synonym – *Euastrum binale* Ralfs, *Euastrum dubium* var. *Triquetrum*]

Pal and Santra 1993, p. 154, pl. 3, Figure 3.

Cells solitary, green, longer than broad, semicells trapezoid, margin smooth, undulated, apical margin of polar lobe with a "u-shaped" invagination in the middle, sinus narrow and linear, cells 18.8 – 20.6 µm long, 14.7 – 16 µm broad, isthmus is 3 – 4 µm broad.

Planktonic in Central sector (4079, 4068, 4730) and Southern sector (4067, 4064, 4730).

Genus – *Cosmarium* **Ralfs**

29. *Cosmarium awadhense* **Prasad *et* Mehrotra (Plate 6.1, Figure 29)**

Misra and Srivastava 2003, p. 87, pl. 1, Figure 21.

Cells solitary, green, slightly longer than broad, sinus deeply constricted, linear, closed, semi cells hemispherical with truncated apex, margin smooth, slightly wavy, chloroplast axial, cells 30 – 34 µm long, 26 – 29 µm broad, isthmus 5 – 6 µm broad.

Planktonic in Northern sector (4756) and Central sector (4076, 4722, 4072, 4730).

30. *Cosmarium decoratum* **W.** *et* **G.S. West (Plate 6.1, Figure 30)**

Misra and Srivastava 2003, p. 87, pl. 1, Figure 27.

Cells solitary, longer than broad, deeply constricted, sinus narrowly linear, semicells bilipped, semi elliptic, apex flat truncate, crenations emarginated, having two distinct pyrenoids, H-shaped ridge at the isthmus region, cells 39.9 – 47 µm long, 26.6 – 30 µm broad, isthmus 8 µm broad.

Epiphytic in Northern sector (2942) and Planktonic in Cenral sector (4066).

31. *Cosmarium lundellii* **Delponte var.** *ellipticum* **W.** *et* **G.S. West (Plate 6.1, Figure 31)**

[Synonym – *Cosmarium lundelii* f. *minus* Strom]

Prasad and Misra 1992, p. 164, pl. 22, Figure 10.

Cell medium size, a little longer than broad, deeply constricted, sinus closed, semi cells sub semicircular with relatively broader and rounded apices, cell wall minutely punctate, cell 51 – 65 µm long, 39 – 44 µm broad, isthmus 8 – 10 µm broad.

Epilithic in Southern sector (4426), Epiphytic in Northern sector (4039) and Planktonic in Central sector (4066) and Southern sector (2968, 4735).

32. *Cosmarium miscellum* **Skuja (Plate 6.1, Figure 32)**

Misra and Srivastava, 2003, p. 87, pl. 2, Figure 3.

Cells longer than wide, constriction deep, sinus broad, closed, semi cells with broad base, narrow towards apex, cell wall with fine granulation, semi cells 40 – 44 µm long, 35 – 37 µm broad, isthmus 16.6 µm broad.

Epilithic in Central sector (4753), Epiphytic in Southern sector (2949) and Planktonic in Outer channel sector (4091).

33. *Cosmarium punctulatum* **Brçbisson (Plate 6.1, Figure 33)**

Brock 2002, p. 544, pl. 134, Figure S.

Cells medium size, sinus narrow, linear, slightly dilated inside, semi cells trapeze form with rounded basal angles, cell wall granulated, semi cells 30 – 34 µm long, 22 – 25 µm broad and isthmus is 7.7 – 8.2 µm broad.

Epilithic in Northern sector (4760), Planktonic in Central sector (4055, 4753), Southern sector (4062) and Outer channel sector (4091, 4714).

Class – Ulvophyceae

Order – Ulotrichales

Family – Microsporaceae

Genus – *Microspora* **Thuret**

34. *Microspora willeana* **Lagerheim (Plate 6.1, Figure 35)**

Ramanathan 1964, p. 122, pl. 32, fig. E-L; pl. 34, Figure A-O.

Filaments cylindrical, cells 23 – 27 µm long and 10.7 µm broad, H-piece clearly visible, chloroplast variable.

Epiphytic in Northern sector (2065).

Order – Ulvales

Family – Ulvaceae

Genus – *Enteromorpha* **Link**

35. *Enteromorpha usneoides* **(Bonnemaison) Agardh (Plate 6.1, Figure 36)**

Rath and Adhikary 2006, p. 51, Figure 5.

Thallus filliform, compressed, wrinkled, branched, dark green in colour, branches 30 – 40 µm broad, become wider when mature, branches more in basal region than the apical area, cells arranged in longitudinal rows, cell wall 1 – 1.8 µm thick.

Floating in Central sector (4720) and Southern sector (4048).

Class – Chlorophyceae

Order – Chlorococcales

Family – Hydrodictyaceae

Genus – *Pediastrum* **Meyen**

36. *Pediastrum simplex* **var.** *simplex* **Komárek (Plate 6.1, Figure 34)**

Komárek and Jankovska 2001, p. 32, Figure 12 A.

Coenobia 16 celled, 126 – 130 µm in diameter, large intercellular spaces or a central space with the cells arranged in a ring at the periphery, inner side of the marginal cells concave, outer surface tapered into a long process, sides of marginal cells concave or straight, internal cells similar to marginal cells with shorter processes, cells 35 – 37 µm long and 16 – 17.5 µm broad.

Planktonic in Southern sector (4744) and Central sector (4071, 4728, 4080, 4722).

37. *Pediastrum tetras* **(Ehrenberg) Ralfs (Plate 6.2, Figure 1)**

[Basionym – *Micrasterias tetras* Ehrenberg]

[Synonym – *Pediastrum ehrenbergii* (Corda) A. Braun]

Komárek and Jankovska 2001, p. 68, Figure 43.

Coenobia 8 celled, circular, 22 – 24.6 µm in diameter, cells without intercellular spaces, marginal cells divided into 2 lobes with a deep single linear incision, inner cell 4 - 6 sided with a single linear incision, cells 6.3 – 6.6 µm in diameter.

Planktonic in Central sector (2933, 4057, 4071, 4729).

Family – Chlorellaceae

Genus – *Chlorella* **Beijernick**

38. *Chlorella protothecoides* **Krüg (Plate 6.2, Figure 2)**

Komárek and Fott 1983, p. 597, pl. 169, Figure 3.

Cells solitary, sometimes aggregated to form clusters, spherical, deep green in colour, chloroplasts cup shaped to wedge shaped, cells 2.5 – 3 µm in diameter.

Epiphytic and also Epilithic in Central sector (4718, 4060, 4755, 4071, 4730).

Family – Scenedesmaceae

Genus – *Scenedesmus* **Meyen**

39. *Scenedesmus bijugatus* **(Turpin) Kützing (Plate 6.2, Figure 3)**

Philipose 1967, p. 252, Figure 164.

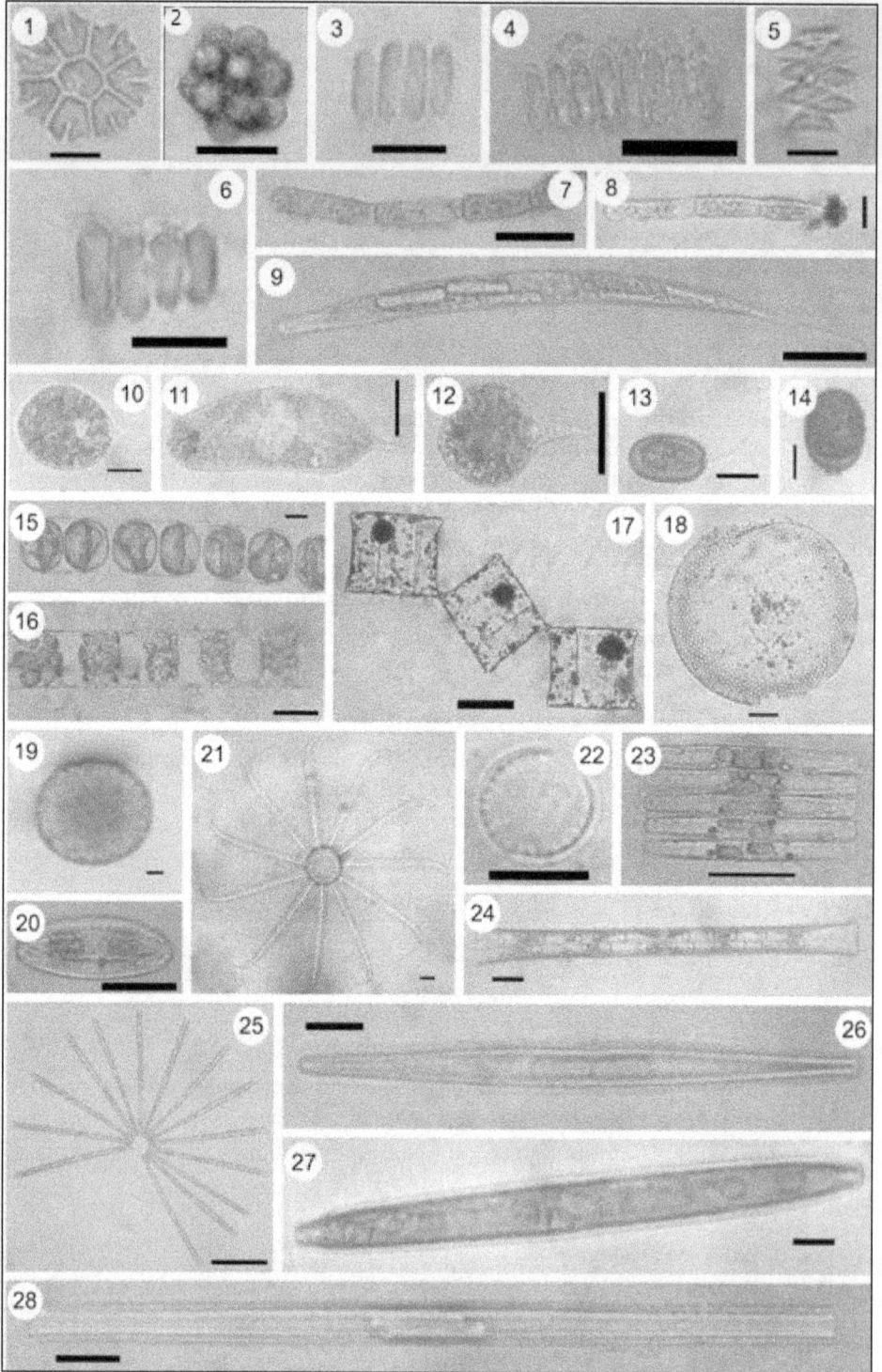

Plate 6.2

1. *Pediastrum tetras*, 2. *Chlorella protothecoides*, 3. *Scenedesmus bijugatus*, 4. *Scenedesmus calyptratus*, 5. *Scenedesmus dimorphus*, 6. *Scenedesmus protuberans*, 7. *Uronema confervicolum*, 8. *Uronema elongatum*, 8. *Euglena acus* var. *acus*, 10. *Euglena agilis*, 11. *Euglena caudata*, 12. *Lepocinclis playfairiana*, 13. *Trachelomonas abrupta* var. *abrupta*, 14. *Trachelomonas hispida* var. *crenulatocollins*, 15. *Melosira decussata*, 16. *Chaetoceros decipiens*, 17. *Odontella polymorpha*, 18. *Coscinodiscus marginatus*, 19. *Coscinodiscus subtilis*, 20. *Cyclotella maxima*, 21. *Bacteriastrum hyalinum*, 22. *Cyclotella meneghiniana*, 23. *Fragilaria crotonensis*, 24. *Synedra crystallina*, 25. *Synedra radians*, 26. *Synedra tabulata*, 27. *Synedra ulna* var. *aequalis*, 28. *Synedra ulna* var. *amphirhynchus*.

(Scale bar: figure 25 = 50 μm; figure 17 = 40 μm; figure 24 = 20 μm; figure 1 – 16, 18 – 23, 26 – 28 = 10 μm).

Coenobia 4 celled, flat, cells arranged in a linear series, cells oblong-ellipsoid to ovoid with the ends broadly rounded, cells 7.5 – 8.5 µm long and 2 – 3 µm broad.

Epilithic in Northern sector (4756) and Planktonic in Central sector (2947, 4071).

40. *Scenedesmus calyptratus* Comas (Plate 6.2, Figure 4)

Komárek and Fott 1983, p. 832, pl. 226, Figure 5.

Coenobia 8 celled, arranged linearly, green in colour, outer cells shorter, 4.5 – 6 µm long and 1 – 2 µm broad, chloroplasts entire.

Planktonic in Central sector (2961, 4728, 4072, 4048, 4750) and also in Southern sector (4064).

41. *Scenedesmus dimorphus* (Turpin) Kützing (Plate 6.2, Figure 5)

[Synonym - *Achnantes dimorphus* Turpin, *Scenedesmus antennatus* Brébisson in Ralfs, *Scenedesmus obliquus* var. *dimorphus* (Turpin) Hansgirg, *Scenedesmus costulatus* Chodat, *Scenedesmus acutus* var. *dimorphus* (Turpin) Rabenhorst, *Scenedesmus acutus* var. *obliquus* Rabenhorst]

Philipose 1967, p. 249, Figure 160.

Colonies 7 - 8 celled, with the cells arranged in linear way, outer cells more or less lunate with the apices attenuated, cells 10 – 11 µm long and 2 – 4 µm broad.

Planktonic in Southern sector (4061, 4731, 4050) and also in Central sector (4729).

42. *Scenedesmus protuberans* Fritsch *et* Rich (Plate 6.2, Figure 6)

[Synonym – *Scenedesmus protuberans* f. *minor* Li]

Hegewald *et al.* 1990, p. 45, pl. 122, Figure 1.

Coenobia is 4 celled, cells are attached laterally except at the ends, cells 7.5 – 8.5 µm long and 2.5 - 3 µm broad, very long spines present at each pole of the terminal cell, the spines much longer in comparison to the length of the cell, spines 13.5 µm long.

Planktonic Central sector (2933, 4060, 4075).

Order – Chaetophorales

Family – Chaetophoraceae

Genus – *Uronema* Lagerheim

43. *Uronema confervicolum* Lagerheim (Plate 6.2, Figure 7)

Ramanathan 1964, p. 50, pl. 13, Figure A-G.

Filaments many celled, straight, 128 - 136 µm long, attached by a disc formed by basal cell, cells 8.3 - 9 µm broad and 10 – 12 µm long, terminal cell pointed, 28 – 30 µm long, cells cylindrical, chloroplast extending the full length of the cell, parietal, containing 1 – 3 pyrenoids.

Epilithic in Northern sector (2933, 4044), Central sector (4754) and also in Outer channel sector (4092, 4726).

44. *Uronema elongatum* **Hodgetts (Plate 6.2, Figure 8)**

Ramanathan 1964, p. 51, pl. 14, Figure A-G.

Simple, unbranched, filamentous, terminal cell of the filaments tapering to an acuminate tip, terminal cell is long, nucleus is located almost centrally in the hyaline protoplasm, enclosed by the chloroplast, apical cell is slightly swollen, and basal cell is gradually attenuated and fixed to the substratum by a small cushion of colourless mucilage. Cells are 100 – 140 μm long and 14 μm broad.

Epilithic in Northern sector (2054).

Phylum – Euglenophyta

Order – Euglenales

Family – Euglenaceae

Genus – *Euglena* **Ehrenberg**

45. *Euglena acus* **var.** *acus* **Ehrenberg (Plate 6.2, Figure 9)**

Wo³owski and Hindák 2005, p. 28, Figure 5-8.

Cells green, long and thin, 77 – 81 μm long and 15.4 – 17.3 μm broad, elongate, truncate at apex of anterior end, gradually tapering and passing into a long, hyaline, sharp tail piece, chloroplasts numerous, small, discoid, parietal, without pyrenoids.

Planktonic in Northern sector (4943) and also in Central sector (2933, 4060).

46. *Euglena agilis* **Carter (Plate 6.2, Figure 10)**

Wo³owski and Hindák 2005, p. 30, Figure 88-101, 118, 119.

Cells variously shaped, usually short cylindrical or spindle shaped, 19 μm long and 7 μm broad, small, with 1 – 4 chloroplasts, situated along the cell axis, parietal, each with a bilateral pyrenoid, paramylon bodies small, rod like, pellicle very finely spirally striated.

Planktonic in Northern sector (2066).

47. *Euglena caudata* **Hübner (Plate 6.2, Figure 11)**

Wo³owski and Hindák 2005, p. 31, Figure 48, 49.

Cell ovate, palmella stage, anterior end rounded, posterior end suddenly ends into a small pointed end or tapering to a tail piece, chloroplast numerous, paramylon bodies small, cell 13.5 μm broad, 30 μm long.

Planktonic in Northern sector (4039, 4756), Central sector (4076, 4409).

Genus – *Lepocinclis* **Perty**

48. *Lepocinclis playfairiana* **Deflandre (Plate 6.2, Figure 12)**

Wo³owski and Hindák 2005, p. 38, Figure 258.

Cells broadly spindle shaped, 43 – 48 μm long, and 20.7 – 24.3 μm broad, anterior end conically narrowed into a slender tip or point, posterior end with a cauda, pellicle smooth, paramylon bodies 2, long, circular or oval ring shaped, chloroplasts numerous small, disc shaped.

Epilithic in Northern sector (4757) and Southern sector (4048, 4430).

Genus – *Trachelomonas* Ehrenberg

49. *Trachelomonas abrupta* var. *abrupta* Swirenko (Plate 6.2, Figure 13)

Wo³owski and Hindák 2005, p. 45, Figure 298, 299, 365.

Lorica ellipsoid to sub cylindrical, 18.2 µm long and 9.7 µm broad, slightly truncate at the anterior end, apical pore small, red in colour.

Epiphytic in Northern sector (2065).

50. *Trachelomonas hispida* var. *crenulatocollins* Lemmermann (Plate 6.2, Figure 14)

Wo³owski and Hindák 2005, p. 44. figure 292, 294-297, 372.

Lorica broadly oval, brown, 25.5 µm long and 19.1 µm broad, spines not clearly visible, apical pore 4.5 µm in diameter.

Epiphytic in Northern sector (2057).

Phylum – Heterokontophyta
Class – Bacillariophyceae
Order – Biddulphiales
Family – Melosiraceae
Genus – *Melosira* C. Agardh
51. *Melosira decussata* (Ehrenberg) Kützing (Plate 6.2, Figure 15)

Kützing 1865, p. 56, pl. 3, Figure VII.

Frustules spherical, joined continuously to form a chain, frustules 8 – 9.5 µm in diameter.

Planktonic in (4043) and Outer channel sector (4086, 4405, 4711).

Family – Triceratiaceae

Genus – *Odontella* Agardh

52. *Odontella polymorpha* Kützing (Plate 6.2, Figure 17)

Kützing 1865, p. 138, pl. 29, Figure 90.

Valves rectangular, intermediated in a zig-zag chain like manner, wall thin, elevations, mantle sloped, labiate processes present on the corners of the valves.

Epilithic in Northern sector (4760) and Outer channel sector (4753).

Family – Coscinodiscaceae
Genus – *Coscinodiscus* Ehrenberg
53. *Coscinodiscus marginatus* Ehrenberg (Plate 6.2, Figure 18)

Desikachary 1988, p. 8, pl. 536, Figure 1–6.

Spherical valve, 66 – 72 µm in diameter, straition tangentially straight, aereolae coarse, more dense towards the periphery, 4 in 10 µm.

Epilithic in Central sector (4717) and Northern sector (4044).

54. *Coscinodiscus subtilis* **Ehrenberg (Plate 6.2, Figure 19)**

Kützing 1865, p. 132, pl. 1, Figure XI.

Vale spherical, 60 – 70 µm in diameter, aereolae equally distributed throughout the valve, striation is not clearly visible in fresh materials.

Epiphytic in Southern sector (4767) and Northern sector (4044, 4756).

Family – Chaetocerotaceae

Genus – *Chaetoceros* **Ehrenberg**

55. *Chaetoceros decipiens* **Cleve (Plate 6.2, Figure 16)**

Tabassum and Saifullah 2010, p. 1143, Figure 8.

Frustules united together to form straight chains, frustules narrow rectangular to lanceolate, setae perpendicular to main axis, 33 – 37 µm long and 8 – 10 µm broad, frustules fuse with each other a short distance outside the margin of the chain, long bristles present at the corners of the frustules.

Planktonic in Outer channel sector (4407, 4711) and Epilithic in Central sector (4058).

Genus – *Bacteriastrum* **Shadbolt**

56. *Bacteriastrum hyalinum* **Lauder (Plate 6.2, Figure 21)**

Desikachary 1988, p. 4, pl. 461, Figure 1- 4.

Spherical valve of diameter 22- 24 µm fringed with about 11 bristles, the ends of the bristles slightly curved backwards, spirally undulate, 77 – 80 µm long.

Epilithic in Northern sector (4039, 4739) and Southern sector (4420).

Order – Bacillariales

Family – Stephanodiscaceae

Genus – *Cyclotella* **(Kützing) Brébisson**

57. *Cyclotella maxima* **Kützing (Plate 6.2, Figure 20)**

Kützing 1865, p. 50, pl. 1, Figure V.

Discoid valve, appear rectangular to elliptical in girdle view, 20 – 24 µm in diameter, striation coarse, radial. 6 – 7 in 10 µm.

Epilithic outer channel sector (2935, 4080) Northern sector (4759) and Southern sector (4744).

58. *Cyclotella meneghiniana* **Kützing (Plate 6.2, Figure 22)**

Kützing 1865, p. 50, pl. 30, Figure 68.

Frustules discoid in valve view, rectangular and undulated in girdle view, margin well defined, coarsely striated and striae wedge shaped, 8 – 10 in 10 µm, diameter 10.7 – 12.3 µm.

Epiphytic in Central sector (4740) and Epilithic in Outer channel sector (4937, 4093).

Family – Fragilariaceae

Genus – *Fragilaria* Lyngbye

59. *Fragilaria crotonensis* Kitton (Plate 6.2, Figure 23)

[Synonym – *Nematoplata crotonensis* (Kitton) Kuntze]

Rath and Adhikary 2005, p. 81, pl. 12, Figure 73.

Cells linear in girdle view, slightly swollen, so that adjacent cells linked at the centre but the ends free, 48.3 – 50.6 µm long and 5.3 – 5.6 µm broad.

Epilithic in Northern sector (2938, 4044, 4760) and Outer channel sector (2975, 4092).

Genus – *Synedra* Ehrenberg

60. *Synedra crystallina* Kützing (Plate 6.2, Figure 24)

[Synonym - *Diatoma crystallinum* Agardh]

Kützing 1865, p. 69, pl. 16, Figure (I). 2.

Valve slender, long, linear, straight, end rounded, obtuse, apices truncate, striation not clearly visible, longer than broad, 75.5 µm long, 11.7 µm broad.

Epilithic in Northern sector (2054).

61. *Synedra radians* Kützing (Plate 6.2, Figure 25)

Kützing 1865, p. 64, pl. 14, Figure VII.

Frustules long, slender, arranged at one end to give a stellar appearance, 108 – 115 µm long and 5 – 8 µm broad, striation transverse, 7 – 8 in 10 µm.

Epilithic in Central sector (4718), Epiphytic in Northern sector (2939, 4039, 4758).

62. *Synedra tabulata* Kützing (Plate 6.2, Figure 26)

[Synonym – *Diatoma tabulatum* Kützing]

Kützing 1865, p. 68, pl. 15, Figure X (1-3).

Frustules linear, slender, slightly attenuated towards apices, rotundous obtuse, two frustules conjugated, striae not distinct at the ends but clear at the middle, 84 µm long and 7.3 µm broad.

Planktonic in Northern sector (2054).

63. *Synedra ulna* (Nitzsch) Ehrenberg var. *aequalis* (Kützing) Hustedt (Plate 6.2, Figure 27)

Huber-Pestalozii 1942, p. 461, Figure 542.

Valve slender, linear, straight, end attenuated, striation linear, parallel almost through out the valve, striation uniformly placed, 7 – 9 in 10 µm, many times longer than broad, 214 µm long and 14 µm broad

Epiphytic in Northern sector (2057).

64. *Synedra ulna* (Nitzsch) Ehrenberg var. *amphirhynchus* (Ehrenberg) Grunow (Plate 6.2, Figure 28)

[Synonym – *Synedra amphirhynchus* Ehrenberg]

Huber-Pestalozii 1942, p. 462, Figure 545.

Valve slender, linear, straight, at the end narrow and suddenly constricted to form capitate end, striation distinct, parallel, absent at the middle, 9 – 12 in 10 µm, many times longer than broad, 114 – 126 µm long and 13.3 – 14.8 µm broad.

Planktonic in Central sector (4718, 4717), Northern sector (4040, 4760) and Outer channel sector (2941, 4092, 4407, 4724).

Family – Tabellariaceae

Genus – *Tabellaria* Ehrenberg

65. *Tabellaria flocculosa* (Roth) Kützing (Plate 6.3, Figure 1)

[Basionym – *Conferva flocculosa* Roth]

[Synonym – *Bacillaria flocculosa* (Roth) Leiblein, *Candollella flocculosa* (Roth) Gaillon, *Striatella flocculosa* (Roth) Kuntze, *Bacillaria tabelaris* Ehrenberg, *Tabellaria ventricosa* Kützing, *Tabellaria flocculosa* var. *ventricosa* (Kützing) Grünow, *Tabellaria fenestrata* var. *intermedia* Grünow].

Huber- Pestalozzi 1942, p. 430, pl. CXXVIII, Figure 522.

Rectangular frustules, flat tip, with keel and punctae, punctuations present on both the sides in a alternative matter, frustules 75 – 79 µm long and 29 – 32 µm broad.

Epiphytic in southern sector (4747), Epilithic in Central sector (4059, 4737) and Planktonic in Central sector (2949, 4411).

Family – Cocconeidiaceae

Genus – *Cocconeis* Ehrenberg

66. *Cocconeis pediculus* Ehrenberg (Plate 6.3, Figure 2)

[Synonym – *Cocconeis communis* f. *pediculus* (Ehrenberg) Chemielevski, *Encyonema caespitosum* var. *pediculus* (Ehrenberg) De Toni]

Kützing 1865, p. 71, pl. 5, Figure IX (1).

Frustules ovoid to ellipsoid, with marginal bend, lanceolate outline, rounded end, 20 – 25 µm long, 10 – 17 µm broad, striation not visible in fresh material.

Planktonic in Northern sector (2053).

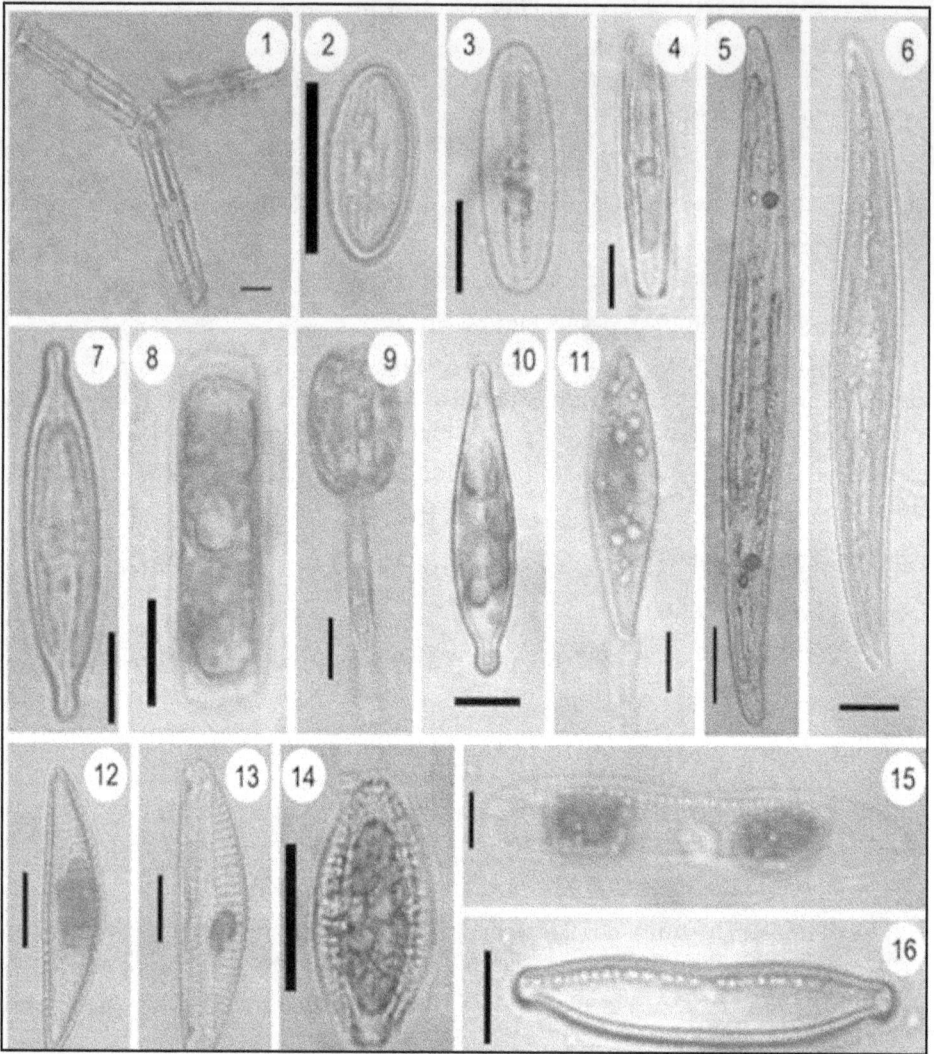

Plate 6.3

1. *Tabellaria flocculosa*, 2. *Cocconeis pediculus*, 3. *Pinnularia nodosa*, 4. *Pinnularia subsimilis*, 5. *Pleurosigma javanicum*, 6. *Pleurosigma normanii*, 7. *Navicula amphirhynchus*, 8. *Navicula major*, 9. *Gomphonema micropus*, 10. *Gomphonema olivaceum*, 11. *Gomphonema sphaerophorum*, 12. *Cymbella affinis*, 13. *Amphora elliptica*, 14. *Nitzschia acuta*, 15. *Hantzschia amphioxys*, 16. *Epithemia gibberula* var. *producta*.

(Scale bar: figure 1 – 16 = 10 *μm*).

Family – Pinnulariaceae

Genus – *Pinnularia* **Ehrenberg**

67. *Pinnularia nodosa* **(Ehrenberg) W. Smith (Plate 6.3, Figure 3)**

[Basionym – *Navicula nodosa* Ehrenberg]

[Synonym – *Pinnularia mesolepta* var. *nodosa* (Ehrenberg) Brun, *Schizonema nodosum* (Ehrenberg) Kuntze, *Navicula mesolepta* var. *nodosa* (Ehrenberg) Gutwinski]

Ettl and Gärtner 1995, p. 102, Figure 18 c.

Valve linear – lanceolate with parallel margins having slight undulations, end broadly rounded, central area slightly widened, raphe thin, straight, with distinct polar nodule, 43 – 70 µm long and 12 – 19 µm broad, striae smooth, transverse, 10 – 12 in 10 µm area.

Epilithic in Northern sector (2938, 4044) and Southern sector (4047, 4746).

68. *Pinnularia subsimilis* **Gandhi (Plate 6.3, Figure 4)**

Gandhi 1970, p. 789, Figure 116 – 118.

Valves linear, lanceolate, slightly attenuated towards the apices, rounded ends, raphe thin, median, axial area linear, narrow, gradually widening towards the centre, central area broad, 55 – 60 µm long and 7 – 9 µm broad, striation not clearly visible.

Planktonic in Southern sector (4417) and Outer channel sector (4406, 4715).

Family – Pleurosigmataceae

Genus – *Pleurosigma* **W. Smith**

69. *Pleurosigma javanicum* **Grunow (Plate 6.3, Figure 5)**

Desikachary 1989, p. 5, pl. 677, Figure 2.

Broadly lanceolate frustules, sigmoid with acuminate obtuse ends, raphe median, towards apices marginal, striae not distinct, 88 – 90 µm long and 9 – 10 µm broad.

Epilithic in Central sector (4721) and Planktonic in Outer channel sector (2934, 4091).

70. *Pleurosigma normanii* **Ralfs (Plate 6.3, Figure 6)**

[Synonym – *Pleurosigma affine* Grunow]

Desikachary 1987, p. 8, pl. 310, Figure 7.

Frustules broadly lanceolate, sigmoid with obtuse end, transverse striae, not clearly visible in fresh material, 107.2 µm long and 14.3 µm broad.

Epiphytic in Northern sector (2065).

Family – Naviculaceae

Genus – *Navicula* **Bory de Saint-Vincent**

71. *Navicula amphirhynchus* **Ehrenberg (Plate 6.3, Figure 7)**

Kützing 1865, p. 95, pl. 4, Figure XIII.

Frustules elliptical, lanceolate, with narrowly rostrate apices, raphe thin, central area slightly wide, 33 – 42.3 μm long and 9.6 – 10.8 μm broad, striation barely visible in fresh material.

Epiphytic Northern sector (2938) and Planktonic in Outer channel sector (2975, 4092, 4714).

72. *Navicula major* Kützing (Plate 6.3, Figure 8)

Kützing 1865, p. 97, pl. 4, Figure XX.

Frustules rectangular in valve view, elongated, striation transverse, less visible striae, 10 – 12 in 10 μm, middle or axial portion granular, 53 – 68 μm long, 7 – 10 μm broad.

Epilithic in Northern sector (2939) and Outer channel sector (2975, 4089).

Family – Gomphonemataceaea

Genus – *Gomphonema* Ehrenberg

73. *Gomphonema micropus* Kützing (Plate 6.3, Figure 9)

Kützing 1865, p. 84, pl. 8, Figure XII.

Frustules small, linear, cuneate, asymmetrical, end truncate, base obtuse, striation marginal, parallel, 8 – 10 in 10 μm, 31.5 – 35.5 μm long, 3.6 – 7.8 μm broad.

Epilithic in Northern sector (2942) and Outer channel sector (2975, 4080).

74. *Gomphonema olivaceum* (Lyngbye) Kützing (Plate 6.3, Figure 10)

[Basionym – *Ulva olivacea* Hormemann]

[Synonym – *Gomphonema olivacea* (Hornemann) Dawson ex Ross and Sims]

Kützing 1865, p. 85, pl. 7, Figure XIII, XV.

Frustules lanceolate clavate with broad, rounded apices, base attenuated, middle wide, striation coarse and distinct, parallel, 8-10 in 10 μm, 39 μm long and 11 μm broad.

Epilithic in Northern sector (2054).

75. *Gomphonema sphaerophorum* Ehrenberg (Plate 6.3, Figure 11)

Pal and Santra 1990, p. 75, pl. 1, Figure 25.

Frustules clavate-lanceolate, one end broad and another end gradually tapering towards apices, apex rostrate, forming astigma like structure, central unilateral, axial area narrow, raphe not seen, striation not clear in fresh material, sometimes having big stalk and attached to aquatic plant, 24 μm long, 7.4 μm broad.

Epiphytic in Northern sector (2057).

Family – Cymbellaceae

Genus – *Cymbella* C. Agardh

76. *Cymbella affinis* Kützing (Plate 6.3, Figure 12)

[Synonym – *Cymbella excisa* Kützing]

Kützing 1865, p. 80, pl. 6, Figure XV.

Frustules biraphid, asymmetrical, elliptic, oblong, end obtuse, dorsal side convex, ventral margin slightly gibbous, raphe arcuate or towards the ventral margin, striation distinct, transverse, parallel, striae 9 – 10 in 10 µm, longer than broad, 49 – 56 µm long and 11.8 – 13 µm broad.

Planktonic in Outer channel sector (2974, 4091, 4714) and Central sector (4060, 4424).

Family – Catenulaceae

Genus – *Amphora* Ehrenberg ex Kützing

77. *Amphora elliptica* Kützing (Plate 6.3, Figure 13)

[Basionym – *Frustulia elliptica* Agardh]

Kützing 1865, p. 107, pl. 5, Figure XXXI.

Frustules in girdle view elliptic lanceolate, slightly biconvex, apices slightly attenuated, obtuse truncate, striation distinct, transverse at both the sides, striae 7 – 8 in 10 µm, 68.6 – 73.3 µm long and 15.7 – 18.2 µm broad.

Planktonic in Outer channel sector (4093, 4406, 4713), Southern sector (4046, 4426) and Central sector (4056, 4424).

Family – Bacillariaceae

Genus – *Nitzschia* Hassal

78. *Nitzschia acuta* Cleve (Plate 6.3, Figure 14)

Desikachary 1989, p. 3, Figure 6, 7.

Frustules linear, lanceolate, deppressed in the middle portion, obtuse apices, 78 – 84.5 µm long and 15 – 17 µm broad, striation clear, 4 – 5 in 10 µm.

Planktonic in Southern sector (4065), Outer channel sector (2973, 4090, 4713) and Epilithic in Northern sector (2939, 4039, 4758).

Genus – *Hantzschia* Grunow

79. *Hantzschia amphioxys* (Ehrenberg) Grunow in Cleve and Grunow (Plate 6.3, Figure 15)

[Basionym-*Eunotia amphioxys* Ehrenberg]

[Synonym- *Nitzschia amphioxys* (Ehrenberg) W. Smith, *Homeocladia amphioxys* (Ehrenberg) Kuntze, *Hantzschia amphioxys* var. *genuina* Grünow, *Hantzschia abundans* Lange- Bertalot]

Ettl and Gärtner 1995, p. 114, Figure 20 a.

Valves small, linear, dorsal side convex, ventral side concave with depression in the middle, ends bluntly rounded, 44 – 50 μm long and 9 – 12.4 μm broad, keel punctae, striation not clearly visible.

Epilithic in Northern sector (2939, 4044), Central sector (2963, 4055, 4753) and Planktonic in Outer channel sector (2973, 4089, 4715).

Family – Rhopalodiaceae

Genus – *Epithemia* Brébisson

80. *Epithemia gibberula* var. *producta* Grunow (Plate 6.3, Figure 16)

Prasad and Srivastava 1992, p. 278, pl. 32, Figure 10.

Frustules broadly elliptical, with sub truncate ends, dorsal margin strongly acute, ventral margin with a very slight curvature, ends prostrate, costae distinct, radiate, slightly dilated below the apices, 33.6 – 47.7 μm long and 23.2 – 27.7 μm broad, striae fine, 2 – 8 in two consecutive costae.

Discussions

We for the first time recorded the algal species occurring at several locations covering all the four sectors after opening of the new mouths to Bay of Bengal. A total of eighty one algal species were documented from four different sectors of Chilika lagoon during different seasons from May 2011 to April 2012 (Plates 6.1–6.3). These belonged to Cyanophyta (24 species), Rhodophyta (2 species), Chlorophyta (19 species), Euglenophyta (6 species) and Bacillariophyceae under Heterokontophyta (30 species). Algal flora of Chilika lake and its distribution had been studied by several workers in the past from 1932 to 2000. Biswas (1932) reported for the first time 22 algal species comprised of 11 species of Cyanophyta, 5 species of Chlorophyta and 6 species of Rhodophyta from the entire lake. Subsequently Roy (1954) reported 33 species of Bacillariophyta in the lake. Ahmed (1966) reported 13 algal species constituting 1 species of Cyanophyta, 5 species of Cholorophyta, 1 species of Bacillariophyta and 6 species of Rhodophyta. Patnaik (1972 and 1978) reported 57 algal species comprising 6 Cyanophyta, 8 Chlorophyta, 40 Bacillariophyceae, 7 Dinophyta, 1 Xanthophyta and 4 Rhodophyta. Raman *et al.* (1990) reported 4 species of Cyanophyta, 3 species of Chlorophyta, 10 species of Bacillariophyceae and 1 species of Dinophyta. Adhikary and Sahu (1992) reported 9 Cyanophyta, 8 Chlorophyta, 1 Xanthophyta, 6 Bacillariophyceae, 3 Dinophyta, and 1 Rhodophyta. Rath and Adhikary (2005) reported the diversity of algal forms in the lake just before opening of the new mouth in 2000 at different seasons and recorded 102 algal species comprising 12 species of Cyanophyta, 23 species of Chlorophyta, 58 species of Bacillariophyceae, 5 species of Dinophyta, and 4 species of Rhodophyta. A comparative account of occurrence of all the algal species so far recorded in the lake by different authors for the period from 1932 to 2000 is given in Table 6.1. Of the 81 algal taxa recorded in the lagoon in 2011-12 after opening of the new mouths, 61 algal species comprised of 20 Cyanophytes, 16 Chlorophytes, 6 Euglenophytes and 19 Bacillariophycae under Heteokontophyta appeared for the first time in the lagoon. These were *Cyanobacterium*

Table 6.1: Comparative Account of Occurrence of Algal Species in different Sectors of Chilika Lagoon Recorded in Present Work with those Occurring before Opening of New Mouth Connecting Bay of Bengal from 1932 up to 2000 (1932 Biswas; 1954 Roy; Ahmed 1966; Patnaik 1973, 1978; Raman 1990; Adhikary and Sahu 1992; and 2000, Rath and Adhikary published in 2005a).

Algal Taxa	1932	1954	1966	1973	1978	1990	1992	1999	2000	2011
Cyanophyta/Cyanoprokaryota										
Cyanobacterium diachloros										+
Synechocystis aquatilis						+				
Aphanocapsa marina										+
Merismopedia elegans									+	
Merismopedia glauca									+	
Merismopedia punctata										+
Merismopedia warmingiana										+
Microcystis aeruginosa										+
Microcystis wesenbergii										+
Chroococcus limneticus										+
Chroococcus turgidus									+	
Pseudanabaena limnetica										+
Pseudanabaena minima										+
Geitlerinema earlei										+
Spirulina labyrinthiformis										+
Spirulina major										+
Spirulina subtilissima									+	+
Arthospira platensis									+	
Phormidium ambiguum										+
Phormidium submembranaceum	+								+	

Contd...

Table 6.1–Contd...

Algal Taxa	1932	1954	1966	1973	1978	1990	1992	1999	2000	2011
Oscillatoria limosa										+
Oscillatoria perornata										+
Oscillatoria princeps									+	+
Oscillatoria proteus										+
Oscillatoria sancta										+
Oscillatoria simplicissima										+
Lyngbya aestuarii	+		+	+	+		+		+	+
Anabaena torulosa	+								+	
Anabaena variabilis						+				+
Anabaena flos-aquae										+
Fischerella sp.									+	
Rhodophyta										
Ceramium diaphanum	+		+		+				+	+
Gracilaria verrucosa	+		+		+		+	+	+	+
Grateloupia filicina	+		+		+				+	
Polysiphonia subtilissima	+		+		+				+	
Chlorophyta										
Spirogyra sp.				+	+	+	+		+	+
Chara sp.							+		+	
Nitella sp.							+		+	
Closterium venus						+				+
Euastrum dubium										+

Contd...

Table 6.1–Contd...

Algal Taxa	1932	1954	1966	1973	1978	1990	1992	1999	2000	2011
Cosmarium awadhense										+
Cosmarium decoratum										+
Cosmarium impressulum									+	+
Cosmarium lundellii										+
Cosmarium miscellum										+
Cosmarium punctulatum										+
Xanthidium sexmamillatum									+	+
Chaetomorpha linum								+	+	
Cladophora glomerata							+	+	+	
Ulva lactuta Linn.					+		+	+	+	
Microspora willeana										+
Enteromorpha compressa	+		+		+		+	+	+	+
Enteromorpha intestinalis	+		+		+		+	+	+	+
Enteromorpha usneoides										+
Eudorina elegans										+
Pediastrum duplex									+	
Pediastrum simplex									+	
Pediastrum tetras									+	+
Chlorella protothecoides										+
Actinastrum hantzschii									+	
Coelastrum cambricum									+	
Scenedesmus bijugatus										+

Contd...

Table 6.1–Contd...

Algal Taxa	1932	1954	1966	1973	1978	1990	1992	1999	2000	2011
Scenedesmus acuminatus										+
Scenedesmus calyptratus									+	+
Scenedesmus dimorphus										+
Scenedesmus protuberans										
Scenedesmus qudricauda										
Selenastrum gracille									+	
Tetraedron gracile									+	
Tetraedron trigonum									+	
Uronema confervicolum									+	+
Uronema elongatum										+
Euglenophyta										
Euglena acus										+
Euglena agilis										+
Euglena caudata										+
Leptocylindrus danicus									+	
Lepocinclis playfairiana										+
Trachelomonas abrupta										+
Trachelomonas hispida										+
Heterokontophyta (Bacillariophyceae)										
Melosira borreii									+	
Melosira decussata										+
Odontella polymorpha										+

Contd...

Table 6.1–Contd...

Algal Taxa	1932	1954	1966	1973	1978	1990	1992	1999	2000	2011
Biddulphia heteroceros						+			+	
Thalassiosira subtilis							+		+	
Lauderia annulata				+					+	
Skeletonema costatum				+			+		+	
Coscinodiscus centralis		+							+	
Coscinodiscus gigas									+	
Coscinodiscus marginatus									+	+
Stephanophyxis turris		+							+	
Chaetoceros affinis		+		+			+		+	
Chaetoceros decipiens										+
Chaetoceros diversus				+						+
Chaetoceros eibenii		+							+	
Chaetoceros lorenzianus		+		+					+	
Chaetoceros paradoxus									+	
Bacteriastrum furactum									+	
Bacteriastrum hyalinum		+		+					+	
Rhizosolenia setigera		+		+					+	
Cyclotella maxima										+
Cyclotella meneghiniana										+
Fragilaria crotonensis									+	+
Grammatophora undulate									+	
Synedra crystalline										+

Contd...

Table 6.1–Contd...

Algal Taxa	1932	1954	1966	1973	1978	1990	1992	1999	2000	2011
Synedra radians										+
Synedra tabulata										+
Synedra ulna									+	+
Leptocylindrus danicus									+	
Licmophora abbreviate										
Asterionallopsis glacialis		+		+					+	
Tabellaria fenestrata					+		+		+	+
Tabellaria flocculosa										
Bacillaria paradoxa		+		+					+	
Cylindrotheca closterium										
Thallassionema nitzschioides		+		+					+	
Cocconeis pediculus									+	
Pinnularia alpine									+	
Pinnularia nobilis									+	
Pinnularia nodosa										+
Pinnularia subsimilis										+
Pleurosigma javanicum										+
Craticula normanii									+	
Gyrosigma acuminatum									+	
Navicula amphirhynchus										+
Navicula lanceolata									+	
Navicula major										+

Contd...

Table 6.1–Contd...

Algal Taxa	1932	1954	1966	1973	1978	1990	1992	1999	2000	2011
Navicula minuscula									+	
Navicula protracta									+	
Navicula salinarum									+	+
Gomphonema micropus										+
Gomphonema olivaceum										
Gomphonema sphaerophorum										
Stauroneis pusilla									+	
Cymbella affinis									+	+
Diatoma elongate									+	
Amphora elliptica									+	+
Amphora ovalis									+	
Amphiprora gigantean									+	
Guinardia flaccid									+	+
Nitzschia acuta										
Nitzschia obtuse									+	
Nitzschia panduriformis									+	
Nitzschia sigma									+	
Hantzschia amphioxys										+
Epithemia gibberula										+
Aulicus sculptus									+	
Climacosphaenia moniligera						+				

diachlloros, Aphanocapsa marina, Merismopedia punctata, Merismopedia warmingiana, Microsystis aeruginosa, Microsystis wesenbergii, Chroococcus limneticus, Pseudanabaena limnetica, Pseudoanabena minima, Geitlerinema earlei, Spirulina labyrinthiformis, Spirulina major, Phormidium ambiguum, Oscillatoria limosa, Oscillatoria perornata, Oscillatoria proteus, Oscillatoria sancta, Oscillatoria simplicissima, Anabaena variabilis and *Anabaena flos-aquae* under Cyanophyta, *Euastrum dubium, Cosmarium awadhense, Cosmarium decoratum, Cosmarium lundellii, Cosmarium miscellum, Cosmarium punctulatum, Microspora willena, Enteromorpha usneoides, Eudorina elegans, Chlorella protothecoides, Scenedesmus bijugatus, Scenedesmus calyptratus, Scenedesmus dimorphus, Scenedesmus protuberans, Uronema confervicolum, Uronema elongatum* under Chlorophyata, *Euglena acus, Euglena agilis, Euglena caudata, Lepocinclis playfairiana, Trachelomonas abrupta* and *Trachelomonas hispida* under Euglenopyta and *Melosira decussata, Odontella polymorpha, Chaetoceros decipiens, Cyclotella maxima, Cyclotella meneghiniana, Synedra crystallina, Synedra radians, Synedra tabulata, Tabellaria flocculosa, Pinnularia nodosa, Pinnularia subsimilis, Pleurosigma javanicum, Navicula amphirhynchus, Navicula major, Gomphonema micropus, Gomphonema olivaceum, Nizschia acuta, Hantzschia amphioxys* and *Epithemia gibberula* under Bacillariophyceae of Heterokontophyta.

In the present work Northern sector showed maximum algal diversity followed by Central, Southern and Outer channel sectors. The Cyanophytes were rich in Northern (16), Central (13) Southern (8) and Outer channel sector (6). Central sector showed maximum occurrence of Chlorophytes (14) followed by Northern sector (10), Southern sector (7) and Outer channel sector (3). Likewise the Euglenophytes were found occurring abundantly in Northern sector due to anthropogenic activity in the nearby area. However, due to tidal impact and change in the hydrological parameters of the lagoon, the Outer channel sector showed highest occurrence of Bacillariophyceae members. Some of the algal species *e.g. Pediastrum simplex, Pediastrum tetras, Nitzschia obtusa* in the Northern sector, *Lyngbya aestuarii, Pediastrum simplex, Pediastrum tetras* in the Central sector, *Lyngbya aestuarii, Ceramium diaphanum, Enteromorpha usneoides, Bacteriastrum hyalinum, Nitzschia acuta* in the Southern sector and *Melosira borreii, Coscinodiscus marginatus, Synendra ulna, Nitzschia acuta* in the Outer channel sector documented in the present work have also been recorded earlier in the lake in 2000 before the opening of new mouth to Bay of Bengal (Rath and Adhikary 2005 b).

Apart from this phytoplankton few seaweed species also occurred in the estuarian regime of Chilika lake. Four species of seaweeds, *e.g.* three species under Chlorophyta *e.g. Chaetomorpha linum, Enteromorpha compressa, Enteromorpha intestinalis* and one species under Rhodophyta, *Gracilaria verrucosa* were recorded in different sectors of Chilika lagoon. Of these *Chaetomorpha linum* occurred in abundance in all the sectors of the lagoon. *Enteromorpha compressa* was recorded in the Central sector at moderate salinity. *Enteromorpha intestinalis* was found occurring in the lake except the Northern sector showing that it preferred low to high salinity level of the lake. *Gracillaria verrucosa* earlier recorded in the Southern and Central sectors, presently also appeared in the Outer channel sector at higher salinity with consistent wave action due to influence of the tides after opening of new mouths to the Bay of Bengal.

Acknowledgment

We are thankful to the Department of Science and Technology, Govt. of India for financial support through a DST- SEED project. We thank the authority of Visva-Bharati for providing laboratory facilities.

References

Adhikary, S.P. and Sahu, J.K. (1992): Distribution and seasonal abudance of Algal forms in Chilika lake, East coast of India, Jpn. Linnol, 53: 197-205.

Ahmed, M.K. (1966): Studies on Gracilaria Greu of the Chilika lake. Bull. Orissa Fish Res. Invest. **1:** 46-53.

Biswas, K. (1932): Algal flora of the Chilika lake. Mem. Asiat. Soc. Bengal, 11: 65 – 198.

Brock, A.J. (2002): Suborder: Closterinieae – In. John, D.M., Whitton, B.A. and Brock, A.J. (eds.) – *The Freshwater algal flora of British Isles.* An identification guide to freshwater and terrestrial algae, The Natural History Museum and British Phycological Society, Cambridge University Press, Cambridge, p. 516 – 593.

Desikachary, T.V. (1959): *Cyanophyta.* I.C.A.R. monograph on Algae. New Delhi. p: 686.

Desikachary, T.V. (1987): *Atlas of Diatoms* (Diatoms from the Bay of Bengal), Madras Science Foundation, Madras. Vol. III and IV: 3-10. Pl. 222 – 400.

Desikachary, T.V. (1988): *Atlas of Diatoms* (Marine diatoms of the Indian Ocean region). Madras science foundation, Madras. Vol. V: 1-3. pl. 401-621.

Desikachary, T.V. (1989): *Atlas of Diatoms* (Marine diatoms of the Indian Ocean region), Madras Science Foundation, Madras. VI: 1-13, Pl. 622 – 809.

Ettl, H. and Gärtner, G. (1995): *Syllabus der Boden-, Luft- und Flechtenalgen.* Stuttgard, New york, p. 699.

Gandhi, H.P. (1970): A further contribution to the diatom flora of the Jog-falls, Mysore state, India. Beihefte, zur Nova Hedwigia Heft. 31: 757-813.

Hegewald, E., Hindák, F. and Schnepf, E. (1990): Studies on the genus *Scenedesmus* (Chlorophyceae, Chlorococcales) from South India, with special reference to the cell wall ultrastructure. Bibliotheca Phycologica, 99, J. Cramer, Berlin, Stuttgart, p. 75.

Huber-Pestalozii, G. (1942): *Das phytoplankton des Süßwassers.* 2. Teil, 2. Hälfte, Schweizerbart'she Verlagsbuchhandlung, Stuttgart, p. 545.

Komárek, J. and Anagnostidis, K. (1998): Cyanoprokaryota I. Teil: Chroococcales. In: Herausgegeben von H. Ettl, G. Gärtner, H. Heynig, D. Mollenhauer (eds.), SüßWasserflora. Von Mitteleuropa, Gaustav Fischer 19, p. 548.

Komárek, J. (2005): Cyanoprokaryota II. Teil: Oscillatoriales. In: Büdel B, Gartner G, Krienitz L and Schagerl M (eds.), SüßWasserflora. Von Mitteleuropa, Elsevier, 19, p. 759.

Komárek, J. and Anagnostidis, K. (1989): Modern approach to the classification system of Cyanophytes 4 – Nostocales. Archiv für Hydrobiologie (Algological Studies), Suppl. 3, 56: 247 – 345.

Komárek, J. and Fott, B. (1983): *Das phytoplankton des Süßwassers,* 7. Teil: - E. Schweizerbart'sche Verlagsbuchhandlung, Stuttgart. p. 1001.

Komárek, J. and Jankovská, V. (2001): Review of the algal genus *Pediastrum;* implication or pollen analytical research. *Bibliotheca Phycologica* J. Cramer, Stuttgart. Band 108: 127.

Kützing, F.T. (1865): Bracillarian order Diatomeen, Verlag von Ferd Förstemann, Nordhausen. P. 152.

Misra, P.K. and Srivastava, A.K. (2003): Some desmids (Chlorophyta) from north eastern Uttar Pradesh, India. Journal of Indian Botanical Society, 82: 85 – 92.

Mohanty, R.K., Mohapatra, A. and Mohanty, S.K. (2009): Assessment of the impacts of a new artificial lake mouth on hydrobiology and fisheries of Chilika lake, India. *Lakes and Reservoirs:* Research and Management, 14: 231 – 245.

Nayak, B.K., Acharya, B.C., Panda, U.C., Nayak, B.B. and Acharya, S.K. (2004): Variation of water quality in Chilika lake, Indian jou. Mar. Sc. 33: 164 – 169.

Nayak, L. and Behera, D.P. (2004): Seasonal variation of some physicochemical parameters of Chilika lagoon (east coast of India) after opening the new mouth, near Sipakuda. *Ind. J. Mar. Sci.* 33: 206 – 208.

Pal, U.C. and Santra, S.C. (1990): Algae of Midnapore, West Bengal II Bacillariophyceae. Phykos, 29: 73 – 81.

Pal, U.C. and Santra, S.C. (1993): Algal flora of Midnapore, III Desmidiaceae. Phykos, 32: 147 – 158.

Panigrahi, S., Wikner, J., Panigrahy, R.C., Satapathy, K.K. and Acharya, B.C. (2009): Variability of nutrients and phytoplankton biomass in a shallow brackish water ecosystem (Chilika Lagoon, India). Limnology, 10: 73 – 85.

Patnaik, S. (1973): Observation on the seasonal fluctuating of plankton in the Chilika lake. Ind. J. Fish., 20: 43-45.

Patnaik, S. (1978): Distribution and seasonal abudance of some algal forms in Chilika lake. J. Inl. Fish. Soc. India. 10: 56-67.

Patra, A.P., Patra, J.K., Mahapatra, N.K., Das, S. and Swain, G.C. (2010): Seasonal variation in physicochemical parameters of Chilika lake after opening of new mouth near Gabakunda, Orissa, India, World J. Fish. Mar. Sci., 2: 109-117.

Philipose, M.T. (1967): *Chlorococcales,* I.C.A.R. Monographs on Algae, New Delhi, p: 365.

Prasad, B.N. and Mishra, P.K. (1992): *Freshwater algal flora of Andaman and Nicober Islands.* Vol. II. Bishen Singh Mahendra Pal Singh, Dehra Dun, p. 284.

Prasad, B.N. and Srivastava, M.N. (1992): *Freshwater algal flora of Andaman and Nicober Islands.* Vol. I. Bishen Singh Mahendra Pal Singh. Dehra Dun. p. 369.

Raman, A.V., Satyanarayan, C.H., Adiseshadri, K., and Prakash, K.P (1990): Phytoplankton characteristic of Chilika lake, a brakish water lagoon along the east coast of India. Indian J. Mar.Sci., 19: 274 – 277.

Ramanathan, K.R. (1964): *Ulotrichales*. ICAR, New Delhi, p. 188.

Rath, J. and Adhikary, S.P. (2005a): *Algal Flora of Chilka Lake*. Daya Publisher New Delhi, p. 206.

Rath, J. and Adhikary, S.P. (2005b): A checklist of algae from Chilika lake, Odisha. Bull. Bot. Surv. Ind., 47: 101-114.

Rath, J. and Adhikary, S.P (2006): Marine macro-algae of Orissa, east coast of India. *Algae*, 21: 49 – 59.

Roy, J.C. (1954): Periodicity of plankton diatoms of the Chilika lake for the years 1950-51. J. Bom. Nat. Hist. Soc., 52: 112 – 123.

Sahoo, D., Sahu, N. and Sahoo, D. (2003): A critical survey of seaweed diversity of Chilika lake, India. Algae, 18: 1 – 12.

Sahu, J. and Adhikary, S.P. (1999): Distribution of seaweeds in Chilika lake. Seaweed Res. Utln., 21: 55 – 59.

Tabassum, A. and Saifullah, S.M. (2010): The planktonic diatom of the genus *Chaetoceros* Ehrenberg from Northwestern Arabian sea bordering Pakistan. Pakistan J. Botany, 42: 1137 – 1151.

Wo³owski, K. and Hindák, F. 2005. *Atlas of Euglenophytes*. VEDA, Publishing House of the Slovak Academy of Sciences, p. 136.

Chapter 7

Sea Level Rise Due to Climate Change

S.N. Das

Retired Scientist,
CSIR-Institute of Minerals and Materials Technology,
Bhubaneswar, Odisha, India

ABSTRACT

Sea level is rising at an increasing rate mainly due to global warming, which is a result of excessive presence of greenhouse gases in the atmosphere emitted from anthropogenic sources. The permanent snow covers of Greenland and Antarctica are melting at a faster rate than whatever amount of ice gets deposited in polar winter months as a result of which the freshwater resources are getting fast depleted and the volume of water in the oceans is increasing with time. The floating ice in the oceans is also melting contributing to volume expansion. All these factors contribute to sea level rise. This might result in inundation of a number of thickly populated coastal cities and jeopardize the lives and livelihoods of millions of people living along the coastlines. A case study undertaken at one of the vulnerable sites located in the east coast of India shows multifaceted adverse effects like loss of fertile agriculture land, loss of coastal mangroves acting as a barrier to cyclonic winds, storm surge and a breeding ground for fishes and other marine animals, salt water incursion into inland freshwater bodies etc. Majority of people opted to move out of the sporadic island areas in search of alternate sources of livelihood. Shifting and rehabilitation efforts by the local and central Governments would need adequate funds and finding alternate homestead and cultivable land is also proving to be difficult. The only alternative is to arrest or reduce the emission of global warming gases and freeze it at a reasonable level (earlier agreed to be 1994 under the UNFCCC platform) urgently with massive international efforts and then think of adopting techno-economically proven carbon capture/sequestration technology in commercial scale to reduce the atmospheric level with a concerted global action plan.

Keywords: *Greenhouse gases, Climate change, Radiative forcing, Sea level rise, Vulnerability and adaptation, Relief/rehabilitation.*

Introduction

The physical interface of sea and land is popularly known as the shore line. It is a dynamic entity, which changes spatially over time due to various reasons. The natural causes are erosion of the coast line due to lateral current of the sea as well as deposition of silts and wash offs carried in by the rivers. Reclamation of land due to population pressure, especially in urban and industrial areas, is the common anthropogenic reason for shoreline change. But the most important reason, of late, is the climate change. It has engaged attention of many scientists working in dynamics of coastal morphology, storm surge and sea level rise. The problem is growing increasingly difficult with time since there are many socio-economic dimensions associated with it. Loss of life and life support systems due to frequent weather extremes like flood, cyclone, tsunami etc. are poorly recorded and responded to, especially in developing countries. More important is the loss of livelihood of people like fishery, agriculture, forestry, bio-diversity, migration/resettlement, saline water incursion into sweet water sources etc, which ultimately affects the food security of the coastal dwellers.

Accurate demarcation and monitoring of shoreline changes are necessary for understanding and deciphering the coastal processes operating in an area. It is essential for the purpose of coastal zone management planning, hazard zoning, erosion/accretion etc. Application of latest techniques like remote sensing with geographical information system (GIS) is a useful approach for such studies due to synoptic and regular data availability with required accuracy/resolution, ease of multispectral database generation and cost competitiveness in comparison to the conventional processes like survey and mapping though laborious and costly physical ground-truthing. However, the later is a must for accuracy and dependable results.

Sea Level Rise

The sea level practically remained unchanged all over the globe during 1000 B.C. to 1900 A.D. with slight fluctuations now and then. The rate of sea level rise during these almost 3,000 years was a mere 0.1 to 0.2 mm per year. Since 1900 AD, however, sea level has been rising rapidly at a rate of 1 to 2 mm per year. Actual observations at many different locations through tide gauge (Ekman *et al.*, 1998) show that the global sea level has been rising at a rate of 1.7–1.8 mm/year over the last century and the rate has increased to 3 mm/year in the last decade (Church *et al.*, 2006). These numbers per annum do not appear to be large, but when accumulated over a long period, they certainly assume a menacing proportion. The sea level has risen by 20 cm in the last 100 years, which is significant and a matter of concern. Kumar *et al.* (2010) discussed vulnerability of Orissa coast to the rising sea level.

The major coastal cities of the world were established during the last few thousand years, a period during which the global sea level did not change much. Coastal and ocean activities, such as marine transportation of goods, offshore energy drilling, resource extraction, fish cultivation, recreation, and tourism are integral to the nation's economy. Satellite measurements taken over the past decade, however, indicate that the rate of increase has gone up to about 3.1 mm/year, which is significantly higher than the average rate for the last century (Nayak *et al.*, 1991). Model projections suggest that the rate of sea level rise is likely to increase during the present century, although there is considerable controversy about the likely extent of the increase. Controversy arises mainly due to uncertainties of three major processes responsible for sea level rise: thermal expansion of sea water, the extent of glacier and ice cap melting, and the loss of ice from the Greenland and West Antarctic ice sheets under different warming scenarios. However, even the minimal effects expected under best possible scenario is alarming (Nicholls R. *et al.*, 2007). The deltas in south and south-east Asia are vulnerable to sea level rise due to their excessive population and dependence on the coastal resources (Saito *et al.*, 2001).

Causes of Sea Level Rise

Before going into the major factors contributing to climate change, it should be understood that the melting back of sea ice will not directly contribute to sea level rise because this ice is already floating in the ocean. However, the melting back can lead to indirect contributions to sea level rise, *i.e.*, a reduction in sea surface reflectivity (albedo) and better absorption of solar radiation leading to a rise in sea surface temperature (SST). More solar radiation being absorbed will accelerate warming, thus increasing the melting back of snow and ice on land as well. In addition, the ongoing breakup of the floating ice shelves will allow a faster flow of ice on land into the oceans, thereby providing an additional contribution to sea level rise.

There are three major processes by which human-induced climate change directly affects sea level. As the climate change increases sea surface temperature (SST), thermal expansion contributes to sea level rise, which has contributed to ~2.5 cm during the second half of the last century. This is likely to triple during the end of present century. The fourth assessment of the Inter-Governmental Panel on Climate Change projected that the thermal expansion may singularly contribute an estimated 17 to 28cm (±50 per cent). This is stated to be a conservative estimate, which might go up further with the present rate of greenhouse gas emission. In fact, their warnings were grim predicting that both the past and future anthropogenic carbon dioxide emissions will continue to contribute to warming and sea level rise for more than a millennium, due to the timescales required for removal of this gas from the atmosphere'. (IPCC Fourth Assessment Report on Climate Change, Working Group I: The Physical Science Basis, 2007)

The second major contributor to sea level rise could be the melting of glaciers and ice caps. Melting of mountain glaciers and ice caps led to ~ 2.5 cm rise during 1951-2000. This is higher than the loss of ice from the Greenland and Antarctic ice sheets, which added ~1 cm to the sea level. This factor may contribute by10 to 12 cm (±30 per cent). This would represent a melting of roughly a quarter of the total amount of ice tied up in mountain glaciers and small ice caps.

The third process may be the loss of ice mass from Greenland and Antarctica. In case all the ice deposits on Greenland melt, which is likely within approximately a thousand years, the sea level would go up by ~7 m. The West Antarctic ice sheet holds about 5 m of sea level equivalent ice and is particularly vulnerable as much of it is immersed below sea level. On the other hand, the East Antarctic ice sheet, which is less vulnerable to climate change, holds about 55 m of sea level equivalent of ice mass. IPCC (2007) estimates that the Greenland ice melt would induce ~2 cm rise whereas Antarctica would induce a fall in sea level by roughly the same amount due to more accumulation of ice than melting assuming global warming effects to be lower in this region. However, these model projections seem to be grossly inaccurate since recent satellite observations indicate that both the Greenland and Antarctica are currently losing ice mass at a steady rate for the last several years.

Mechanism of Sea Level Rise due to Climate Change

Sea level can rise by two different mechanisms with respect to climate change. First, as the oceans warm due to an increasing global temperature, the volume of seawater increases taking up more space in the ocean basin and causing a rise in water level. The second mechanism is the melting of ice over land, which then adds water to the ocean. The Intergovernmental Panel on Climate Change (IPCC 4th Assessment Report, 1997) predicts that the total global-average sea level rise from 1990 to 2100 will be 110 to 770 millimeters (4.3 to 30.3 inches) under different warming scenarios. The expansion will be different for different greenhouse gas emission scenarios.

Unwarned and temporary sea level rise might occur due to tsunami and tectonic plate slip as per latest findings. The tragic loss of life associated with the Indian Ocean tsunami of December 2004 followed by sea level rise in certain coasts is the latest example. The nature and extent of impact due to a tsunami is related to the mechanism responsible for its generation like seismic event, mass movement, etc. Consequently, the materials deposited and the events responsible for these deposits assume significance. This requires the linkage of terrestrial and marine records, and involves observation of palaeo-environmental reconstruction and computer model simulation. Model predictions for the Paradip coast of Orissa in east coast was analyzed for the area 19° to 22° N and 84.9° to 88.5° E with its southern open boundary located at latitude 19° N due to natural hazards like cyclones. Although the principal component is surge, the sea-surface elevation may either be decreased or increased with respect to surge value, depending on its phase with astronomical tide. However, the mutual interaction of tide and surge is nonlinear, and both processes must be considered simultaneously. Paradip and Chandabali shore lines experienced exceptionally high tide levels due to storm surge resulting out of the super cyclone of 1999 that inundated land as far away as 20 km from the shoreline (Sinha *et al.*, 2008).

Effects of Sea Level Rise

The physical effects of sea level rise are categorized into five types: inundation of low lying areas, erosion of beaches and bluffs, salt water intrusion into aquifers and surface waters, higher inland water tables and increased flooding/storm damage (Nichols *et al.*, 2007). The tide gauge records at five coastal locations in India such as Mumbai, Kolkata, Cochin, Kandla and Sagar Islands have been documenting the rise in sea level for the last several years. The change in sea level appears to be higher in the eastern coast compared to western coast due to their inherent coastal dynamics.

The main reason is the flatness of the east coast due to deposition through several rivers meeting the sea after flowing through lengthy coastal plains. The global average sea level rise has been reported as 2.5 mm/year (0.04 to 0.1 inches) since 1950s and this holds good for India as well. Sea level rise is contributing to coastal erosion in many places of the world.

A small increase in sea level can have dramatic impacts on many coastal environments. Over 600 million people live in coastal areas that are <10 m above sea level, and >60 per cent of the cities with >5 million population are located in these at-risk areas. With sea level projected to rise at an accelerated rate for at least several years, most of these locations are going to be inundated and so the vulnerable people are to be relocated. Other detrimental effects are coastal erosion, wetland and coastal plain flooding, salinization of aquifers and soils and loss of habitat for fish, birds, and other wildlife as well as the plants. The US Environmental Protection Agency (EPA) estimates that 26,000 square kilometers of land would be lost if the sea level rises by 0.66 m, while the IPCC notes that as much as 33 per cent of coastal land and wetland habitats are likely to be lost in the next hundred years if the level of the ocean continues to rise at its present rate. As a result, very large numbers of wetland and swamp species are likely at serious risk. In addition, species that rely upon the existence of sea ice to survive are likely to be especially impacted as the retreat accelerates, posing the threat of extinction for polar bears, seals, and some breeds of penguins as well.

Unfortunately, many of the vulnerable countries do not have the required resources to prepare for it. Low-lying coastal regions in developing countries such as Bangladesh, Vietnam, India, and China have large population living in risky coastal areas such as estuaries and deltas. Certain island nations like the Philippines, Indonesia, Tuvalu and Vanuatu are at severe risk since they do not have enough land at higher elevations to rehabilitate the displaced coastal population. Another risk for some island nations is the loss of their freshwater supplies as sea level rise pushes saltwater into their aquifers. Therefore, those living on several small islands in the Indian Ocean and Pacific would very likely be forced to evacuate by the end of this century.

Sea Level and Tsunami

Tsunami results in deposits that are out of place, such as marine sand layers within a terrestrial peat or freshwater lake (Bondevik *et al.*, 2003). These attributes are not unique, since storm surges can also deposit marine sediments many meters above normal tidal levels (Dawson and Shi, 2000). Ongoing research is investigating the processes and impacts of modern tsunamis to develop a clearer picture of the distinctive anatomy of these events. A 2004 study in New Zealand examines sediments from a 15th century tsunami and a large storm that occurred in 2002. The results show clear differences in their sedimentology, continuity and extents of deposits. This type of information enables the development of a set of diagnostic criteria for finding their possible origin (Smith *et al.*, 2004).

For example, in the case of a fault-related tsunami, run-up height rarely exceeds twice the fault slip. Consequently, given maximum water levels of 25–30 m recorded in Sumatra suggest the Indian Ocean event was triggered by a fault displacement of between 12 and 15 m. Tectonic mass movements have the potential to produce even stronger waves, but at present these are hard to quantify, particularly when the slide or slump is submarine in nature. This picture is further complicated when mass movements are triggered by seismic events. Investigations of past tsunamis can be used to develop and test models of tsunami generation, and identify regions at higher risk of future events.

Reconstructing run-up height from a palaeo-tsunami requires information on the maximum altitude attained by the waves, and the altitude of sea level at the time of the event. Run-up heights based on the inland limit of sand layers are treated as minimum estimates since water levels commonly exceed the height of sediment deposition (Dawson and Shi, 2000). Many studies rely on estimates of former relative sea levels derived from geophysical models of the glacial isostatic adjustment process (Bondevik *et al.*, 2003). In an interesting development, Smith *et al.* (2004) used the tsunami deposits as a time horizon by locating the inland limit of intertidal sediments capped by this marker, the shoreline position at the time of the tsunami can be reconstructed and subsequent patterns of land movement identified.

Sea Levels and Seismic Events

In addition to tsunamis, earthquakes are associated with relative sea-level changes driven by vertical land movement. Evidence for cycles of land uplift and subsidence has been known in coastal deposits from the Pacific coast of North America for some years (Atwater, 1987). A period of pre-seismic subsidence occurs in some areas which could potentially be used as an indicator of forthcoming large earthquakes (Shennan *et al.*, 2005). Distinguishing pre-seismic subsidence from co-seismic change requires detailed investigations employing precise indicators of relative sea level. Quantitative palaeo-environmental reconstructions employing microfossils have the potential to provide these records. Sawai *et al.*, 2004 used diatoms and pollen to examine relative sea-level changes at two sites in Cook Inlet (Alaska) due to 1964 earthquake. Their data, with a maximum precision of 0.06 m, indicate pre-seismic subsidence of around 0.15 m at both the locations. In two related papers, Zong *et al.* (2003) use a diatom-based transfer function for tide level to reconstruct changes associated with a series of seismic events in the area. Their results show evidence for pre-seismic subsidence of similar magnitude to that associated with the 1964 earthquake, although these subtle changes are at the upper limit of their transfer function resolution. Others used a combination of foraminifera and diatoms to identify sudden elevation changes in three Holocene sedimentary sequences from Ohiwa Harbour, New Zealand. Their records indicate potential seismic-related subsidence of around 2 m, although the reconstructions are of lower precision than the transfer functions from North America.

Climate Change and Sea Level Rise

The earth's climate has been changing under normal conditions with cycles of snow and thaw periodically due to natural reasons. But too much of human interference has resulted in visible changes in weather and climate, which are well documented. It has now been proved beyond doubt and with a reasonable degree of accuracy that industrial revolution followed by relative material abundance and over-exploitation of natural resources has led to certain non-amendable changes in our climate, which are growing faster with time due to increasing abundance of air pollutants in our atmosphere.

One of the most pronounced effects of climate change has been melting of masses of ice around the world. Glaciers and ice sheets are large, slow-moving assemblages of ice that cover about 10 per cent of the world's land area and exist on every continent except Australia. They are the world's largest reservoir of freshwater, holding approximately 75 per cent of available freshwater on earth surface. Most of the mountain glaciers and the ice sheets in both Greenland and Antarctica have lost huge mass over the past century. Retreat of these ice sheets occurs when the mass balance, *i.e.*, the difference between accumulation of ice in the winter against melting in the summer, is negative such that more ice melts each year than is replaced. Thus climate change affects the mass balance of glaciers and ice sheets by affecting the temperature and precipitation of a particular area, both of which are key factors in the ability of a glacier to replenish its volume of ice. The glaciers and ice sheets lose mass when the temperature exceeds a particular level and lasts for a long period or there is insufficient precipitation in the region (Thompson *et al.*, 2002). Similar glacial melt-backs have been observed in other massive ice deposits like Alaska, the Himalayas, and the Andes.

Not only loss in ice cover but also the rate of these melting is important. Recent studies show that the movement of ice towards the ocean from both the major polar region ice sheets has increased significantly. As the speed increases, the ice streams flow more rapidly into the ocean, too quick to be replenished by snowfall. For example, the rate of the ice streams originating from the Greenland Ice Sheet has doubled in just a few years and the mass balance has become negative in the past few years. Estimates put the net loss of ice at anywhere between 82 and 224 cubic kilometers per year, which is a colossal loss of permanent ice cover.

Recent estimates show a sharp contrast between the East and West Antarctic Ice Sheet loss. The acceleration of ice loss from the West Antarctic Ice Sheet has doubled in recent years, which is similar to what has happened in Greenland. The main reason for this increase is the quickening pace at which glacial streams are flowing into the ocean. The loss of ice from the West Antarctic ice sheet is estimated to be 47 to 148 km^3/yr. On the other hand, recent measurements indicate that the East Antarctic ice sheet is gaining mass because of increased precipitation. However, this gain in mass by the East Antarctic ice sheet is negligible in comparison to the loss in West Antarctic ice sheet. Therefore, the mass balance of the entire Antarctic Ice Sheet is negative.

The melting back of the glaciers and ice sheets has two major impacts. First, areas that rely on the runoff from the melting of mountain glaciers are very likely to experience severe water shortages as the glaciers disappear. Less runoff will lead to a reduced capability to irrigate crops as freshwater dams and reservoirs more frequently go dry. Water shortages could be especially severe in parts of South America and Central Asia, where summertime runoff from the Andes and the Himalayas, respectively, is crucial for freshwater supplies. In addition, the melting of glaciers and ice sheets adds water to the oceans, contributing to sea level rise.

Mangrove Ecosystem

Natural ecosystems perform fundamental life-support services without which human civilization would cease to thrive (Chand *et al.*, 2010). Ecosystem functions are often defined in terms of fluxes of matter and energy. According to Field *et al.* (1998) ecosystem function covers three major areas - Biogeographical, ecological and anthropocentric, with overlaps between all three. In some cases a single ecosystem service is product of one and more ecosystem functions whereas in other cases a single ecosystem function contributes to two or more ecosystem services (Costanza, 1997). The values of mangrove ecosystem functions have been estimated to be very high in terms of social and economic returns for the immediate surroundings (Gilbert *et al.*, 1998).

The inundation due to sea level rise along Indian coasts will be different depending on the elevation of the coast with respect to the seawater level. It will be felt much more in flat surfaces like that of east coast, particularly north Orissa coast as compared to other shorelines. Similarly, model predictions by scientists from National Institute of Oceanography, Goa and Indian Institute of Tropical Meteorology, Pune have predicted different sea level rises at different shore locations of Indian coast. Net sea level rise is expected to be 1.20 mm/yr in Mumbai coast while that of Kochi would be 1.75 mm/yr, Vishakhapatnam 1.90 mm/yr, Diamond Harbor (Kolkata) 5.74 mm/yr, which might lead to sinking of the some of the islands in deltaic regions of Sundarbans. This has been predicted from tide gauge data over the past 50-100 years after making glacial isostatic correction at Diamond harbor (Talley *et al.*, 2007).

Functions of Coastal Mangrove

The undisturbed and natural mangrove forests or ecosystems act as seaward barrier and check the coastal erosion and minimize the tidal thrust or storm hit arising from the sea considerably. The degree of protection varies with the width of mangroves. Mangrove root systems retard water flow. Resistance to water flow serves to dissipate the energy of floodwaters, of particular service during cyclone. A super cyclone in the month of October in 1999 hit the Orissa coast near Saharabedi, a village lying about 1.5 km away from the seacoast in Ersama Block of Jagatsinghpur district had a wind speed of around 260 km/h and a storm surge of about 10 m. This super storm traveled more than 250 km inland and within a period of 36 hrs ravaged more than 200 lakh hectares of land, devouring trees and vegetation, leaving behind a huge trail of destruction. This cyclone affected around 15 million people in 12 districts and caused the deaths of about 20,000 people and over 4 lakh cattle. The loss to property, crops and plantations, communication and transportation networks was colossal, the value of which is estimated to be over Rs 10,000 crores. The most severely affected districts were Balasore, Bhadrak, Cuttack, Ganjam, Jagatsinghapur, Jajpur and Kendrapara affecting around population a population of around 11 million people.

Case Study on the Effects of Sea Level Rise in Bhitarkanika Estuarine Region

The study area selected was the Bhitarkanika coastal mangrove region, a severely depleted mangrove forest and protected wild life sanctuary. It covers an area of 672 sq. km of mangrove forest and wetland extending between 20° 17' 32" to 20° 48' 00" N Long. and 86° 38' 51" to 86° 17' 36" E Lat. (Figure 7.1). Bhitarkanika covers a total area of 2154.26 km^2 of which Bhitarkanika Wildlife Sanctuary and National Park covers 672 km^2, the Gahirmatha (Marine) Wildlife Sanctuary covers 1435 km^2 while the buffer zone in the Mahanadi delta covers 47.26 km^2. The natural boundaries of the Sanctuary are defined by Dhamara river to the north, Maipura river to the south, Brahmani river to the west and the Bay of Bengal in the east. The coastline from Maipura to Barunei forming the eastern boundary of the Sanctuary is an ecologically crucial habitat. Bhitarkanika is a wetland of International importance and 2672 km^2 area of wetland habitat has been declared as a Ramsar site in the year 2002. The general elevation above mean tide level is between 1.5 and 2 m (Dani and Kar, 1999). Higher ground extends up to 3.4 m. The river flow is influenced twice daily by high and low tides at approximately six hourly intervals. The maximum and minimum tide level varies according to lunar days and seasons. Siltation is a common phenomenon in the river systems. Soil erosion is taking place on the banks of the Baitarani, Honsua, Bramhani and Dhamra rivers. This area experiences tropical warm and humid climate, with no distinct season. Rain occurs due to the southwest monsoon from May to September, and the nor. The average rainfall is about 1642 mm, bulk of which is received during June to mid October. The maximum temperature recorded is 41°C and the minimum is 9°C during May and January respectively.

Figure 7.1: Location Map of Bhitarkanika Area.

Mean relative humidity ranges from 70 to 85 per cent throughout the year. The most important weather phenomenon is the prevalence of tropical cyclones. These open wetlands are influenced by monsoon rain and regular tidal inundation.

Bhitarkanika and Dangmal Bocks constitute the core area. These sites experience tide of semi-diurnal type. The mean sea level in the region is about 1.66 m. The area has about 200 km of water body inside the sanctuary and falls in the deltaic region of the river Brahmani, Baitarani, and their tributaries. The estuarine rivers like Brahmani, Baitarani, Kharasrota, Dhamra, Pathasala, Maipura, Hansua, and Hansina flow into the Bay of Bengal, which are further crisscrossed by numerous creeks, channels, and nallahs. Thus the area provides the peculiar ecological conditions for the growth and development of rich and varied mangrove life forms, both flora and fauna, along with their associates. There are many villages within the sanctuary as well as surrounding it. The population in these villages has been growing very fast during the last few decades. Part of the population rise is because of the heavy influx of refugees from neighbouring areas and habitations were started by clearing mangrove forests. A total of 81 villages are adjacent to the mangrove forests. The population increase is attributed as one of the reasons for decreasing mangrove of the area.

The region comes under the tropical monsoon climate with three pronounced seasons: winter (October to January), summer (February to May) and rainy (June to September). The maximum temperature is recorded during April to June and the minimum temperature in winter during the December to January. The relative humidity ranges from 75-85 per cent throughout the year. Wind speed during non-monsoon months is ~20 km/h mostly from S/SW directions. Rainfall is 1550-1620 mm/yr, most of it is received during monsoon. The most important weather phenomenon is the tropical cyclones formed in Bay of Bengal during pre-monsoon months. Rainfall conditions decide the sequence of mangrove distribution in the different zones in the tidal region. Baunsagada River experiences 1.5-2.5 m of tides during summer, which goes up to 3-5 m during monsoon. The creeks like Khola receive tidal water from both the sides, which reaches 3-4 m in summer to 5-6 m during the rainy season.

The intertidal zones near mouth of rivers Maipura and Dhamra has prolific mangrove forests. The vegetation of Bhitarkanika is broadly classified into mangrove and salt bush formations. The salt bush formation is found along the littoral tracts of Satbhaya and Gahirmatha sea shore where the soil is sandy and is rarely inundated under normal conditions except for high tides during flood and cyclones. The coastline here is characterized by sand dunes. There is a huge sand dune adjacent to the estuary called Babubali.

Rhizophora and Avicennia are the most commonly used species as they have very high calorific value. Stilt roots of Rhizophora are also used as roof material and the branches of Avicennia are used as fencing material and fishing poles. Mangrove forests also serve as resource for other faunal elements such as source for fish and prawns. Oysters clinging on stilt roots of Rhizophora are gathered by poachers and used as an ingredient for making whitewash for buildings. Mangrove stilt roots are also used for mosquito repellent and ayurvedic medicines. The ecosystem functions of the area have been brought out clearly by Badola and Hussain (2005).

The sea level mapping was carried out in the region and a social survey carried out in the most vulnerable villages like Prabhati, Bhitarkanika, Gahirmatha etc., in which >90 per cent of people are engaged in agricultural and fishing activities while only a sizeable population have alternate sources of income like business or service elsewhere in small towns. The land utilization pattern of all the villages in sanctuary area shows that more than 50 per cent of total land area is being utilized for agricultural purpose of which most of the land depends on rain for the rice cultivation. Animal husbandry also provides some engagement. All these activities are centered in and around the villages and therefore flooding or inundation by salt water during cyclones or storm surge affects their life directly. Vulnerability of the Asian coastal population to the rising sea levels and adaptation mechanisms has been discussed by Cruz *et al.* (2007). Aquaculture, especially shrimp and fish farming, is widely practiced in the mangrove area. Many shrimp and fish-ponds have been built by clearing the mangrove forest near the semi-haline zone, which is rarely replenished. These activities will also be severely affected by the sea level rise.

Acknowledgements

The author is thankful to the CSIR-IMMT Bhubaneswar for having provided opportunity to participate in this program earlier funded by different external agencies. However, the results provided are not a part of the reports submitted to the agencies concerned.

References

Atwater, B.F. (1987): Science 236, 942-944, Evidence for great halocene earthquakes along the outer coast of Washington state.

Badola, R. and Hussain, S.A. (2005): Environmental Conservation Valuing ecosystem functions: an empirical study on the storm protection function of Bhitarkanika mangrove ecosystem, India. 32(1): 85–92.

Bondevik, S., Mangerud, J., Dawson, S., Dawson, A. and Lohne, O. (2003): Record-breaking height for 8000-year-old tsunami in the North Atlantic. Eos 84(31): 289–300.

Chand, P. and Acharya, P. (2010): Int. J. Geomatics and Geosc, Shoreline change and sea level rise along coast of Bhitarkanika wildlife sanctuary, Orissa: An analytical approach of remote sensing and statistical techniques. 1(3): 437-455.

Church, J.A. and White, N.J. (2006): Geophys. Res. Lett., 33, L01602, 20th century acceleration in global sea level rise.

Costanza, R., d'Arge, R., de Groot, R., Farber, S. and Grasso, M. (1997): The value of the world's ecosystem services and natural capital. Nature (387): 253–60.

Cruz, R.V., Harasawa, H., Lal, M., Wu, S., Anokhin, Y., Punsalmaa, B., Honda, Y., Jafari, M., Li, C. and Huu Ninh, N. (2007): Asia Climate Change - Impacts, Adaptation and Vulnerability. Contribution of Working Group II to the Fourth Assessment Report of the Intergovernmental Panel on Climate Change, M.L. Parry, O.F. Canziani, J.P. Palutikof, P.J. van der Linden and C.E. Hanson, Eds., Cambridge University Press, Cambridge, UK, 469-506.

Dani, C.S. and Kar, C.S. (1999): IOSEA Indian Ocean - South-East Asian Marine Turtle Memorandum of Understanding, Conservation of sea turtles and environmental relationship of arribadas of Olive Ridley [*Lepidochelys olivacea*] in relation to Bhitarkanika mangrove ecosystem of Orissa coast.

Dawson, A.G. and Shi, S. (2000): Pure and Applied Geophys., 157: 875-897, Tsunami deposits.

Ekman, M. (1998): The world's longest continued series of sea level observations, Pure and Appl. Geophys, 127: 73–77.

Field, C.B., Osborn, J.G., Hoffman, L.L., Polsenberg, J. Ackerly, D.D., Berry, J.A., Bjorkman, O., Held, A.A., Matson, P.A., Mooney, H.A. and Vitousek P.M. (1998): Mangrove biodiversity and ecosystem function, Global Ecology and Biogeography Letters 7(1): 3–14.

Gilbert, A.J. and Janssen, R. (1998): Use of environmental functions to communicate the values of a mangrove ecosystem under different management regimes, Ecological Economics, 25: 323-346.

Horton, R., Herweijer, C., Rosenzweig, C., Liu, J., Gornitz V. and Ruane, A.C. (2008): Sea level rise projections for current generation CGCMs based on the semi-empirical method, Geophysical Res. Lett., 35: L02715.

IPCC Fourth Assessment Report: Climate Change (2007): Working Group I: The Physical Science Basis - Projections of Future Changes in Climate.

Kumar, T.S., Mahendra, R.S., Nayak, S.K., Radhakrishnan, K. and Sahu, K.C. (2010): Coastal Vulnerability Assessment for Orissa State, East Coast of India, Journal of Coastal Research, 26: 523–534.

Nayak, S.R. (1991): Manual for mapping of coastal wetlands/landforms and shoreline changes using satellite data - Technical Note, Space Applications Centre, Ahmedabad, IRSUP/SAC/MCE/TN/32/91.

Nicholls, R.J., Wong, P.P. Burkett, V.R., Codignotto, J.O. Hay, J.E. McLean, R.F. Ragoonaden, S. and Woodroffe, C.D. (2007): Coastal systems and low-lying areas, in *Climate Change, Impacts, Adaptation, and Vulnerability.*

Saito, Y. (2001): Deltas in Southeast and East Asia: Their evolution and current problems. In Mimura, N. and Yokoki, H., eds. (2000) Global Change and Asia Pacific Coasts Proceedings of APN/SURVAS/LOICZ Joint Conference on Coastal Impacts of Climate Change and Adaptation in the Asia-Pacific Region, APN, Kobe, Japan, November 14(16): 185-191.

Shennan, I., Hamilton, S., Hillier, C. and Woodroffe, S. (2005): Quaternary International 133– 134, 95-106, A 16000-year record of near field relative sea-level changes, northwest Scotland, UK.

Sinha, P.C., Jain, I., Bhardwaj, N., Rao, A.D. and Dube, S.K. (2008): Numerical modeling of tide-surge interaction along Orissa coast of India, Nat Hazards, 45: 413–427.

Smith, D.E., Shi, S., Cullingford, R.A., Dawson, A.G., Dawson, S., Firth, C.R., Foster, I.D.L., Fretwell, P.T., Haggart, B.A., Holloway, L.K. and Long, D. (2004): The

Holocene Storegga Slide tsunami in the United Kingdom. Quaternary Science Reviews 23: 2291– 321.

Talley, C.K. and Unnikrishnan, A. (2007): Journal of Coastal Research Special Issue Observations: Oceanic Climate Change and Sea Level, Coastal Vulnerability Assessment Database: Vulnerability to sea-level rise in the U.S. Southeast.

Thompson, L.G., Mosley-Thompson, E., Davis, M.E., Henderson, K.A. Brecher, H.H., Zagorodnov, V.S., Mashiotta, T.A. and Lin, P.N. (2002): Science 298 (5593): 589– 593.

Zong, Y., Shennan, I., Combellick, R.A., Hamilton, S.L. and Rutherford, M.M. (2003): Microfossil evidence for land movements associated with the AD 1964 Alaska earthquake. The Holocene 13: 7–20.

Chapter 8

Harvesting Nutrients from Water using Bio-Coagulant

Narendra Kumar Sahoo, Sumedha Nanda Sahu,
S. Sharma and S.N. Naik

Centre for Rural Development and Technology,
Indian Institute of Technology, Delhi, Hauz Khas,
New Delhi – 110 016, India

ABSTRACT

The efficiency of a phytoplankton community in nutrient (nitrogen and phosphorus) removal from water column and performance of *Moringa oleifera* seed extract as a biocoagulant in harvesting the algal biomass has been studied here. For this purpose nitrogen (N) and phosphorus (P) was added to water in presence of an algal community and analyzed regularly for residual nutrient in the water column. A parallel set was maintained for two weeks to produce algal biomass for the bio-coagulation study using deoiled seed powder of *M. oleifera*. The phytoplankton community at a higher algal density is able to remove around 90 per cent of N and P within 2-5 days and the bio-coagulant provides around 93-98 per cent harvest of biomass depending on algal density. In lieu of the higher-cost and secondary pollution from conventional treatment methods, utilization of micro-algae for treatment of wastewater is an alternative which can also ensure environmental security and continuous biomass supply.

Keywords: Bio-coagulant, Micro-algae, Nutrient, Moringa oleifera.

Introduction

Unmanaged anthropogenic activities and demophoric expolsion (*i.e.* the rapid increasing growth of populations and technology) are indirectly and directly adding nutrient to aquatic ecosystems. This extra nutrient in the waters lead to a process called Eutrophication. Due to nutrient enrichment of waterbodies, eutrophication has become an acute problem (UNEP, 1999) everywhere in the world with a few exceptions. The negligence or unwillingness to treat its cause and symptoms are rendering the scarce water resources unusable day by day. The increase in nutrient content of water bodies leads to a corresponding increase in primary production by autotrophs *e.g.*, algae, macrophytes etc. in the aquatic system. When the consumption in and from the systems is less than this production, the natural balance in the systems get disturbed to stimulated an array of undesirable changes both in terms of health of the ecosystem as well as human uses. The secondary consequences like surface-scums, anoxia, loss of biodiversity, toxic algal blooms, bad odour, production of low value fish, fish kills, etc. thus follows. Toxic blue-green algae and other algal groups producing unpleasant odours and taste in drinking water (*e.g.* some Crysophyceae) are particularly of major concern (Ryding and Rast, 1989). In the contrary, algae (phytoplankton) are very efficient at removing nutrients from the water column and can exhibit very high growth rates (Ryding and Rast, 1989).

Micro-algae are generally very efficient in nutrient scavenging (Lardon *et al.*, 2009, Sahoo 2010) and biomass production (*e.g.* 2 g/l/day dry biomass; Kong *et al.*, 2010) compared to terrestrial as well as other aquatic plants. Algae are up to several hundred times more productive per unit of land than other crops (Singh and Gu 2010, Mandal and Mallick 2009) depending on the species and growth conditions. Lardon *et al.* (2009) reported that photosynthetic efficiencies of algae range from 3 to 8 per cent, compared with 0.5 per cent for many terrestrial crops. They can double their biomass within 24 hr; as short as 3-4 hours during the exponential growth (Banerjee *et al.*, 2002). Micro-algae (*e.g. Chlorella* sp., *Scenedesmus* sp., *Cosmarium* sp. etc.) can be cultivated on otherwise non-productive lands or in brackish, saline, and wastewater that has little competing demand. These advantageous characteristics of micro-algae have brought them into treatment of wastewater otherwise called as phycoremediation (John 2000).

Compared to physical and chemical treatment processes, algae based treatment can potentially achieve nutrient removal in a less expensive and ecologically safer way with the added benefits of resource recovery and recycling (Graham *et al.*, 2009, Oswald 2003). However, sustainable harvesting of microalgae is a major challenge (Grima *et al.*, 2003) due to their small size, low specific gravity (equivalent to water), and low concentration in the culture medium. Hence, efficient and cost-effective methods of biomass harvest for micro-algal need more research inputs.

There is a wide array of harvesting methods that can be employed to harvest microalgal biomass such as centrifugation, flocculation, sedimentation and micro-filtration and any combination of these (Munoz and Guieysse 2008, Mutanda *et al.*, 2011, Danquah 2009, Grima *et al.*, 2003). Chemical flocculants such as alum, lime, $FeCl_3$, cationic polyelectrolytes, and $Ca(OH)_2$ (de la Noue *et al.*, 1992) are in use for removing algae from water but they causes secondary pollution in water and produce low grade biomass. In general, use of methods like straining, centrifugation, using chemical coagulant, auto-flocculation, flocculation by air-bubbling, electroflocculation, ultrasonic flocculation etc. for harvesting of micro-algae have limitations in one or other aspects such as cost, versatility, efficiency, energy consumption, quality and utility of harvested material, etc.

Use of bioflocculants (Vandamme *et al.,* 2010, Kim *et al.,* 2011) overcome many of the problems by making its requirements lower in terms of dosage, cost, energy, space, etc. ensuring a quality harvest. Use of one such bioflocculant namely *Moringa olefera* seed kernel extract (Ghebremichael 2004, Ali *et al.,* 2010) for efficient and low cost harvesting (Folkard 1999, Sutherland *et al.,* 1995) of micro-algal biomass will be a promising idea.

With this background the study envisaged to see the efficiency of a phytoplankton community in nutrient (nitrogen and phosphorus) removal from water column and performance of *Moringa oleifera* seed extract as a biocoagulant in harvesting the algal biomass.

Materials and Methods

Nutrient Removal

Microalgae laden water from different waterbodies of Delhi was mixed followed by filtration through a GF/F filter paper and washing in distilled water to prepare a working stock of 5.8×10^7 cells/lt. The initial community composition of phytoplankton was noted using high power (45x × 10x magnification) compound microscope. 50ml of the stock water was taken in nine conical flasks of 3lt capacity each. 500 µg/l P (as KH_2PO_4) and proportionate amount of NO_3-N (as KNO_3) were added to the flasks so as to prepare three ratios of N: P *viz.* 4:1, 8:1 and 16:1 in triplicates (Table 8.1). The three ratios have been noted in the text ahead as R4, R8 and R16 respectively. Micronutrients were also added in accordance with the algal media as noted in APHA, 1998. On 16th day, again 250 µg/l P and proportionate amount of NO_3-N was added to the culture.

Table 8.1: Experimental Set-up for Nutrient Uptake Study

Treatment	Nutrient Addition on 0th Day (Study period: 0 - 16th day)		Nutrient Addition on 16th Day (Study period: 16th - 21st day)	
	P (µg/l)	N (µg/l)	P (µg/l)	N (µg/l)
R4	500	2000	250	1000
R8	500	4000	250	2000
R16	500	8000	250	4000

Biomass Harvest

To each beaker of one liter capacity, with one liter de-chlorinated tap water, 20 ml of the stock microalgae is added followed by Nitrogen (N) and phosphorus (P) in nine combinations (Table 8.2). Different combinations of P and N are taken so as to get eutrophied water samples with a range of microalgal biomass (*i.e.* a range of optical density (OD) of water samples). The treatments are run in triplicates at room temperature-light condition. The microalgae are grown for two weeks with one time stirring each day and growth measurement is taken in terms of OD678 (Sorokin, 1975) at the end of experiment.

Table 8.2: Treatments Used in Algal Cultivation for Bio-coagulation Study

P Input (µg/l)	N Input (µg/l)	Treatment Code#
100P	400N	100R4
100P	800N	100R8
100P	1600N	100R16
250P	1000N	250R4
250P	2000N	250R8
250P	3000N	250R16
750P	3000N	750R4
750P	6000N	750R8
750P	9000N	750R16

#The prefix and suffix to 'R' denotes concentration of P (µg/l) and ratio of N:P respectively

Coagulant Extraction from the Seed

Dried *Moringa oleifera* (MO) seeds (Figure 8.1) were collected from IIT Delhi campus and stored at room temperature. The seeds were shelled just before the extraction and the kernel was powdered using a kitchen blender. Oil was removed by cold extraction (AOAC, 1984) with petroleum ether and solids were dried at room temperature. From the dried samples, 10 per cent (w/v) solution were prepared using distilled water, stirred for 30 minute and filtered through GF/F Whatman (0.45 µ) to prepare the crude extract. 1 ml of this extract is equivalent to 100 mg of dry seed powder.

Microalgae Harvest with MO Coagulant

Coagulation study was done using 1 lt capacity beakers coupled with stirrer with speed control. To all beakers 1 ml of the extract was added. Rapid mixing and slow mixing were done at 250 rpm for 20s and 40 rpm for 2 minutes, respectively (Ghebremichael, 2004) followed by a settling period of 1 hour (Muyibi and Evison, 1995).

After treatment with natural coagulant (*M. oleifera* seed powder extract) the microalgae in the beakers formed flocks and settled down to the bottom. The supernatant was decanted later on carefully avoiding any biomass loss and the bottom material was filtered on pre-weighed Whatman GF/F filter paper (APHA, 1998). After drying the filtrate along with the filter paper in hot-air oven at 70°C for 24 hours, final weight was noted down. The decanted water was filtered and weight was taken in the same way. Ash was assessed in the harvested biomass following incineration in a muffle furnace at 525°C for 6 hours (APHA, 1998).

Figure 8.1: *Moringa oleifera* **Plant and its Different Parts**
(Leaf, flower, green pod, dried pod showing seed, seed and kernel).

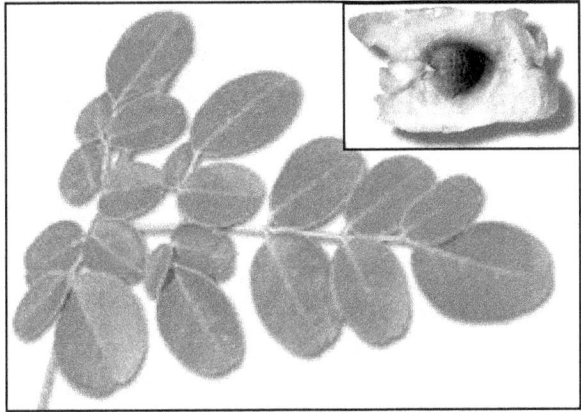

Figure 8.1–*Contd...*

Results

More than 20 species of micro-algae occurred in the algal stock solution (Table 8.3). Out of the 20 species representing their respective genus the number is highest under the class chlorophyceae. This list consists of the frequently occurring genera and the not so frequent ones have been left out deliberately. The diversity helped in successfully culturing the community at different ratios of nitrogen and phosphorus. The whole community has shown good growth response to addition of nutrient as mentioned in Table 8.5.

Table 8.3: Community Composition of Phytoplanktons in Stock

	Class	Genera		Class	Genera
1	Chl.	Ankistrodesmus	11	Chl.	Tetraedron
2	Chl.	Chlamydomonas	12	Cyn.	Aphanothece
3	Chl.	Chlorella	13	Cyn.	Merismopedia
4	Chl.	Cosmarium	14	Cyn.	Synechocystis
5	Chl.	Cylindrocystis	15	Blc.	Amphora
6	Chl.	Kirchneriella	16	Blc.	Cymbella
7	Chl.	Oocystis	17	Blc.	Fragilaria
8	Chl.	Peridinium	18	Blc.	Navicula
9	Chl.	Staurastrum	19	Eug.	Euglena
10	Chl.	Scenedesmus	20	Eug.	Phacus

Nutrient Removal

Study of percentage removal of the initial amount (Table 8.4) of phosphorus is showing that around 50 per cent removal on first day highest being for R4 reaching up to 49.5 per cent. By 2nd day up to 81 per cent (R8) of phosphorus was removed from the media. Almost up to 90 per cent (R8 and R16 ratio treatment) removal was achieved by 5th day from the systems. The maximum of phosphorus removal was 98 per cent (R16) and that took 16 days to occur. In case of the second event of nutrient addition, the trends were bit different from the 1st event; very high per cent removal of 70.6 per cent was noted for R4 ratio-treatment within first 4.5 hours but for R8 and R16 the values were 55.8 per cent and only 36.7 per cent respectively and around 90 per cent (R8) removal was recorded by 2nd day (day 18 is 2nd day when day 16 is 0th day). In general when the amount of phosphorus availability is high the removal is also very high; the removal becomes still more if the initial phytoplankton density is high. The residual phosphorus approach zero but never reaches to zero. Although 90 per cent of the phosphorus get removed within 3-5 days it takes another 5-10 days to achieve 98 per cent removal.

Unlike phosphorus, nitrogen (Table 8.4) does not have higher values in the initial days. For 50 per cent removal, R4 took 5-6 days and R8 took 6-7 days while R16 took almost 11 days. By the 16th day maximum per cent removals were 87, 94 and 96 respectively for R4, R8 and R16. In contrary, on the 2nd event of nutrient addition the values were high on the 1ˢᵗ day itself *i.e.* 54.7 per cent, 51.8 per cent and 43.7 per cent for R4, R8 and R16 ratio-treatments respectively and the highest per cent removal values were 79, 90 and 93 respectively for the trio by 5ᵗʰ day after the nutrient addition event.

Table 8.4: Pattern of P and N Removal from Water as Percentage of the Initial Amount

Days of Incubation	PO$_4$-P			NO$_3$-N		
	R4	R8	R16	R4	R8	R16
0	0.0	0.0	0.0	0.0	0.0	0.0
1	49.5	45.5	46.5	6.8	4.7	3.2
2	76.2	81.3	74.6	11.8	9.5	6.0
3	83.8	87.5	78.7	17.0	13.6	7.5
4	86.4	89.3	81.5	23.9	18.7	9.8
5	87.3	90.4	90.8	45.5	37.0	12.5
6	88.1	91.1	91.2	69.3	42.8	15.0
7	89.1	91.7	92.2	84.1	61.8	17.3
8	89.6	92.3	92.6	84.6	86.2	20.7
9	90.5	92.8	93.5	84.9	90.5	28.8
10	91.9	93.7	95.5	85.2	91.7	37.6
11	**	**	**	85.5	92.5	49.2
12	**	**	**	85.6	92.7	62.1
14	92.9	94.5	97.0	85.9	93.0	84.7
16m	95.1	95.6	97.7	86.0	93.1	95.7
16n	0.0	0.0	0.0	0.0	0.0	0.0
16e	70.6	55.8	36.7	38.2	31.8	28.2
17	74.3	72.7	62.0	54.6	53.9	43.6
18	77.9	89.1	84.6	65.6	66.0	55.2
21	90.9	92.2	94.6	78.5	89.5	93.0

**: Not measured due to the very low rate of removal per day.

Biomass Harvest

As evident from Table 8.5, with increasing input of phosphorus and nitrogen concentration in water the OD678 value in the water samples has increased indicating increased production in the systems. The optical density of 750P:9000N is almost 16 times that of 100P:400N. These treatment systems resulting from a wide range of P-N input can be considered as natural systems with various degrees of eutrophication and hence laden with different quantities of microalgal biomass.

From the biomass harvesting experiment with natural coagulant it was found that almost all the biomass in suspension in the water column has got flocculated and settled down to the bottom of the beakers. As evident from the biomass harvest values, one liter of highly productive system has produced around 0.5 g of algal biomass and that of a moderate one is 0.12-0.25 g. Taking in to account, the loss of biomass in the decanted water, this biomass harvest is almost 93-98 per cent of the total biomass. Moreover, the ash content in some selected sample (where harvest is good) the samples were around 2 per cent only, indicating that the biomass has a very good resource potential.

Table 8.5: Growth of Microalgae in Response to Nutrient Inputs and Performance of *Moringa oleifera* Seed Extract as a Coagulant

P:N Input (µg/l)	Mean OD678	Mean Dry Wt. (mg)	Yield (per cent)
100P:400N	0.0431	0.023±0.001	93
100P:800N	0.0898	0.057±0.003	93
100P:1600N	0.1261	0.084±0.003	94
250P:1000N	0.1679	0.113±0.004	95
250P:2000N	0.2000	0.140±0.005	96
250P:3000N	0.2314	0.163±0.006	96
750P:3000N	0.3416	0.240±0.011	97
750P:6000N	0.4865	0.341±0.014	98
750P:9000N	0.6837	0.472±0.019	98

When the per cent of biomass harvested is analyzed, it was found that the yield has increased with increasing OD678 of the samples. The lowest yield of 93 per cent has occurred at lowest OD678 and highest of 98 per cent has occurred at highest OD678.

Thus, it is evident that natural coagulants are very much efficient in harvesting microalgal biomass from eutrophied (productive waters) waters by forming flocks out of micro-algal cells. The algal cells were settled down to the bottom of the container making the process of harvesting easy and in turn resulting in clear water (free from algal cells).

Discussions

Nutrient Removal

Similar to these nutrient uptake results, Gonzalez *et al.* (2008) found an 80 per cent phosphate removal by a consortium of algae-bacteria from swine manure within 3 weeks of inoculation. Density driven nutrient uptake has been reported by Lau *et al.* (1995) which shows that nutrient uptake is influenced by algal density. Lafarga-De la Cruz *et al.* (2006) got upto 99 per cent P removal and 92 per cent N removal from the culture media, similar to the present study. Similar kind of uptake trends in case of Nitrogen and Phosphorus has also been reported by Varkitzi *et al.* (2010). Sorokin and Dallocchio (2008) reported in the lagoon of Vincia that the DIP (dissolved inorganic phosphate) uptake depends upon bloom density.

This Phycoremediation is one of the most effective methods of wastewater treatment available (Park and Craggs 2010) which removes nutrients and heavy metals, discourages growth of pathogens (due to aeration and increased pH by photosynthesis), furnish O_2 to heterotrophic aerobic bacteria to mineralize organic pollutants, and sequestration of CO_2 in turn (Munoz and Guieysse 2006). The use of microalgae in wastewater treatment (phycoremediation) has long been promoted (Oswald and Gotaas 1957) and continued to be studied widely for several decades (Golueke and Oswald 1965, Oswald 1995, Pagand *et al.*, 2000, Shelef and Azov 1987). However, the application of microalgae in the wastewater industry is still fairly limited and mainly either through the use of conventional oxidation (stabilization) ponds or the more developed suspended algal pond systems such as high-rate algal ponds.

Biomass Harvesting

The observation that biomass harvest is higher when the turbidity (OD) of the water is high, hints towards easy harvesting of hypertrophic waters. This observation have happened, must be due to availability of less number of cells for formation of flock of optimum size in case of water samples with lesser algal biomass (low OD678). On the contrary, water samples with high optical density have higher number of algal cells to participate in the formation of flock of optimum size and weight to settle down easily to the bottom of the container. *Moringa oleifera* (or other non toxic natural coagulants) as evident from the results, proves to be a very good option for using as a coagulation substance. In combination with sodium bentonite as weighting agent it produces a final water clarity equivalent to that produced using the conventional chemical coagulants (Folkard *et al.*, 1995), simultaneously ensuring a better water quality (Ndabigengesere and Narasiah, 1998). The sludge thus produced is significantly more compact and represents a potentially useful output as a soil conditioner/fertilser (Folkard *et al.*, 1995). Though algae with bioaccumulated toxins and heavy metals may render the algae unfit for human or animal consumption (Gordon *et al.*, 1982) the same could easily be used for the generation of biogas or for organic chemical extraction. With the use of *Moringa* seed protein, algae, particularly spirullina, has successfully been harvested in Mexico, Israel etc (Price, 2007).

The various chemical coagulants (*e.g.*, alum, Ferric salts, and synthetic polymers etc.) for precipitating algae co-precipitate the inorganic P too. Aluminum salts are so far the widely used coagulants in water and wastewater treatment (Bratby, 1980). However, problems such as Alzheimer's disease and similar health hazards are associated with residual aluminum in treated waters (APHA, 1998; Miller *et al.*, 1984; Letterman and Driscoll, 1988; Qureshi and malm, 1985), besides production of large volumes (James and O'Melia, 1982) of sludge during treatment. The water quality degrades in terms of decreased pH, increased EC, reduced alkalinity, and residual chemicals and need further treatments (studies cited in Ndabigengesere and Narasiah, 1998) for making it usable. In addition, this treated water sometimes lead to encrustation in supply systems. Alum flock carry over from the clarifiers cause "filter blinding". The sludge from the clarifiers is voluminous and difficult to dewater and present pollution problems on discharge to the receiving water. Besides that the treated water needs further treatment as mentioned previously.

Conclusion

Algae and particularly phytoplankton are very fast growing autotrophs and act as efficient scavengers of nutrients. Hence algae can be employed in removing nutrients from wastewater which on harvest inturn would curtail the process of eutrophication. The harvesting is better done with non-toxic natural coagulant instead of chemical coagulants as the former yield more intact and useful sludge. The non-toxic natural coagulants (such as *M. olefera* seed protein) will be promising in the process of cleaning waterbodies without any net investment as the harvested biomass ensures economy in terms of bio-crude production or other industrial/domestic/agricultural uses. Thus, process could be developed as a remedial measure to eutrophication as well to ensure a continuous biomass feedstock for various applications.

Acknowledgement

Authors duly acknowledge the financial assistance in the form of a sponsored project from Department of Science and Technology for carrying out the study.

References

Ali, E.N., Muyibi, S.A., Salleh, H.M., Alam, M.Z. and Salleh, M.R.M. (2010): Production of Natural Coagulant from Moringa Oleifera Seed for Application in Treatment of Low Turbidity Water J. Water Resource and Protection 2: 259-266.

AOAC. (1984): Official methods of analysis, 14th ed. Washington, DC: Association of Official Analytical Chemists.

APHA, AWWA, and WEF. (1998): Standard methods for the examination of water and wastewater. Published jointly by American Public Health Association, American Water Works Association and Water Environment Federation. 20th edition. Inc. Boltimore, Maryland, USA.

Banerjee, A., Sharma, R., Chisti, Y. and Banerjee, U.C. (2002): *Botryococcus braunii*: a renewable source of hydrocarbons and other chemicals. Critical Reviews in Biotechnology 22(3): 245-279.

Bratby, J. R. (1980): Coagulation and Flocculation, with emphasis on Water and Wastewater Treatment. Uplands Press Ltd., Croydon.

Danquah, M.K., Gladman, B., Moheimani, N. and Forde, G.M. (2009): Microalgal growth characteristics and subsequent influence on dewatering efficiency. Chemical Engineering Journal 151: 73-78.

de la Noüe, J. and de Pauw, N. (1988): The potential of microalgal biotechnology: A review of production and uses of microalgae. Biotechnology Advances 6: 725-770.

Folkard, G.K., Sutherland, J.P. and Shaw, R. (1999): Water clarification using *Moringa oleifera* seed coagulant. Pages 109-112 in Shaw R, ed. Running Water. Pub: Intermediate Technology Publications, London, ISBN 1-85339-450-5.

Folkard, G. K., Sutherland, J. P. and Al-Khalili, R. (1995): Natural Coagulants - A Sustainable Approach. In; Sustainability of Water and Sanitation Systems: Proceedings of the 21st WEDC Conference", Kampala, Uganda, 4-8 September, 1995 (Eds. Pickford, J. *et al.,)* WEDC Publications, pp. 263-266.

Ghebremichael, K. A. (2004): *Moringa* seed and pumice as alternative natural materials for drinking water treatment. ISSN 1650-8602, ISRN KTH/LWR/PHD 1013-SE, ISBN 91-7283-906-6.

Golueke, C.G., Oswald, W.J. and Gee, H.K. (1965): Harvesting and processing sewage-grown planktonic algae. Journal of the Water Pollution Control Federation 37: 471-498.

Gonzalez, C., Marciniak, J., Villaverde, S., Leon, C., Garcia, P.A. and Munoz, R. (2008): Efficient nutrient removal from swine manure in a tubular biofilm photo-bioreactor using algae-bacteria consortia. Water Science and Technology-WST 58(1): 95-102.

Gordon, M. S., Chapman, D. L., Kawasaki, L.Y., Tarifino-Silva, E., and Yu, D. P. (1982): Aquacultural approaches to recycling of dissolved nutrients in secondarily treated domestic wastewaters. IV: Conclusions, design and operational considerations for artificial food chains. Water Res. 16, 509-16.

Graham, L.E., Graham, J., Graham, J.M. and Wilcox, L,W. (2009): Algae. Benjamin Cummings. 616 p.

Grima, M.E., Belarbi, E.H., Fernandez, F.G.A., Medina, A.R. and Chisti, Y. (2003): Recovery of microalgal biomass and metabolites: process options and economics. Biotechnology Advances 20(7-8): 491-515.

James, C. and O'Melia, C. R. (1982): Considering sludge production in the selection of coagulants. J. Am. Wks. Ass., 74, 158-251.

John, J. (2000): A self-sustainable remediation system for acidic mine voids. In: 4th International conference of diffuse pollution 506-511.

Kim, D.G., La, H.J., Ahn, C.Y, Park, Y.H. and Oha, H.M. (2011): Harvest of *Scenedesmus* sp. with bio flocculant and reuse of culture medium for subsequent high-density cultures. Bioresource Technology 102: 3163-3168.

Kong, Q.X., Li, L., Martinez, B., Chen, P. and Ruan, R. (2010): Culture of microalgae *Chlamydomonas reinhardtii* in wastewater for biomass feedstock production. Applied Biochemistry and Biotechnology 160: 9-18.

Lafarga-De la Cruz, F., Valenzuela-Espinoza, E., Millan-Nunez, R., Trees, C.C., Santamarýa-del-Angel, E. and Nunez-Cebrero, F. (2006): Nutrient uptake, chlorophyll-α and carbon fixation by *Rhodomonas* sp. (Cryptophyceae) cultured at different irradiance and nutrient concentrations. Aquacultural Engineering 35: 51-60.

Lardon, L., Hlias, A., Sialve, B., Steyer, J.P. and Bernard, O. (2009): Life-cycle assessment of biodiesel production from microalgae. Environmental Science and Technology 43: 6475–6481.

Lau, P.S., Tam, N.F.Y. and Wong, Y.S. (1995): Effect of algal density on nutrient removal from primary settled wastewater. Environmental Pollution 89: 59-66.

Letterman, R. D., and Driscoll, C. T. (1988): Survey of Residual aluminium in filtered water. J. Am. Wat. Wks. Assoc., 80, 154-158.

Mandal, S. and Mallick, N. (2009): Microalga Scenedesmus obliquus as a potential source for biodiesel production. Applied Microbiology and Biotechnology 84: 281-291.

Miller, R. G., Kopfler, F. C., Kelty, K. C., Strober, J. A., and Ulmer, N. S. (1984). The Occurrence of aluminium in drinking water. J. Am. Wat. Wks. Assoc., 76, 84-91.

Munoz, R. and Guieysse, B. (2006): Algal–bacterial processes for the treatment of hazardous contaminants: a review. Water Research 40: 2799-2815.

Mutanda, T., Ramesh, D., Karthikeyan, S., Kumari, S., Anandraj, A. and Bux, F. (2011): Bioprospecting for hyper-lipid producing microalgal strains for sustainable biofuel production. Bioresource Technology 102(1): 57-70.

Muyibi, S. A. and Evison, L. M. (1995): Optimizing physical parameters affecting coagulation of turbid water with *Moringa oleifera* Seeds. Wat. Res. 29(12), 2689-2695.

Ndabigengesere, A., and Narasiah, K. S. (1998): Quality of Water Treated by Coagulation Using *Moringa Oleifera* Seeds. Wat. Res., 32, 781-791.

Oswald, W.J. (1995): Ponds in the twenty-first century. Water Science and Technology 31: 1-8.

Oswald, W.J. (2003): My sixty years in applied algology. Journal of Applied Phycology 15: 99-106.

Oswald, W.J. and Gotaas, H.B. (1957): Photosynthesis in sewage treatment. Transactions of the American Society of Civil Engineers 122: 73-105.

Pagand, P., Blancheton, J.P., Lemoalle, J. and Casellas, C. (2000): The use of high rate algal ponds for the treatment of marine effluent from a recirculating sh rearing system. Aquaculture Research 31: 729-736.

Park, J.B.K. and Craggs, R.J. (2010) : Wastewater treatment and algal production in high rate algal ponds with carbon dioxide addition. Water Science and Technology 61: 633–639.

Price, M. L. (2007): The *Moringa* Tree. An ECHO Technical Note.

Qureshi, N., and Malmberg, R. G. (1985): Reducing Aluminium residuals in Finished Water. J. Am. Wat. Wks Ass., 77, 101-108.

Ryding, S.O. and Rast, W. (1989): The Control of Eutrophication of Lakes and Reservoirs. The Parthenon Publishing Group Inc. New Jersey, USA.

Sahoo, N.K. (2010) : Nutrient removal, growth response and lipid enrichment by a phytoplankton community. Journal of Algal Biomass Utilization 1 (3): 1-28.

Shelef, G. and Azov, Y. (1987): High-rate oxidation ponds: the Israeli experience. Water Science and Technology 19(12): 249-255.

Singh, J. and Gu, S. (2010) : Commercialisation potential of microalgae for biofuels production. Renewable Sustainable Energy Review 9: 2596-2610.

Sorokin, Y.I. and Dallocchio, F. (2008): Dynamics of phosphorus in the Venice lagoon during a picocyanobacteria bloom. Journal of Plankton Research 30(9): 1019-1026.

Sorokin, C. (1975): Dry weight, packed cell volume and optical density. In; Handbook of Phycological Methods : Culture Methods and Growth Measurements (Ed. Stein, J. R.). Cambridge University Press, pp. 321-343.

Sutherland, J.P., Folkard, G.K. and Al Khalili, R.S. (1995): Preliminary investigations of alternative coagulant - flocculent dosing regimens to treat the Morton Jaffray source water, Report to Construction Associates (PVT), Harare, May 1995.

UNEP. (1999): Technology Needs for Lake Management in eutrophication of Rawa Danau and Rawa Pening, Downloaded: http://www.unep.or.jp/ietc/Publications/Tech Publications.

Vandamme, D., Foubert, I., Meesschaert, B. and Muylaert, K. (2010): Flocculation of microalgae using cationic starch. Journal of Applied Phycology 22: 525-530.

Varkitzi, I., Pagou, K., Granéli, E., Hatzianestis, I, Pyrgaki, C., Pavlidou, A., Montesanto, B. and Economou-Amilli, A. (2010): Unbalanced N:P ratios and nutrient stress controlling growth and toxin production of the harmful dinoflagellate Prorocentrum lima (Ehrenberg) Dodge Original Research Article. Harmful Algae 9(3): 304-311.

Chapter 9

Predicting Discolouration Potential in Water Distribution Networks: An Index-Based Approach

Paul Chadwick[1], Alex Francisque[2],
Andrew Heather[1], Philip Selby[1] and Rehan Sadiq[2]

[1]*Mott MacDonald, Demeter House, Station Road, Cambridge UK CB1 2RS*
[2]*School of Engineering, University of British Columbia (Okanagan Campus)*
Kelowna, British Columbia, Canada, V1V 1V7

ABSTRACT

Distribution Operation and Maintenance Strategies are now widely established in the United Kingdom water industry and actively used by water companies to manage water quality in distribution systems. The main objective of this research was to develop tools for the DOMS forward-looking approach, for use by water companies in developing their long-term drinking water quality plans. It was achieved through (1) developing models for estimating future discolouration in drinking water quality, which can be tested and applied by companies; (2) assessing the discolouration relative importance compared with other DOMS water quality issues; and (3) developing a methodology for assessing what interventions are appropriate in different circumstances and what benefits are obtained. Previous modelling attempts were generally hindered by insufficient data or quantification of the underlying processes. This research aimed to take a simpler approach that would work with current data and knowledge. An index-based approach was developed using data collected by water companies to generate three indices: water quality, vulnerability, and hydraulic discolouration potential. A system discolouration index was calculated from the three indices at water quality zone level. The proposed distribution discolouration Index (D2I) was generally able to identify troublesome zones, where most customer complaints about discolouration were received.

Keywords: *Distribution networks, Discolouration, DOMS, Index-based approach, Water quality models.*

Introduction

Water companies in United Kingdom (UK) have traditionally been required to report their compliance with drinking water standards on an annual basis, with compliance being assessed at customers' taps. The compliance reporting has always been retrospective – *i.e.* it reports compliance over the period that has been monitored. Although it can show if levels of service have changed, it does not in itself show that service will change (or continue a trend) in the future. Water companies generally report high levels of compliance (99.6 per cent) with drinking water standards. However of 0.4 per cent of samples that fail to meet the standard, a significant proportion (~60 per cent) of fail parameters are related to the water discolouration (Drinking Water Inspectorate (DWI, 2007). Discolouration events occur when accumulations are mobilized from within the network and re-suspended due to systems changes (Vreeburg and Boxall 2007). The water may appear to be darkly coloured black, brown or red. White discolouration is generally associated with dissolved air or carbon dioxide being released from the water leaving the pressurised network. The material causing discolouration is of various origins, *e.g.* the source water (Lin and Coller 1997, South East Water 1998, Kirmeyer *et al.* 2000, Slaats *et al.* 2002, Ellison 2003), the treatment works (Vreeburg *et al.* 2008), the distribution network through various reactions and mechanisms such corrosion, erosion, biological growth, external contamination during repair, intrusion and backflow (Le Chevallier *et al.*1987, Stephenson 1989, Sly *et al.*1990, Walski 1991, Clark *et al.* 1993, Lin and Coller 1997, Ruta 1999, Prince *et al.* 2001, Meches 2001, Gauthier *et al.* 2001, Clement *et al.* 2002, Slaats *et al.* 2002, Boxal *et al.* 2003, Seth *et al.* 2004, Vreeburg and Boxall 2007) and is accumulated within the network due to many processes and mechanisms (Vreeburg and Boxall 2007). Discolouration events result ofcomplex, poorly understood and interactive mechanisms. The processes may be understood through the concept that particles attached by some means to the pipe are the cause of discolouration (Vreeburg and Boxall 2007).

Distribution Operation and Maintenance Strategies (DOMS) were officially introduced in 2002 by the DWI Information Letter 15/2002 (DWI, 2002) to provide a mechanism for companies to demonstrate to the DWI that they had in place a strategy for maintaining levels of service delivered to customers. The introduction of the UKW ater Industry Research (UKWIR) common framework for capital maintenance planning (UKWIR, 2002) reminded companies to consider the future effects on service of changes in the asset base or external factors, increasing the emphasis on understanding deterioration and its impacts on service to customers. The water supply network deteriorates from the time of installation until its eventual replacement. The rate of deterioration may be influenced by the water chemical properties, temperature, age and velocity, and the pipe material (Kleiner 1998; Sadiq *et al.* 2006; 2007; Francisque *et al.* 2009a; Yamini and Lence 2010). This deterioration may affect the water quality at any time. Overall water quality changes result of a combination of physical, chemical and microbial processes (Hendrickson 1996; Kirmeyer *et al.* 2000; Payment *et al.* 2003; Sadiq and Rodriguez 2004) which are complicated and not yet fully understood. Discolouration is not frequently recorded in routine water quality monitoring because when it occurs it tends to be transient. Because discoloured water events often occur over short duration for unpredictable reasons they are difficult to study (Vreeburg and Boxall 2007).Continuous monitoring in many locations around a network would provide more information about the discolouration nature but the cost of providing and maintaining sufficient equipment to monitor whole networks is likely to be unacceptable to customers. Of the relatively few samples failing water quality tests (508 samples, representing 0.4 per cent across England and Wales), 53 per cent exceeded the prescribed concentration or value (PCV) for discolouration-related parameters (DWI, 2007). Discolouration is an important cause of customer contact to water companies (~80 per cent of contacts about water quality), and is one of the main reasons for customers to complain to the DWI (DWI, 2007). Consumers may contact the DWI if they feel that their complaint about water quality has not been sufficiently dealt with by, or if they have a lack of confidence in, their water supplier. An analysis of the 130 complaints to the DWI about water quality in 2006 showed that the majority relate to aesthetic parameters, with appearance the greatest cause of complaint.

Discolouration is one of the main causes of customer service failure. Discoloured water incidents greatly affect customer confidence in tap water quality and the quality of service provided by water companies (Vreeburg and Boxall 2007). The specific extent of discolouration is difficult to quantify because the different parties involved measure and describe it in different ways, but the proportion of water quality samples failing for discolouration-related parameters is similar to the proportion of water quality complaints to the DWI about appearance. Hence tools that help companies to reduce the incidence of discolouration are likely to have direct customer service benefits. The most common intervention to control discolouration is flushing of water pipes, from wash-out valves and fire hydrants. This increases the flow in the pipes above normal and consequently scouring forces and shear stress increase (Boxall *et al.* 2005, 2001), then mobilize and re-suspend material accumulated within the pipes (Vreeburg and Boxall 2007). By definition flushing causes discolouration and hence the timing and management of the activity is important. Flushing is a short-term way of removing deposits and will have to be repeated as long as particle accumulation continues. Some companies have developed targeted flushing programmes, with the return period varied according to the particle accumulation rate or in response to rising customer contacts. Flushing has the advantage that it is relatively cheap and non-intrusive, but it does not address any underlying deterioration of the network.

Previous studies (UKWIR 2006a) have reviewed the linkages between asset parameters and discolouration but no strong causal relationship was confirmed.

UKWIR (2006b,c) produced a guidance manual for water companies in building their DOMS. The study reviewed service indicators and examined the use of customer contact data about interruptions to the water supply as a measure of service. Customer contacts provide a 'first-hand', continuous and in some cases near instantaneous indication of the water acceptability to consumers, but there was some variability in contact rates between different groups of customers. For example, some population groups were much more (or less) likely to contact their water undertaker to report an interruption to their water supply than others. Therefore the potential variability in customer contact rates should be taken into account when using customer contacts as a success or failure measure in service delivery, meaning that differences in the absolute number of contacts might not directly be related to the service failure magnitude.

The Drinking Water Regulations (DWI, 2001) set the minimum standards and monitoring frequencies for water supplies. Consequently, all English and Welsh water companies have a common dataset of statutory monitoring results, with networks divided into water quality zones (WQZs). However, the main limitation of statutory monitoring is that it is designed to audit long-term compliance rather than provide detailed performance data. To monitor network performance each water company sets its own data collection (and storage) regime according to the perceived performance monitoring requirements of individual networks. Companies also have hydraulic models, built for investment planning, but the letters typically consider steady-state or diurnal flow ranges, rather than exceptional states such as the velocity generated by a burst main or fire hydrants operation. Customer contacts to water companies are recorded, including a categorisation of the contact reason. Discolouration is categorised as either brown/black/red water, or white water. Although there may be a relatively long data history, the meaning of the records may change, reducing the usefulness of time-series analysis. Although customers are not likely to contact the company if there has not been discolouration, there might be varying rates of contact when discolouration does occur, depending on the day time and if customers accept the level of service. However, published studies of differences in customers' attitudes to discolouration or likelihood of contacting their water supplier about it could not be found.

A major limitation to combining the various data sets is that they are often stored in different formats and without common reference data. Thus water quality results cannot be related to asset registers or to network model results. Data about hydraulic performance tend not to be reported from models other than at the time of investment planning, and hence there is no common format of storage of these data. In some cases a geographic information system may be used to combine data sets (Francisque *et al.* 2009b), but often there are no geo-referable data or the areas to which the data relate are not clear. Often the proportion of water supplied to a WQZ from a source is estimated based on average conditions, but in practice there might be significant variation throughout the year, as the individual source output may vary with changing demand. Different data sets cover different historical periods: hydraulic models tend to be one-off data sets, whereas water quality monitoring and customer contact data tend to be on-going. The period of data retention may also vary between companies, but even where data have been retained for statutory purposes some may be in archived format and effectively not available unless a business case justifies its retrieval. Whilst many types of data are collected by water companies, the linkages between them are weak and there is no requirement for a common sample other than the statutory water quality monitoring. Therefore it is important for this research to adopt a method that makes best use of common data, but allows additional data sets to be included for more detailed analysis. Where long-term data sets were available, companies reported that there were changes in the networks during the data collection period, reducing the meaning of long-term comparison. For example, networks were relined or pipes replaced, DMAs and water quality zones were reconfigured, either to meet new regulations on zone size or to improve leakage control, and water treatment works received new processes. Therefore even when data were retained over long periods, they refer to a dynamic network and as such, long-term analysis does not generally offer an advantage over short-term data. Where the main features of a network have remained unchanged for the duration of the data period, then long-term analysis would be meaningful, but even then the results would be specific to the individual portion of network being studied.

The general objective of this research was to develop tools for the DOMS forward-looking approach, for use by water companies in developing their long-term drinking water quality plans. The specific objectives were to: *i*) Develop models for estimating future deterioration in drinking water quality owing to the distribution system state, that can be tested and applied by companies; *ii*) Assess the relative importance of discolouration compared with other DOMS water quality issues; and *iii*) Develop a methodology for assessing what interventions are appropriate in different circumstances and what benefits are obtained. Previous studies have applied traditional statistical techniques such as regression analysis, and detailed process modelling, to forecast water networks discolouration potential. Such models have generally been hindered by insufficient data or quantification of the underlying processes. Therefore, this research aimed to take a simpler approach that would work with current data and knowledge. Hence, an index approach was developed using data currently collected by water companies. The next sections describe the methodology used, the model testing process including a case study, the intervention and rehabilitation strategies proposed followed by the conclusions.

Methodology

Modelling Approach

The limited number and type of data available do not support meaningful regression analysis and other statistical techniques for which sample populations would have to be representative of the wider asset base. For such techniques to work it would be necessary to have a detailed understanding of the water abstraction, treatment, transfer, storage and distribution system so that causes and effects may be ascertained. The techniques based on continuous monitoring of turbidity (Vreeburg and Boxall 2007) are limited by the prohibitive cost to maintain expensive equipment throughout the network at all times. Boxall*et el. al.* (2001) developed a cohesive transport model involving the hydraulic forces generated within distribution networks (Vreeburg and Boxall 2007). However, most distribution networks lack a hydraulic model or it is usually not reliable when available. Some companies involved in this research have hydraulic models, built for investment planning, but these models consider typically steady-state or diurnal flow ranges, rather than exceptional conditions such as the velocity generated by a main burst or fire hydrants operation. An index-based approach to assessing discolouration potential was proposed (Figure 9.1). It allows limited data to be combined with expert knowledge and hence can cope with limited or missing data. The approach is flexible in that additional parameters may be added where companies have more data. The technique can be adapted to make use of statutory monitoring data or of more targeted investigative samples. It is suitable for use at different levels of detail in the network, *e.g.* whole systems, a trunk main, treatment and distribution network various points, individual zone, DMA, or pipe. The proposed index approach identifies the main factors contributing to discolouration and hence indicates not only the propensity of the study area to cause discolouration, but also the most appropriate interventions. By examining the change in index values over time it will be possible to indicate if discolouration is becoming more or less likely over time. It therefore lends itself to the estimation of future expenditure requirements to deal with the discolouration problem in a targeted way. The benefits of an index-based approach are:

☆ The technique is transparent and hence auditable;

☆ It can be built from a range of data sources;

☆ It can be designed to still work when some data is missing, also so that it benefits from additional or improved data;

☆ It can combine expert knowledge with data to provide a network state clear indication.

Typical indices take account of several factors when considering the impact of contributing factors on a level of performance (CCME, 2001). For example, for a given level of performance, the magnitude of failure and the number of samples failing will all be taken into account, although recent failures may be more important than historical failures. Hence the important measures (sub-indices) are the number or proportion of samples failing (non-compliance), the frequency, magnitude, and failure regency.

Water supply systems comprise of sources, treatment, storage and pipes, which may have a relatively complex interaction and affect water quality in many ways. Within the distribution network (the part of the water supply system from the treatment works to the customer), the propensity to cause discolouration is represented by aggregating three indices:

1. Water Quality Discolouration Index (WDI) – an assessment of the discolouration potential arising from the quality of water in the network. This indicates the extent to which water entering the network contributes to discolouration.

2. Vulnerability Discolouration Index (VDI) – an assessment of the discolouration potential arising from the distribution network's integrity. This indicates the extent to which the pipes contribute to discolouration.

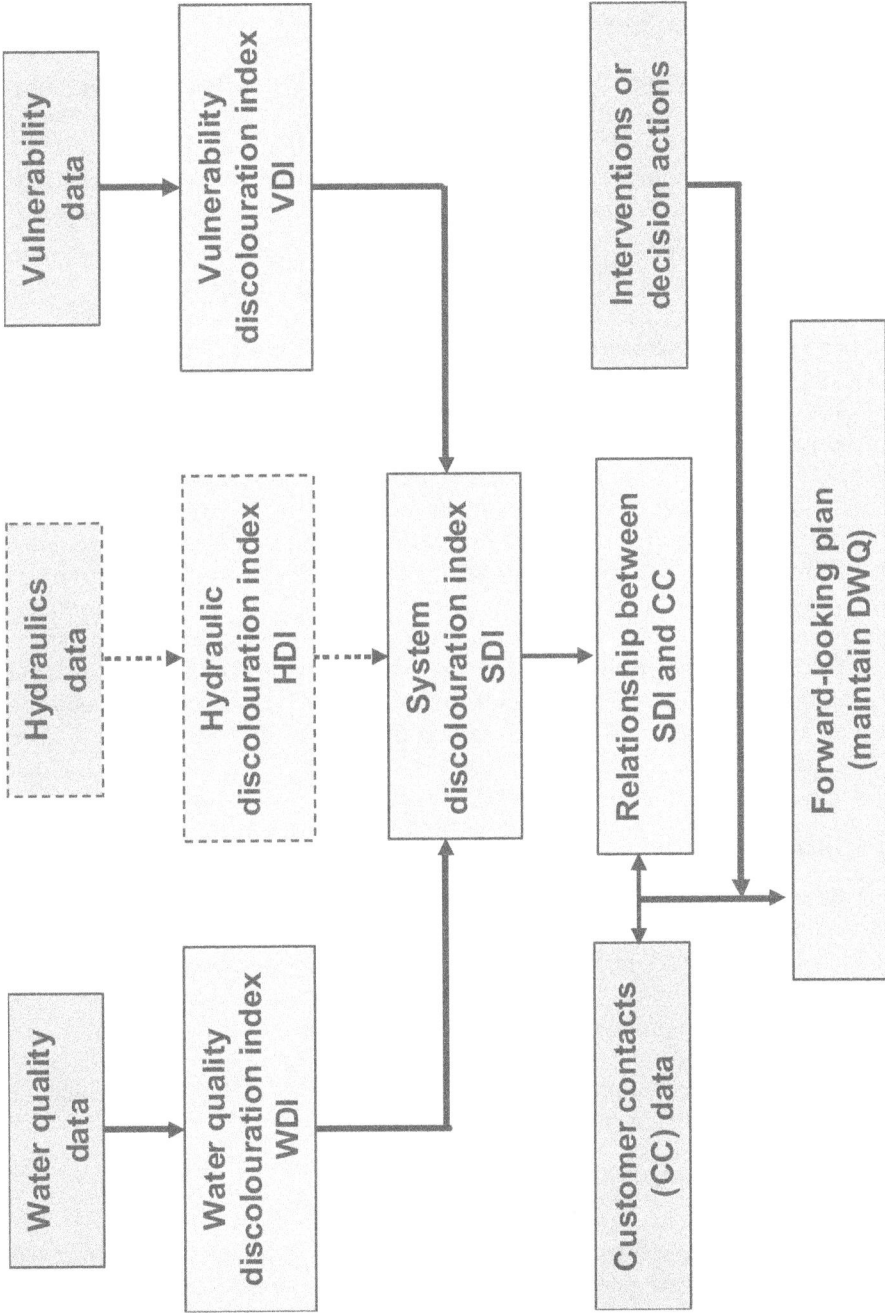

Figure 9.1: Conceptual Water Quality Model.

3. Hydraulic Discolouration Index (HDI) – an assessment of the discolouration potential arising from the system's hydraulic conditions. This indicates the extent to which the network hydraulic features, such as flow in the pipes, contribute to discolouration.

Since these factors are related to root causes, they also help to indicate which interventions are likely to control discolouration potential in the network part being studied. The indices are aggregated to create a system discolouration index (SDI), which accounts for the total discolouration potential of the network part under study. Data requirements and the detailed development of the indices are described thereafter.

A simple logical approach would count the number of samples failing (non-compliant with) a set threshold and use that value as the input to the index. However, in reality there is a broader scale of contribution to discolouration, rather than a fixed cut-off point. For example, as iron concentrations rise so the mass of material available to cause discolouration increases. There is no fixed value representing 'failure' in terms of discolouration potential, as opposed to the maximum prescribed concentration or value (PCV) in the drinking water standards. Hence, it can be said that the PCV represents ultimate failure, but any concentration above the detection limit makes a progressively greater contribution to failure. Determinations falling into the positive detection range but within the PCV therefore pass the standard but also contribute to its later failure – they belong to both the 'pass' and 'fail' group in varying proportions, rather than having a binary pass/fail cut-off. This approach is illustrated in Figure 9.2. The 'ramp' functions defining the progression from pass to fail may slope in either direction or be curved. For some parameters, such as pH, the 'fail' conditions are both sides of an ideal value, in which case the ramp could be triangular or parabolic. This approach based on 'fuzzy logic' (Zadeh 1965; Ross 2004; Sadiq *et al.* 2004) is an important component of the method: it enables the index to differentiate between samples that nearly-fail and those which are clear passes.

Data Collection

As mentioned above, the index-based approach has been developed to make the best use of data that are already collected by water companies. Companies were easily able to provide data for the water quality discolouration and network vulnerability indices, but abstraction of data for the hydraulic discolouration index was more complex. The parameters selected for each index and their justification are listed in Table 9.1. The unit of measure is given, including, where appropriate, threshold values, those that were used in model development. Users may consider varying the threshold values in developing and testing their own models.

Step I - Transformation

Transformation is an important step as it converts all observed measurements into a commensurate scale, thus allowing the parameters' contributions to be compared with each other and aggregated to form a single index. In this research a 'pollution index' type scale [0, 1] has been used, where '0' is the best value (no contribution to discolouration) and '1' the worst value (full contribution to discolouration). The start

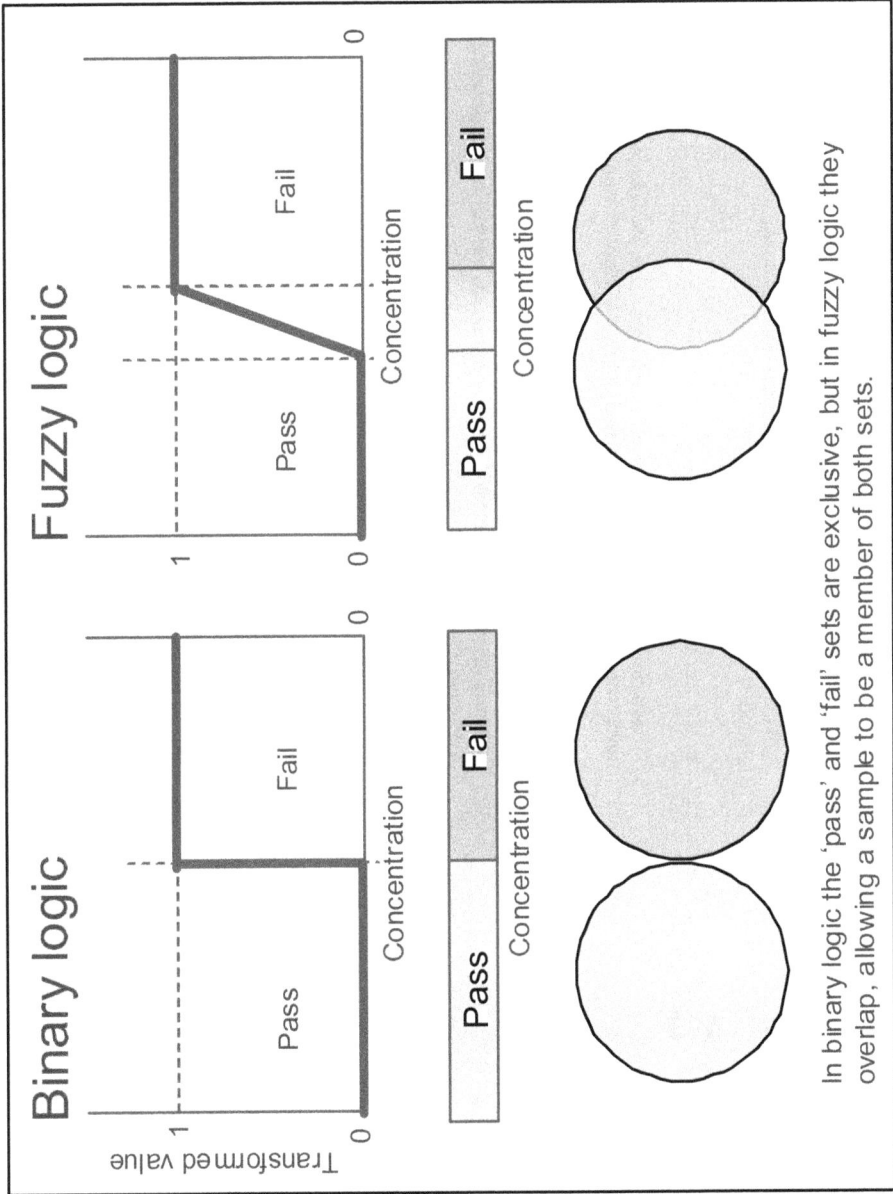

In binary logic the 'pass' and 'fail' sets are exclusive, but in fuzzy logic they overlap, allowing a sample to be a member of both sets.

Figure 9.2: Comparison of Transformations using Binary and Fuzzy Logic.

Table 9.1: Parameters Selected for Each Index and a Brief Justification of their Use

Parameter	Unit	Justification
Water quality discolouration index (WDI): Individual sample results or monthly averages where the sample population was too large for use in a spreadsheet.		
Aluminium	$\mu g.L^{-1}$ Al	These metals may be present in source water or coagulants (Fe and Al). They are commonly found in discoloured water.
Iron	$\mu g.L^{-1}$ Fe	
Manganese	$\mu g.L^{-1}$ Mn	
Turbidity	NTU	These are characteristics of discoloured water and also related to total suspended solids as evidenced by previous work carried out by "WRcplc" company.
Colour	Hazen	
pH	pH	Water companies at the stakeholder workshop stated that pH had a significant effect on the observed discolouration rates.
Hardness	$\mu g.L^{-1}$ CaCO₃	Water companies at the stakeholder workshop stated that hardness had a significant effect on observed discolouration rates.
Vulnerability Discolouration Index (VDI): Summary data about the asset base, which indicate the nature of its construction and overall condition, hence the *likelihood* of it contributing to discolouration.		
Pipe material	Per cent of pipe length of cast iron (CI).	Cast iron is prone to corrosion and may cause discolouration through release of rust. Data may be restricted to *unlined* cast iron.
Pipe diameter	Per cent of pipe length ≤θ4" *or* per cent of pipe length ≤θ6"	Small diameter pipes are more prone to sudden changes in velocity and to bursting than larger or trunk mains.
Burst frequency	Bursts per km, for individual months	The burst frequency indicates the overall condition of the network, and hence the likelihood that CI pipes will have corroded. It also indicates the likelihood of sudden changes in velocity which will re-suspend sediments.

Contd...

Table 9.1–*Contd...*

Parameter	Unit	Justification
Hydraulic discolouration index (HDI): Summary data about the hydraulic characteristics of the system that indicate the likelihood that sedimentation or re-suspension will occur.		
Velocity	Per cent pipe length having a modelled mean velocity of >1.0 m.s⁻¹	Where velocity is unusually high, it is likely to re-suspend sediment that has formed in low-velocity parts of the network. Adjacent areas of significant velocity gradient will be prone to sedimentation and re-suspension as velocity fluctuates with demand.
Total leakage	Customer night flow *or* per cent measured leakage (unaccounted for water in a given zone)	Leaks indicate poor network condition and together with burst rate indicate the likelihood of corrosion and sudden velocity changes.
Water age	Per cent network length in which the water is >1 day old	Older water age indicative of long networks (or poor design which lack circulation of water) causing low velocities, which increase the opportunity for deposits to form and precipitate.
Pressure	per cent WSZ >60m head	Where pressure is high it can easily cause scouring velocities in the vicinity of bursts, demand fluctuations, leaks, or re-suspending sediments that have formed.

Distribution Discolouration Index (D2I): Having prepared the data described above, the index was created through a four-stage approach's) Transforming the values of individual parameters into a consistent commensurate scale; *i)* Calculating sub-indices for discolouration;*iii)* Calculating indices for discolouration; and *iv)* Calculating the system or distribution discolouration index. The steps are illustrated in Figure 9.3.

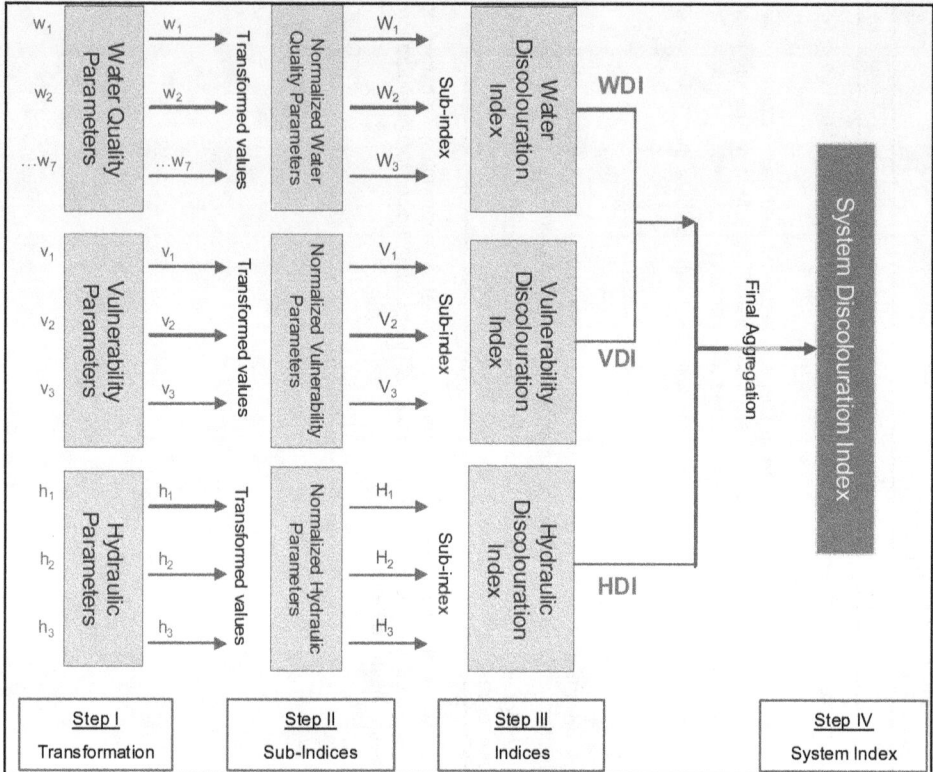

Figure 9.3: Steps in Developing the Index.

and end-points of ramp functions are important factors in determining the output of the index, in the same way as a threshold (cut-off point) would be in the case of a binary logical approach. Three functions were used: increasing ramp, where higher sample results convert to higher index values; a parabolic, where deviation either above or below a preferred value contributes to discolouration; and a decreasing ramp, where higher sample results convert to lower index values (a decreasing contribution to discolouration). The ramp functions are illustrated in Figure 9.4, and their assignment to the data types is listed. The regulatory PCV was initially used as the upper limit for all parameters; because where determinations exceed the PCV the water has failed a statutory standard/guideline. However, these standards are not based on propensity to cause discolouration and so other thresholds were tested.

Step II – Calculate Sub-indices

Each of the three discolouration indices is founded on three sub-indices. The transformed data for all parameters in each category (water quality, vulnerability and hydraulic) from Step I are used to derive the sub-indices, which measure: 1) The *number* of parameters that do not meet the objectives; 2) The *frequency* with which the objectives have not been met; and 3) The *recency* of events in which objectives have not been met. These sub-indices will later be combined to indicate the potential for

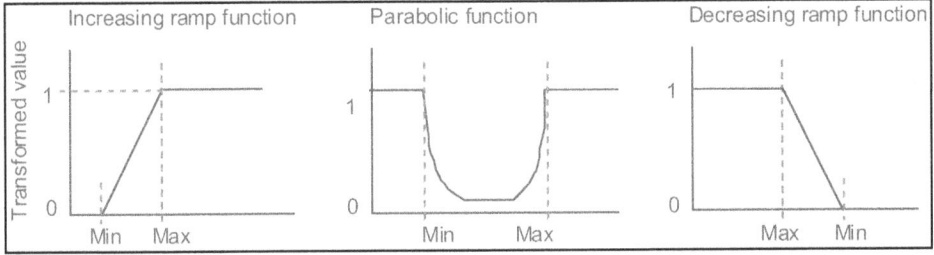

Name	Parameter	Ramp function
W_1	Aluminium	Increasing
W_2	Iron	Increasing
W_3	Manganese	Increasing
W_4	Turbidity	Increasing
W_5	Colour	Increasing
W_6	pH	Parabolic (low centre)
W_7	Hardness	Decreasing
V_1	Pipe material	Increasing
V_2	Pipe diameter	Increasing
V_3	Burst frequency	Increasing
H_1	Velocity	Increasing
H_2	Total leakage	Increasing
H_3	Water age	Increasing
H_4	Pressure	Increasing

Figure 9.4: Illustration of Transformation Functions Assigned to Input Parameters.

discolouration in the area being studied. For example, the conditions in which discolouration occur are more likely to be met in areas where, recently, many parameters have frequently failed to meet objectives. The inclusion of regency also facilitates an indication of the change in discolouration index. By comparing recent and early 'failures' to meet objectives, it is possible to identify improvement or deterioration in the potential to cause discolouration. Although this does not provide a deterioration forecast, it does indicate the rate and direction of recent change. The sub-indices are briefly described as followed:

Sub-index 1 (or factor F_1)

Number of parameters that have not met the objectives: Expressed as a fraction of the number of parameters considered in the calculation (CCME, 2001). It is based on the available data over the predefined time period (the time period being studied, or for which data are available). Its nomenclature is W1 for water quality, V1 for vulnerability, and H1 for hydraulic.

Sub-index 2 (or factor F₂)

Frequency with which the objectives have not been met: Expressed as a fraction of events of non-compliance for all parameters over a given period of time (CCME, 2001). The factor F_2 is calculated by adding all non-compliant events, and dividing them by the total number of observations. Its nomenclature is W2 for water quality, V2 for vulnerability, and H3 for hydraulic.

Sub-index 3 (or factor F3)

Regency of events in which objectives have not been met: It is a comparison of the early and recent failure rate for each parameter, expressed as the ratio of the average of the last 25 per cent of data (by date) to the average of the first 75 per cent of data for each parameter. A second calculation which reverses the ratio, *i.e.* compares the average of the last 75 per cent of samples with the average of the first 25 per cent, was also produced, so the index could be produced to assess recent and historical failure to meet objectives. This is used later to indicate improvement or deterioration. The 25:75 split selection was a matter of judgement; it could be varied if required, for example to increase the emphasis on the more recent events, but the sample population size will be a limiting factor if focusing on more recent results. Its nomenclature is W3 for water quality, V3 for vulnerability, and H3 for hydraulic.

Step III – Calculate Discolouration Indices

Each of the three discolouration indices – water quality discolouration index (WDI), vulnerability discolouration index (VDI), and hydraulic discolouration index (HDI) – is calculated by combining three sub-indices. A range of mathematical approaches to combining the sub-indices was considered (Ott 1978; Silvert 2000; Somlikova and Wachowiak 2001; Kumar and Alappat 2004; Sadiq and Tesfamariam 2007). The method chosen is known as *'root mean additive'* and its variants are commonly used models for rating lakes and streams in Canada (CCME, 2001). The formulation for each index is:

$$WDI = \left(\frac{\left(W_1^2 + W_2^2 + W_3^2\right)}{3}\right)^{0.5} \text{ and } VDI = \left(\frac{\left(V_1^2 + V_2^2 + V_3^2\right)}{3}\right)^{0.5} \text{ and } HDI = \left(\frac{\left(H_1^2 + H_2^2 + H_3^2\right)}{3}\right)^{0.5}$$

(1)

Step IV – Calculate System Discolouration Index

During the initial model development it became evident that hydraulic data would be difficult to extract from system models or monitoring studies in a form that could be related to the other parameters. Therefore, it became important to be able to calculate the system discolouration index (SDI) with or without hydraulic discolouration index availability. In order to make it easier to include or exclude the HDI, the SDI is first calculated on the basis of WDI and VDI, with HDI being added if it is available, as shown in Figure 9.5. The indices are aggregated using compositional rules to form the SDI, by following three steps: 1) Mapping to sets, 2) Composition, and 3) Scoring, described briefly thereafter.

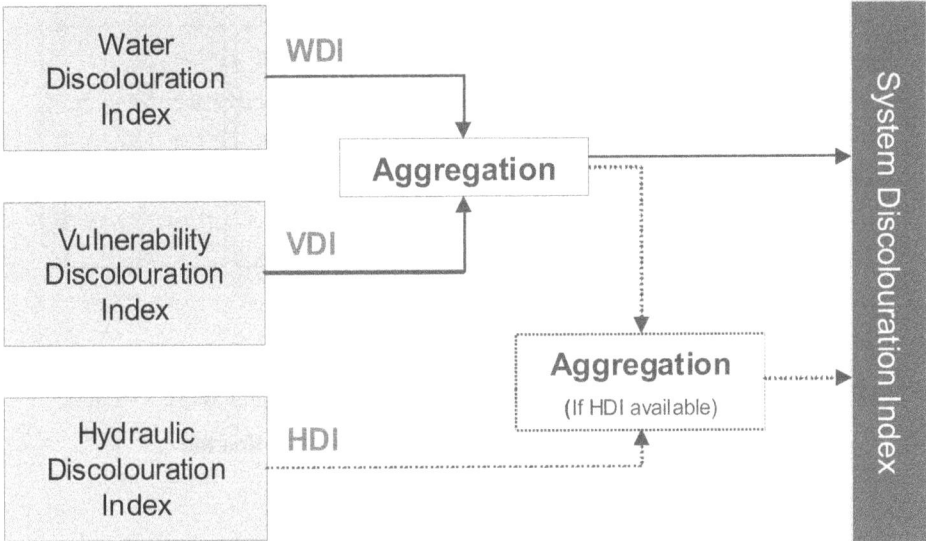

Figure 9.5: Process for Calculating the System Discolouration Index (SDI).

Mapping to Sets (Discretization)

This step apportions the individual indices into three qualitative sets: Low, Medium, and High. The index value is apportioned between the three sets so that total is always 1 (unity). It may belong to two contiguous states ('L' and 'M', or 'M' and 'H' but not all three and not 'L' and 'H' at the same time. The step is similar to the transformation step applied to the original input data, in that it converts the index value from one scale to another, but in this case the process determines how much of the index value belongs to each of the three sets.

Composition

Having mapped the indices onto the high, medium, and low sets, rules are used to decide what value to assign to the system discolouration index, SDI. The rules illustrated in Figure 9.6 shows a rule-base (consisting of 9 rules) defined using judgement, with the following general principles: i) *If all input indices are low, then the SDI is low*; ii) *If all input indices are medium, then the SDI is medium*, and iii) *If all input indices are high, then the SDI is high*. Since the composition stage is a matrix calculation, the resulting SDI is in the same form as the mapped (discretised) indices, apportioned between the low, medium and high sets. Whereas if hydraulic data are available, a second composition step is taken, in which the matrix result of the SDI is then aggregated with the hydraulic index HDI (figure 5) using similar type of rule sets.

Scoring

This step transforms the composition result from its apportionment among the low, medium and high sets to a single value on a continuous scale [0, 1] as followed:

$$SDI = 0.01 \times d_{SDI}^{L} + 0.50 \times d_{SDI}^{M} + 0.99 \times d_{SDI}^{H} \tag{2}$$

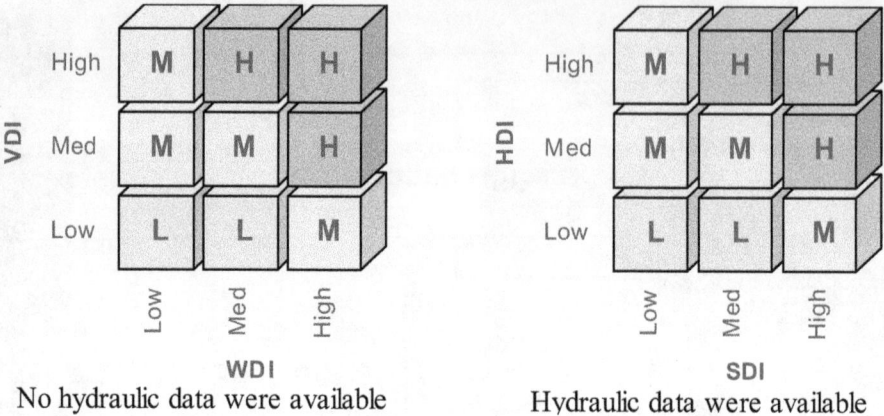

Figure 9.6: Illustration of the Rules Used to Define the System Discolouration Index (SDI).

The choice of weighting is important to the result that is obtained. We chose to weight the sets symmetrically about the medium set, with the greatest weight applied to the high set. This means that the SDI will be strongly influenced by the proportion of the composed indices in the high set, and least influenced by the proportion in low. Hence any study area whose indices are primarily in the low set will receive a low SDI score. Having established the principle of emphasising the high set, the assignment of weighting values was a matter of judgement, informed by previous experience. The result of the scoring step is a system discolouration index in which a high value means that the conditions for discolouration are more likely to have been met. Therefore the SDI indicates that there is a greater possibility of discolouration occurring in those study areas having a high SDI.

Model Testing

Members of a constituted steering group were invited to supply data for a selection of water quality zones, for use in model development. A data set that provided full coverage of the main parameters was chosen and distributions of the sample results for each parameter were plotted, to show the natural cut-off points that defined the typical data range. The median value was initially chosen as the lower cut-off.

Sensitivity to Transformation Function

The water quality, vulnerability, and hydraulic discolouration indices (WDI, VDI and HDI) are a measure of frequency with which the water quality, network vulnerability and hydraulic parameters exceed precise values. Ramp functions define the minima and maxima between which these values must lie. The minima and maxima are set on the hypothesis that the parameters only contribute to discolouration if they exceed certain values. For most of the parameters the initial assumption was that it only contributed to discolouration if values exceeded the 'Most Likely' value. It was also surmised that the discolouration would more likely be evidenced for high values with low probabilities. In general it was assumed that correlation between the parameters and discoloration would be positive and therefore the Maxima were set to the limit of the data. For each of the water quality, network vulnerability and hydraulic parameters a frequency plot of the sample data was drawn. The most likely value was extrapolated and set as the lower limit from which to select as the ramp function minimum and the maximum value was taken as the ramp function maximum. Scenarios were then run for with the ramp function minimum between the lower limit and the maximum.

Over 15,000 scenarios were run with the ramp function minima of each parameter set independently and at random. For each outcome a correlation between the SDI index and the customer contacts about discoloration (CCD) was determined. The outcomes with high correlations were filtered and the data were examined for patterns in the ramp function minima. The data showed very low correlations with R^2 values ranging between 0.01 and 0.10. The data were filtered for correlations with R^2 greater than 0.03 but the resultant data set proved to be too small to analyse with any degree of confidence. As each of the parameters is equally weighted, any correlation will be obscured by the noise of the other parameters and hence high R^2 values are unlikely. If patterns can be spotted with high values of R^2 it is possible to deduce that certain parameters are more likely to correlate than others, in which case weighting them may give a good correlation.

Another simulation with 500,000 scenarios was run where the lower limit for each of the ramp functions was set to zero. The outcomes were filtered for records with $R^2 > 0.025$; this produced a set of 16,000 values. This set was progressively filtered and the ramp function minima of the parameters were analysed. The ramp function minima were calculated as a percentage of the ramp function maxima and were plotted along with their confidence limits. For low R^2 values each of the parameters shows the minima at approximately 50 per cent, which represents the arithmetic average of the scenarios (Figure 9.7a). As the results are progressively filtered for higher R^2 values the ramp function minima move away from the 50 per cent value but this is largely due to the statistical variation in the scenarios which is indicated by the increase in confidence limits. Figure 9.7b shows an example of the variation in the minima when the results are filtered for $R^2 > 0.135$. Here the analysis would suggest that only Leakage and per cent <d4 parameters indicate a move away from the 50 per cent value that is not accounted for in the increased confidence limits. For this particular analysis it may suggest that these parameters have a strong correlation with customer contacts about discolouration.

Case Study

One water company was selected to employ the D2I methodology. This company had submitted the most relevant data (Table 9.2) to allow testing with a reasonable number of Water Supply Zones (83 WSZs). For this study, only regulatory compliance sampling data was used for distribution water quality analysis. We recommend the use of detailed investigational sampling results if they are available, although the impact of using such data was not investigated. Company D was selected because in recent years it had completed a large 'Section 19 programme' and we wished to see how the model coped with a mix of data from rehabilitated and non-rehabilitated

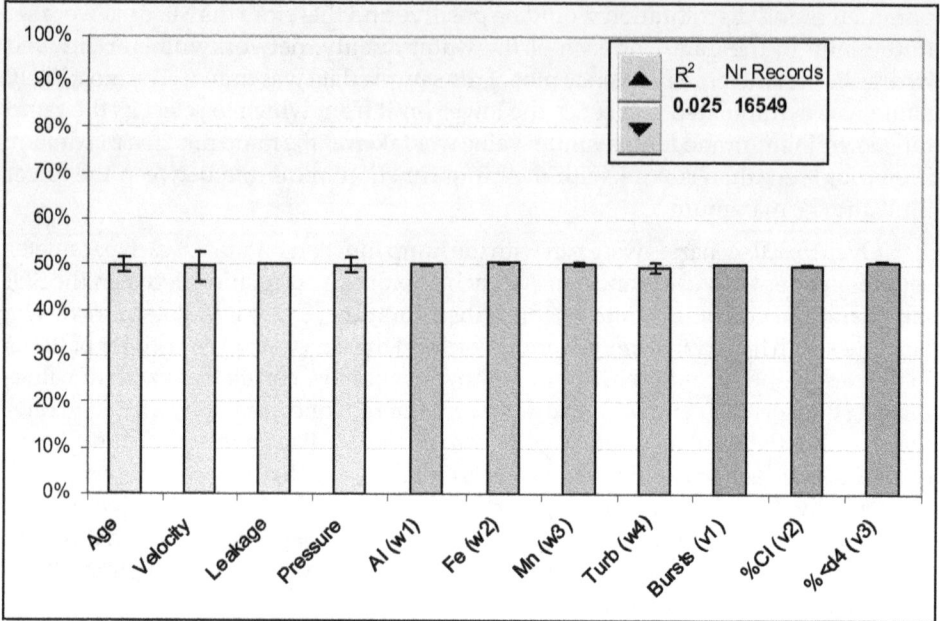

Figure 9.7a: Transformation Function Minima as a Percentage of
Maxima for each Parameter.

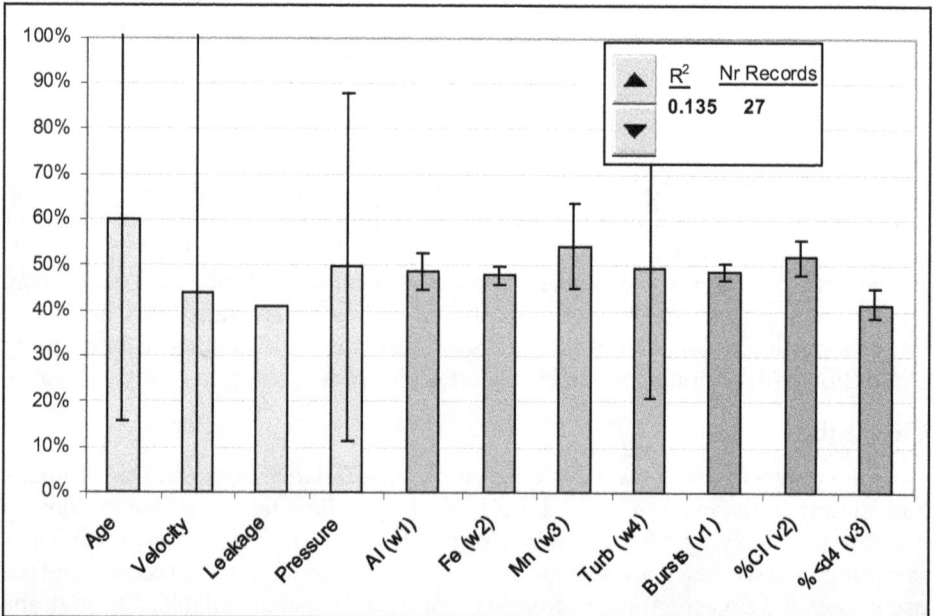

Figure 9.7b: Variation in Ramp Function Minima when Filtering Results for $R^2>0.135$.

zones. The case study was initially conducted without hydraulic data, which were unavailable at the time. It was also carried out using a one-off spreadsheet to calculate the indices for each zone, since the tool had not been developed. The calculations were the same but in the tool the data loading and index calculation has been automated. Note that the D2I methodology is not designed to be a 'one-size fits all' model. It can be altered to suit all data restrictions/availability; however, the more data available the more reliable an outcome will be. The case study attempts to identify problem zones and sub-zones and understand where the water quality failures are occurring before identifying a suitable intervention, as illustrated in Figure 9.8.

Table 9.2: Data Availability for Case Study

Water quality parameters used for estimating WDI	
Al (µg/L)	☆ Data were available for years 2004-2007
Fe (µg/L)	☆ Monthly averages were calculated for each WSZ
Mn (µg/L)	☆ pH and colour data were not available
Turbidity (NTU)	
Vulnerability parameters used for estimating VDI	
Pipe material	**Per cent CI pipe** = (Length of CI pipe in WSZ/Total WSZ length) × 100
Pipe diameter	**Per cent pipe ≤θ4"** = (Length pipe ≤θ4" in WSZ/Total WSZ length) × 100
Bursts	**Bursts** = (Sum of monthly bursts in WSZ/Total WSZ length)
Customer's complaints (CC)	
Yearly counts of customer's complaints in each WSZ.	

In the approach shown in Figure 9.8, the appropriate indices would be calculated up to seven times. The results would provide a clear indication of the discolouration potential source and hence of the most appropriate intervention to address it. Not all indices can be calculated at all points in the network: for example, it is unlikely that VDI or HDI will be appropriate at point 1, the source water abstraction, but WDI will indicate whether the problem is source related. Whereas a high WDI result in distribution, for example at points 5, 6 or 7, might indicate poor quality of the water, incomplete removal of suspended solids at the treatment works, addition of particles by the treatment (Gauthier *et al.* 2001, Vreeburg *et al.* 2004), intrusion 'contaminant'; it might also be due to 'iron pick-up' from the network. This can be confirmed by the results obtained from points 2 (the treated water input to the network) and 1. The result of the analysis will inform the choice of intervention between, for example, improving treated water quality and improving the network condition (through flushing, relining, or replacement). Likewise, the effect of water storage and trunk mains may be assessed by comparing points 2, 3 and 4.

Data Preparation

Data for Al, Fe, Mn and Turbidity were used to determine WDI as Colour and pH data were not available for Company D. Data relating to pipe material, diameter, WSZ length and bursts were used to determine the VDI, as these take into account the pipe network general state for a particular WSZ, as it takes into account the pipe

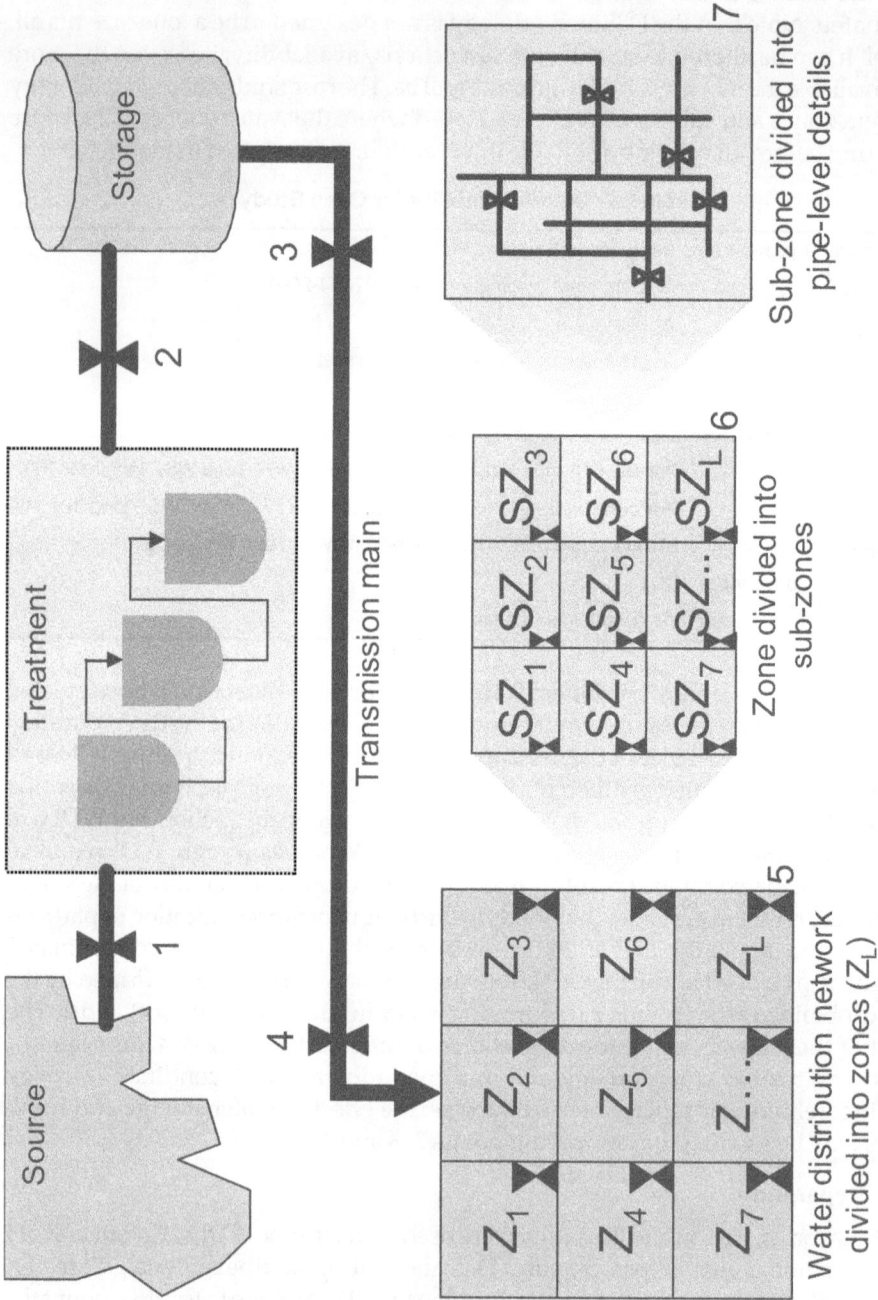

Figure 9.8: Schematic Diagram Showing Application of D2I Model at Seven different Points in the Network to Identify the Cause of Discolouration.

physical failures and its composition. The root mean additive model was used to calculate WDI and VDI. It was assumed that recent trends in discolouration would continue if there were no 'external' changes during the period being studied. However, the source data do not include information about what changes have been made during the study period – such as mains rehabilitation, flushing, or improvements to treatment works – and these should be recorded to help explain the results that are obtained.

The data were arranged in a suitable format so that the various indices could be calculated. A Pivot Table was used to calculate the monthly averages for each water quality parameter for individual WSZs. The results for the W_1, W_2, W_3 and WDI values were copied into a summary spreadsheet. This step was repeated for each WSZ. The same process was repeated with the input data for VDI (bursts, per cent CI pipe, per cent pipe $\leq 4\phi$"), however for the VDI data, only single values for per cent CI pipe and per cent pipe $<= 4\phi$" were required. Once WDI and VDI were calculated, the SDI values were also copied into the summary spreadsheet, to finally provide a summary spreadsheet similar to Table 9.3.

Table 9.3: Example Summary Spreadsheet

WSZ	W_1	W_2	W_3	WDI*	V_1	V_2	V_3	VDI*	SDI	CC
1	0.25	0.07	0.25	0.21	0.38	0.02	0.83	0.53	0.33	91
2	0.09	0.02	0.00	0.05	0.19	0.22	0.00	0.17	0.09	8
3	0.20	0.06	0.00	0.12	0.48	0.03	0.82	0.55	0.26	41
4	0.09	0.03	0.00	0.06	0.88	1.00	0.00	0.77	0.21	3
...	...	...	...	...	...	...	...	...	...	...

* The results are provided only for the RMA calculation approach

Identification and Analysis of Problem Zones

Problem zones were identified using a 'classification' approach. This identifies the zones where there is broad agreement between the model and observed customer behaviour, as well as those areas where the monitoring data used to generate the index do not support the level of customer contacts received. To classify the zones, the results of each index were assigned to one of three categories based on 25th and 75th percentiles:

☆ Class 1: $x < 25$[th] centile;

☆ Class 2: 25th centile $< x < 75$[th] centile;

☆ Class 3: $x > 75$[th] centile.

The same approach was then taken for customer contact data, categorising the number of customer contacts into the same three percentile categories. The first two rows of Table 9.4 show the distribution of the indices between the classes. Where the model forecasts customer contact frequency, then it would be expected that an upper quartile zone in customer contact terms (*i.e.* class 3, $x > 75$[th] centile) would also be the upper quartile of index results. Table 9.4 shows the percentage of zones where the classes matched and where they did not. The indices matched reasonably well because only a small proportion were two classes different. The W_2 sub-index (frequency with which water quality parameters were not met) provided the best match with customer contacts.

Table 9.4: Identification of Problem Zone – An Example

Percentile	Calculated Results for Each Sub-Index, Index, and Customer Contacts									
	W_1	W_2	W_3	WDI	V_1	V_2	V_3	VDI	SDI	CC
Class 1 threshold (25th centile)	0.117	0.048	0.001	0.073	0.467	0.045	0.001	0.313	0.179	28
Class 2 threshold (75th centile)	0.280	0.065	0.25	0.192	0.712	0.231	0.385	0.572	0.274	305

Percent Matching Classes *

CC**	Showing the Count of Zones in Each Index Class that Match the Associated Class for Customer Contacts									
	W_1	W_2	W_3	WDI	V_1	V_2	V_3	VDI	SDI	CC
0-Diff	51	64	24	45	33	28	31	41	36	**100**
1-Diff	41	35	52	48	48	45	57	39	52	0
2-Diff	8	1	24	7	19	28	12	20	12	0

* Class 1 = less than 25th percentile value.

Class 2 = between 25th and 75th percentile values

Class 3 = more than 75th percentile value

**0-Diff: exact match between customer contact and index classes; 1-Diff: Difference of one class between customer contacts and index; 2-Diff: Difference of two classes between customer contacts and index.

The process above may be used to develop a strategy for investigation and more detailed analysis. For example, 'problem zones' would be those where both the index value and customer contacts are high; 'at risk' zones would be those where there are fewer customer contacts than might be expected. The 'at risk' zones would be subject to more detailed investigation such as flushing and sampling, whereas the problem zones would be put forward for intervention planning.

To assess these zones in more detail, the indices were calculated from the original sample results, rather than from monthly averages as done initially, because the full data set could be handled for the short-listed zones. (Now that the tool has been developed it would be feasible to calculate the index from individual samples for all zones, subject to the limit of 63,000 rows in Microsoft Excel.) This more detailed investigation allowed the individual sample points to be reviewed, where sufficient data were available. At the sub-zone level pipe data were not available; therefore it was not possible to calculate the VDI. Hence only the WDI is used to compare with the distribution water quality results. In the case of the development data set the data were for many different sample points within the zone. It is likely that different materials and flows are present in different parts of the zone and this might explain some of the variability in results. Only W_1 and W_2 were calculated, because *recency* (W_3) cannot be calculated at an individual sample level. The results show which parts of the zone are contributing to high index values, although individual sample results are likely to depend on the conditions prevailing at the time of sampling. A further development of the approach would be to use geographic analysis to cluster samples according to the part of the zone they represent, so that a larger, but related, sample is built. A suitable intervention could be identified at this level, to remedy the problem for that part of the zone.

The WDI was calculated for the treated (finished) water using monthly averaged data collected from the treatment plant. At this stage there is no VDI to be calculated, because there are no pipes. These results can then be compared with the distribution water quality results at various stages along the distribution system to highlight the discolouration problems caused by old water mains or at the water treatment works (Figure 9.8).

Relationship among Various Indices and Customer Contacts

The estimated values for various discolouration indices (sub-indices such as $W_1, W_2, W_3, V_1, V_2, V_3$; indices such as WDI and VDI, and system index SDI) were used to develop an empirical model for customer contacts. The indices were estimated based on monthly averaged data derived from Company D's database. The data from 83 zones were divided into two sets, namely training and testing, by randomly selecting 58 values (*i.e.* 70 per cent) for training and the remaining 25 values for testing. A 'hit and trial' method similar to step-wise regression was used. Two goodness-of-fit criteria, namely mean absolute error (MAE) and coefficient of determination (R^2) were used to select the best-fitted model (Table 9.5). The following empirical model was obtained after numerous trials:

$$CC = 9e^{\left(5.6W_1 - 2.3\,W_2 + 74.9\,W_3 - 0.4\,V_1 - 1.7V_2 - 0.6V_3\right)} \tag{3}$$

where

CC: Customer contacts

W_1: Number of WQ objectives not met Î [0, 0.6]

W_2: Number of times the WQ objectives were not met Î [0, 0.1]

W_3: Recency of non-compliance of WQ objectives Î [0, 0.25]

V_1: Number of vulnerability objectives not met Î [0, 1]

V_2: Number of times the vulnerability objectives were not met Î [0, 1]

V_3: Recency of non-compliance of vulnerability objectives Î [0, 1]

The approximate ranges for sub-indices are also indicated (Equation 3). While developing empirical models the available data only varied within these ranges, therefore the use of this model should be very specific to those ranges. Figure 9.9a shows the model fitness based on training and testing data sets respectively. Equation

3 is also used to compare the prediction of customers' complaints based on the same classification as described previously. For both the training and testing data sets, more than 75 per cent of the time the classes were exact matches. There were only a few incidences of the class difference between the actual and predicted CC being of one or more classes. It is also worth mentioning that 0-Diff results are significantly improved by fitting the empirical model (compare Tables 9.4 and 9.5).

Table 9.5: Goodness-of-Fit Criteria Used for Selection of Empirical Model

Data Set	Number of WSZ	R2	MAE	0-Diff	1-Diff	2-Diff
Training	58	0.55	70	77	8	15
Testing	25	0.43	91	76	19	5

R^2 = Coefficient of determination.

MAE = Mean absolute error.

0-Diff: Percent exact match.

1-Diff: Difference of one class level expressed in percentage.

2-Diff: Difference of two classes expressed in percentage.

At the end of the study hydraulic data were received and used to calculate the HDI for the zones being studied. The approach was identical to that used for the WDI and VDI. SDI composition with hydraulic data included was carried out as shown earlier in Figure 9.5, by calculating SDI on the basis of VDI and WDI, and then the extended index, XDI, from SDI and HDI. The HDI inclusion did not have a significant effect on the results. Further work would be necessary, to review the choice of parameters and transformation functions applied in calculating HDI. The class-difference approach described earlier could be used to identify zones for further investigation in order to improve the understanding of the hydraulic effects.

Sensitivity Analysis

Sensitivity analysis is the process of estimating the degree to which the output of a model changes, as values of input parameters are changed (Cullen and Frey 1999). The empirical model sensitivity is linked to sub-indices through the governing Equation 3. The sub-indices may be represented by given ranges as provided (Eq. 3) and identification of inputs significantly contributing to output variance gives the analyst an awareness of which sub-index is controlling the outputs. The scatter plot, partial and rank correlation coefficients, multivariate regression, and contribution to variance and probabilistic sensitivity analysis are some among various methods of identifying key input variables from model outputs (Cullen and Frey 1999). Figure 9b shows the percent contributions of six sub-indices towards the variability of customer complaints. It appear that more than 50 per cent contribution is for number of times (frequency) water quality parameters are not met (W_2). All other sub-indices contributions roughly range from 6 to 10 per cent. These results emphasise the need to control repetitive violations or higher values of selected water quality parameters.

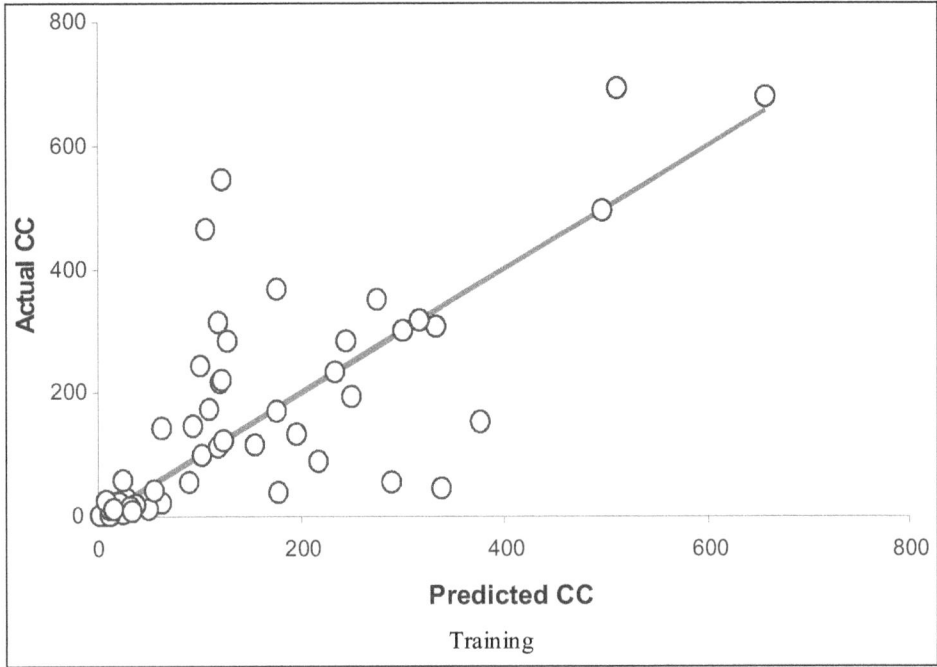

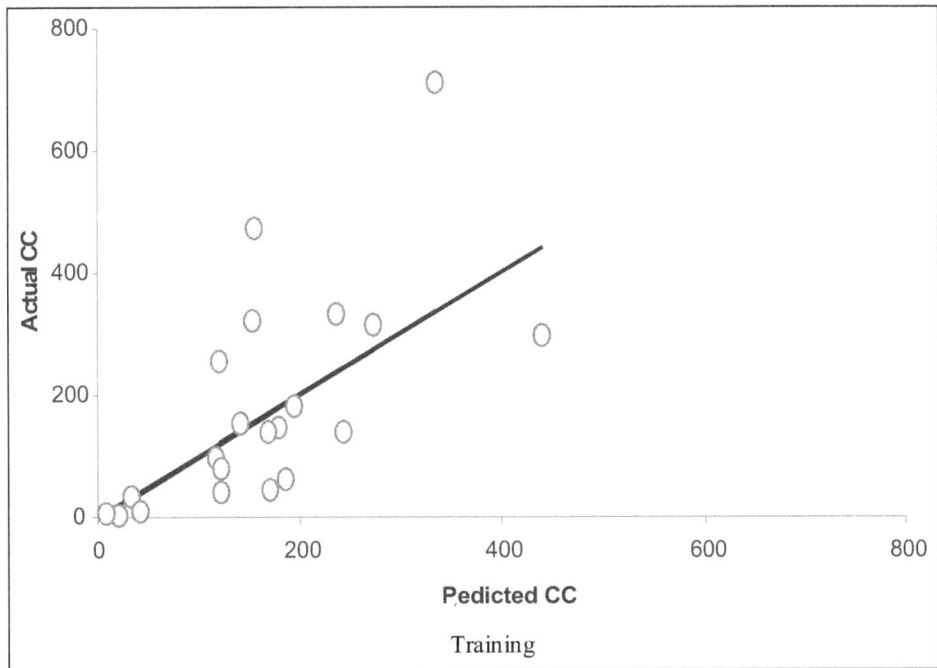

Figure 9.9a: Fitting of the Empirical Model for Customer Contacts as a Function of Sub-indices of Discolouration.

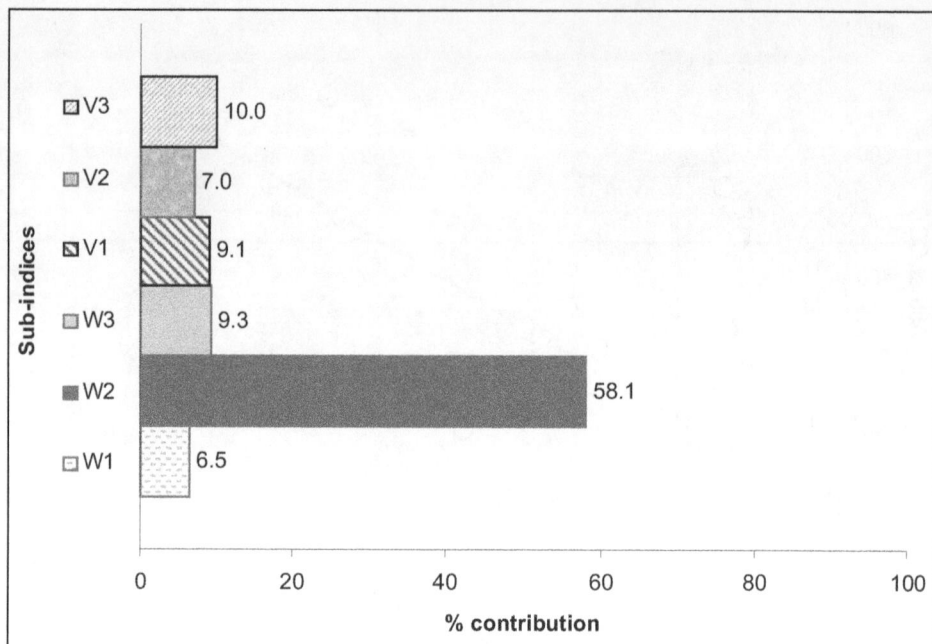

Figure 9.9b: Studying the Sensitivity of Sub-Indices for CC.

Intervention and Rehabilitation Strategies

When problem zones (DMAs, treatment works, service reservoirs, or pipes) have been identified, an intervention can be planned. The intervention choice should balance cost and efficacy in protecting the level of service to customers. Some interventions that reduce discolouration will also provide other customer service benefits. For example, pipe replacement may also reduce bursts, leakage, interruptions to supply and low pressure. The inclusion of other benefits is not reviewed in this study but should be considered in compiling a business plan. This section reviews the calculation of intervention lifespan and effects, and a simple intervention matrix derived from a stakeholder workshop held.

Assessment of Interventions

Many factors affecting the appropriateness of an intervention are uncertain and lead to uncertainties in the cost and benefits of interventions. Most interventions are unlikely to eliminate future discolouration without any subsequent interventions being required. The extent of the effect and its duration – the time acquired between an intervention and discolouration recurring – is a factor of its impact on the discolouration root cause in the studied area. Figure 9.10 illustrates the main steps in calculating the acquired time (AT) and the impact of uncertainty. Uncertainty affects the whole process since neither the starting value ($W_1 T_0$) nor the current deterioration rate is likely to be certain and therefore, at time T_1, the water quality will be uncertain. The figure shows how an intervention is implemented after time T_1, but as a result of the uncertainties in the original estimates and in estimating the impact of the

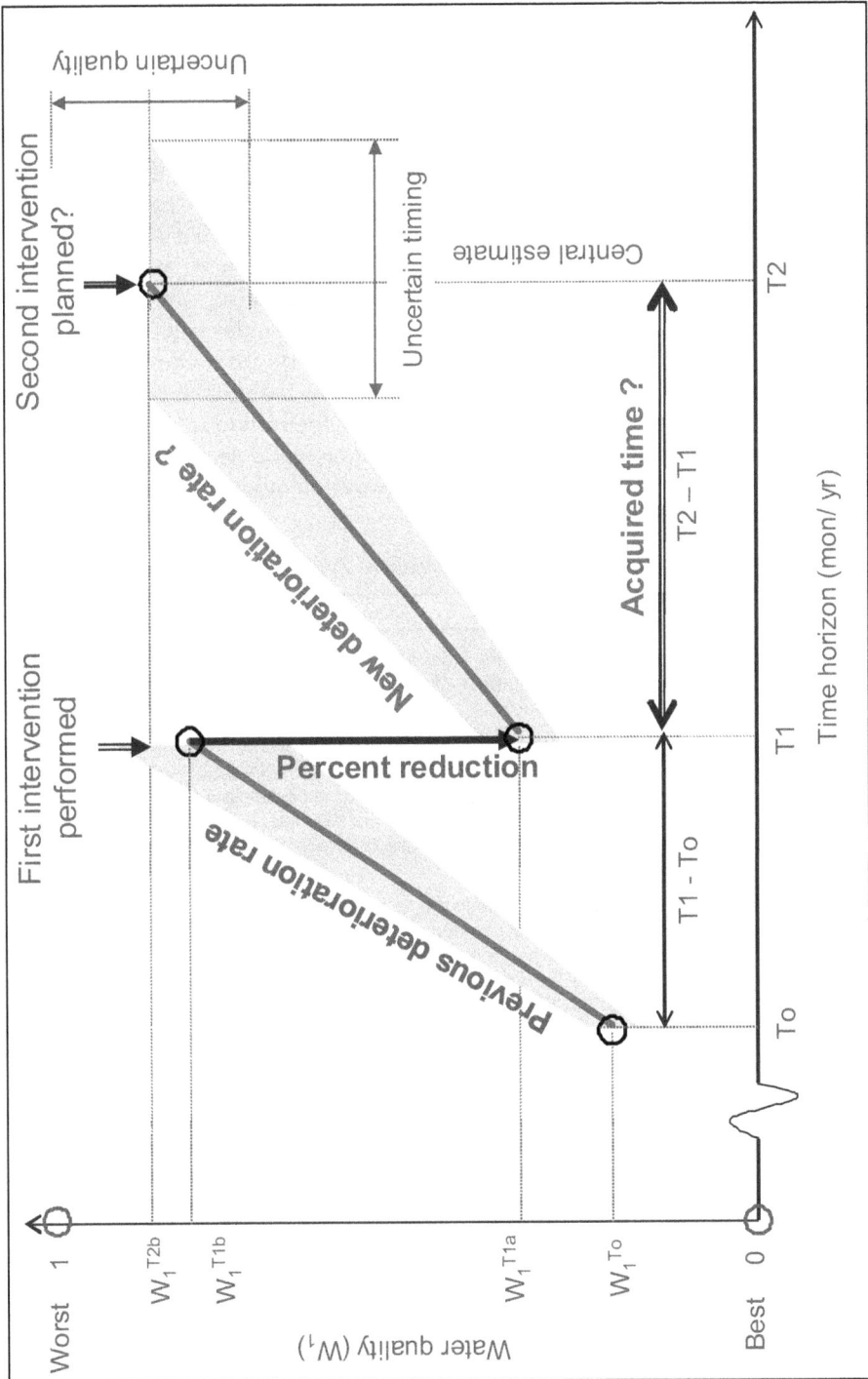

Figure 9.10: Steps and Uncertainty in Assessing the Time Acquired by an Intervention.

intervention, the new water quality is also uncertain. The deterioration rate post-intervention may differ from the previous deterioration rate, depending on whether or not the intervention addresses the root cause. The result of the sequence of uncertain effects is that there may be considerable uncertainty in the AT by the selected intervention. This may have a significant impact on the whole-life cost assessment or, for a given cost, on the level of service attainable.

Figure 9.10 also shows the parameters, described in Table 9.6, that need to be included in the intervention analysis. When these parameters have been estimated, together with their uncertainties, it is possible to estimate the time that will be acquired by an intervention, and hence the intervention frequency, long-term cost, and level of service. A linguistic approach was built into the forward-looking tool. The impact of an intervention is defined in terms of its effect on the current index value, in categories such as 'large reduction'. Each category has a range of index reductions assigned to it by expert judgement. This enables companies to follow the process of making their time acquired best estimate by different interventions for individual zones, although the detailed data for calculation of the values are not available. In the longer term companies should aim to collate data from deterioration models and other relevant sources to further develop the analysis.

Table 9.6: Essential Parameters to include in the Intervention Analysis

Parameter	Description
Current index value	The current value of the index being modelled (*e.g.* WDI, VDI, HDI, and SDI). It forms the starting point of the analysis and should be based on the most robust data available.
Previous deterioration rate	The rate of deterioration which has prevailed and will continue if no intervention (or other change) takes place. In the case of hydraulic or vulnerability indices, this may be informed by varying the input parameters of the index based on asset deterioration model outputs. For example, burst rate models or demand models may provide data to use in forecasting HDI and VDI from current values.
Reduction achieved by intervention	It is necessary to estimate the reduction in index value that each intervention will deliver. This may be derived from the effect the intervention will have on individual input parameters for each index. For example, a new pipe will reduce the burst rate; changing zone configuration will change the hydraulic performance. The changed parameters may be substituted for the existing values in order to evaluate the impact on the relevant index.
New deterioration rate	The deterioration rate after intervention will not necessarily be the same as that before, depending on the nature of the intervention. For example, deterioration of a new MDPE pipe will not affect discolouration in the same way as that of an unlined iron pipe.
Index before first intervention	The maximum value of the relevant index (WDI, VDI, HDI, or SDI) must be decided, in order to determine when the first intervention will be required (on the basis of the starting index value and deterioration rate). However it will also be necessary to understand the linkage between index values and level of service, for example as discussed previously.
Index before repeat intervention	If the current index value is too high (*i.e.* the level of service is poor) then a lower value may be chosen as the threshold for subsequent interventions. The acquired time will depend on the value achieved by the initial intervention and the future deterioration rate, subject to uncertainty.

Intervention Matrix

A matrix of interventions was constructed, based on the results of the stakeholder consultations. It was then simplified (Table 9.7) through an iterative process, to reduce the overlap between the options, while maintaining a good spread of capital and operating interventions. The interventions (and causes addressed) that were not included were removed on the basis of relevance to discolouration, overlap with other options, or because the relationships were unlikely to be quantifiable in the short to medium term. Apart from the extent to which an intervention directly addresses the discoloration cause, its efficacy will depend on the context in which it is applied, and whether it is applied in combination with other options. For example, relining a pipe will fully address its corrosion, but if upstream pipes accumulate particles, then railing's effect might behave more like that of flushing, as particles move downstream. This could lead to more frequent discolouration events since in the newly lined surfaces layer strength of these particles is weak (Vreeburg and Boxall 2007, Boxall *et al.*, 2001). Pairs of interventions and their effects on discolouration are given in Table 9.8. In some situations, three interventions might be considered together. For example, where discolouration results from pipe corrosion and treated water, it might be necessary to implement vulnerable pipe relining, improved treatment, and mains cleaning to remove the previously accumulated particles. Such combinations will be considered in more detail as rules for selecting interventions in modeling are developed.

It might be necessary to sub-divide some interventions in order to explain their efficacy in different circumstances. For example, customer-reactive flushing (network flushing realized only after a threshold level of complaints is reached) depends on a degree of failure of customer service, *i.e.* there are enough complaints to reach the threshold. But reactive flushing in response to targeted sampling could be arranged so as to remain below the threshold of customer complaints, as could flushing on a fixed timescale dependent on the particle accumulation rate. Table 9.8 does not sub-divide interventions according to the circumstances in which they are applied. This refinement may be introduced as companies develop their modelling approach. The interventions in Table 9.8 have been included in the default settings for the D2I tool and companies may add others as they consider necessary.

Cost Data

Development and application of the D2I method will help companies to identify problem zones where the conditions for discolouration to occur are likely to be met, together with an indication of the likely underlying cause. The ultimate selection of an intervention should follow the UKWIR common framework approach on a whole-life basis, considering the impact on service and cost, and customers' willingness to pay for the service level. It will therefore be necessary to include the intervention cost in the analysis, at the level of detail at which the intervention is being planned. The D2I tool enables interventions to be reviewed in terms of their costs and impacts over 25 time steps.

Table 9.7: Simplified Intervention Matrix

Intervention	Extent to which the Cause of Deteriorating Quality is Improved									Time Scale of Effect
	Pipe Lining	Pipe Material	Pipe Condition Quality	Source Water	Sedimentation/ Deposition	Corrosion	Treated Water	Flow	Bio-film	
Flushing or mains cleaning	⊗	⊗/Δ	⊗	⊗	⊗	⊗	⊗	⊗	⊗	Immediate, short term
Mains replacement/relining	•	•	•	O	O	•	O	O	O	Long term
Water filters	⊗	⊗	⊗	⊗	⊗	⊗	⊗	⊗	O	Immediate
Change (improvement) in source	O	O	•	•	•	•	•	O	•	Long term
Change (improvement) in treatment	O	O	O	•	•	O/•	•	O	•	Long term
Rezoning	O	O	O	O/•	O/•	O/•	•	•	•	Immediate

Key: O No effect; ⊗ Addresses symptoms and cause; • Addresses symptoms only; Δ May increase deterioration effect of the cause.

Table 9.8: Pairs of Interventions and their Effects on Discolouration

Option	Flush/ Clean Mains	Reline/ Replace Pipes	Water Filters	Change (Improve) Source Water	Change (Improve) Treatment	Rezoning
Flush/clean mains	Removes local deposits. Does not address cause, so re-accumulation will occur. Might expose unprotected metal and lead to 'bleeding' or increased rate of corrosion.	Stops corrosion-related discolouration and removes deposits in non-vulnerable pipes, which have accumulated downstream. Can provide a long-term solution, if no deposits arising from treatment.	Mitigates the effect of discolouration for individual properties or small groups of properties. Flushing/ cleaning controls the rate of discolouration for those customers without filters. Flushing also helps to extend filter life.	Stops source-related discolouration and removes deposits in non-vulnerable pipes, which have accumulated downstream. Can provide a long-term solution if rate of corrosion not so great as to make flushing ineffective or uneconomic.	Stops treatment-related discolouration and removes deposits in non-vulnerable pipes, which have accumulated downstream. Can provide a long-term solution if rate of corrosion not so great as to make flushing ineffective or uneconomic.	Controls flow, addresses water age- or composition-related discolouration, while flushing/cleaning removes deposits that have already accumulated. Cleaning might expose unprotected metal and lead to 'bleeding' or increased rate of corrosion. Does not address treatment issues.
Reline/replace pipes	–	Stops corrosion-derived discolouration. Deposits in untreated sections may still be re-suspended and cause discolouration.	Unlikely combination.	Stops corrosion-derived discolouration. Reduces or stops source-derived deposition. Deposits in untreated sections may still be re-suspended and cause discolouration.	Stops corrosion-derived discolouration. Reduces or stops source-derived deposition. Deposits in untreated sections may still be re-suspended and cause discolouration.	Controls flow, addresses water age- or composition-related discolouration, while relining/replacing stops corrosion-derived discolouration, removes previously accumulated deposits. Can provide a long-term solution but does not address treatment issues.

Contd...

Table 9.8–Contd...

Option	Flush/ Clean Mains	Reline/ Replace Pipes	Water Filters	Change (Improve) Source Water	Change (Improve) Treatment	Rezoning
Water filters	—	—	Mitigates the effect of discolouration for individual properties or small groups of properties. Does not remove existing deposits or address the cause.	Mitigates the effect of discolouration for individual properties or small groups of properties. Changing source quality prevents further deterioration but does not address deposits already in the network (Unlikely combination).	Mitigates the effect of discolouration for individual properties or small groups of properties. Changing source quality prevents further deterioration but does not address deposits already in the network. (Unlikely combination.)	Controls flow, addresses water age- or composition-related discolouration, while filters mitigate the effect of discolouration for individual properties or small groups of properties. Does not remove existing deposits.
Change (improve) source quality	—	—	—	—	Unlikely combination/same effect.	Unlikely combination without flushing to remove accumulated deposits.
Change (improve) treatment	—	—	—	Prevents further deterioration but does not address deposits already in the network.	Prevents further deterioration but does not address deposits already in the network.	Unlikely combination without flushing to remove accumulated deposits.
Rezoning	—	—	—	—	—	Controls flow and reduces water age. Can effectively change the source water. Does not address deposits already in the network.

Overview of the D2I Tool

The D2I tool is an analytical tool developed to automate the process of calculating the indices for each zone in an imported data set, and to enable basic intervention analysis and costing to be carried out. The tool follows a workflow process shown in Figure 9.11. Each step from setting up the tool to reviewing the plan is described in detail in a guidance notes. In summary, the first step, data preparation, is to populate the data import sheet with the appropriate source data for modelling. A flat-file approach (the complete modelling data set is prepared on a single worksheet) has been adopted. The users may modify some settings, including the transformation functions and data headings for each imported data column. This provides flexibility but also transparency in that the functions may be viewed and edited directly. The tool settings should be changed, if required, before importing data. Once the data are imported the D2I is calculated immediately and the user may then review the individual zones (or other study areas). The index results are displayed on a zone-by zone basis as shown in Figure 9.12a. The bar chart shows the SDI calculations, using the two regency values (biased to first 25 per cent, and last 75 per cent of samples). Whilst this does not constitute a full deterioration analysis it does indicate the extent to which the index has improved or deteriorated during the study period and this change is categorised into five bands from deteriorating quickly to improving quickly.

An intervention option, or combination of interventions, may be identified for each zone (Figure 9.12b). Each intervention cost and lifespan is entered and the tool

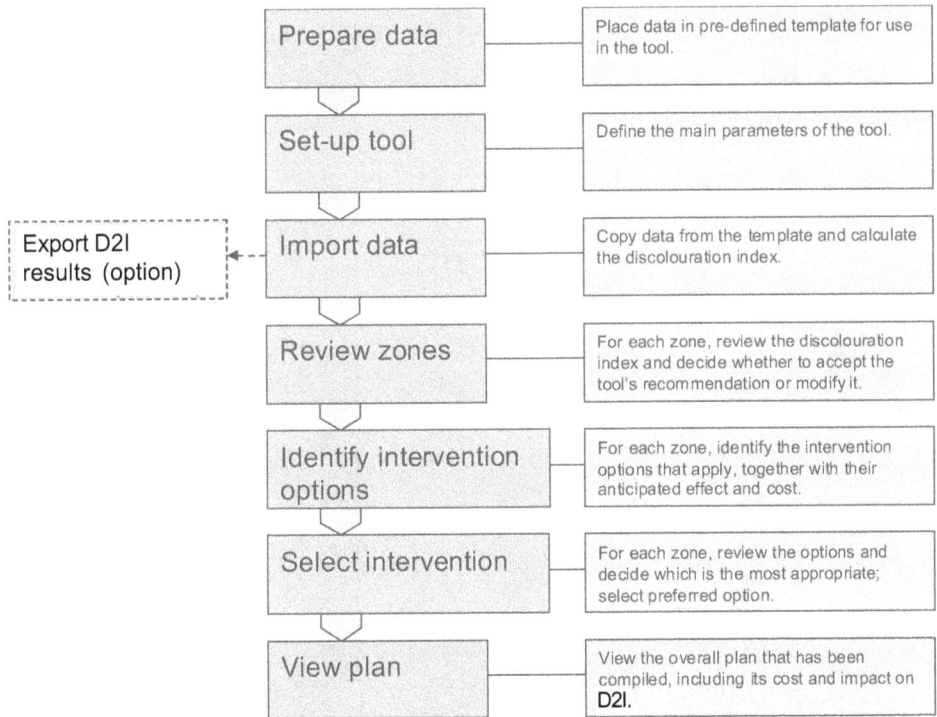

Figure 9.11: DOMS Tool Workflow Process.

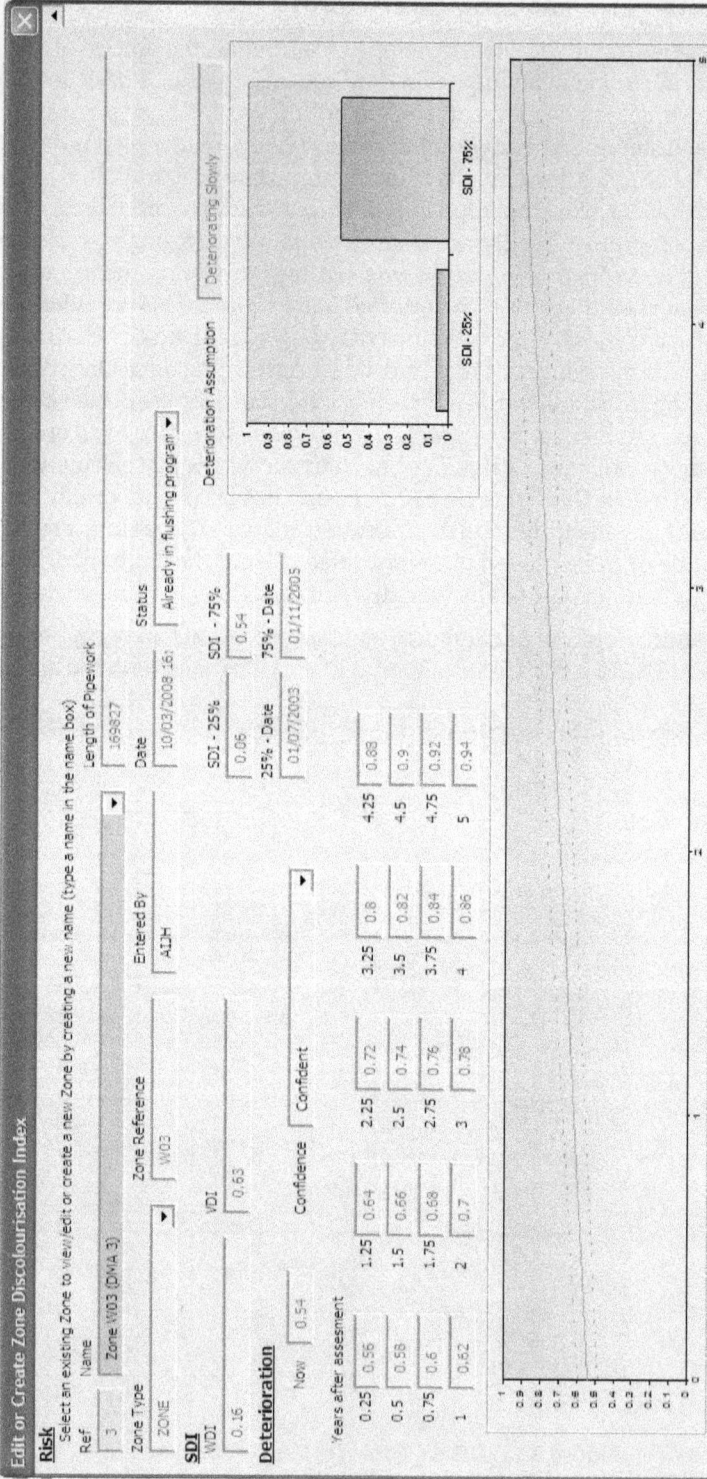

Figure 9.12a: D2I Results for an Example Water Supply Zone.

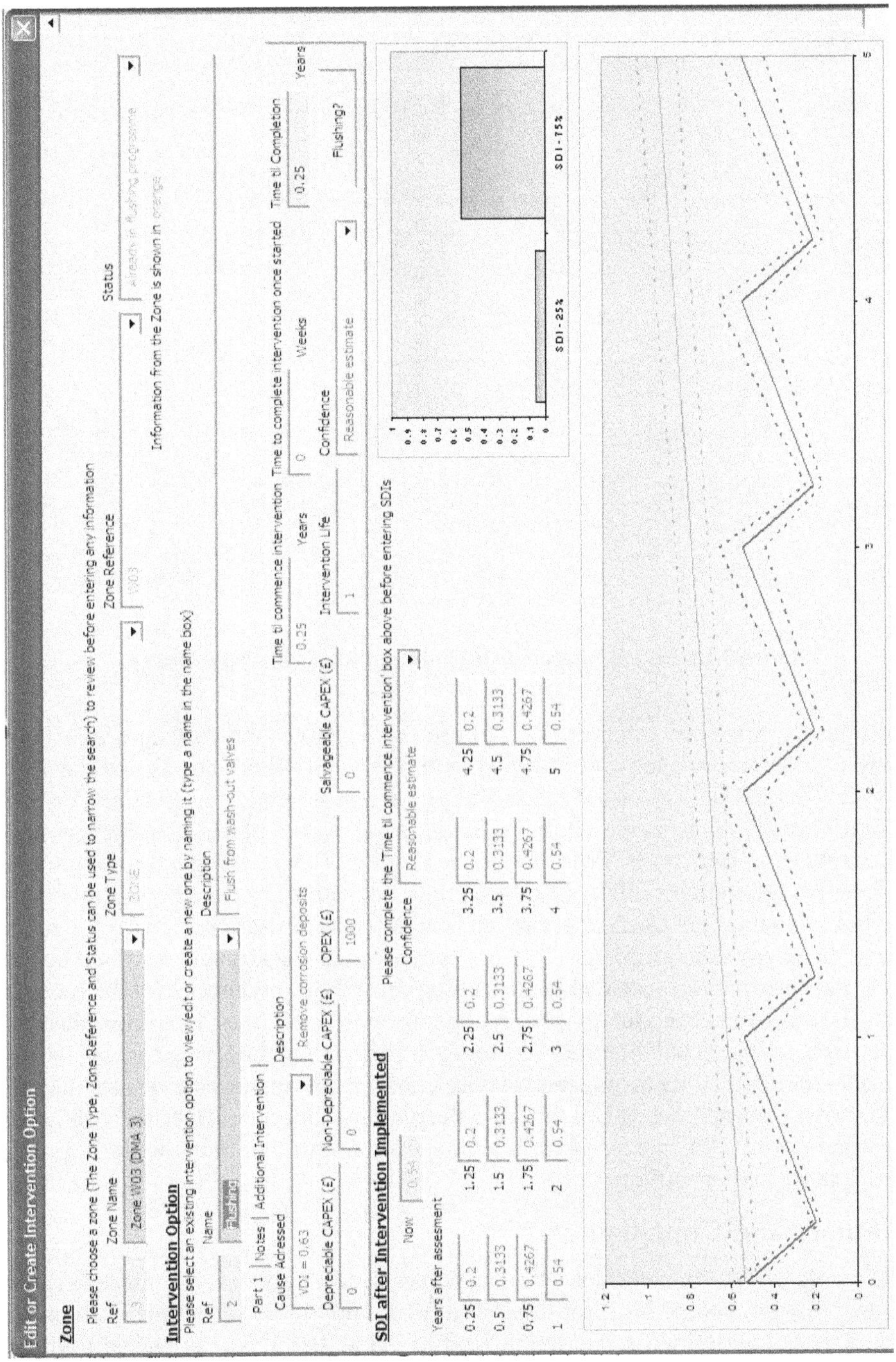

Figure 9.12b: Specifying Flushing Parameters in the Tool.

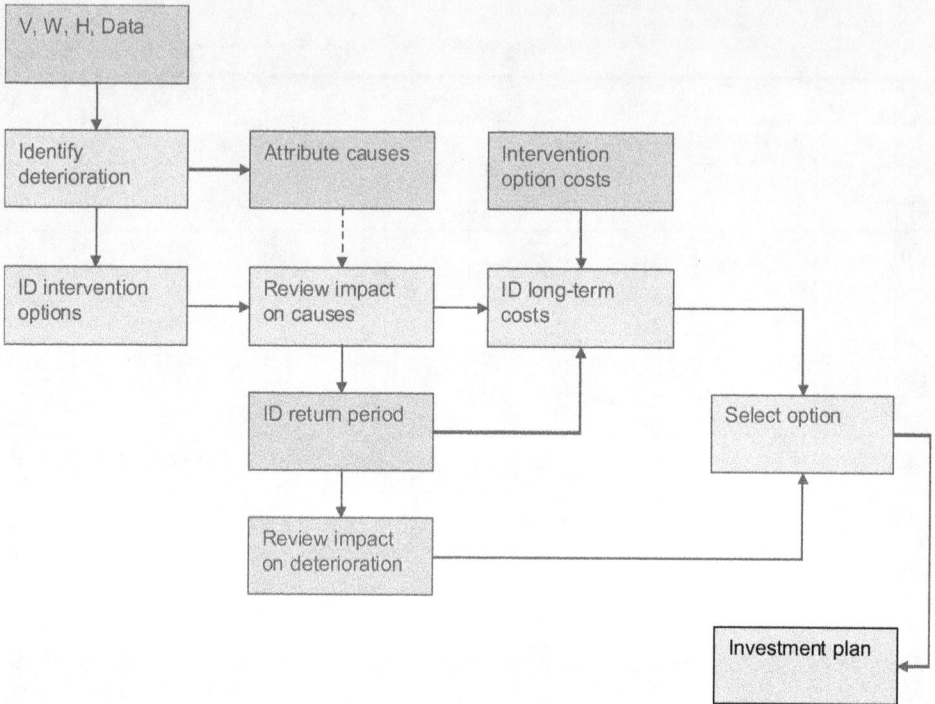

Figure 9.13: DOMS Analysis and Intervention Planning Process.

illustrates the specified intervention pattern. This enables the costs and effects of potential interventions for each study area to be visualised on the basis of the data entered. The cost data are used for assessing each intervention normalised or Net Present Value costs, from which the tool builds the overall plan. Each intervention may have up to four parts. This enables real options to be assessed, for example treatment improvement followed by flushing; or temporary flushing followed by network rehabilitation. Users select an intervention from those specified for the zone. To help decide which option to select, the tool presents data the user entered about the capital and operating costs, effects and lifespan of the interventions for the specific zone. The model has been set up for a five year forward look. The plan is compiled by the tool from the selected interventions for each zone. It is in the form of a table listing the zones (or other study areas), index values, chosen interventions and costs. Figure 9.13b shows the approach to constructing the plan. Assuming no overlap with other investment needs, this process would produce an investment plan for maintaining water quality in distribution.

Summary and Conclusions

During this study a number of areas where the analysis could be improved or the techniques developed further were identified. Thus, some recommendations for further work are defined and summarized in Table 9.9. It also appeared that the companies collect sufficient data to construct a distribution discolouration index,

Table 9.9: Principal Recommendations for Future Work

Item	Recommendation
National database	The UKWIR water mains failure database has been valuable in providing a dataset for the review of process and model development. A similar data set relating to discolouration would facilitate a wider review of the relationship between water, pipe and flow characteristics and the impact on water quality. To construct a database of value it would be necessary to collate customer service, water quality, and asset and performance data into a database with common references. Whilst relatively straightforward, the time and scheduling burdens hindered data availability to this project. Therefore we recommend the creation of a single dataset for further development of the D2I approach. In the absence of a national data set companies should collate their own data into a common-reference format, so that all the relevant parameters may be joined at asset level.
Further review of hydraulic data	Due to the difficulty of extracting hydraulic data in a format suitable for use in the hydraulic discolouration index, little development of HDI was possible. One company's data were available at the end of the study and some initial modelling was possible, but further development will be required in order to make best of the data.
Intervention efficacy	Further work is required in order to quantify the efficacy of interventions. There is a high level of expert knowledge of intervention impacts as a result of the 'Section 19' rehabilitation programmes, but the data are now largely obsolete because of subsequent changes to the distribution networks.
Water monitoring	The D2I model was developed using routine water quality monitoring data, because all companies carry out water quality monitoring to the same standard. But routine regulatory monitoring is not well suited to discolouration modelling because its randomised sampling pattern prevents time-series analysis of individual points in the network. Where companies have investigational monitoring data for single locations, they should be used for further development of the modelling approach. The results could then be used to help determine whether investment in continuous monitoring of that part of the network would be beneficial.
Level of service data	Customer contacts about discolouration were used as the indicator of the level of service, and therefore as the measure of success of the D2I index in forecasting discolouration. However, customer contact data are unlikely to be the best data for assessing the model results. Firstly, they represent a failure in service and hence interventions based on rising complaints are too late. Secondly, the number of contacts might not be comparable between incidents, depending on the time of day and location of the incident, meaning that two incidents with the same number of customer contacts might not be of the same scale. Therefore, companies should seek other data as a measure of the success of the model in forecasting discolouration. This could include samples taken during flushing, which show whether particles are present; or frequent monitoring of individual points in the network. This will be particularly important for indicating where conditions leading to discolouration events have been met, even though incidents have not yet been reported.

but in several different databases and in different formats, without a common referencing system. This hindered the modelling approach development and meant that the full index based on the water quality, network vulnerability, and hydraulic data could only be created for one of the companies that supplied data to this study.

The D2I index approach could be used to indicate where conditions for discolouration have been met, based on a range of data routinely collected by water companies, but to quantify this in terms of risk to service would require more detailed investigation and more targeted data than is provided by compliance monitoring. The networks are constantly changing in terms of input water quality, configuration, pipe materials, and hydraulics. Therefore, it is unlikely to be beneficial (or even feasible) to obtain a long-term time series data set covering all input parameters as this would not reflect the network current configuration. However, if such data were available together with information about changes to the network, they might have some use in isolating causal factors.

Although customer contacts provide good evidence of a failure in the level of service – and hence confirm where interventions are needed – the number of contacts does not necessarily reflect the event magnitude. Therefore, the number of customer contacts is not a reliable comparator with discolouration potential indicated by the D2I model. Flushing colour samples may be a useful comparator for testing the D2I result since it provides physical evidence of the potential for discolouration. This has the advantage that it will confirm areas where the conditions for discolouration have been met, even if there have been few customer contacts about discolouration, and would hence benefit a forward-looking approach. Finally, we were able to develop a simple discolouration modelling and intervention planning tool based on the D2I approach.

Acknowledgements

This study was primarily done when the last author was working as a research officer at National Research Council (NRC) Canadain Collaboration with Mott MacDonald (Cambridge, UK). The research was funded by UK Water Industry Research (UKWIR).

References

Boxall, J.B. and Saul, A.J. (2005): Modeling discoloration in potable water distribution systems. Journal of Environmental Engineering *ASCE*, 131(5): 716–725.

Boxall, J.B., Skipworth, P.J. and Saul, A.J. (2001A): Novel approach to modeling sediment movement in distribution mains based on particle characteristics. In: Proceedings of the Computing and Control in the Water Industry Conference, De Monfort University, UK.

Boxall, J.G., Skipworth, P.J. and Saul, A.J. (2003): Aggressiveness flushing for discolouration event mitigation in water distribution networks. Water Science and Technology, 3(1/2): 179-186.

CCME (Canadian Council of Ministers of the Environment) (2001): Canadian water quality guidelines for the protection of aquatic life.CCME water quality index 1.0 Technical Report. Available at: http://www.ccme.ca/assets/pdf/wqi_techrprtfctsht_e.pdf

Clark, R.M., Grayman, W.M., Males, R.M. and Hess, A.F. (1993): Modeling contamination propagation in drinking water distribution systems. Journal of Environmental Engineering ASCE, 119 (2): 349–354.

Clement, J., Hayes, M., Sarin, P., Kriven, W.M., Bebee, J., Jim, K., Beckett, M., Snoeyink, V.L., Kirmeyer, G.J. and Pierson, G. (2002): Development of red water control strategies. Awwa Research Foundation and American Water Works Association, USA.

Cullen, A. C. and Frey, H. C. (1999): Probabilistic techniques in exposure assessment: a handbook for dealing with variability and uncertainty in models and inputs. Plenum Press, New York.

DWI (2001): The Water Supply (Water Quality) Regulations 2001. Water, England and Wales, London. Available at: http://dwi.defra.gov.uk/stakeholders/legislation/ws_wqregs2001.pdf

DWI (2002): DWI Information Letter *15/2002*. DWI, London, UK. Available at: http://dwi.defra.gov.uk/stakeholders/information-letters/2002/15_2002.pdf

DWI (2007): Drinking water (2006): Drinking water in England and Wales.DWI, London, UK.

Ellison, D. (2003): Investigation of pipe cleaning methods. Report No. 90938, Awwa Research Foundation and American Water Works Association, Denver, USA.

Francisque, A., Rodriguez, M. J., Miranda-Moreno, F. L., Sadiq, R. and Proulx, F. (2009a): Modeling of heterotrophic bacteria counts in a water distribution system. Water Research 43(4): 1075–1087.

Francisque, A., Rodriguez, M. J., Sadiq, R., Miranda-Moreno, F. L. and Proulx, F. (2009b): Prioritizing monitoring locations (zones) in a water distribution network: a fuzzy risk approach. Journal Water Supply Research and Technology – AQUA 58(7): 488–509.

Gauthier, V., Barbeau, B., Millette, R. and Prévost, M. (2001): Suspended particles in the drinking water of two distribution systems. Water Supply Research and Technology: Water Supply, 1 (4): 237–245.

Hendrickson, H. (1996): Loss of chlorine residual: what do you do when it happens: cross connection case histories.In: Proc. of the 1996 Annual AWWA Conference. Denver, Co, USA.

Kirmeyer, G. J., Friedman, M., Martel, K., Howie, D., Clement, J., Sandvig, A., Noran, P. F., Smith, D., LeChevallier, M., Volk, C., Dyksen, J. and Cushing, R. (2000): Guidance manual for maintaining distribution system water quality. AWWARF, Denver, CO, USA.

Kleiner, Y. (1998): Risk factor in water distribution systems. In: British Columbia Water and Waste Association 26th Annual Conference, Whistler, BC.

Kumar, D. and Alappat, B.J. (2004): Selection of the appropriate aggregation function for calculating leachate pollution index. ASCE Practice Periodicals of Hazardous, Radioactive and Toxic Wastes, 8(4): 253-264.

Le Chevallier, M.W., Babcock, T.M. and Lee, R.J. (1987): Examination and characterization of distribution system biofilms. Applied Environmental Microbiology. 53(12): 2714–2724.

Lin, J. and Coller, B.A. (1997): Aluminum in a water supply, Part 3: domestic tap waters. Water. Journal of Australian Water Association, 11–13.

Meches, M. (2001): Biofilms in drinking water distribution systems. Controlling disinfection by-products and microbial contamination in drinking water. Clark, R.M., Boutin, B.K. (Eds.), National Risk Management Research Laboratory, Office of Research and Development, US Environmental Protection Agency, Cincinnati, Ohio, USA, p. 13.1–13.10.

Ott, W.R. (1978): Environmental indices: theory and practice. Ann Arbor Science Publishers, Michigan, USA.

Payment, P., Waite, M. and Dufour, A. (2003): Introducing parameters for the assessment of drinking water quality. In: Assessing microbials safety of drinking water: improving approaches and methods (A. Dufour, M. Snozzi,W. Koster, J. Bartram, E. Ronchi and L. Fewtre, eds). WHO and OECD, IWA Publishing, London, UK.

Prince, R., Goulter, I. and Ryan, G. (2001): Relationship Between Velocity Profiles and Turbidity Problems in Distribution Systems. World Water and Environmental Resources Congress, American Water Works Association, Orlando, Florida, pp. 9.

Ross, T. J. (2004): Fuzzy Logic with Engineering Applications. John Wiley and Sons, New York.

Ruta, G. (1999): Assessing Water Quality at Customers Internal Taps. In: Australian Water and Wastewater Association, 18th Federal Convention Proceedings.

Sadiq, R. and Rodriguez, M. J. (2004): Disinfection by-products (DBPs) in drinking water and the predictive models for the occurrence: a review. The Science of the total environment, 321(1-3): 21-46.

Sadiq, R., Rajani, B. and Kleiner, Y. (2004): A fuzzy based method of soil corrosivity evaluation for predicting water main deterioration. Journal of Infrastructure Systems, 10: 149–156.

Sadiq, R. and Tesfamariam, S. (2007): Probability density functions based weights for ordered weighted averaging operators (OWA): An example of water quality indices. European Journal of Operational Research, 182: 1350-1368.

Sadiq, R., Kleiner, Y. and Rajani, B. (2007): Water quality failures in distribution networks - risk analysis using fuzzy logic and evidential reasoning. Risk Analysis, 27(5): 1381–1394.

Sadiq, R., Najjaran, H. and Kleiner, Y. (2006): Investigating evidential reasoning for the interpretation of microbial water quality in a distribution network. Stochastic Environmental Research and Risk Assessment, 21: 63–73.

Seth, A., Bachmann, R.T., Boxall, J.B., Saul, A.J. and Edyvean, R. (2004): Characterization of materials causing discolouration in potable water systems. Water Science and Technology, 49 (2): 27–32.

Silvert, W. (2000): Fuzzy indices of environmental conditions. Ecological Modelling, 130 (1-3): 111-119.

Slaats, P.G.G., Rosenthal, L.P.M., Sieger, W.G., van den Boomen, M., Beuken, R.H.S. and Vreeburg, J.H.G. (2002): Processes involved in generation of discoloured water. Report No. KOA 02.058, American Water Works Association Research Foundation/Kiwa, The Netherlands.

Sly, L.I., Hodgkinson, M.C. and Arunpairojana, V. (1990): Deposition of manganese in drinking water distribution system. Applied Environmental Microbiology, 56(3): 628–639.

Somlikova, R. and Wachowiak, M. P. (2001): Aggregation operators for selection problems. Fuzzy Sets and Systems, 131: 23-34.

South East Water (1998): Water Quality Report 97/98. South East Water, Melbourne, Australia.

Stephenson, G. (1989): Removing loose deposits from water mains: Operational guidelines. Water Research Centre, Swindon, UK.

UKWIR (2006a): DOMS - What can we learn? Research Report.Ref. No. 06/WM/18/3.

UKWIR (2006b): DOMS Guidance Manual Volume 1: Non-Technical Overview. Ref. No.06/WM/18/1.

UKWIR (2006c): DOMS Guidance Manual Volume 2: Guidance. Ref. No. 06/WM/18/2.

UKWIR (2002): Capital maintenance planning: a common framework. UKWIR, London, UK.

Vreeburg, J.H.G. and Boxall, J.B. (2007): Discolouration in potable water distribution systems: A review. Water Research, 4(6): 519-529.

Vreeburg, J.H.G., Schaap, P.G. and van Dijk, J.C. (2004): Particles in the drinking water system: from source to discolouration. Water Science and Technology, 4(5–6): 431–438.

Vreeburg, J.H.G., Schippers, D., Verbecrk, J.Q.J.C. and van Dijk, J.C. (2008): Impacts of particles on sediment accumulation in a drinking water distribution system. Water Research, 42(16): 4233-4242.

Walski, T.M. (1991): Understanding solids transport in water distribution systems. In: Proceedings Water Quality Modelling in Distribution Systems, AWWA Research Foundation, USA, pp. 305–309.

Yamini, H. and Lence, B. J. (2010): Probability of failure analysis due to internal corrosion in cast-iron pipes. Journal of Infrastructure Systems, 16(1): 73-80.

Zadeh, L. A. (1965): Fuzzy sets. Information Control, 8: 338-353.

Chapter 10

Nutrient Removal from Domestic Wastewater: An Environmental Biotechnology Approach

Prangya Ranjan Rout, Rajesh Roshan Dash
and Puspendu Bhunia

School of Infrastructure, Indian Institute of Technology,
Bhubaneswar, Odisha, India

ABSTRACT

Excessive nutrient loading is a major ongoing threat to water quality. The chapter presents a review on the impact of nutrient discharges from domestic wastewater and analysis of existing information on different technologies used for nutrient removal from domestic wastewater. While effluent from wastewater treatment plants (WTPs) are major contributors of nutrients from point source category, domestic wastewater, urban and agricultural land uses are considered as significant nonpoint nutrient contributors. Non-point source pollution of closed water bodies such as lakes, reservoirs, inner bays, ponds and parts of rivers is a global concern. The current deterioration of water quality of above mentioned enclosed watersheds is primarily due to pollution from domestic wastewater. One of the major concerns regarding constituents in domestic wastewater is the concentration of nutrient compounds, in particular nitrogen and phosphorus. The discharge of nutrients into bodies of water can cause many problems in the environment as well as for human health. Reclaim, recycle and reuse can be adjudged as an alternative to eliminate direct nutrient discharges to receiving waters and allowing the beneficial use of treated wastewater at the same time. However, nutrients in reclaimed water can still be a concern for reuse applications, such as agricultural and landscape irrigation. At the present time the increased concentration of nutrients is a significant water quality concern in many of the nation's waters and a leading cause of impairment of designated uses. Therefore current and impending world legislation on effluent discharge has led to the need for improved wastewater treatment for nutrient removal. More or less improved wastewater treatment systems are focused towards biological nutrient removal (BNR). The technologies based on application of biological processes for nutrient removal. This is nothing but the implementation of biotechnology for environmental protection, which can be coined as Environmental Biotechnology.

Keywords: Nutrient removal, Domestic wastewater, Denitrifying bioreactor, Biological nitrogen removal, Biophysical phosphorous removal.

Introduction

Wastewater often contains large amounts of nitrogen and phosphorus. Both nitrogen and phosphorus along with carbon are essential elements for growth of microorganisms, plants and animals, thus known as nutrients or biostimulants. When discharged into the aquatic environment, these nutrients can lead to the growth of undesirable aquatic life. When discharged in excess amount on land, they can also lead to the pollution of groundwater (Metcalf and Eddy, 2003). Many aquatic systems have very low ambient nutrient concentrations and small shifts in the nutrient load can result in dramatic changes in community structure (Dodds and Welch 2000; Rabalais 2002). Sustained inputs of phosphorus and/or nitrogen to aquatic environments lead to increased rates of eutrophication, the nutrient enrichment of water bodies causing excessive growth of aquatic plants like algae, cyanobacteria, rooted aquatic vegetation and duckweed, a widespread problem throughout the world affecting the quality of domestic, industrial, agricultural and recreational water resources. Some of the ill effects of eutrophication include low dissolved oxygen, fish kills, murky water, and depletion of desirable flora and fauna (USEPA, 1993). Nutrients from wastewater have also been linked to ocean "red tides" that poison fish and cause illness in humans. Nitrogen in drinking water may contribute to miscarriages and is the cause of a serious illness in infants called methemoglobinemia or "blue baby syndrome". Higher amounts of these nutrients can also stimulate the activity of microbes, such as *Pfisteria*, which can be potentially harmful to human health in short excessive amounts of these nutrients lead to a range of undesirable effects, including impairment of human health (algal toxins), reduced biodiversity of aquatic species, reduction in amenity value, and increased costs of treatment for drinking water (Withers *et al.*, 2009).

The potential eutrophic influence a nutrient will have on an aquatic system is related to its bioavailability. Biologically available forms of nitrogen are nitrate (NO_3-N) and ammonia (NH_3-N) but organic nitrogen (organic-N) can also be made available through conversion to NH_3-N. Similarly For phosphorus, soluble orthophosphate (PO_4-P) is the only form that can be assimilated directly by autotrophs (Correll 1998) but phosphorus is lost from soils in both particulate-which is more dominant and dissolved forms (Bennett *et al.*, 2001). To support algal growth, some forms of particulate phosphorus are converted through mineralization reactions and desorption to PO_4-P. The elements of nitrogen and phosphorus required for the growth of phytoplankton are mostly derived from human activities. The principal source of nutrient compounds is the nitrogen and phosphorous containing compounds of plant and animal origin. Thus the primary input of nutrients into nature occurs through its use in agriculture and from domestic wastewater (constituents shown in Table 10.1), (Bennett *et al.*, 2001; Camargo and Alonso, 2006). Whereas nutrients can be regulated by balanced fertilization in agricultural use (Sharpley and Tunney, 2000), wastewater needs proper treatment for the removal of nutrients up to the levels that can be acceptable for natural systems (Kadlec and Knight, 1996). The contribution of nitrogen from human wastes to the environment is estimated at 25×10^{12} g N/year. While only about 17 per cent of that is from agro ecosystems, it is nevertheless significant in its impacts on aquatic ecosystems and public health (Galloway *et al.*, 2003). Likewise households are responsible for more than 45 per cent of the total phosphorus load discharged into surface waters. Thus measures of controlling emission sources such as domestic wastewater are important in a restoration program for aquatic environment since domestic wastewater accounts for a large part of the pollution load (Sudo, 2000). Therefore, nutrient inputs must be reduced and measures for effective reduction of nutrient concentrations in domestic wastewater must be developed to protect these ecosystems from deterioration.

Table 10.1: Major Constituents of Typical Domestic Wastewater

Constituent	Concentration, mg/l		
	Strong	*Medium*	*Weak*
Total solids	1230	720	390
Total Dissolved Solids (TDS)	860	500	270
Suspended solids	400	210	120
Nitrogen (as N)	70	40	20
Phosphorus (as P)	12	7	4
Chloride	90	50	30
Sulfate	50	30	20
Alkalinity (as $CaCO_3$)	200	100	50
Oil and grease	100	90	50
BOD_5	350	190	110
COD	800	430	250
TOC	260	140	80
VOCs	>400	100-400	<100

Source: Metcalf and Eddy, 2003.

Background of Nutrient Removal

The removal of organic matter has long been considered as the main target of wastewater treatment systems. WTPs have been used worldwide to remove the organic mater and to improve the quality of wastewater before it is discharged to surface or groundwater and re-enters water supplies. Regardless of the methods used at any particular WTP, almost all treatment processes fall into three main categories: (1) primary, (2) secondary, and (3) advanced tertiary treatment. Primary treatment removes both inorganic and organic suspended solids. Secondary treatment is responsible for removal of dissolved organics and a negligible amount of dissolved nutrients. So some suspended and dissolved material along with a major portion of nutrient may still remain in the effluent after typical secondary treatment as depicted in Figure 10.1. That is why advanced or tertiary treatment comes into picture to meet regulatory requirements protecting receiving waters. The conventional WTPs are devoid of tertiary treatment facilities, so the secondary effluents from wastewater treatment plants enriched with nutrients (NH^+_4, NO^-_3 and PO_3^{-4}) may become a dominant contributor to nutrient inputs into aquatic systems. Besides the discharge of untreated or poorly treated domestic wastewater directly to water bodies, particularly in rural areas of developing countries is solely responsible for many watercourses, reservoirs and lakes not meeting their quality objectives. Compounds containing available nitrogen and phosphorous have received considerable attention since the mid-1960s. Initially, nitrogen and phosphorous in wastewater discharges became important because of their effects in accelerating eutrophication of lakes and promoting aquatic growth. More recently, nutrient control has become a routine part of treating wastewater use for the recharge of groundwater supplies. Therefore, more and more stringent wastewater discharge standards have been set in the world for effluent containing nutrients and this has been prompted researchers for the development of new technologies. In the present scenario nutrient removal is one of the basic objectives in wastewater treatment systems and biological, physical, chemical or combinations of these methods are used to remove wastewater nutrients. Available several unit processes for nutrient removal include Biological Nutrient Removal (BNR), biological filtration combined with biological nitrogen removal, adsorption, chemical precipitation, membrane filtration, reverse osmosis, ion exchange, air stripping, breakpoint chlorination and biological nitrification and denitrification (Metcalf and Eddy, 2003).

Existing Nutrient Removal Processes

Nutrients can be removed from wastewater through advanced or tertiary treatment systems. Several chemical and biological nutrient removal (BNR)

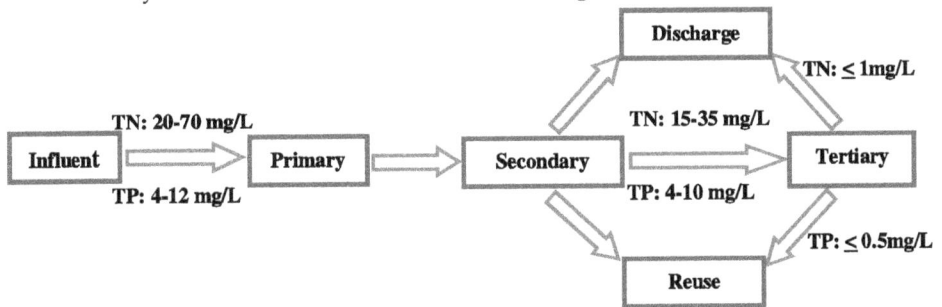

Figure 10.1: Nutrient Concentration Ranges for Untreated to Tertiary Treated Wastewater.
Source: **Richard** *et al.*, **2009.**

technologies are available to lower effluent nutrient concentrations before discharge (Pagilla *et al.*, 2006). Out of multiple alternatives BNR is considered as an economical and effective wastewater treatment process for removing nitrogen and phosphorus (Fan *et al.*, 2009). BNR emerged approximately thirty years ago and during the past several decades, a great deal of biological treatment systems with the function of simultaneous removal of nitrogen and phosphorus have been developed including the sequencing batch reactor (SBR), the Bardenpho process, the University of Cape Town (UCT) system, and the anaerobic-anoxic-aerobic (A_2O) process (Tchobanoglous *et al.*, 2003). BNR technologies, basically use microorganisms under aerobic or anaerobic conditions to remove nutrients via suspended and/or attached growth processes. So here the role of living organism, the role of some bioprocesses and biosystems for removal of nutrients from wastewater can be realized as the direct application of biotechnology for the protection of environment, and the interdisciplinary approach can be termed technically as environmental biotechnology. Membrane technology, in recent times also evolved as one of the innovative advanced methods for nutrient removal which combines biological with physical processes (Bitton, 2005). Some of the case studies based upon above mentioned technologies are discussed in the following section.

Limited studies have been reported in the literature in the context of nutrient removal from domestic wastewater. The effectiveness of existing technologies as reported by various reseachers at different operating paramentes has been presented in Table 10.2. Llabres *et al.* (1998) demonstrated the feasibility of using the biodegradable organic fraction of municipal solid waste (BOF-MSW) as an easily biodegradable C source for nitrogen and phosphorous removal. The experiment was carried out in a continuous UCT (University of Cape Town) - bench scale plant by taking synthetic wastewater. The percentage removal rate of total nitrogen and total phosphorous were 89.3 and 88.7 respectively. Ahn *et al.*, [2003] proposed the innovative process SAM (sequencing anoxic/anaerobic membrane bioreactor) to enhance biological phosphorus removal (EBPR). A laboratory-scale experiment was performed to treat the household wastewater including toilet-flushing water. The aerobic zone with the submerged membrane was continuously aerated for nitrification and phosphorous uptake as well as fouling control. The design was based upon the Modified Luzack-Ettinger (MLE) type Membrane Bio-Reactor (MBR) process. The phosphorous removal was much better in the SAM process, yielding 93 per cent removal efficiency. The nitrogen removal efficiency of the SAM was about 60 per cent, which was slightly lower than that of the MLE-type MBR process. However, the hydraulic retention time of the SAM process in the anoxic condition was 2.3 times shorter than that of MLE-type MBR process. As reported by Yang *et al.* (2003) an entrapped mixed-microbial cell (EMMC) with and without humic substances for both fixed and moving carrier reactors and conventional suspended growth culture (*i.e.* conventional activated sludge process) were investigated for removal of organics, total nitrogen, total phosphorous and sulfides from domestic wastewater simultaneously. Both synthetic and actual domestic wastewater were investigated under operational conditions of 12 h of hydraulic retention time (HRT) with 1 h of aeration and 1 h of non-aeration, and 6 h of HRT with continuous aeration, at a room

Table 10.2: Details of Existing Nutrient Removal Processes

Reactor Type	HRT	MLSS	Per cent Removal of TN	Per cent Removal of TP	Reference
Anaerobic-Anoxic (A$_2$N) SBR and Nitrification SBR.	24 hours	Not Reported	88	99	Kuba et al., 1996
Sequencing Batch Reactor (SBR)	12 hours	1875 mg/L	73	54	Carucci et al., 1997
Anaerobic-Aerobic-Anoxic (AO)$_2$ SBR.	12 hours	1480 mg/L	88	100	Lee et al., 2001
Sequencing Anoxic/Anaerobic (SAM) Membrane Bioreactor	Not Reported	10000 mg/L	93	60	Ahn et al., 2003
Vertical Submerged Membrane Bioreactor (VSMBR)	8 hours	4830 mg/L	75	71	Chae et al., 2006
Membrane Bioreactor (MBR)	2-3 days	Not Reported	50-90	25-70	Abegglen et al., 2008
Soil Infiltration Technology (SIT)	24 hours	Not Reported	68–75	94	Chun et al.,2008
Anaerobic Bioreactor (ABR) and hybrid Constructed Wetland (CW)	1.2 days	Not Reported	96	74	Singh et al., 2009
Anaerobic Anoxic Aerobic (A$_2$O) Membrane Bioreactor	1-6 hour	7500 mg/L	60-67	74-82	Banu et al., 2009
Constructed soil filter (CSF)	0.5-1hours	Not Reported	35	67	Kadam et al., 2009
Hybrid constructed wetland (CW) system	2.2-8 days	Not Reported	78.3	65.4	Vymazala et al., 2011
Anaerobic/Aerobic/Anoxic (AOA) Process	8 hours	3830–4720 mg/L	70.3 ± 2.9	87.3 ± 11.8	Xu et al., 2011
Constructed wetlands and denitrifying bioreactors for on-site and decentralized wastewater treatment	24 hours	Not Reported	58-95	36-65	Tanner et al., 2012

temperature of 25°C. In general, the EMMC associated systems which provide high solids retention time achieve a better removal of chemical oxygen demand (COD), nitrogen, phosphorous and the odour producing substance than the suspended growth system for both HRTs of 6 h (continuous aeration) and 12 h (1 h of aeration and 1 h of non aeration).

Yang *et al.* (2005) developed sludge ceramic from lake sediment and subsequently used them as a medium for biological filtration in a small-scale wastewater treatment system capable of removing nitrogen and phosphorus from wastewater. They found the sludge ceramic superior than the conventional ceramic media and they got very significant result in the form of BOD levels of 10 mg L^{-1} or less, Total Nitrogen (TN) of 10 mg/L or less and Total Phosphorous (TP) of 1mg/L or less. A novel vertical submerged membrane bioreactor (VSMBR) with anoxic (lower layer) and oxic (upper layer) zones were developed to remove organic and nutrients from synthetic wastewater including glucose as a sole carbon source Chae *et al.* (2006). The optimal volume ratio of anoxic and oxic zones was determined as 0.6. In addition, the desirable internal recycle rate and HRT were found to be 400 per cent and 8h, respectively. Average Mixed Liquor Suspended Solids (MLSS) concentrations of the anoxic and the oxic zones were 4830 and 2970 mg/L, respectively. Under these conditions, the average removal efficiencies of TN and TP were 75 per cent and 71 per cent, respectively. The conceptualization of a carrier anaerobic baffled reactor (CABR), to treat domestic wastewater was done by Feng *et al.* (2008). The performance of the reactor was satisfactory in case of total chemical oxygen demand (TCOD) removal but the reactor displayed nutrient removal efficiency of only 19.22- 24.85 per cent for TN and 30.86- 35.58 per cent for TP throughout the experiment, when the influent nitrogen and phosphorus concentrations were 61.95- 67.95 mg TN/L and 7.25- 7.77 mg TP/L, respectively. The study performed by Abegglena *et al.* (2008) showed the biological nutrient-removal potential of an on-site Membrane bioreactor (MBR) located in the basement of a four-person house. This treatment plant differs from other conventional MBRs by a highly fluctuating influent water flow and a lack of pretreatment. The reactor consists of two tanks. When using the first reactor as an anaerobic/anoxic reactor by recycling activated sludge and mixing the first reactor, nitrogen and phosphorus removals of over 90 per cent and 70 per cent were achieved, respectively.

A full-scale, two-stage anaerobic tank and soil trench system was designed and developed by Chun *et al.* (2008) to evaluate the feasibility and performances in treating domestic wastewater in a decentralized approach. The raw sewage was prepared and fed into the first anaerobic tank and the second tank by 60 per cent and 40 per cent, respectively. This novel process could decrease chemical oxygen by 89 per cent 96 per cent, suspended solids by 91 per cent–97 per cent, and total phosphorus by 91 per cent–97 per cent. The denitrification was satisfactory in the second stage soil trench, so the removals of TN as well as ammonia nitrogen reached 68 per cent– 75 per cent and 96 per cent–99 per cent, respectively. In order to investigate the nutrient removal efficiency and the governing internal dynamics of the most widely used wetland type, the horizontal subsurface flow reed bed, in receiving domestic septic tank and secondary effluent in a temperate climate such as Ireland, Luanaigh *et al.* (2010) designed, constructed and rigorously monitored two systems for a period of over 2 years and they reported that nitrogen removal was poor across both reed beds, with only 29 per cent removal of TN across the secondary treatment bed and 41 per cent removal across the tertiary treatment bed, with little distinctive seasonal change. Removal of PO_4-P from the secondary and tertiary treatment beds was equally poor at rates of 45 per cent and 22 per cent, respectively. Vymazal *et al.* (2011) proved the three-stage hybrid constructed wetland consisting of saturated vertical flow, free-drained vertical-flow and horizontal sub-surface flow wetlands, proved to be very effective in reducing organics, suspended solids and nitrogen from domestic wastewater. Removal of BOD_5 amounted to 78.1 per cent and 94.5 per cent, respectively with outflow concentrations of 16 and 10 mg/L, while the inflow average NH_4-N concentration of 29.9 mg/L was reduced to 6.5 mg/L with the average removal efficiency of 78.3 per cent. Also phosphorus was removed efficiently despite the fact that the system was not aimed at P removal and therefore no special media was used. Phosphorus removal efficiency was 65.4 per cent but the average outflow concentration of 1.8 mg/L was still high. Wang *et al.* (2010) demonstrated the performance of a novel three-stage vermin filtration (VF) system using the earthworm, *Eisenia fetida*, for domestic wastewater. In their study, the researchers showed that the tower VF efficiently removed water pollutants compared to that of traditional VF. The average removal efficiencies of the tower VF planted with *Penstemon campanulatus* were chemical oxygen demand, 81.3 per cent ; ammonium, 98 per cent ; total nitrogen (mainly in the form of nitrate), 60.2 per cent ; total phosphorus, 98.4 per cent. They justified the result by stating that soils played an important role in removing the organic matter whereas the three-sectional design with increasing oxygen demand concentration in the effluents, and the distribution of certain oxides in the padding was likely beneficial for ammonium and phosphorus removal, respectively.

Kadam *et al.* (2009) developed constructed soil filter (CSF) and monitored the performance of two CSF units located in Mumbai, India for contaminant removal from municipal wastewater. The system operated with hydraulic retention time of 0.5-1.0 h, hydraulic loading of 0.036-0.047 $m^3\,m^{-2}\,h^{-1}$, and the result showed elevated dissolved oxygen (DO) levels, removal of COD (136-205 to 38-40 mg/L) and BOD (80-125 to less than 12 mg/L) suspended solids from 135-203 to 1318 mg/L and turbidity from 84-124 to 8-11 NTU, bacterial removal of 2.4-3.1 log order for total coliform and Fecal coliform. Total Nitrogen 10.8±6.4 to 7.0±6.3 mg l^{-1} with a removal efficiency of 34.8 per cent while Total Phosphorous 1.4±0.7 to 0.4±0.2 mg/L with removal efficiency of 67.3 per cent. This work illustrates the potential of a novel vermi-biofiltration system in treatment of urban wastewater. Tomar *et al.* (2011) constructed a small-scale vermi-biofiltration reactor using vertical subsurface-flow constructed wetlands (VSFCWs) aided with local earthworms *Perionyx sansibaricus*. The wastewater was treated through this system for a total of eight repetitive cycles. Vermi-biofiltration caused significant decrease in level of TSS (88.6 per cent), TDS (99.8 per cent), COD (90 per cent), NO_3^- (92.7 per cent) and PO_4^{3-} (98.3 per cent). However the author suggested some key issues *e.g.*, loading rate, flow alternation impacts and earthworm stocking density need to be explored further. Balachandran *et al.* (2012) focused on the development of a high rate process for simultaneous removal of COD, NH_4-N and

Phosphorous from synthetic wastewater having COD to NH_4-N ratio equal to 1. A continuous down-flow fixed film reactor was used by the researcher for the development of the process. The reactor operated at COD and NH_4-N loading rate of 5.4 g/L/d at 3 h HRT showed the removals of COD, NH_4-N, total nitrogen and phosphorous varied in the range of 91-39 per cent, 51-53 per cent, 47-50 per cent and 26-58 per cent respectively. The investigator studied high rate nitration-denitration process with phosphorous removal at low dissolved oxygen concentration of 0.2-0.3 mg/L and suggested that the developed process can be used for economical biological nutrient removal from wastewater.

After a brief analysis of some of the above mentioned processes or systems for domestic wastewater nutrient removal we can infer that more or less almost all the methods based upon either the mechanism of Biological Nutrient Removal (BNR) or biological filtration combined with biological nitrogen removal and physicochemical phosphorous removal processes are more efficient than their counterparts (Inamori *et al.,* 1993). The nutrient removal efficiency of some of the discussed systems is compared in Figure 10.2. Anaerobic-anoxic sequence batch reactor achieved 99 per cent phosphorous removal while anaerobic bioreactor and hybrid constructed wetland showed a significant 96 per cent nitrogen removal.

Mechanism for Nutrient Removal

As discussed earlier the BNR process is very common in wastewater treatment. There exist a number of biological nutrient removal treatment process configurations. Some biological nutrient removal systems are designed to treat only Total Nitrogen or

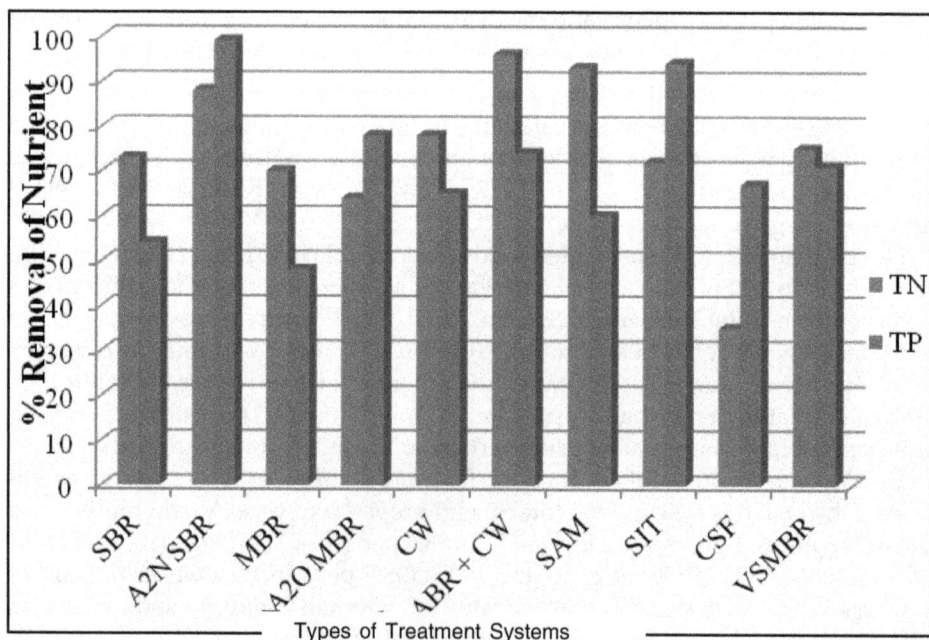

Figure 10.2: Comparison of Nutrient Removal of different Wastewater Treatment Systems.

Total Phosphorus, while other processes are meant for the treatment of both. Though there will be variability in the configurations of each system, still biological nutrient removal systems designed to treat Total Nitrogen must have an aerobic zone for nitrification and an anaerobic zone for de-nitrification, while biological nutrient removal systems designed to treat Total Phosphorus must have an anaerobic zone free of dissolved oxygen and nitrate. In the following two sections we will focus on the removal mechanism of phosphorous and nitrogen from wastewater.

Phosphorus Removal

Biological phosphorus removal (BPR) from wastewater utilizes activated sludge with specific microorganisms, categorized as Phosphorus Accumulating Organisms (PAOs), which can assimilate excess phosphorous as polyphosphates (Wentzel *et al.*, 1991). Barnard *et al.* (1984) suggested that, the activated sludge should be exposed first to anaerobic and then to aerobic zones to achieve effective phosphorus removal. Anaerobic contact time is a key component of BPR as secondary phosphorus release can occur if contact times are too long. The PAOs work best in optimal pH, DO, and temperature ranges and therefore he process parameters need to be maintained to maximize BPR. The (PAOs) are very sensitive to fluctuation in process parameters and may sometimes be attacked by another group of microbes called Glycogen Accumulating Organisms (GAOs). Thus the conventional biological phosphorus removal process is very complex in nature. Therefore a lot of process control is needed for biological phosphorous removal. Hence the process may not be reliable and may not achieve high levels of phosphorus removal. Similarly, phosphorus removal through the dosing of chemicals such as aluminum, calcium or iron salts, is a common method used to effectively reduce effluent concentrations of nutrients released to surface waters. The fundamental concept of chemical precipitation is that the above mentioned salts, when added to wastewater, produce insoluble phosphates that can be separated from the discharged effluent through sedimentation processes. But the method results in higher production of sludge, thus significantly increasing the operational and maintenance cost. The chemicals are expensive and their handling and storage is too dangerous. Therefore, many more other alternative removal methods have been investigated. Out of many other alternatives for phosphorous removal like flocculation, sedimentation, adsorption, precipitation, iron contacting and iron electrolysis (Takai *et al.*, 2002; Haruta *et al.*, 1992 Moriizumi *et al.*, 1999), sorption *e.g.* adsorption and/or precipitation mechanisms of phosphorus to the substratum has been recognized as one of the most important removal mechanisms (Richardson, 1985). Chemical composition of the substrate has been reported to affect the P-sorption capacity. Since the sorption is a finite process, it is an important factor to consider when selecting substrates for potential use in filter-based systems or in constructed wetland systems meant for phosphorus removal. Therefore filter bed having substrates capable of phosphorus absorption can be assumed as a suitable alternative for phosphorus removal from wastewater.

Nitrogen Removal

On the other hand nitrogen removal can be accomplished by physical, chemical and biological processes. The physical and chemical processes for nitrogen removal have proven to be costly, unreliable, and problematic. Among the other biological (Van der star *et al.*, 2007) and non-biological processes (Metcalf and Eddy, 2003) that have been used, biological nitrification/denitrification is the principal process that has been demonstrated to be feasible, both economically and technically, for nitrogen removal in both centralized and decentralized systems (USEPA, 1993, 2003). Nitrification is defined as a two-stage biological process, which occurs under aerobic conditions in the presence of oxygen. During nitrification, ammonia is oxidized to nitrite by one group of autotrophic bacteria, known as Nitrosomonas. The nitrite is then further oxidized to nitrate by another group of autotrophic bacteria, known as Nitrobacter. De-nitrification occurs under anaerobic conditions, in the absence of oxygen, and involves the biological reduction of nitrate to nitric oxide, nitrous oxide, and nitrogen gas which is released to the atmosphere. De-nitrification can be accomplished by a host of heterotrophic bacteria in the absence of dissolved oxygen. These bacteria use the oxygen in the nitrate, instead of dissolved oxygen, to digest organic material, thereby releasing nitrogen gas as a waste by-product. The commonly used nitrogen removal process configurations are preanoxic, postanoxic, and simultaneous nitrification–denitrification as given in Figure 10.3 (Metcalf and Eddy, 2003). Simultaneous configurations are most often used in suspended-growth designs whereas preanoxic and postanoxic are most commonly in attached-growth or combinations of suspended/attached-growth (Metcalf and Eddy, 2003; US EPA, 1993). Both nitrification and denitrification can be mediated by suspended-growth or attached-growth processes. While suspended-growth processes are typically used for both nitrification and denitrification in centralized systems, attached growth denitrification is occasionally seen (US EPA, 2008a, b; US EPA, 1993). Attached-growth processes are predominantly used for nitrification in on-site wastewater treatment systems while both attached and suspended-growth systems have been used for denitrification (Oakley, 2005). So a filter bed which favors attach growth process and a post positioned attached growth denitrifying bed can be adjudged as an appropriate technology for biological nitrogen removal.

Impact of Nutrient on Reusability

In recent times, water reclamation, recycle and reuse programs have been implemented in both developed and developing countries because of increasing freshwater demand and the difficulty of complying with nutrient discharge standards. (USEPA 2004b). Reuse programs provide an opportunity to reduce or eliminate direct nutrient discharges to receiving waters while allowing for the beneficial use of reclaimed water. The main reuse categories of reclaimed water are agricultural and landscape irrigation, industrial processing, groundwater recharge, environmental and recreational uses, nonpotable urban uses, and indirect potable reuse (Tchobanoglous *et al.*, 2003). Risk assessment of water reuse is necessary which includes public health concerns due to direct or indirect exposure to pathogens (Ongerth *et al.*, 1982) and irrigation issues such as salinity, sodium adsorption ratios, and metal content in soils (Qian *et al.*, 2005). Keeping the risk factor in view, treatment system is decided for a particular type of wastewater and also for a particular type of reuse. Typical nutrient concentration in reclaimed water varies depending on the level of treatment and reusability as depicted in Table 10.3. Nutrient removal may not

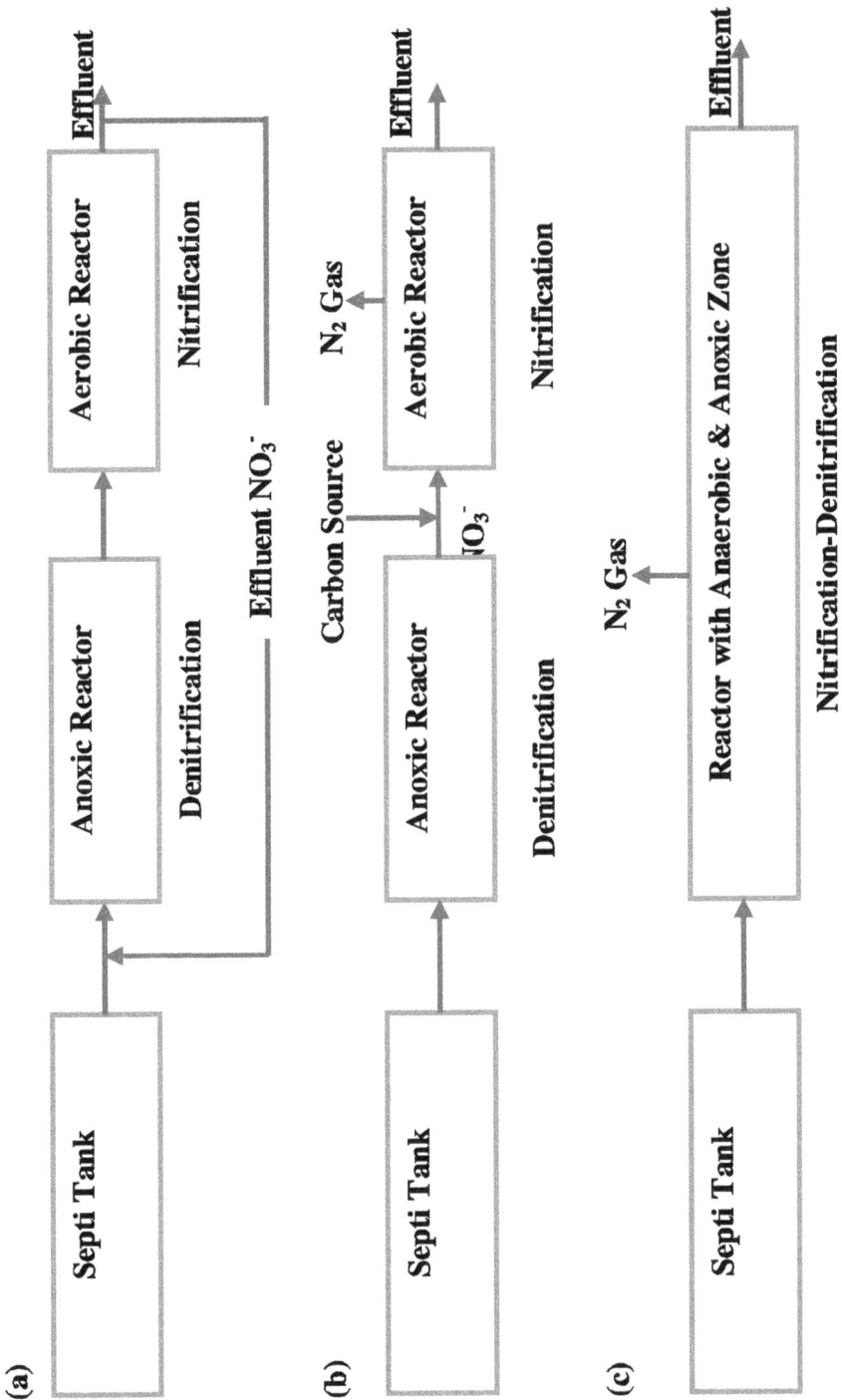

Figure 10.3: Schematic Diagram of (a) Pre-anoxic Process, (b) Post-anoxic Process, (c) Simultaneous Nitrification-Denitrification Process.

be that much necessary for treated water intended for irrigational reuse, since crops require nutrients for their growth and development, but for effluents discharged to water bodies stringent nutrient removal method should be followed. However the long term effect of using reclaimed water for irrigation may have a detrimental effect. Nutrient rich water may results in increased soil microbial activity, which can eventually lead to the biological clogging of pores and reduced hydraulic conductivity (Magesan *et al.*, 2000). Phosphorus inputs from reclaimed water beyond plant nutrient requirements leads to phosphorus accumulation in soils, the potential for leaching increases and this poses a water quality threat. Groundwater recharge via percolation or direct injection using reclaimed water poses a contamination risk from ammonia and nitrate as well. Therefore continual monitoring of groundwater recharge systems using treated effluent, is required (Fryar *et al.*, 2000). An interesting aspect of wastewater reuse is the use of both natural and artificial wet-lands to improve the quality of treated water. Constructed wet-land can provide nutrient removal services equivalent to tertiary treatment processes while providing suitable wildlife habitat. Thus nutrient removal efficiency in treatment wet-lands provide direct ecosystem benefits.

Table 10.3: Nutrient Concentration after Treatment through different Unit Processes

Components (mg/L)	Untreated Wastewater	ASP	ASP+ BNR	ASP+ BNR+ MF+RO
NH_3-N	12-45	1-10	1-3	≤0.1
NO_3-N	0-trace	10-30	2-8	≤1
Total Nitrogen (TN)	20-70	15-35	3-8	≤1
Total Phosphorus (TP)	4-12	4-10	1-2	≤0.5

Source: Richard *et al.*, 2009

ASP: Activated Sludge Process; BNR: Biological Nutrient Removal; MF: Membrane Filtration; RO: Reverse Osmosis.

Conclusion

There is a need to evaluate and improve the treatment efficiency of existing wastewater treatment facilities for significant removal of nutrients from domestic wastewater. The literature review indicates that, stimulated by the leading world legislations like European legislation, American legislation, almost all developed countries are implementing nutrient removal technologies in domestic wastewater treatment systems. But the scenario in developing countries is totally different. In many of the developing countries, the potential harmfulness of domestic wastewater is yet to be realized. Though very few wastewater treatment processes with nutrient removal facilities are already in action in some of the developing countries, they are not efficient enough to meet the discharge standard. Also literature analysis leads to the inference that most of the wastewater treatment systems with nutrient removal options operated in developed countries are mechanized devices based on the activated sludge process (ASP). Many of the systems are energy-intensive, scaled-down models of centralized plants and less reliable. Owing to the above shortcomings, further study may be carried out on the recommended idea detailed in section-8 of this article. The recommendation device may consists of single pass biofilter with denitrifying bioreactor. This will be of an attached growth process and attached growth process may show resistance towards variable wastewater flow and characteristics of domestic wastewater compare to ASP. This may be a natural wastewater treatment system emphasizing on passive, robust, ecologically engineered designs for nutrient removal from domestic wastewater.

Recommendations

The extensive literature study related to nutrient removal from domestic wastewater gave an insight into the current technologies employed and the common practices followed to tackle the issue. It was interpreted that many of the systems of the present time are less reliable. The lower reliability is due to the inherent variability of domestic wastewater characteristics and the challenges of operationally controlling nutrient removal processes at the level of residences. In order to deal with the shortcomings the further study may focus on a hybrid system with two components (*a*) multi stage filter bed and (*b*) denitrifying bioreactor. For simultaneous removal of nutrients such as nitrogen and phosphorus, the system may be designed to provide various environmental conditions of aerobic and anoxic type. The specific environmental condition may favour specific type of mechanism like nitrification and denitrification. The formulated filter media for the filter bed may serve the purpose of phosphorous removal. During the operation the wastewater will be allowed to be gravity-fed over the entire system. Wastewater percolation through the filter media and wastewater dropping down between the different stages of filter bed will result in passive oxygenation. So the aerobic condition in the filter media will the desirable condition for nitrification. The use of a filter bed plus a post-positioned denitrifying bioreactor with a carbon source will be the other ideal condition for denitrification to take place.

References

Abegglena, C., Ospelta, M. and Siegrista, H. (2008): Biological nutrient removal in a small-scale MBR treating household wastewater. Water research 42: pp. 338-346.

Ahn, K.H., Song, K.G., Cho, E., Cho, J., Yun, H., Lee, S. and Kim, J. (2003): Enhanced biological phosphorus and nitrogen removal using a sequencing anoxic/anaerobic membrane bioreactor (SAM) process. Desalination 157: pp. 345-352.

Balachandran, P., Sabumon, P.C., Gopakumar, A. and Malliyekkal, S.M. (2012): International Symposium on Southeast Asian Water Environment,10: Part II.

Banu, J.R., Uan, D.K. and Yeom, I.T. (2009): Nutrient removal in an A2O-MBR reactor with sludge reduction, Bioresource Technology 100: pp. 3820-3824.

Barnard, J.L. (1984): Activated primary tanks for phosphate removal. Water SA 10: pp. 121- 126.

Bennett, E.M., Carpenter, R. and Caraco, N.F. (2001): Human impact on erodable phosphorus and eutrophication: a global perspective. BioScience 51: pp. 227-234.

Bitton, G. (2005): Wastewater microbiology, Wiley-Liss, John Wiley and Sons, New Jersey, USA, pp. 766.

Camargo, J. and Alonso, A. (2006): Ecological and toxicological effects of inorganic nitrogen pollution in aquatic ecosystems: a global assessment. Environmental International 32: pp. 831-849.

Carey, R.O. and Migliaccio, K.W. (2009): Contribution of Wastewater Treatment Plant Effluents to Nutrient Dynamics in Aquatic Systems: A Review. Environmental Management, 44: pp. 205-217.

Carucci, A., Majone, M., Ramadori, R. and Rossetti, S. (1997): Biological phosphorus removal with different organic substrates in anaerobic/aerobic sequencing batch reactor. Water Science Technology 35 (1): pp. 161-168.

Chae, S.R., Kang, S.T., Watanabe, Y. and Shin, H.S. (2006): Development of an innovative vertical submerged membrane bioreactor (VSMBR) for simultaneous removal of organic matter and nutrients. Water research 40: pp. 2161-2167.

Chun, Y.E., Zhan-Bo, H.U., Hai-Nan, KONG., Xin-Ze, WANG. and Sheng-Bing, H.E. (2008): A New Soil Infiltration Technology for Decentralized Sewage Treatment: Two- Stage Anaerobic Tank and Soil Trench System. Pedosphere 18: pp. 401-408.

Correll, D.L. (1998): The role of phosphorus in the eutrophication of receiving waters: a review. Journal of Environmental Quality 27: pp. 261-266.

Dodds, W.K. and Welch, E.B. (2000): Establishing nutrient criteria in streams. Journal of the North American Benthological Society 19: pp. 186-196.

Fan, J., Tao, T., Zhang, J. and You, G.L. (2009): Performance evaluation of a modified anaerobic/anoxic/oxic (A2/O) process treating low strength wastewater. Desalination 249 (2): pp. 822-827.

Feng, H., Hu, L., Mahmood, Q., Qiu, C., Fang, C. and Shen, D. (2008): Anaerobic domestic wastewater treatment with bamboo carrier anaerobic baffled reactor. International Biodeterioration and Biodegradation 62: pp. 232-238.

Fryar, A.E., Macko, S.A., Mullican, W.F., Romance, K.D. and Bennett, P.C. (2000): Nitrate reduction during ground-water recharge, southern High Plains, Texas. Journal of Contaminant Hydrology 40: pp. 335-363.

Galloway, J.N., Aber, J.D., Erisman, J.W., Seitzinger, S.P., Howarth, R.W., Cowing, E.B., and Cosby, B.J. (2003): The nitrogen cascade. Bioscience 53: pp. 341-356.

Haruta, S., Takahashi, T. and Nishiuchi, T. (1992): Experimental study on dephosphorization technique using the aerating submerged iron contractor process (II) —long-term stabilization on dephosphorization and the effect of water temperature and flow rate. Journal Japanese Society of Irrigation, Drainage and Reclamation Engineering 158: pp. 57-63.

Inamori, Y., Takai, T. and Sudo, R. (1993): Recent aspects of nitrogen and phosphorus removal technology. Journal of Resource and Environment. 29 (8): pp. 12-23.

Kadam, A.M., Nemade, P.D., Oza, G.H. and Shankar, H.S. (2009): Treatment of municipal wastewater using laterite-based constructed soil filter. Journal of Ecological Engineering 35: pp. 1051-1061.

Kadlec, R.H. and Knight, R.L. (1996): Treatment Wetlands, Lewis Publishers, CRC Press, Boca Raton, pp. 893-901.

Kuba, T., van Loosedrecht, M.C.M. and Heijnen, J.J. (1996): Phosphorus and nitrogen removal with minimal COD requirement by integration of denitrifying dephosphatation and nitrification in a two-sludge system. Water Research. 30 (7): pp. 1702-1710.

Lee, D.S., Jeon, C.O. and Park, J.M. (2001): Biological nitrogen removal with enhanced phosphate uptake in a sequencing batch reactor using single sludge system. Water Research 35 (16): pp. 3968-3976.

Llabres, P., Pavan, P., Battistioni, P., Cecchi, F. and Malta-alvarez, j. (1998): The use of organic fraction of municipal Solid waste hydrolysis products for Biological nutrient removal in wastewater treatment plants, pii: s0043-1354; 00179-1.

Luanaigh, N.D., Goodhueb, R. and Gill, L.W. (2010): Nutrient removal from on-site domestic wastewater in horizontal subsurface flow reed beds in Ireland, Journal of Ecological Engineering 36: pp. 1266-1276.

Magesan, G.N., Williamson, J.C., Yeates, G.W. and Lloyd-Jones, A.R. (2000): Wastewater C:N ratio effects on soil hydraulic conductivity and potential mechanisms for recovery. Bioresource Technology 71: pp. 21-27.

Metcalf and Eddy Inc. (2003): Wastewater Engineering: Treatment, Disposal and Reuse 4th edn., McGraw-Hill Co., New York.

Moriizumi, M., Fukumoto, A., Yamamoto, Y. and Okumura, S. (1999): Basic studies on the characteristics of phosphorus removal by the electrochemical elution of iron. Journal of Japan Society on Water Environment 22 (6): pp. 459-464.

Oakley, S.M., Gold, A.J. and Oczkowski, A.J. (2010): Nitrogen control through decentralized wastewater treatment: Process performance and alternative management strategies, Journal of Ecological Engineering, 36: pp. 1520-1531.

Ongerth, H.J., Ongerth and J.E. (1982): Health consequences of wastewater reuse. Annual Review of Public Health 3: pp. 419-444

Pagilla, K.R., Urgun-Demirtas, M. and Ramani, R. (2006): Low effluent nutrient technologies for wastewater treatment. Water Science and Technology 53: pp. 165- 172.

Qian, Y.L. and Mecham, B. (2005): Long-term effects of recycled wastewater irrigation on soil chemical properties on golf course fairways. Agronomy Journal 97: pp. 717-721.

Rabalais, N.N. (2002): Nitrogen in aquatic ecosystems. Ambio 31: pp. 102-112.

Richardson, C.J. (1985): Mechanisms controlling phosphorus retention capacity in freshwater wetlands. Science 288: pp. 1424-1427.

Singh, S., Haberl, R., Moog, O., Shrestha, R.R., Shrestha, P. and Shrestha R. (2009): Performance of an anaerobic baffled reactor and hybrid constructed wetland treating high-strength wastewater in Nepal-A model for DEWATS. Ecological Engineering 35: pp. 654-660.

Sharpley, A.N. and Tunney, H. (2000): Phosphorus research strategies to meet agricultural and environmental challenges of the 21st century. Journal of Environmental Quality 29: pp. 176-181.

Sudo, R. (2000): The development of countermeasures for lake and sea environment, Journal of Japan Society on Water Environment 23 (10): pp. 608-613.

Takai, T., Miyasaka, A. and Inamori, Y. (2002): Phosphorus removal and recovery technique using zirconium-ferrite adsorbent. Journal of Water Waste 44 (7): pp. 54-60.

Tanner, C.C., Sukias, J.P.S., Headley, T.R., Yates, C.R. and Stott R. (2012): Constructed wetlands and denitrifying bioreactors for on-site and decentralized wastewater reatment: Comparison of five alternative configurations. Ecological Engineering 42: pp. 112-123.

Tchobanoglous, G., Burton, F.L. and Stensel, H.D. (2003): Wastewater Engineering: Treatment Disposal and Reuse, 4th ed. New York: Metcalf and Eddy Inc., McGraw Hill Science Engineering.

Tomar, P. and Suthar, S. (2011): Urban wastewater treatment using vermi-biofiltration system, Desalination, 282: pp. 95-103.

USEPA, (1993): Manual: Nitrogen Control. Office of Water, Washington, D.C. EPA/625/R-93/010.

USEPA, (2004b): Guidelines for water reuse. EPA-625-R-04-108, Washington, DC, USA.

USEPA, (2008 a,b): Memorandum: Development and Adoption of Nutrient Criteria into Water Quality Standards.

Van der Star, W., Abma, W.R., Blommers, D., Mulder, J.W., Tokutomi, T., Strous, M., Picioreanu, C. and van Loosdrecht, M. (2007): Startup of reactors for anoxic ammonium oxidation: experiences from the first full-scale anammox reactor in Roterdam, Journal of Water Research 41: pp. 4149-4163.

Vymazal, J. and Kropfelova, L. (2011): A three-stage experimental constructed wetland for treatment of domestic sewage: First 2 years of operation, Journal of Ecological Engineering 37: pp. 90-98.

Wallace, S.D. and Knight, R.L. (2006): Small-scale Constructed Wetland Treatment Systems: Feasibility, Design Criteria, and O&M Requirements. Water Environment Research Foundation, Alexandria, VA.

Wang, S., Yang, J., Lou, S.J. and Yang, J. (2010): Wastewater treatment performance of a vermifilter enhancement by a converter slag–coal cinder filter, Journal of Ecological Engineering, 36: pp. 489-494.

Wentzel, M. C., Lotter, L. H., Ekama, G. A., Loewenthal, R. E. and Marais, G. (1991): Evaluation of biochemical models for biological excess phosphorus removal. Water Science and Technology 23: pp. 567-576.

Withers, P.J.A., Jarvie, H.P., Hodgkinson, R.A., Palmer-Felgate, E.J., Bates, A., Neal, M., Howells, R., Withers, C.M. and Wickham, H.D. (2009): Characterization of phosphorus sources in rural watersheds. Journal of Environental Quality 38: pp. 1998-2011.

Xu, X., Liu, G. and Zhu L. (2011): Enhanced denitrifying phosphorous removal in a novel anaerobic/aerobic/anoxic (AOA) process with the diversion of internal carbon source, Bioresource Technology, 102: pp. 10340-10345.

Yang, P.Y., Su, R. and Kim, S. J. (2003): EMMC process for combined removal of organics, nitrogen and an odor producing substance. Journal of Environmental Management, 69: pp. 381-389.

Yang, Y., Inamori, Y., Ojima, H., Machii, H. and Shimizu, Y. (2005): Development of an advanced biological treatment system applied to the removal of nitrogen and phosphorus using the sludge ceramics. Water Research 39: pp. 4859–4868.

Chapter 11

Adsorption of Reactive Red 120 from Aqueous Solution by Locally Available Modified Adsorbents

Pramit Sarkar, Akshaya Kumar Verma,
Puspendu Bhunia and Rajesh Roshan Dash

School of Infrastructure, Indian Institute of Technology,
Bhubaneswar – 751 013, Odisha, India

ABSTRACT

In this chapter, the use of locally available and waste material for the removal of dyes from wastewater was proposed as an alternative substitution over activated carbon and other expensive adsorbents. The study investigates the adsorption potential of different locally available materials and their chemical modified forms. Chemical modification of sand and flyash was carried out using ferric nitrate and manganese dioxide and thus, adsorbent prepared from sand and flyash successfully used to remove Reactive Red 120, a diazo textile dye from the aqueous solution in batch mode. Out of six different alternative adsorbents namely sand, sand coated with manganese dioxide (SCM), sand coated with ferric nitrate (SCF), flyash, flyash coated with manganese dioxide (FCM) and flyash coated with ferric nitrate (FCF), FCF was appeared as the best adsorbent giving complete dye removal at a dose of 50 g/L within contact time only. Equilibrium study showed that the Langmuir isotherm best describes the adsorption process with correlation coefficient 0.9973 and giving adsorption capacity of 2.08 mg/g.

Keywords: Adsorption, Dye, Flyash, Reactive red 120, Sand, Textile wastewater.

Introduction

Water pollution resulting from dyestuff has been a brooding issue for scientists over the world. Industries like textile, paper, plastic, tanning, where dyeing and printing units utilise complex synthetic dyes, are the major source of dye wastewater (Sharma *et al.*, 2009; Verma *et al.*, 2012; Yousefi *et al.*, 2011). These dyes exhibit a spectrum of structures besides their complex molecular arrangement owing to which many of them are impervious to heat, light and bio-degradation (Sharma *et al.*, 2009; Yousefi *et al.*, 2011). They are commonly detected in trace amounts in industrial effluents at the discharge sites. The characteristic colouration of water due to the presence of dyes makes it easily detectable even at levels as low as 1 ppm (Yousefi *et al.*, 2011). Additionally, they may inhibit photosynthesis and thus induce negative impacts on the aquatic ecosystem. Thus the removal of colouring dyes from the industrial effluents and more importantly from that of the textile industries has become a major area of concern (Khan *et al.*, 2004).

Physico-chemical methods such as adsorption, coagulation, precipitation and electro-chemical processes have been widely used for the removal of these dyes from the industrial effluents. The chemical methods however offer the problem of disposing the accumulated sludge besides the requirement of additional electrical energy, thereby adding on to the cost of the process (Khan *et al.*, 2004). On the other hand, biological processes of treatment are handicapped with the requirement of large open area, sensitivity to diurnal variation and rigidness of design and operation amongst others. Almost all advanced oxidation processes are associated with high cost of operation and may produce the toxic secondary products. Nonetheless, adsorption process seems to be more effective in removal of colour from dyestuff effluents (Amin, 2008; Amin, 2009; Gupta and Suhas, 2009).

Considering the expenses of activated carbon, which is otherwise most effective in this regard, many economical adsorbents have been experimented in recent times for the removal of these dyes, *viz.* orange leaves, neem leaves, banana peel (Velmurugan *et al.*, 2011), *Polyalthia longofolia* (Ashoka) seed powder (Mundhe *et al.*, 2012), sugarcane bagasse (Abdullah *et al.*, 2005). Quite a bit of literature is available on the use of low-cost adsorbents on the removal of synthetic dyes from textile wastewater. To the best of our knowledge, no study has been observed in the literature on the removal of reactive dyes from the textile wastewater employing various low cost adsorbents and their modified forms. Therefore, the present chapter has been focused on the comparative evaluation of the efficiencies of sand, flyash and their modified forms for the removal of reactive red 120 (RR 120), a diazo dye from aqueous solution.

Materials and Methods

Preparation of Adsorbents

Sand was procured from the nearby places, where it is available in abundance. The sand was first sieved to a specific size range of 0.180 mm to 0.220 mm and washed with distilled water followed by drying in hot air oven to a temperature of

150°C for further use. For modification of adsorbents, the dried sand or was mixed with the chemical solution of $Fe(NO_3)_3$ $9H_2O$ or MnO_2 solution and kept in the hot air oven for drying at a temperature of 150°C, with constant stirring at regular intervals. After 24±2 hrs of mixing, when the mixture had dried completely, it was run through another cycle of washing and oven-drying before being subjected to experimentation. The similar modifications were also provided to the flyash, which was obtained from a nearby industry, already graded in accordance to a specific size of 0.02 mm.

For preparing $Fe(NO_3)_3$ $9H_2O$ (FINAR reagents) solution, 30g of the chemical was diluted with 100 mL of distilled water. The resulting solution was added to 100 grams of sand and/or flyash each to produce sand coated with $Fe(NO_3)_3$ $9H_2O$ and fly-ash coated with $Fe(NO_3)_3$ $9H_2O$, hereafter denoted as SCF and FCF, respectively. As for coating with MnO_2 (CDH Chemicals), 13 grams of the chemical was diluted with 100 mL of distilled water and then separately added to 100g of sand and/or 100g of fly-ash to produce sand coated with MnO_2 and fly-ash coated with MnO_2, hereafter denoted as SCM and FCM, respectively.

Preparation of Dye Solution

Initially a stock solution of 50 mg/L was prepared with RR 120 in distilled water. Dye was procured form SIGMA Chemicals and the characteristics of which was summarised in Table 11.1. All the chemicals used in this study were of analytical grade and all the experiment were performed at room temperature (25±5°C).

Table 11.1: Characteristics of RR 120

Molecular Formula:	$C_{44}H_{24}Cl_2N_{14}O_{20}S_6Na_6$
Molecular Weight:	1469.58 g/mol
λ_{max}:	535 nm
Type of dye:	Reactive, diazo
Molecular Structure:	

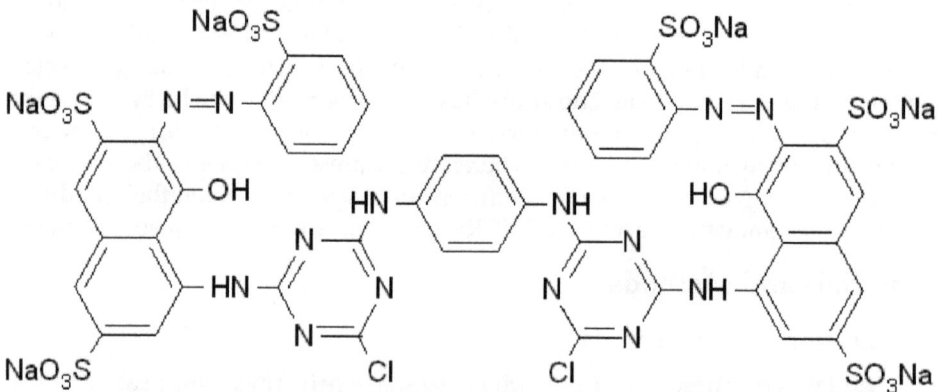

Adsorption Studies

The batch experiments were carried out to study the adsorption of RR 120 on sand, SCF and SCM as well as flyash, FCF and FCM as adsorbents. Sorption experiments were carried out in 250 mL of Erlenmeyer flask containing 100mL of dye solution with desired concentrations (10-100 mg/L). For the studies, 50 mg/L of dye solution was treated with the varying dosage of adsorbents using orbital shaker at 100 rpm and at desired temperature for various contact times. Aliquot from the flask were taken at predetermined intervals and subjected to filtration with Whatman 42 filter paper. The filtered solutions thus collected were finally neutralised to a pH 7+0.2 before measuring the absorbance. A digital pH meter (HACH) was used for all pH measurements. Colour of the samples was determined by absorbance measurement using UV-VIS Spectrophotometer (Perkin-Elmeyer, Lambda 25). The residual dye content in the test flasks were measured at the maximum absorbance wavelength (λ_{max} = 535 nm) of reactive red 120. Dye removal efficiency was calculated using following equation:

$$\text{Dye removal of efficiency (\%)} = \frac{(Ao - At)}{Ao}$$

where,

Ao = Absorbance of the blank/untreated solution, At = Absorbance of treated solution.

Results and Discussions

Effect of Adsorbent Dosage on Dye Removal Efficiency

Adsorbent dose is one of the main parameters in efficiency of absorption process. With increasing adsorbent dose, dye removal efficiency increased that enhancement in dye removal is due to increases in adsorption sites (Hameed, 2009). To study the effect of adsorbent dosage on dye removal, a series of adsorption experiments were carried out using different adsorbents namely sand, SCF, and SCM as well as flyash, FCF, and FCM. The specified amounts of above adsorbents were added in Erlenmeyer flasks containing 100 mL dye solution of 50 mg/L concentration. These flasks were subjected to shaking at 100 rpm for 30 min at room temperature. The results of sorption study was summarised in Figure 11.1a and 11.1b.

This can be observed from the findings that the dye removal efficiency continuously increased with increasing dosage of adsorbents. Further, FCF was found to be the most effective adsorbent out of all other adsorbent, producing 99.56 per cent dye removal efficiency even at very less dose (5g/100mL) (Figure 11.1a). Approximately same degree of dye removal efficiency was also observed in case of SCF at higher dose (50g/100 mL) (Figure 11.1b). Considerable dye removal efficiency of more than 80 per cent was also found in case of flyash at its extreme dose of 30g/100 mL. FCM did not appear as a significant adsorbent as compared with even flyash but found superior as compared to the sand. The superiority of FCM over sand as well as SCM may be due to dual adsorption properties *i.e.* flyash provides larger

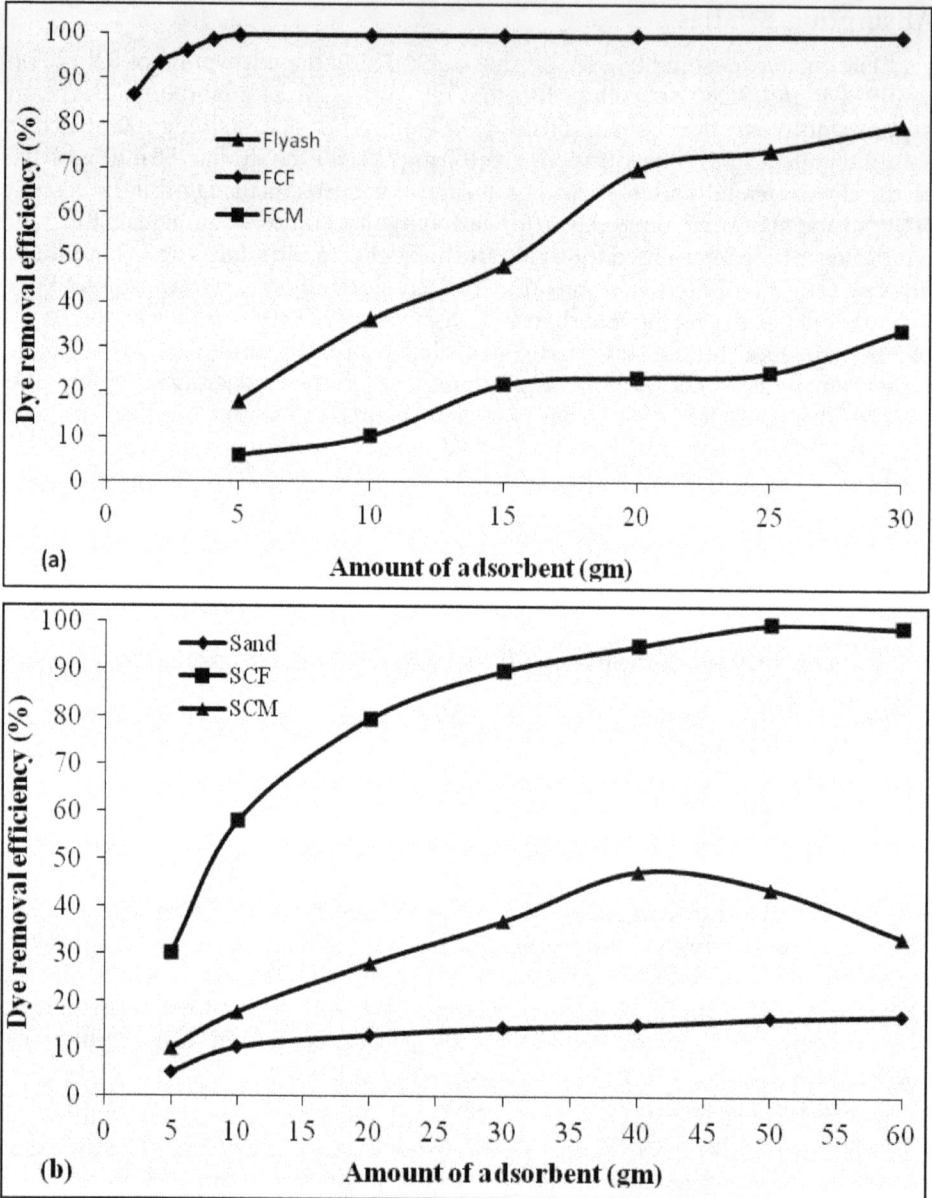

Figure 11.1: Effect of Adsorbents Dosage on Reactive Red 120 Removal
(Contact time = 30 minutes, Temperature= 25 °C).
(a) Effectiveness of flyash and its modified form,
(b) Effectiveness of sand and its modified form.

surface area and coating of magnesium oxide improves the adsorption sites for dye adsorption and therefore shows better dye removal efficiency. Similarly, magnesium oxide coating on sand further improves the adsorption ability by providing active adsorption sites for dye adsorption. Highest dye removal efficiency at very less dosage using FCF can be attributed by the fact that flyash provides the large surface area and the coating of $Fe(NO_3)_3$ $9H_2O$ in the form of ferric hydroxide provide greater number of activated sorption sites to the adsorbent.

Effect of Contact Time

Equilibrium study was carried out to determine the adsorption capacity of adsorbents and kinetic models were used to examine the rate of adsorption process. Sorption experiments were conducted using 100 mL of 50 mg/L dye solution with optimised dosage of different adsorbents. The flask containing dye solution and adsorbents were shaken at 100 rpm for 40 min. The aliquots were then taken out from the shaker at different time intervals (5, 10, 15, 20, 25, 30, 35 and 40 min). The collected samples were filtered and neutralised to pH 7 and finally analysed for residual concentration of dye.

For Sand and SCM, dye removal efficiency was consistently increased with the contact time till 40 min, which was considered as an extreme value. However, only 5 min contact time was enough to give dye removal efficiency of close to 100 per cent with SCF as adsorbent (Figure 11.2a). Similar trends of increase in dye removal efficiency with contact time up to certain duration were observed using Flyash and FCM as an adsorbent. Complete dye removal efficiency was also produced by FCF even at 5 min of contact time (Figure 11.2b). Reduction of dye removal efficiency after a certain duration was observed in almost all the cases. This can be explained by the fact that desorption of the dye molecule might be taking place at greater contact times during shaking. The results of the present study are in good agreement with the findings reported by Saha, 2010 and Velmurugan *et al.*, 2011 who used Tamarind shell and orange peel respectively for adsorption of different types of dye in aqueous solution.

Effect of Temperature

To study the effect of temperature, sorption experiments were carried out in six sets as per the six different adsorbent forms at five different temperatures 20°C, 25°C, 30°C, 35°C and 40°C. In this study dye removal efficiency for various adsorbent was investigated at their optimised dosage and contact time. It can be seen for almost all the cases that dye removal decreases with increase in temperature. However, adsorption using optimised adsorbent (based on the adsorbent dosage and contact time) namely SCF and FCF virtually appears as independent of temperature since reduction in dye removal efficiency is very nominal (Figure 11.3).

The study indicates that there is no appreciable effect of temperature on dye removal efficiency using both the optimised adsorbent SCF and FCF, which in turn reflects the suitability of the process for all the climatic conditions. Further, FCF was observed as the effective and optimised adsorbent since flyash is a waste material and modified adsorbent based on it give slightly improved efficiency over SCF and also shown absolute independency with the temperature increase (Figure 11.3).

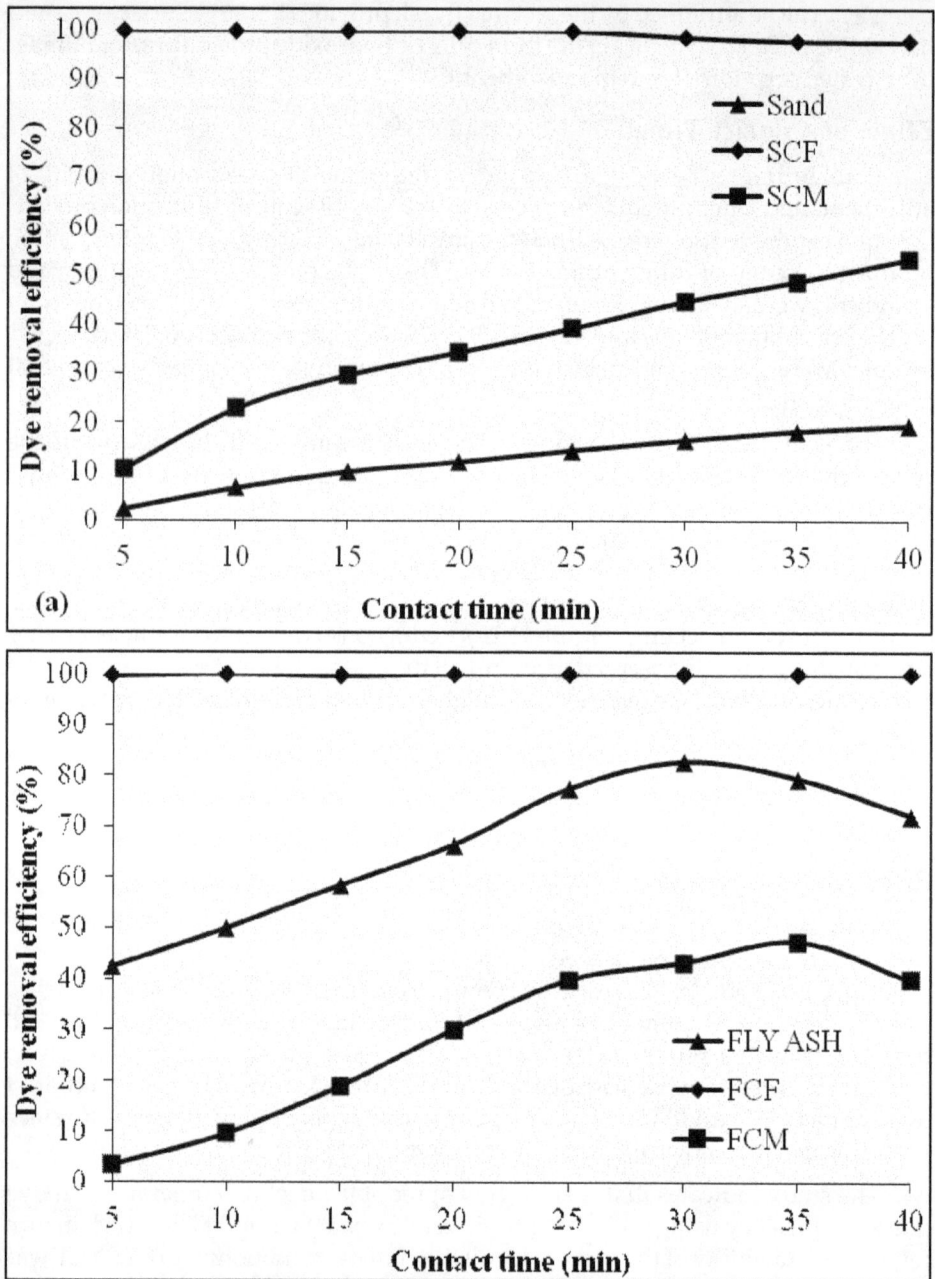

Figure 11.2: Effect of Contact Time on Dye Removal Efficiency
Utilising Optimised Dosage of Adsorbent,
(a) Sand and its modified form, (b) Flyash and its modified form.

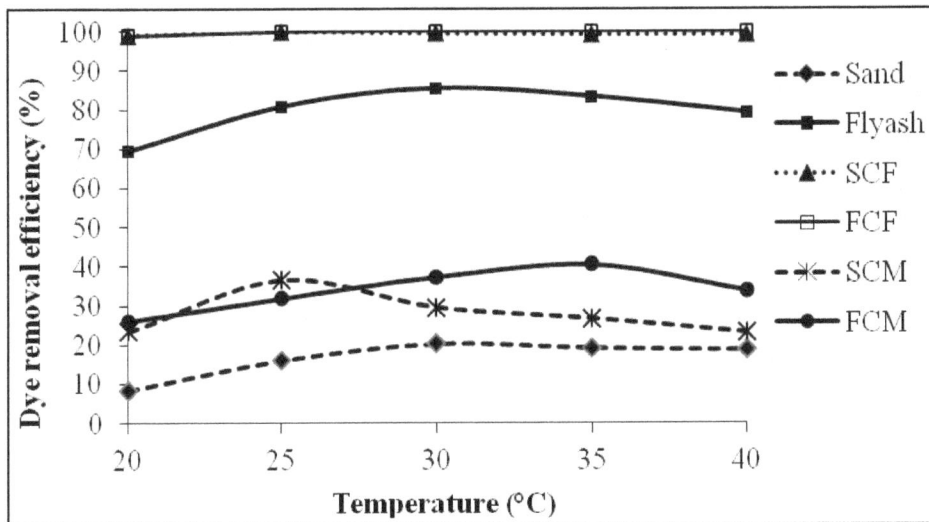

Figure 11.3: Effect of Temperature on Dye Removal Efficiency for different Forms of Adsorbents.

Effect of Initial Dye Concentration

In batch adsorption studies, the initial concentration of adsorbate in solution plays crucial role as a driving force which overcome mass transfer resistance of adsorbate between solid and aqueous phase (Gulnaaz *et al.*, 2006). A series of adsorption experiments were carried out at different initial concentrations of RR 120 (varying from 10 mg/L to 100 mg/L) with FCF at its optimised dose (50g/L) and contact time (5 min). The effect of dye concentration on the removal efficiency for the optimised adsorbent as FCF was shown in Figure 11.4.

It can be seen from the Figure 11.4 that increase in the initial dye concentration to 100 mg/L, resulted a decrease of its removal to 97.58 per cent. This can be attributed to the saturation of the available adsorption sites on the adsorbent (Mezenner and Bensmaili, 2009).

Adsorption Isotherm Analysis

A number of mathematical models have been applied for analysis of adsorption studies. Selection of isotherm depends upon the system and adsorbent types. For dye adsorption onto the solid surface, Frendulich and Langmuir isotherm equation have been used more frequently (Sharma *et al.*, 2009; El-sayad *et al.*, 2011). Therefore, the results of the experiments carried out at different dye concentrations were analysed using the Freundlich and Langmuir isotherms.

Langmuir Isotherm

According to the Langmuir isotherm, sorption of pollutants on solid surface can be described linear equation:

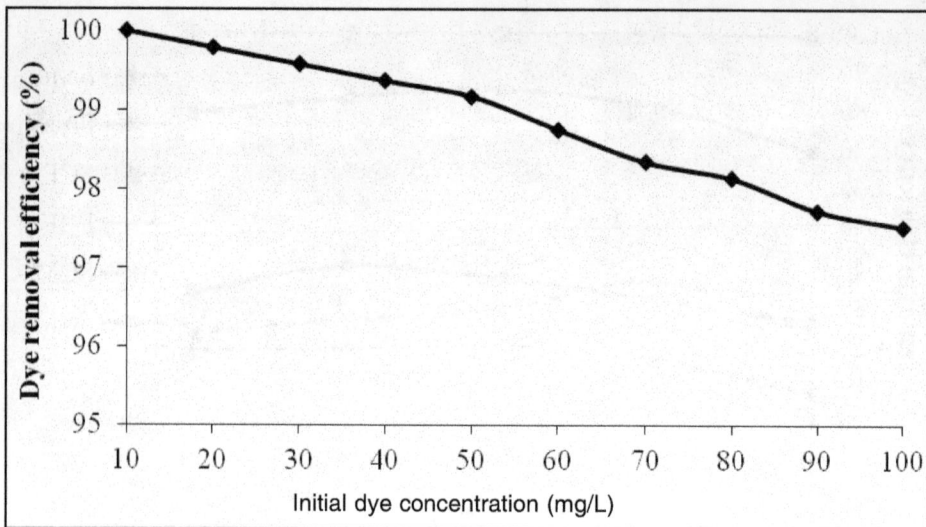

Figure 11.4: Effect of Initial Concentration on Adsorption of Dye on FCF.

$$\frac{Ce}{qe} = \frac{Ce}{Q_o} + \frac{1}{Q_o b}$$

where,

Ce: Equilibrium concentration of adsorbate

q_e: Adsorptive capacity, mg/g

Q_o: Langmuir constant related to the adsorption capacity, mg/g

b: Langmuir constant related to the energy of adsorption

The Langmuir plot for the adsorption of RR 120 onto the FCF is shown in Figure 11.5.

Freundlich Isotherm

The equilibrium adsorption data at different dye concentrations were also fitted with Freundlich isotherm model.

$$\log qe = \{\log Kf\} + \left\{ \left(\frac{1}{n}\right) \times \log Ce \right\}$$

where,

n and K_f (mg/g) are Freundlich constants related to the intensity of adsorption and adsorption capacity, respectively. The plot of log q_e against log C_e is shown in Figure 11.6.

The adsorption study revealed that the adsorption of RR 120 onto FCF can be described by both Langmuir as well as Freundlich isotherm model with correlation

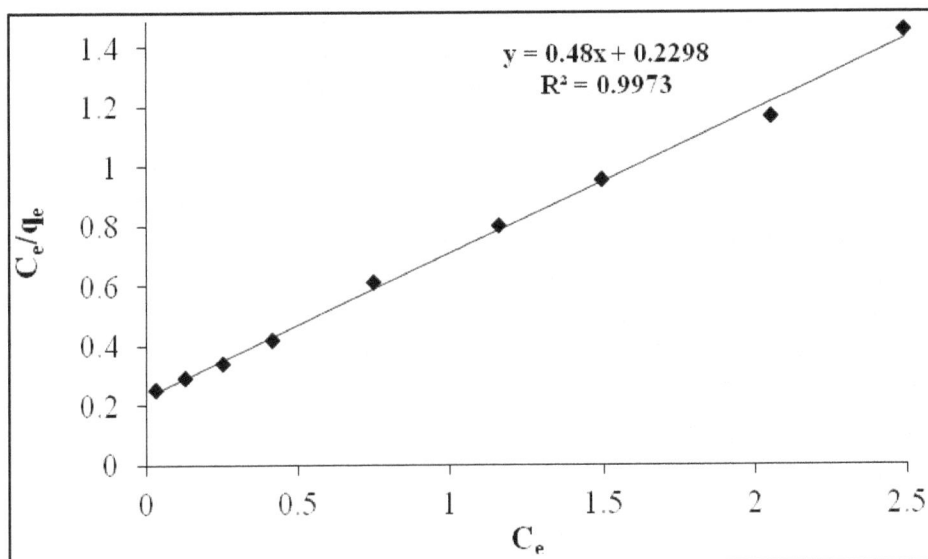

Figure 11.5: Langmuir Isotherm Plot for Adsorption of RR 120 onto FCF.

$y = 0.48x + 0.2298$
$R^2 = 0.9973$

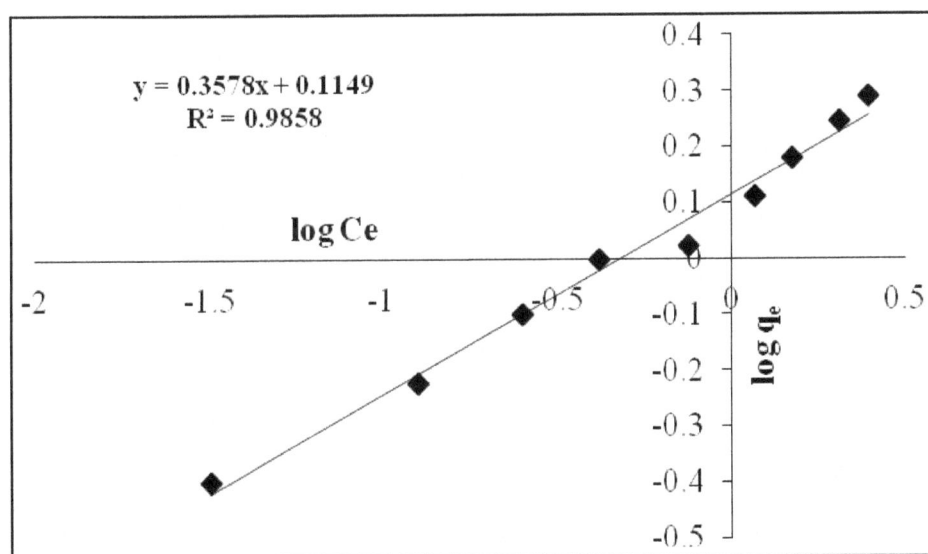

Figure 11.6: Freundlich Isotherm Plot for Adsorption of RR 120 onto FCF.

$y = 0.3578x + 0.1149$
$R^2 = 0.9858$

coefficients (R^2) of 0.9858 and 0.9973. The values of Langmuir and Freundlich isotherm constant are given in the Table 11.2. Freundlich isotherm is generally applicable to the system where the calculated value of $1/n$ lies between 0 and 1, which in turn indicate that favourable adsorption onto the utilised material (Khan *et al.*, 2004). Maximum adsorption capacity (Qo) of 2.08 mg/g was found with R^2 value of close to unity using Langmuir isotherm model. Therefore, it confers that Langmuir isotherm model better describe the adsorption process of RR 120 onto FCF. The results of present study are in good agreement with the adsorption study reported by different researcher using different adsorbents for dye removal (Santhy and Selvapathy, 2006; Khan *et al.*, 2004; Sharma *et al.*, 2009; Velmurugan *et al.*, 2011).

Table 11.2: Adsorption Isotherm Constants

Isotherm	Constants	R^2
Freundlich	1/n = 0.358, Kf = 1.128	0.9858
Langmuir	Q_o = 2.08, b = 2.09	0.9973

A comparison was made for the adsorbents used in this work and those already reported for dye removal (Table 11.3). For comparison, the maximum adsorption capacity was considered for all adsorbents. It is clear from this table that being a locally available and modified form of flyash as FCF as adsorbent came out with promising adsorption capacity.

Table 11.3: Comparison of Adsorbent Capacities of different Adsorbents

Adsorbent	Adsorption Capacity	Reference
Blast furnace sludge	1.3 mg g^{-1}	Jain *et al.* (2003)
Fly ash	0.7428 mg g^{-1}	Gupta and Suhas (1996)
Diatomite	0.42 mmol g^{-1}	Shawabkeh and Tutunji (2003)
Fly ash (treated with H_2SO_4)	0.0021 mmol g^{-1}	Lin *et al.* (2008)
Peat	0.43-0.91 mg g^{-1}	McKay *et al.* (1981)
Coal	0.223-0.798 mg g^{-1}	Khan *et al.* (2004)
Neem sawdust	3.42 mg g^{-1}	Khattri *et al.* (2000)
Raw Coir pith	1.65	Namasivayam *et al.*, 2001
Kaolnite	0.6506	Gupta and Suhas,1996
Wallastonite	0.6957	Gupta and Suhas,1996
FCF	2.08 mg g^{-1}	Present study

Conclusion

The results of present chapter revealed that the adsorption of RR 120 on the various adsorbents greatly depends upon the adsorbent dosage, contact time and initial dye concentration. All locally available adsorbents and their modified forms showed considerable removal of dye from aqueous solution; however $Fe(NO_3)_3$ $9H_2O$ coated forms of sand and flyash as SCF and FCF was appeared to depict the appreciable dye removal efficiency over other forms. Higher percentages of adsorption were observed at the lower concentrations of RR 120. Almost all the adsorption process was almost independent of temperature variation within in the selected temperature range. Based on adsorbent dosage and contact time, FCF was appeared as the best adsorbent giving almost 100 per cent dye removal at the dose of 50g/L within only 5 min. The adsorption equilibrium study showed that Langmuir models best describes the adsorption of RR 120 onto FCF producing 2.08 mg/g of adsorption capacity with R^2 value close to unity.

References

Abdullah, A.G. L., Salleh, M.A.M., Siti Mazlina, M.K., Noor, M.J. M.M., Osman, M.R., Wagiran, R. and Sobri, S. (2005): Azo dye removal by adsorption using waste biomass: sugarcane bagasse. International Journal of Engineering and Technology, 2(1): 8-13.

Amin, N. K. (2008): Removal of reactive dye from aqueous solutions by adsorption onto activated carbons prepared from sugarcane bagasse pith. Desalination, 233: 152-161.

Amin, N. K. (2009): Removal of direct blue-106 dye from aqueous solution using activated carbons developed from pomegranate peel: adsorption equilibrium and kinetics. Journal of Hazardous Materials, 165: 52-62.

El-Sayad, G.O., Aly, H.M. and Hussein, S.H.M. (2011): Removal of acrylic dye Blue-5G from aqueous solution by adsorption on activated carbon prepared form Maize cops. International Journal of Research in Chemistry and Environment,1(2): 132-140.

Gulnaz, O., Kaya, A. and Dincer, S. (2006): The reuse of dried activated sludge for adsorption of reactive dye. Journal of Hazardous Materials, 134 (1-3): 190-196.

Gupta, G.S. and Shukla, S.P. (1996): An inexpensive adsorption technique for the treatment of carpet effluents by low cost materials. Adsorption Science and Technology, 13: 15–26.

Gupta, V. K. and Suhas. (2009): Application of low-cost adsorbents for dye removal-a review. Journal of Environmental Management, 90: 2313- 2342.

Hameed, B. H. (2009): Grass waste: A novel sorbent for the removal of basic dye from aqueous solution. Journal of Hazardous Materials, 166: 233–238.

Jain, A.K., Gupta, V.K., Bhatnagar, A. and Suhas. (2003): Utilization of industrial waste products as adsorbents for the removal of dyes. Journal of Hazardous Materials, 101: 31–42.

Khan, T.A., Singh, V.V. and Kumar, D. (2004): Removal of some basic dyes from artificial textile wastewater by adsorption on Akash Kinari coal. Journal of Scientific and Industrial Research, 63: 355-364.

Khattri, S.D. and Singh, M.K. (2000): Colour removal from synthetic dye wastewater using a bioadsorbent. Water, Air, and Soil Pollution, 120: 283–294.

Lin, J.X., Zhan, S.L., Fang, M.H., Qian, X.Q. and Yang, H. (2008): Adsorption of basic dye from aqueous solution onto fly ash. Journal of Environmental Management, 87: 193–200.

McKay, G., Allen, S.J., McConvey, I.F. and Otterburn, M.S. (1981): Transport processes in the sorption of coloured ions by peat particles. Journal of Colloid and Interface Science, 80: 323–339.

Mezenner, N. Y. and Bensmaili, A. (2009): Kinetics and thermodynamic study of phosphate adsorption on iron hydroxide- eggshell waste, Chemical Engineering Journal, 147: pp. 87–96.

Mundhe, K.S., Gaikwad, A.B.,Torane, R.C., Deshpande, N.R. and Kashalkar, R.V. (2012): Adsorption of methylene blue from aqueous solution using *Polyalthia longofolia* (Ashoka) seed powder. Journal of Chemical and Pharmaceutical Research, 4 (1): 423-436.

Namasivayam, C. and Kavitha, D. (2002): Removal of Congo red from water by adsorption onto activated carbon prepared from coir pith, an agricultural solidwaste. Dyes and Pigments, 54: 47–58.

Saha, P. (2010): Assessment on the Removal of Methylene Blue Dye using Tamarind Fruit Shell as Biosorbent. Springer Science+Business Media 213: 287– 299.

Santhy, K. and Selvapathy, P. (2006): Removal of reactive dyes from wastewater by adsorption on coir pith activated carbon. Bioresource Technology, 97(11): 1329-1336.

Sharma, Y. C., Upadhyay, Uma S. N. and Gode, F. (2009): Adsorptive removal of a basic dye from water and wastewater by activated carbon. Journal of Applied Sciences in Environmental Sanitation, 4(1): 21-28.

Shawabkeh, R.A. and Tutunji, M.F. (2003): Experimental study and modeling of basic sorption by diatomaceous clay. Applied Clay Science, 24: 111–120.

Velmurugan, P., Rathinakumar, V. and Dhinakaran, G. (2011): Dye removal from aqueous solution using low cost adsorbent. International Journal of Environmental Sciences, 1(7): 1492-1503.

Verma, A.K., Dash, R.R. and Bhunia, P. (2012): A review on chemical coagulation/flocculation technologies for removal of colour from textile wastewaters. Journal of Environmental Management, 93: 154-168.

Yousefi, N., Fatehizadeh, A., Azizi, E., Ahmadian, M., Ahmadi, A., Rajabizadeh, A. and Toolabi, A. (2011): Adsorption of reactive black 5 dye onto modified wheat straw: isotherm and kinetics study. Sacha Journal of Environmental Studies, 1(2): 81-91.

Chapter 12

Impact of Mercury Contained Effluent and Reclamation of Mercury Contaminated Environment by a Cyanobacterium

Alaka Sahu and A.K. Panigrahi

Environmental Science Research Division,
Department of Botany, Berhampur University,
Berhampur – 760 007, Odisha, India

ABSTRACT

The present chapter was designed to study the impact of the mercury contained effluent of a chlor-alkali industry on a cyanobacterium and its possible detoxification in the environment. The liquid effluent of the industry was continuously discharged into Rushikulya River Estuary where highest level of mercury to the tune of maximum 2.65 mg of Hg l^{-1} was recorded, which is much more than the stipulated limit of 0.01 mg l^{-1}, set by Pollution Control Board. The alga, *Westiellopsis prolifica,* Janet is more tolerant and less sensitive to toxicants. At lower concentrations, dry weight increased, chlorophyll content increased, GPP and NPP value increased,. At higher concentrations dry weight decreased, chlorophyll level depleted, GPP and NPP values declined, significantly. The exposed alga showed $1.22 \pm 0.09 \mu g$, $2.42 \pm 0.52 \mu g$ and $3.92 \pm 0.88 \mu g$ of Hg/50 ml culture after 15 days of exposure in 0.45 per cent (X), 1.6 per cent (Y) and 3.8 per cent (Z) effluent concentration, respectively. After 15 days of recovery period, 68.5 per cent, 23.97 per cent and 19.89 per cent of mercury removal/excretion was marked in X, Y and Z effluent concentrations, respectively. At lower concentration exposure, higher percent of mercury removal was marked, when compared to higher concentration exposure. Significant increase in chlorophyll content was recorded in 0.45 per cent effluent treatment, when compared to the control value. Significant decrease in chlorophyll content was marked at 3.8 per cent effluent concentration, when compared to the control value. Significant decrease in phaeophytin content was recorded in 3.8 per cent effluent treatment, when compared to the control value. Where as, increase in phaeophytin content was marked at 0.45 per cent and 1.6 per cent effluent concentration, when compared to the control value. No recovery was marked during recovery studies. Carotenoid content decreased in all the exposed concentrations, when compared to the control value. Higher photosynthetic rate (P.R.) was recorded at all exposure periods in X concentration, when compared to the control value. In Y concentration, increase in photosynthetic rate up to 12th day of exposure and then decrease in P.R. value was noted. Significant depletion in P.R. value was recorded in Z concentration at all exposure periods, when compared to the control value. No significant recovery was recorded during recovery studies. *Westiellopsis prolifica*, Janet could volatilize mercury from the medium as seen from the experiment designed by us in our laboratory. The exposed alga could volatilize 1.93 µg, 3.94 µg and 5.14 µg of mercury in X, Y and Z concentrations within a period of 15 days. The exposed alga could accumulate 1.26 µg, 2.62 µg and 3.94 µg of mercury within 15 days in X, Y and Z effluent concentrations. The exposed alga could remove 98.46 per cent, 56.94 per cent and 35.83 per cent of total mercury present in the medium of X, Y and Z effluent concentration either by residual accumulation or by volatilization. In the entire process 0.93 per cent, 2.86 per cent and 1.82 per cent of mercury was lost/unseen/not traceable in X, Y and Z concentrations, respectively.

Keywords: *Chlor-alkali industry, Effluent, Mercury, Cyanobacterium, Pigments, Photosynthesis, Detoxification, Reclamation.*

Introduction

The chlor-alkali industry M/S Jayashree Chemicals Pvt. Ltd., is situated at Ganjam, on the Bank of Rushikulya estuary about 1.5 km. Away from the sea, Bay of Bengal, on the East and 30 km. North of Berhampur city (Figure 12.1) on the south-eastern side of India at 84° 53′E longitude and 19° 16′N latitude. The industry was discharging its effluent containing mercury into the estuary and deposits solid waste on the adjacent land areas. This addition of mercury is the primary contamination. The secondary contamination occurs through the chimney by way of evaporation during electrolysis in the cell house, into the atmosphere and its fall out by the process of precipitation. All these discharges collectively seem to cause a major environmental threat to crop production and also to fisherman engaged in fishing both in the river and also in the estuary.

It is well established that the wastes from chlor-alkali industry contain mercury as the potential pollutant which makes an entry into the paddy fields. In near by paddy fields, nitrogen fixing cyanobacteria *Westeillopsis prolifica,* Janet grows abundantly and thereby, is exposed to toxic stress of mercury. The present piece of work was designed to study the impact of the effluent containing mercury on the growth of a blue-green alga, abundantly available in nearby crop fields, the *Westiellopsis prolifica,* Janet and whether this alga might be used as a biological agent

Figure 12.1: Arc GIS Explore Photograph Showing Positional Location of Chlor-alkali Industry at Ganjam, Rushikulya River, Estuary and Bay of Bengal)

to detoxify the contaminated environments simultaneously fixing atmospheric nitrogen and acting as a biofertilizer.

Materials and Methods

Collection and Storage of the Effluent

Effluent from the discharge channel of the chlor-alkali industry was collected in 5 liter capacity stoppered glass bottles from three different points during the discharge of cell house washings and brought to the laboratory. All the effluent samples were mixed in a bigger glass container and stored for experimental use. Since storage of the effluent for longer period would cause some physico-chemical changes, it was left as such for three months during which any change liable to take place would be completed and after that very little or no change would take place on further storage (Shaw, 1987). The important physico-chemical properties of the effluent was then determined as per the standard procedures followed by Shaw (1987) and Sahu (1987).

Experimental Procedure

Each time the effluent before use was thoroughly hand-shaken and the suspended particles were allowed to settle. The decanted clear supernatant was used for the experiments. Graded concentrations of the effluent were prepared using the culture medium as diluent and were expressed in terms of per cent (v/v). Three concentrations of the effluent *viz.* 0.45 per cent,1.6 per cent and 3.8 per cent were selected from the algal bioassay and were named as treatment 'X', 'Y' and 'Z' respectively in all the experiments. Treatment 'X' was the maximum allowable concentration (MAC), treatment 'Y' was LC_{50} and treatment 'Z' was LC_{90} deduced from toxicity test. Fifty milliliter each from the three concentrations of the effluent amended with nutrient medium and normal culture medium as control, were taken in 100 ml Borosil/Corning conical flasks stoppered with non-absorbent cotton plugs. They were irradiated with UV for 10 minutes and then inoculated with 1 ml of homogeneous suspension of the alga to each flasks. For each concentration of the effluent and control, eight sets of flasks were taken and each one of them was again replicated either 3 times or 5 times, as per the requirement of the experiment. Estimation of different parameters was made after the 3rd, 6th, 9th, 12th and 15th days of exposure. The algal cultures were centrifuged in a refrigerated centrifuge at 20°C and 5000 rpm, for 10 minutes. The algal mass in the pellet was thoroughly washed in double distilled water and centrifuged again before any estimation. After 15 days of exposure, three sets of flasks from each treatment and control, were centrifuged, the algal residues were thoroughly washed in double distilled water, centrifuged again and were re-suspended in freshly prepared sterilized culture medium (50 ml in each 100 ml flask). These flasks were allowed to recover for 15 days. Estimation of different parameters during recovery was made after 5th, 10th and 15th day of recovery. In order to maintain proper concentration of the effluent during the period (15 days) of exposure, the medium in all the treatments was changed at 5 (five) days interval under aseptic conditions, except in case of nitrogen estimation studies. In nitrogen estimation study, the exposed solutions were not changed but at 5 day interval, the exposed solutions were recharged with fresh limited toxicant only.

Methods of Algal Culture

The experimental algal cultures were grown under controlled conditions of lighting, temperature and type of culture vessels (100 ml Borosil conical flasks stoppered with non-absorbent cotton plugs). Culture flasks were kept in series on a culture rack of glass plate with steel frame. The light intensity at the mid level of the culture rack varied between 2200 ± 200 lux. Since the light intensity varied with the location under the light bank, the positions of culture vessels of the culture racks were changed alternating every day in order to compensate for any possible variation. The cultures were illuminated 16 hrs a day throughout the entire period of experiment. Temperature of the culture room was maintained at 26 ± 2°C. The culture flasks were hand shaken twice daily to avoid clumping and adherence of the algal filaments to the walls of the culture vessels (Allen and Arnon, 1955).

Growth Measurements

Growth of the alga was measured by dry weight method. Dry weight of the alga in the culture flasks was estimated after centrifuging in a refrigerated centrifuge at 8000 rpm and 20°C for 10 minutes. The algal pellet was then transferred to pre-weighed glass bottle and dried in an oven and cooled in a desiccators and the weight of the dried algal mass was recorded in a single pan balance.

Pigment Studies

The amount of total chlorophyll and phaeophytin was estimated and calculated by using the formula given by Vernon (1960). The amount of carotenoid was estimated and calculated by using the formula given by Davies (1976).

Total chlorophyll (mg l^{-1}) = 6.45 (E at 665) + 17.72 (E at 649)

Total phaeophytin (mg l^{-1}) = 6.75 (E at 666) + 26.03 (E at 655)

where, E = Extinction value (Optical density)

$$\text{Carotenoid content (in g. } l^{-1}) = \frac{E\,V\,K}{2500 \times 100}$$

where,

E: Extinction at 475 nm

V: Volume of the extract,

K: dilution factor

2500: Specific extinction co-efficient at 475 nm.

Residual Mercury Analysis

The algal mass was digested in a Bethge's apparatus with 10 ml of acid digestion mixture (1:3 Conc. H_2SO_4 and HNO_3). The gas evolved during digestion along with acid fumes was condensed simultaneously in the reflux condenser fitted to the apparatus. To ensure complete digestion the process of reflux condensation was repeated thrice (Wantorp and Dyfverman, 1955). Care was taken to avoid loss of mercury vapour during digestion. In the analytical process a suitable aliquot (1 ml) of blank, standard or sample was stirred for 5 minutes with the help of a magnetic stirrer in acid medium (7 ml 10 per cent HNO_3) along with 2 ml of 20 per cent (w/v) $SnCl_2$ in a closed reaction vessel. Mercury vapour released in the reaction flask was pumped into the mercury analyser, set at 100 per cent transmission. The absorbance was read at 253.7 nm in a cold vapour atomic absorption 'Mercury Analyser' (MA 5800, ECIL, India). The amount of mercury in the samples was computed from the standard graph. The residual mercury contents were expressed in µg of mercury per 50 ml algal culture.

Oxygen Evolution Measurements

The evolution of oxygen due to photosynthesis was measured manometrically with the help of a photowarburg's apparatus (New Paul, India) followed the procedures of Oser (1965) and Sahu (1987). Rate of photosynthesis was expressed in µl of O_2 evolved/hr/50 ml culture of the alga.

Decontamination Studies

An experimental set was designed to study the possible volatilization of mercury by the alga. During the trial and error selection and pilot test, it was observed that the alga Westiellopsis prolifica, Janet, can tolerate and grow better in mercury contaminated environments, when compared to some other heterocystous and non-heterocystous blue-green alga. The alga was collected from these contaminated environments isolated and cultured in the laboratory. Periodic sub-culturing was carried out to obtain a pure axenic culture of the alga. The axenic culture of *Westiellopsis* was grown sufficiently in large quantities for the experimental purposes. Three wide mouth stoppered flasks (Corning Glass) with lateral outlets were selected for the purposes. A glass tube was inserted up to the bottom of each flask through the stopper. Both the outlet and inlet glass tubes were connected with narrow PVC tubes and both were fitted with pinch cocks to make the system air-tight. The entire system was sterilized following the usual laboratory procedures. Experimental culture mix was prepared with the effluent and culture medium, so as to get 0.45 per cent, 1.6 per cent and 3.8 per cent effluent concentrations and kept in separate flasks. The media in all the flasks were irradiated with UV, homogenized axenic culture of *Westiellopsis prolifica*, was inoculated into the flasks under aseptic condition, and kept in the culture room under 2200 ± 100 lux illumination and temperature maintained at $26 \pm 2°C$. The air-tight flasks were hand shaken twice daily. After 5 days interval the outlet was connected to the Mercury analyser through different trappers. Before hand a stream of aseptic air was pumped through the trappers to get clean sterilized air into the culture flasks. The bubbling of air through the culture medium carries the mercury vapour along with any particle, which passes through the trappers and finally enters into the 'Mercury Analyser'. The instrument was a cold vapour atomic absorption spectrophotometer unit. The air from the culture vessel was continuously drawn into the analyser and the instrument was kept in 'Hold' position, so as to record the highest absorption. The optical density obtained in the experiment was recorded. The values for mercury were obtained from the standard curve, prepared earlier with the help of a graded series of concentrations of standard (pure redistilled mercury reacted with pure Hg free concentrated HNO_3, GR). The experiments were repeated several times for confirmation. The efficiency, accuracy and sensitivity of the procedure were quantified by statistical calculations and interpretations (Sahu, 1987). All the obtained data were statistically analyzed.

Results and Discussions

The effluent of the chlor-alkali industry was carefully collected and brought to the laboratory in glass containers and analyzed. After analysis of the mercury content present in the effluent samples brought in different containers were mixed and mean mercury concentration was estimated. The mixed effluent was sterilized by UV irradiation and stored in closed containers, to avoid biomethylation of mercury by microbes. A graded series of concentrations of the mixed effluent (now called as effluent) was prepared in different culture flasks along with the culture medium. Unialgal, pure axenic cultures of *Westiellopsis prolifica*, Janet was inoculated in an inoculation chamber (aseptic). The survival percentage and lethal concentration values were determined after 15 days of exposure. From the toxicity study and toxicity curve, the lethal concentration values were determined (Table 12.1).

Table 12.1: Showing the Per cent Survival Value, Lethal Concentration Value and Effluent Concentration

Lethal Concentration Value (LC)	Effluent Concentration (per cent)	Percent Survival Value (PS)
LC_0 (X)	0.45	PS_{100} (X)
LC_{10}	0.65	PS_{90}
LC_{50} (Y)	1.6	PS_{50} (Y)
LC_{90} (Z)	3.8	PS_{10} (Z)
LC_{100}	4.1	PS_0
MAC	0.45	PS_{100}

The above Table 12.1 indicates the lethal concentration and percent survival values deduced from toxicity study. This alga could tolerate up to 0.45 per cent of the effluent concentration, where hundred percent survivals and no death was recorded. This concentration was selected and named as "X" concentration for the entire period of experimentation. The second selected concentration was 1.6 per cent of the effluent, where 50 per cent mortality and 50 per cent survivability was marked and named as "Y" concentration. The final and third selected concentration was 3.8 per cent of the effluent, where 90 per cent mortality and 10 per cent survivability was marked and named as "Z" concentration. MAC was selected to study the effect of the effluent on the organism without killing the organism. The second selected concentration was 1.6 per cent and the third selected concentration was 3.8 per cent of the effluent where 50 per cent and 90 per cent mortality was observed, respectively. These three concentrations were selected to study the effect of the mercury contained effluent on the physiological and biochemical parameters of the exposed alga and the observed effects were compared with the control alga. The Table also indicated that with the increase in the effluent concentration, the percent survival decreased and the per cent mortality increased. The PS values showed the existence of a negative correlation and LC values showed a positive correlation. With the increase in exposure period, the lethal concentration values decreased, showing a negative correlation. Experiments were conducted exposing the blue-green alga, only for 15 days. This alga showed maximum exponential growth up to 12 days and then the growth was stabilized and after 15 days of exposure, the declining trend in growth started. Once the exposed alga was transferred to toxicant free medium, in recovery studies, with fresh nutrient medium the growth rate was revitalized for another period of 15 days. After which, the aged alga need to be homogenized and recharged. Hence, experiments were planned to complete within 30 days. Fifteen days of exposure and 15 days of recovery was planned to test whether 15 day exposed alga could recover within the same period of recovery.

Experiments were conducted to study the residual mercury accumulation at different exposure periods and excretion/removal of mercury during recovery period. In the control set, no residual mercury accumulation was recorded indicating absence of mercury in the background. No mercury was detected by the Mercury Analyser in the control set and at '0' day exposure (inoculation day). At 0.45 per cent effluent (X) concentration, the residual mercury increased from 0.62 ± 0.14 µg of mercury to 1.22 ± 0.09 µg of mercury/50 ml algal culture within 15 days of exposure. When the exposed alga was transferred to toxicant free medium, after 15 days of recovery, 0.38 ± 0.06 µg of Hg/50 ml culture was recorded. On 5th day of recovery 9.02 per cent mercury removal might be due to excretion, was recorded. On 10th day and 15th day of recovery, 22.95 per cent and 68.8 per cent of mercury removal was recorded, respectively, in the exposed culture.

In Y set (1.6 per cent effluent), higher rate of residual mercury accumulation/ retention was marked, when compared to the X set. Higher values were recorded at all exposure periods, when compared to the X set. A maximum of 2.42 µg of mercury/ 50 ml culture was recorded on 15th day of exposure and when the exposed alga was transferred to toxicant free medium, a maximum of 1.84 ± 0.28 µg of mercury/50 ml algal culture was recorded. A maximum of 23.97 per cent decrease in the residual mercury level was marked on 15th day of recovery, when compared to 15th day exposure value (Figures 12.2A and B). With the increase in exposure period, the residual mercury increased showing the existence of a significant positive correlation. The correlation coefficient analysis indicated the existence of positive correlations ($r = 0.931$, $P \leq 0.01$) in the X set and ($r = 0.973$, $P \leq 0.01$) in the Y set. Data indicated the existence of non-significant correlation up to 6th day of exposure and significant correlation ($P \leq 0.05$) after 6th day of exposure and up to 15th day of exposure. In the Z set (3.8 per cent effluent concentration), the residual mercury accumulation was highest. With the increase in exposure period, the residual mercury accumulation increased showing a positive significant correlation ($r = 0.987$, $P \leq 0.001$). A maximum of 3.92 ± 0.88 µg of mercury/50 ml algal culture was recorded. During recovery period, no significant recovery was recorded. A maximum 20.66 per cent, 0.25 per cent and 19.89 per cent mercury removal was recorded on 5, 10 and 15th day of recovery (Figure 12.2B).

Figure 12.3 indicated changes in dry weight of the control and effluent exposed alga at different exposure and recovery periods. In the control set, significant growth

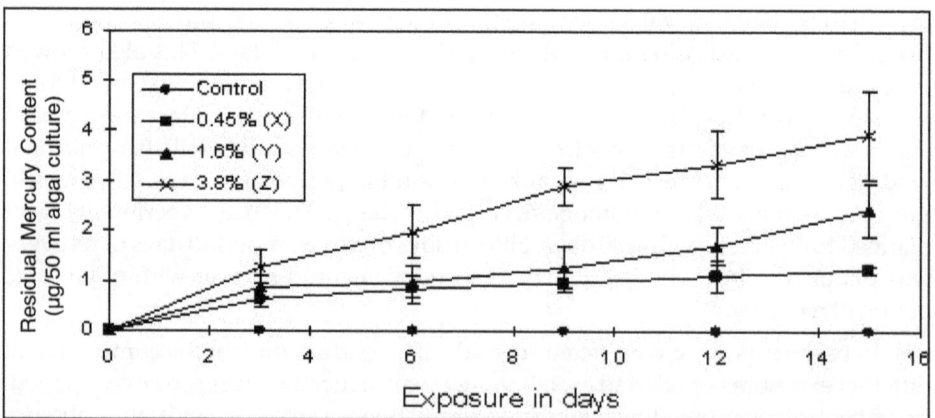

Figure 12.2A: Showing Changes in Residual Mercury Content of Control and Effluent Exposed Alga at different Effluent Concentrations at different Exposure Periods.

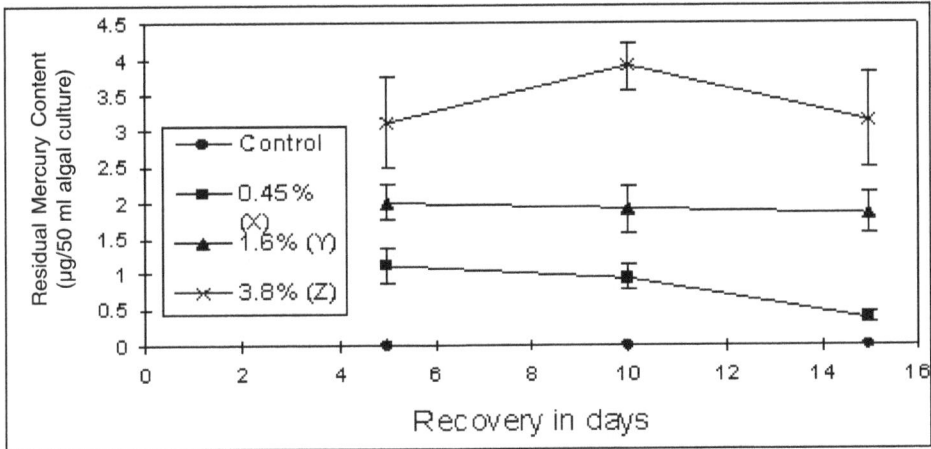

Figure 12.2B: Showing Changes in Residual Mercury Content of Control and Exposed Alga at different Effluent Concentrations at different Recovery Periods.

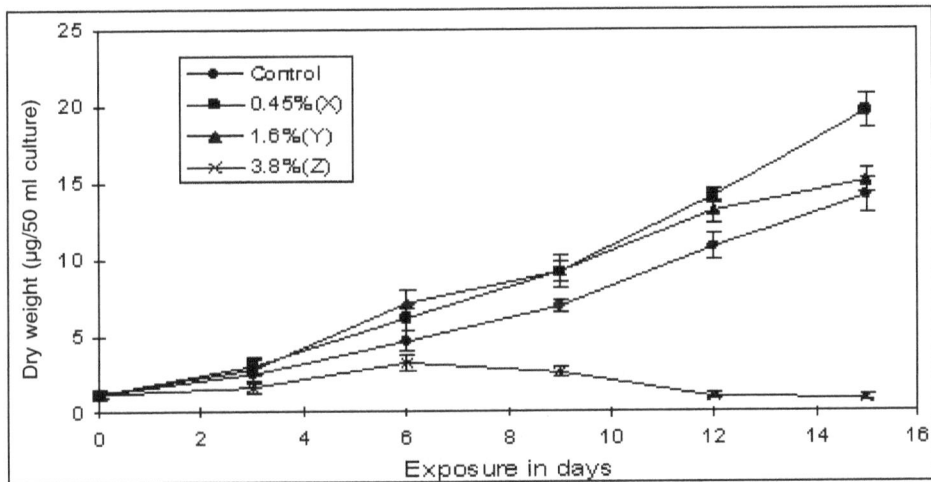

Figure 12.3: Showing Changes in Dry Weight of Control and Exposed Alga at different Effluent Concentrations and different Exposure and Recovery Periods.

of the alga was recorded. The dry weight increased from 1.16 ± 0.22 mg to 14.14 ± 1.06 mg within 15 days of exposure, showing significant growth of the alga. The control set alga, when recharged with fresh nutrient medium, still better growth was recorded during recovery period. At 0.45 per cent effluent concentration, excellent growth of the alga was observed, when compared to the control set. The dry weight increased from 1.16 ± 0.22 mg to 19.66 ± 1.14 mg within 15 days of exposure. When compared to the control set, 39 per cent increase over the control value was recorded. At 1.6 per cent of the effluent concentration, on 6th day of exposure 53.9 per cent increase in dry weight over the control value was recorded. With the increase in exposure period, the percent increase, decreased significantly and 6.8 per cent increase in dry weight was recorded on 15th day of exposure. The 'Y' set exposed alga was transferred to toxicant free medium for recovery studies. An initial recovery followed by significant decrease in the dry weight was marked. On 15th day of recovery, 7.3 per cent decrease in the dry weight was recorded. When we observe the data of Figure 3.3 and Table 3.3, in Y set, the dry weight increased from 15.11 ± 0.73 mg to 29.32 ± 1.32 mg within 15 days but when we compare the obtained value with the control set, decrease in dry weight was marked (Figure 12.4).

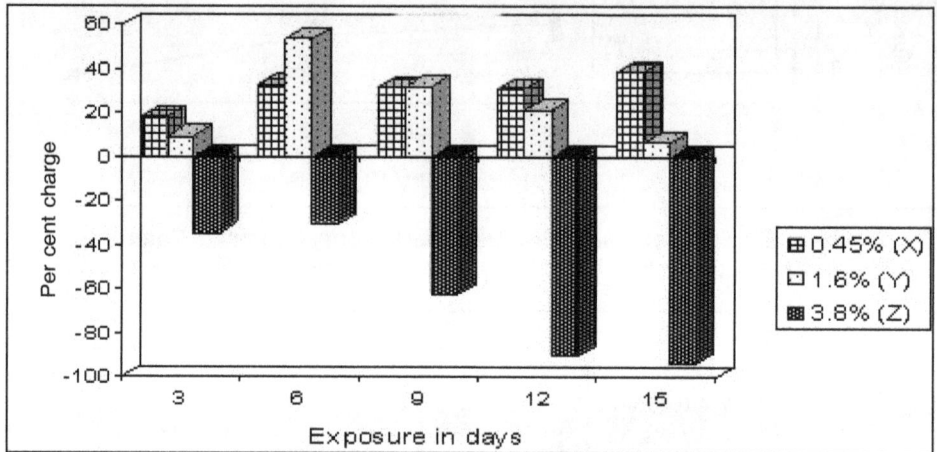

Figure 12.4: Showing Changes in Dry Weight of Exposed Alga at different Concentrations of the Effluent of the Chlor-Alkali Industry, at different Exposure Periods when Compared to Control.

At 3.8 per cent effluent concentration (Z set), the dry weight increased, insignificantly up to 3.22 ± 0.52 mg on 6th day of exposure and then with the increase in exposure period, the dry weight of the alga in exposed flasks declined and 1.01 ± 0.14 mg was recorded on 15th day of exposure. No recovery was marked, when the exposed alga was transferred to toxicant free medium. A maximum of 94.1 per cent decrease in dry weight, when compared to the control value was marked and the percent decrease reached maximum to 99.1 per cent on 15th day of recovery (Figure 12.4). The correlation coefficient indicated the existence of non-significant correlation between the dry weight vs. effluent concentrations.

Figure 12.5 shows the changes in total chlorophyll content in control and effluent exposed alga at different exposure and recovery periods. The chlorophyll content increased from 6.2 ± 1.1 to 26.8 ± 3.4 µg/50 ml culture within a period of 15 days, in the control set. Whereas, at 0.45 per cent effluent concentration (X), the total chlorophyll content significantly increased with the increase in exposure period and the value of each exposure period was more than the control value. Both the control set and X set showed the existence of a significant positive correlation. On 12th day of exposure 27.2 per cent and on 15th day of exposure 19 per cent increase was marked when compared to the control value. At 1.6 per cent effluent concentration, the total chlorophyll content increased significantly, when compared to the control set and X set up to 6th day of exposure. The total chlorophyll content increased with the increase

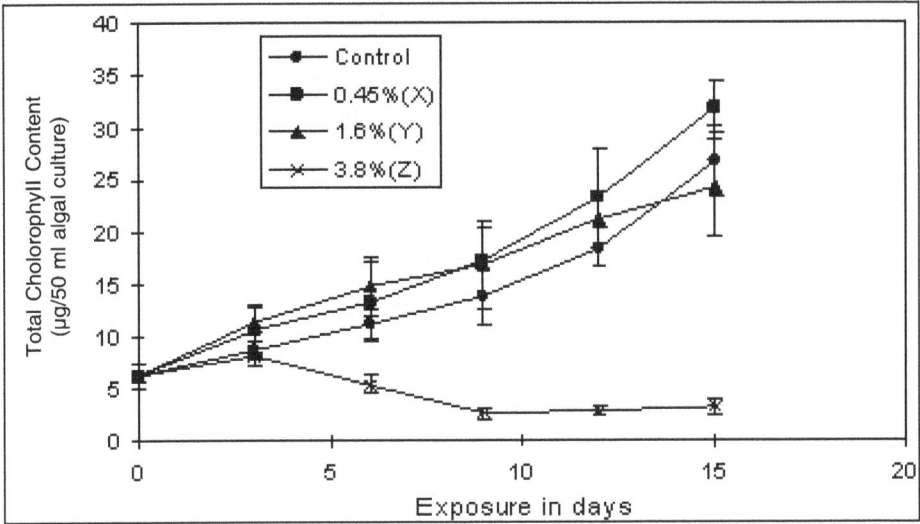

Figure 12.5: Showing Changes in Total Chlorophyll Content in Control and Effluent Exposed Alga at different Concentrations of the Effluent and different Exposure and Recovery Periods.

in exposure period up to 15th day of exposure and also in the recovery period. But the values were less than the control values. On 15th day of exposure 9.7 per cent decrease in the parameter was recorded. In recovery studies, we observed an initial recovery

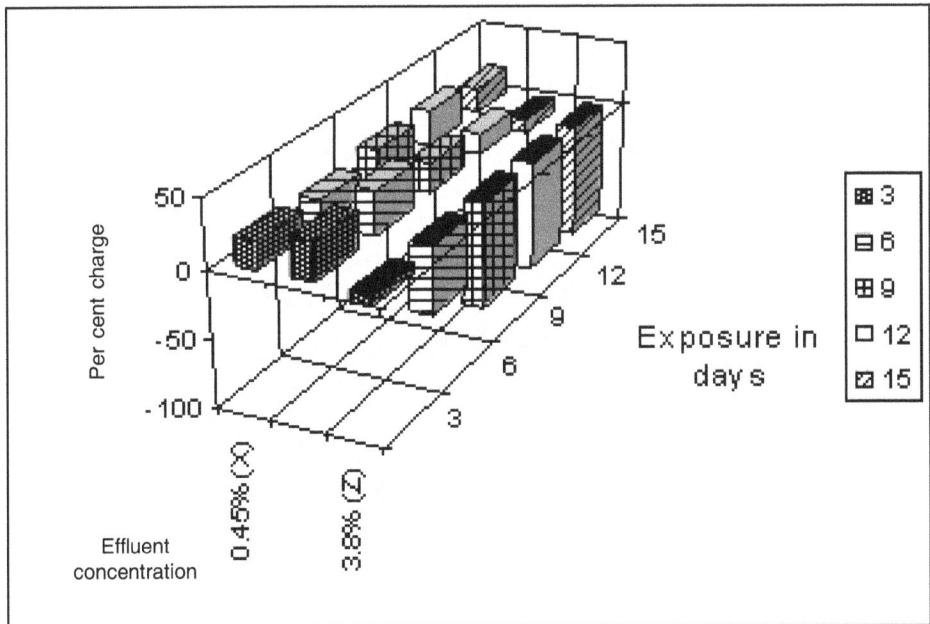

Figure 12.6: Showing Per cent Change in Total Chlorophyll Content in Effluent Exposed Alga at different Exposure Periods when Compared to Control.

by 22.1 per cent then the per cent decrease, increased and a maximum of 13.6 per cent decrease was recorded on 15th day of recovery (Figure 12.6). At 3.8 per cent effluent concentration, an initial insignificant increase followed by decrease on total chlorophyll content was recorded. On 15th day of exposure, 88.1 per cent decrease in total chlorophyll content when compared to the control value was marked. When the exposed alga (Z set) was transferred to toxicant free media, an insignificant recovery by 7.1 per cent was recorded on 15th day of recovery. The two way analysis of variance ratio test conducted for the total chlorophyll content data based on Figure 12.5 indicated the existence of significant difference between rows and columns. The correlation coefficient analysis conducted between the total chlorophyll content and effluent concentration indicated the existence of non-significant (P=NS) negative correlation at all exposure periods. The phaeophytin content increased steadily with the increase in exposure period and recovery period in the control set, showing the existence of a positive correlation between exposure period and phaeophytin content. At 0.45 per cent effluent concentration, the phaeophytin content showed an insignificant variation, when compared to the control alga. The pigment content increased at few exposure periods and decreased on 3rd and 15th day of exposure. During recovery studies, the rate of depletion of the pigment though increased, but the values were significantly less than the control set (Figure 12.7).

In the Y set, interestingly the phaeophytin content increased from 7.6 $\pm$ 1.1 to 34.6 $\pm$ 2.8 µg/50 ml of algal culture within 15 days of exposure period. The 15th day of exposure value was more than double of the respective control value, where 102.3 per cent increase was recorded. But during recovery period, the pigment content

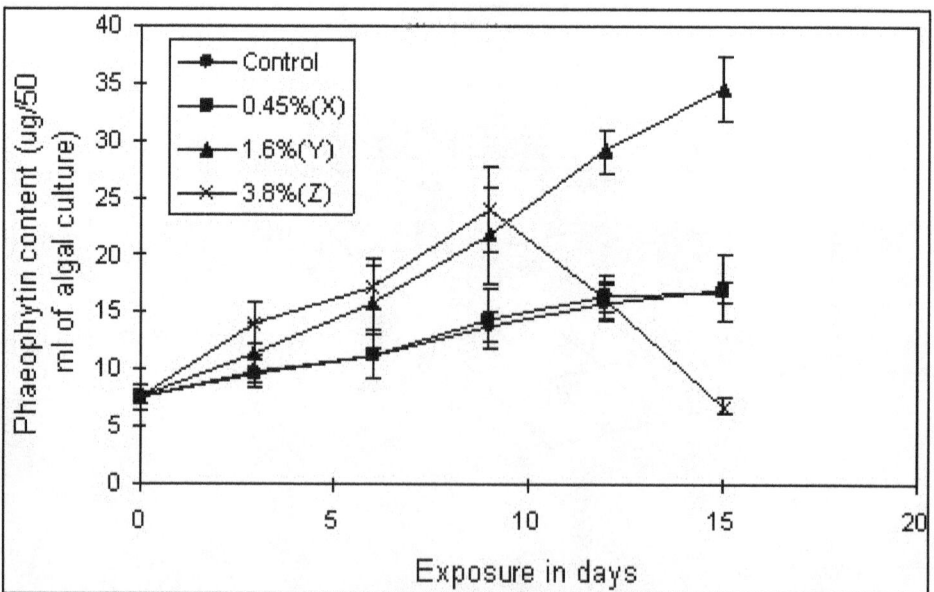

Figure 12.7: Showing Changes in Phaeophytin Content in Control and Effluent Exposed Alga at different Concentrations of the Effluent and different Exposure and Recovery Periods.

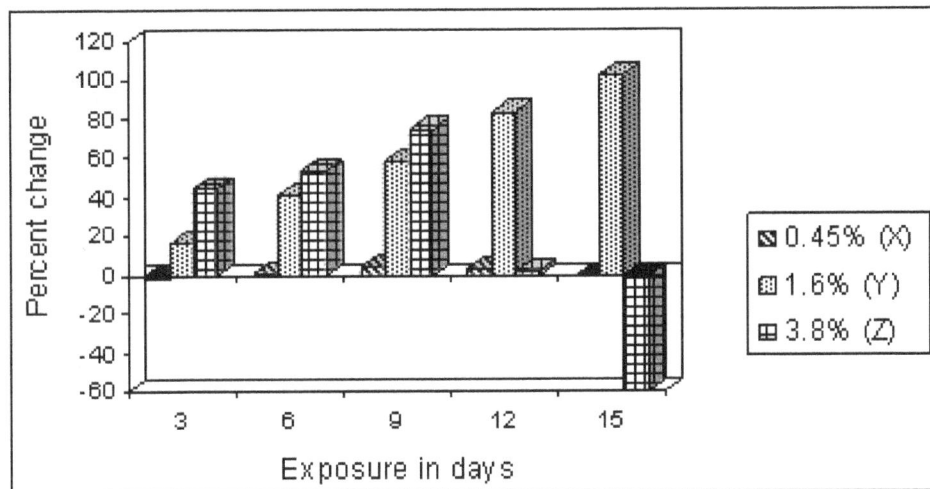

Figure 12.8: Per cent Change in Phaeophytin Content in Effluent Exposed Alga at different Exposure Periods when Compared to Control.

decreased interestingly and on 15th day of recovery, 43.9 per cent decrease was noted. In the Z set, the phaeophytin pigment increased with the increase in exposure period up to 9th day of exposure, then the phaeophytin content decreased significantly and 59.7 per cent decrease was recorded on 15th day of exposure. When the exposed alga was transferred to toxicant free medium 85.8 per cent, 93.3 per cent and 85.1 per cent decrease, when compared to the control value was recorded on 5, 10 and 15th day of recovery, respectively (Figures 12.7 and 12.8). The two way analysis of variance ratio test conducted for the data of Figure 12.7 on the changes of phaeophytin content indicated the existence of significant difference between columns and no significant difference between rows. The correlation coefficient analysis conducted between effluent concentration vs. phaeophytin content indicated the non-existence of significant correlation (P=NS) at all exposure periods except on 3rd day of exposure, where a positive and significant (r=0.990; P≤0.01) correlation was marked. The correlation coefficient analysis between days of exposure vs. phaeophytin content indicated the existence significant positive correlation in the control set (r=0.996, p≤0.001), X set (r=0.986, P≤0.001) and Y set (r=0.994, P≤0.001) and negative non-significant correlation in the Z set (r= –0.080, P=NS). The correlation coefficient analysis conducted between residual mercury concentration vs. phaeophytin content indicated the existence of significant positive correlation in the X set (r=0.927, P≤0.01) and Y set (r=0.964, P≤0.01) and non-significant (P=NS) positive correlation in the Z set.

The carotenoid pigment content increased from 3.8 ± 0.6 to 17.2 ± 0.8 µg/50 ml of algal culture within 15 days of exposure in the control set (Figure 12.9). When the control set was transferred to toxicant free medium for another period of 15 days, the rate of carotenoid content biosynthesis reduced in the control set. However, the value reached to 76.8 ± 1.3 µg/50 ml of algal culture. At 0.45 per cent of effluent concentration, an initial insignificant increase by 4.3 per cent and 5.1 per cent on 3rd and 6th day of exposure and consequent decrease by 9.8 per cent, 16.9 per cent and 23.3 per cent on 9, 12 and 15th day of exposure, when compared to the control value was marked, respectively. The exposed alga was transferred to toxicant free medium for recovery studies. A partial recovery was recorded in the X set (Figure 12.9).

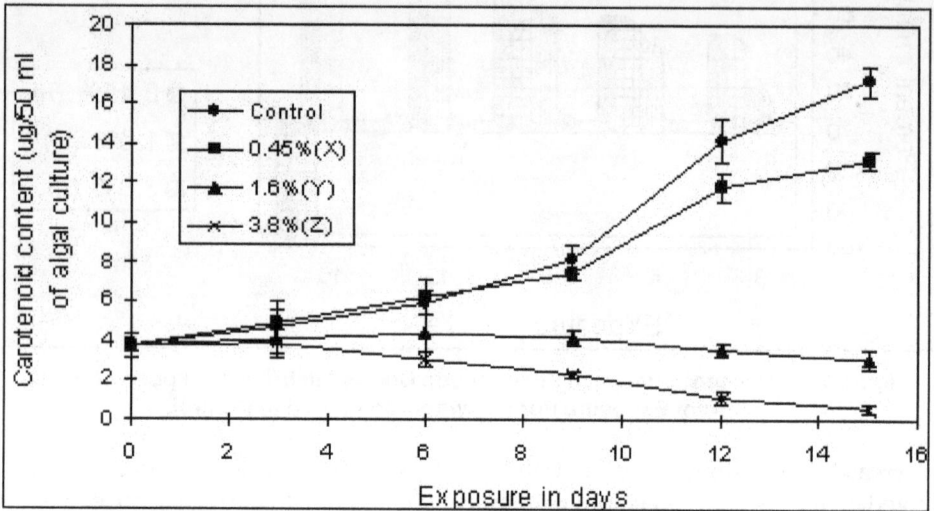

Figure 12.9: Showing Changes in Carotenoid Content in Control and Effluent Exposed Alga at different Effluent Concentrations and different Exposure and Recovery Periods.

At 1.6 per cent effluent concentration, the carotenoid content increased up to 6th day of exposure, and then the carotenoid content significantly declined, when compared to the control value. All the observed values were significantly less than the control values and a maximum 81.9 per cent decrease was noted on 15th day of exposure. When the exposed alga was transferred to toxicant free medium, no recovery was marked. Rather the carotenoid content further depleted and 85.6 per cent, 86.3 per cent and 85.4 per cent decrease was recorded on 5th, 10th and 15th day of recovery, respectively.

At 3.8 per cent effluent concentrations, the carotenoid content significantly declined at all exposure period, showing the existence of a negative correlation. The percent decrease of the pigment content increased with the increase in exposure period and 96.3 per cent decrease was recorded on 15th day of exposure. No significant recovery was marked in the recovery studies, when the exposed alga was transferred to toxicant free medium (Figure 12.10). The two way analysis of variance ratio test conducted based on the data of the Figure 12.9 pertaining to changes in carotenoid content indicated the existence of significant difference between rows and non-significant difference between columns. The correlation coefficient analysis conducted between residual mercury concentration vs. changes in carotenoid content indicated the existence of significant positive correlation (r=0.842; P≤0.05), in the Z set non-significant negative correlation (r= −0.599, P=NS) in the Y set and significant negative correlation (r= −0.926, P≤0.01) in the Z set. The correlation coefficient analysis

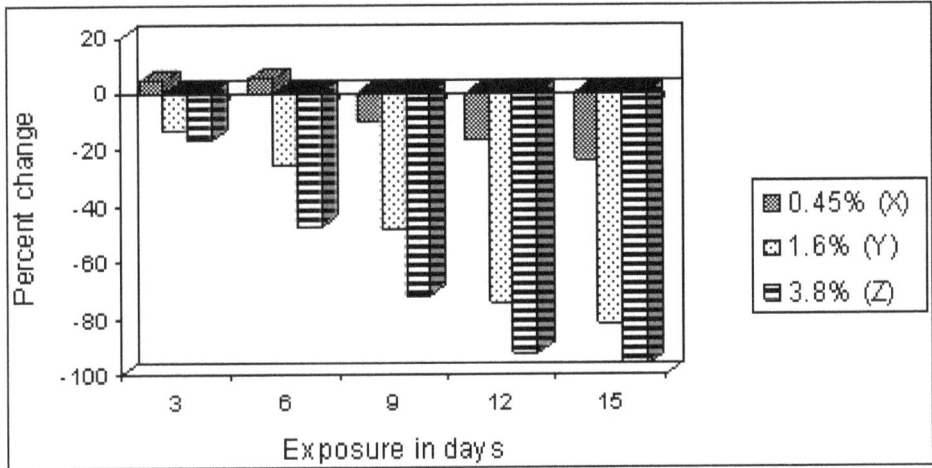

Figure 12.10: Per cent Changes in Carotenoid Content in Effluent Exposed Alga at different Exposure Periods when Compared to Control.

conducted between exposure period vs. carotenoid content indicated the existence of significant positive correlation (r=0.954, P≤0.01) in the control set, in X set (r=0.969, P≤0.01), negative non-significant correlation (r=-0.589, P=NS) in the Y set and significant negative correlation (r= –0.971; P≤0.01) in the Z set. Figure 12.11 shows the changes in the photosynthetic rate in control and effluent exposed alga at different exposure and recovery periods. In the control set the photosynthetic rate increased

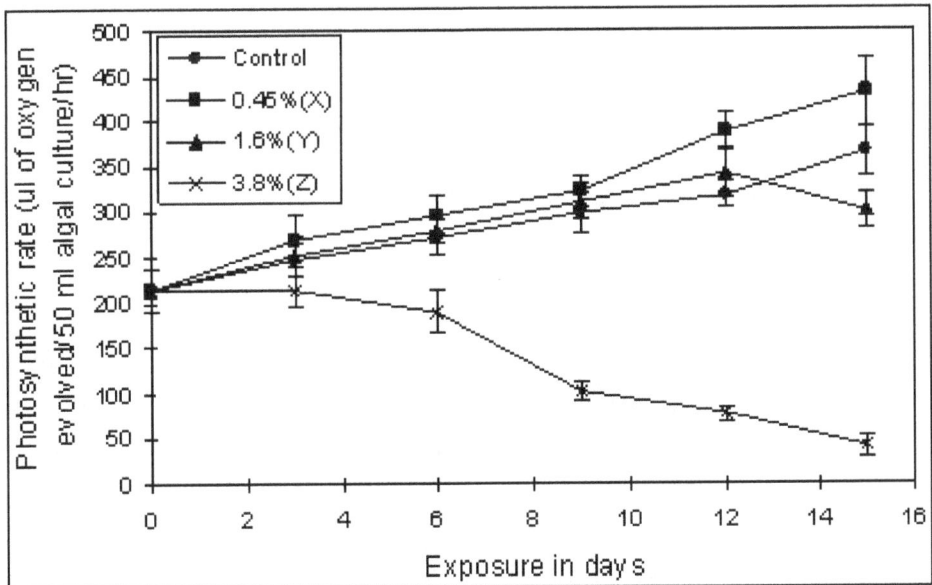

Figure 12.11: Showing Changes in Photosynthetic Rate in Control and Effluent Exposed Alga at different Effluent Concentrations and at different Exposure and Recovery Periods.

from 212.6 ± 24.2 to 365.6 ± 26.4 µl of O_2 evolved/50 ml algal culture/hr within 15 days of exposure. During the next 15 days of exposure, the photosynthetic rate increased to 472.2 ± 31.4 µl of O_2 evolved/50 ml algal culture/hr. In case of 0.45 per cent effluent concentration, the photosynthetic rate increased from 212.6 ± 24.2 to 431.3 ± 37.6 µl of O_2 evolved/50 ml algal culture/hr, after 15 days of exposure. Significant increase in the photosynthetic rate was marked in the X set at all exposure periods and a maximum of 17.9 per cent increase in the photosynthetic rate was recorded on 15th day of exposure.

The photosynthetic rate of the control set, X set and the per cent increase in the photosynthetic rate in the X set showed the existence of significant positive correlation with the exposure period (Figure 12.11). The exposed alga of the X set was transferred to toxicant free medium, where significant increase in the parameter was marked. At 1.6 per cent effluent concentration, the photosynthetic rate did not show any significant variation, when compared to the control set up to 12th day of exposure. Later, on 15th day of exposure, 17.5 per cent decrease in the photosynthetic rate when compared to the control value was marked. During recovery studies, no recovery was marked, rather significant depletion in the photosynthetic rate was marked. During recovery period, 39.1 per cent decrease in the photosynthetic rate was recorded on 15th day of recovery (Figure 12.12). In the Z set (3.8 per cent effluent concentration), the photosynthetic rate declined from 212.6 ± 24.2 µl of O_2 evolved/50 ml algal culture/hr within 15 days of exposure, where a maximum of 88.4 per cent decrease, when compared to the control value was marked. The exposed alga was transferred to toxicant free medium for recovery studies, where a partial insignificant increase

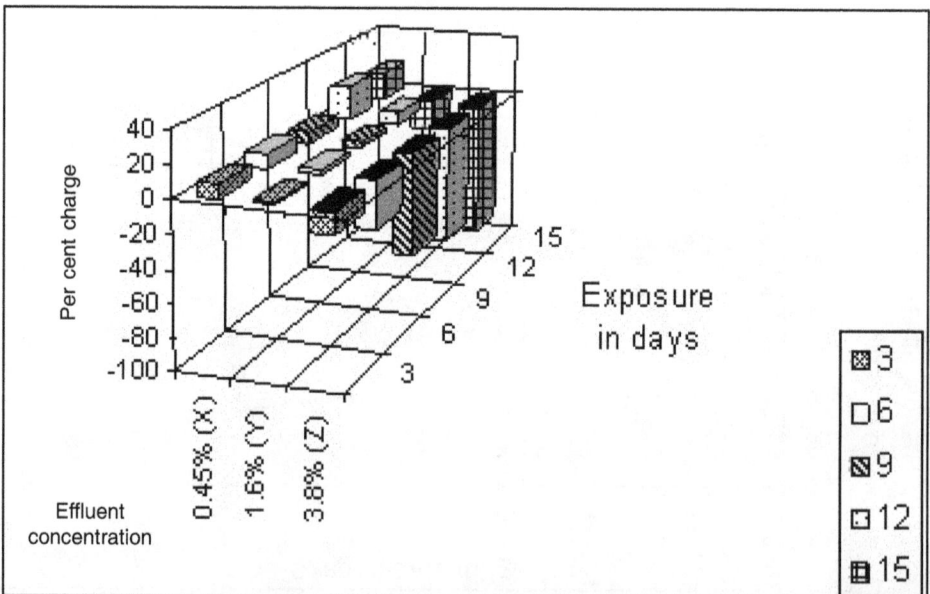

Figure 12.12: Per cent Changes in Photosynthetic Rate in Effluent Exposed Alga at different Exposure Periods when Compared to Control.

was marked, after 15 days of recovery. But the values were much less, when compared to the control values (Figure 12.12).

The correlation coefficient analysis conducted between the photosynthetic rate vs. effluent concentration at different exposure periods indicated the existence of non-significant (P=NS) negative correlation at all exposure periods except on 15th day of exposure, where a significant negative correlation (r= –0.958, P≤0.05) existed. The correlation coefficient analysis conducted between photosynthetic rate vs. exposure period, indicated the existence of highly significant positive correlation in the control set (r=0.993, P≤0.001) and X set (r=0.991, P≤0.001); significant positive correlation (r=0.869, P≤0.05); and negative significant correlation (r= –0.961, P≤0.01) in the Z set. The correlation coefficient analysis indicated the existence of significant positive (r=0.922, P≤0.01) correlation in the X set; non-significant (P=NS) positive correlation in the Y set and negative significant (r= –0.929, P≤0.01) correlation in the Z set. The two way analysis of variance ratio test indicated the existence of significant difference between rows and columns. The test was conducted on the data of the Figure 12.11 related to photosynthetic rate of control and exposed alga.

Figure 12.13 represent the amount of mercury present in the medium and the dynamics and the fate of mercury in the control and exposed medium, in presence of a blue-green alga, *Westiellopsis prolifica*, Janet. The background mercury level in the control set was not traceable, that might be zero. At 0.45 per cent effluent concentration, the amended culture media ('X' set) contained 3.24 µg of mercury/50 ml culture. In 1.6 per cent ('Y' set) effluent concentration, the mercury content was 11.52 µg/50 ml cultures. In 3.8 per cent effluent concentration ('Z' set), 25.34 µg of mercury/50 ml culture was present.

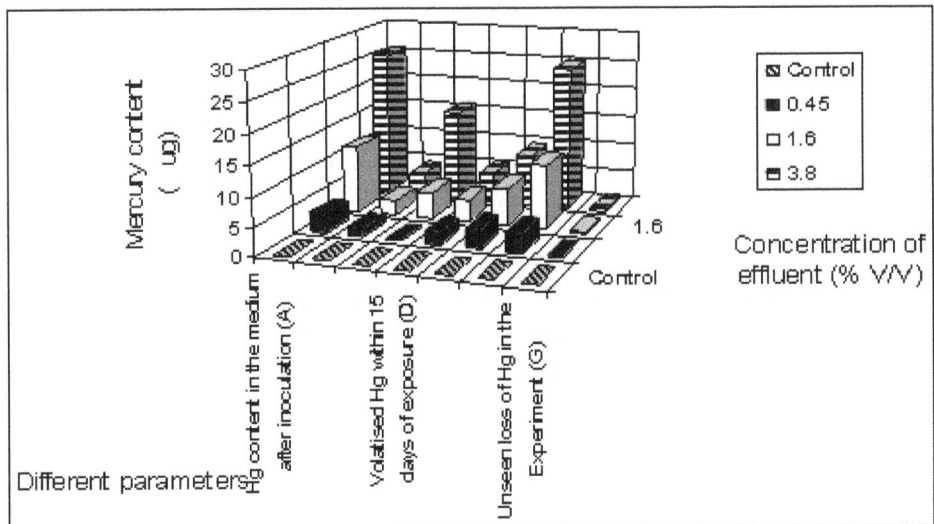

Figure 12.13: Mercury Dynamics in Experimental Media, Showing Mercury Content in the Medium, Residual Accumulation by the Algae, Amount of Mercury Volatilised, and Loss of Mercury during the Experiment. Data are mean of samples.

After 15 days of exposure, the exposed alga could accumulate 1.26 µg and 2.62 µg and 3.94 µg of mercury, absorbed from the medium in X, Y and Z concentrations, respectively. After 15 days of exposure, the X, Y and Z media contained 0.02, 4.63 and 15.8 µg of mercury in the exposed solution after the removal of BGA. During the experiment, 1.93 µg, 3.94 µg and 5.14 µg of mercury was detected in the form of vapour present in the inoculated flasks of X, Y and Z concentrations, respectively (Figure 12.14). In these experiments, 3.19 µg, 6.56 µg and 9.08 µg of mercury was removed from the exposed culture flasks either by way of residual accumulation by the alga and by way of volatilization of mercury from the exposed environment (Figure 12.14). During the experiment, we could record only 3.21 µg, 11.19 µg and 24.88 µg of mercury showing a loss/unseen mercury to a tune of 0.03 µg, 0.33 µg and 0.46 µg of mercury in X, Y and Z experimental sets (Figure 12.14), the sensitivity of the methodology was 81.36 per cent. 93.5 per cent accuracy was noted in the procedure by feed back analysis.

Figure 12.14 shows the per cent accumulation of mercury by alga, percent of mercury volatilized by the alga, percentage of total mercury removal from the medium, and per cent of mercury loss during the experiment. The blue-green alga could accumulate 38.8 per cent, 22.7 per cent and 15.5 per cent of total mercury present in the X, Y and Z concentration media (Figure 12.14). From the X, Y and Z media, 59.56 per cent mercury, 34.2 per cent mercury and 20.28 per cent of total mercury was volatilized, respectively (Figure 12.14). During the experiment, 98.46 per cent of total mercury present was removed from the medium by the alga, leaving around 0.61 per cent of mercury within 15 days in concentration X (0.45 per cent effluent concentration). In case of concentration Y (1.6 per cent effluent concentration), 56.94 per cent of mercury was removed from the medium by the alga. After the experiment 40.2 per cent of total mercury was available in the medium. During the process, 2.86 per cent

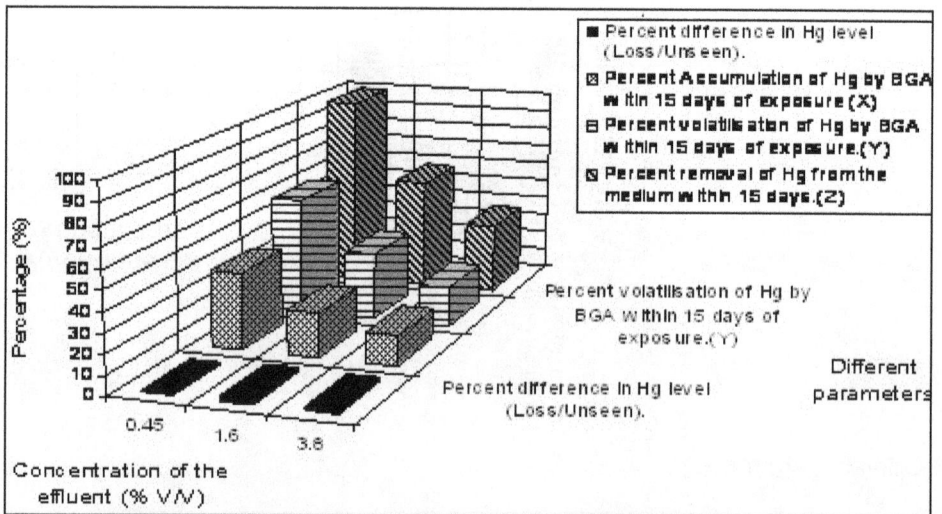

Figure 12.14: Shows the Per cent Accumulation, Volatilization, Removal from the Medium and Loss/Unseen during the Experiment. Data calculated from the mean of the samples.

of total mercury was either lost or unseen (Figure 12.14). In case of Z concentration (3.8 per cent effluent concentration), 35.83 per cent of total mercury was removed by the alga from the medium. During the process 1.82 per cent of total mercury was lost/ unseen. In the media, 62.35 per cent of total mercury was available after the experimental period of 15 days (Figure 12.14). It was observed from the results that at lower concentrations of mercury (X set) the alga, *Westiellopsis* could remove higher amount of mercury to the tune of 98.46 per cent when compared to Y and Z set. Where, only 56.94 per cent and 35.83 per cent of total mercury was removed. This experiment indicated that the blue-green alga could remove higher amount of mercury from the environment either by way of residual accumulation or by way of volatilization at lower/sub-lethal concentrations. The present data agrees with the findings of Sahu (1987), Mishra (2013) and Raut (2013) and disagrees with the findings and interpretations of Mohapatra (1992) and Pradhan (1998). The tolerance of the exposed alga was not genetic but purely due to slow acclimatization/adaptation in the environment, as the presence of resistant genes were neither identified nor confirmed like bacteria. The effluent should be diluted for possible use as a stimulant in the crop fields. But we should not forget the bioconcentration of mercury and biomagnification of mercury in a food chain, in the ecosystem, which can be hazardous. The present stimulatory effect of the toxicant showing higher growth rate, better nitrogen fixing capability will be a curse in future. To remove the mercury toxicant from the environment, the contaminated effluent/solid waste is only to be diluted to proper sub-lethal/maximum allowable concentrations and then the alga is to be inoculated for getting better results. From the present investigation, it can be inferred that in any mercury contaminated site, the effluent or solid waste is to be diluted to concentration X, if not possible than dilute up to concentration-Y first. Then, we can inoculate, a unialgal culture of *Westiellopsis prolifica*, for removal of mercury from the medium. The contaminated environment can be detoxified and the site can well be used for any useful purpose. Hence, proper care should be taken, while handling the issue at early phases. Protection and preservation of the environment is more important then short term benefits for a better and purer environment for the future generations.

Acknowledgement

Authors wish to thank the authorities of Berhampur University for providing necessary laboratory and library facilities.

References

Allen, M. B. and Arnon, D. I. (1955): Studies on nitrogen fixing blue-green algae, growth and nitrogen fixation by *Anabaena cylindrica*, Lemm. Pl. Physiol. Lancaster. 30: 366-372.

Davies, A. G. (1976): An assessment of the basis of mercury tolerance in *Dunaliela tertiolecta*. J. Mar. Biol. Ass., U. K. 56: 39-57.

ECIL (Electronic Corporation of India Limited) (1981): Analytical methods for determination of mercury with mercury analyser, MA 5800 A.

Mishra, A. K. (2013): Status of Environmental Mercury Contamination, physico-chemical analysis, geographic distribution of mercury and bioconcentration of mercury in and around a Caustic-chlorine Industry. Ph. D. thesis, Berhampur University, Orissa, India.

Mohapatra, A. (1992): Eco-physiology, resistance and ecological implications of a mercurial compound on a blue-green alga. Ph. D. thesis. Berhampur University.

Oser, B. L. (1965): Biuret test, Determination of Glycogen and Warburg. In : Hawks Physiological Chemistry, 14th edn., Edited by B. L. Oser, TMH Publishers, New Delhi, reprinted 1979, p. 179, 224 and 444.

Pradhan, P. K. (1998): Eco-toxicological effects of the leached chemicals of the solid waste of a chlor-alkali industry on a cyanobacterium. Ph.D. Thesis, Berhampur University, India.

Raut, P. (2013): Geographical distribution of mercury in and around a chlor-alkali industry at Ganjam, Odisha. Ph.D. Thesis, Berhampur University, India.

Sahu, A. (1987): Toxicological effects of a pesticide on a blue-green alga: III. Effect of PMA on a blue-green alga, *Westiellopsis prolifica*, Janet. and its Ecological Implications. Ph.D. Thesis, Berhampur University, India.

Sharma, D. (2013): Decontamination of mercury contaminated environment by environmental and biological agents. Ph. D. thesis, Berhampur University, Orissa, India.

Shaw, B. P. (1987): Eco-physiological studies of the waste of a chlor-alkali factory on biosystems. Ph. D. thesis, Berhampur University, Orissa, India.

Vernon, L. P. (1960): Spectrophotometric estimation of chlorophylls and phaeophytins in plant extracts. Anal. Chem., 32:114-150.

Wanntorp, H. and Dyfverman, A. (1955): Identification and determination of mercury compounds in fish. Identification and determination. Arkiv, Kemi, 9 (2).

Chapter 13

Heavy Metal Concentration through Idol Immersion Activities: A Case Study of River Budhabalanga, Balasore, Odisha

Tanuja Panigrahi, Bita Mohanty and R.B. Panda

P.G. Department of Environmental Science,
F.M. University, Balasore, Odisha, India

ABSTRACT

Balasore town in the state Odisha, India is situated in the bank of river Budhabalanga. The impact of idol immersion on water quality of Budhabalanga River is discussed, for this purpose Balighat point was selected as sampling station in the year 2012 because huge number of idols are immersed in this Ghat of Budhabalanga River. Water samples were collected at morning hours during pre immersion, during immersion and post immersion in the periods of idol immersions. The immersion of idol of Lord Ganesh, Lord Viswakarma and Goddess Durga during month of August to October is the major source of contamination and sedimentation to the River Budhabalanga. The idols are been made up of clay, plaster of paris, cloth, paper, wood, thermo cool, jute, adhesive materials and synthetic paints etc. Out of the all material used in making the idol, thermo cool is Non-biodegradable while paints contain heavy metals such as Chromium, Lead, Cadmium and Mercury. The present study was under taken to evaluate the concentrations of heavy metals. The findings of the increase heavy metal concentration after immersion may magnify in their concentrations at different tropic levels by food chain. On the basis of these changes it is concluded that the level of water pollution increases in Budhabalanga River due to these religious activities and cause adverse effect to the aquatic life or entire aquatic ecosystem. No one can change or stop these religious activities but awareness among the people and society can reduce the pollution.

Keywords: Budhabalanga river, Heavy metals, Religious activities, Water quality, Water pollution.

Introduction

"WATER" the elixir of life referred as nature, was worshiped since Vedic days. But now a day there is an increasing menace of water pollution throughout the globe. The major riverine systems are getting polluted day by day. It occurs when pollutants are discharged directly or indirectly into water bodies without adequate treatment to remove harmful compound. Water pollution occurs due to the discharge of municipal sewage both domestic and industrial without any treatment which brings considerable changes in the river water quality in addition to many religious activities now became a threat to the ecosystem (Bajpai *et al.*, 2003, Mukherjee *et al.*, 2003). The idols of Lord Viswakarma, Lord Ganesh, Goddess Durga etc. worshipped by Hindus are immersed in the month of August to October respectively every year. Similarly during the Moorum festival, tazias are being immersed by Muslims in the month of May every year (Bibicz *et al.*, 1982). The idol are been made up of clay, plaster of paris, cloth, paper wood, thermocol, jute, adhesive materials and synthetic paints etc. Out of the all material used in making the idol, thermo cool is Non-biodegradable while paints contain heavy metals such as Chromium, Lead, Cadmium and Mercury (Das *et al.*, 2012). The chemical paints used to decorate the idols increases heavy metal concentration and acidity in the water (Bowen *et al.*, 1966). Lead and Chromium, which also adds through SINDUR in the water bodies, are very toxic even in very small quantity for human being through the process known as Bioaccumulation and Biomagnification (Dhote *et al.*, 2001). When immersed, these colours and chemical dissolve slowly leading to significant alternation in the water quality (Ujjania *et al.*, 2011).

Study Area

Balasore is situated on the bank of Budhabalanga River that plays important and major role in its economic and social growth and development. Budhabalanga River is a river of Eastern India and North east Odisha with a length around 175 km. It originates from the Similipal Hill, Mayurbhanj district, Odisha and has a total catchment area of 4840 square kilometres. Its major tributaries are the Sono River, the Gangahar River and the Catra River. The flow of Budhabalanga covers Fuladi, Remuna, Balasore in Odisha and empty into the Bay of Bengal at Balighat (DSHB, 2011). The confluence point of river Budhabalanga with Bay of Bangal is very near to the sampling station Balighat, where due to back water of sea the flow of water is almost nil. Therefore after immersion of idols the pollutants accumulated in the sampling site due to still water current of the river.

In India, a lot of religious activities take place all around the year. Most of the Temples and ritual places are located near the aquatic resources like pond, lakes, river etc. The people of Balasore are always excided for celebration of festivals. Ganesh

Chaturthi and Durga puja (Figure 13.1) are of the important festivals of them. In this festivals number of Ganesh and Durga idols of different sizes are immersed in Budhabalanga River. About 500 or more Ganesh idols and about 150 Durga idols were immersed during 2012 in Balighat immersion point (Figure 13.2). Ganesh idol increases pollution in Hussainsagar Lake, Hyderabad was observed (Reddy *et al.*, 2001). The floating materials released through idol in the river and lake after decomposition result in eutrophication of the river, lake etc. (Leland *et al.*, 1991). The idol immersion is a religious activity which is responsible for adding pollution load in the water bodies. The reservoir can serve as a model for studying heavy metal contamination through idol immersion.

Figure 13.1: Idol of Goddess Durga.   **Figure 13.2: Lord Ganesh Immersion.**

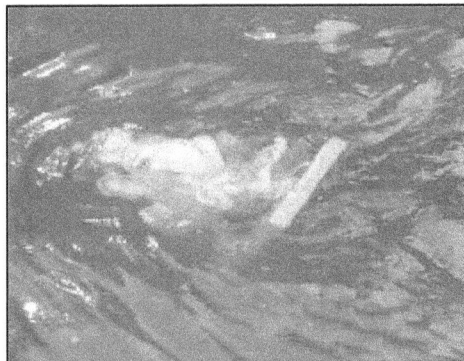

Material and Methods

The water samples was collected from surface layer during morning hours from Balighat idol immersion point and the site of idol immersion at different intervals *i.e.* pre immersion (a week before the commencement of the immersion activities), during immersion and post immersion (10 days after and 45 days after the completion of immersion activities) in the period of Ganesh chaturthi, Viswakarma puja and Durga puja in the month of August, 2012 to October, 2012 respectively. The water samples collected for the heavy metals analysis like chromium and lead were analyzed according to standard methods (APHA, 1995). The heavy metals were preserved by adding 5 ml of 1N HNO_3 and bringing down the pH to near about 4 and analyzed using AAS (Perkin Elmer A Analyst 100). A comparative study of Ca and Mg were carried out to identify their effect on the water quality as well as on aquatic life.

Results and Discussions

The results of the present investigation show that the water of River Budhabalanga is deteriorated due to the immersion of different idols. The concentration of calcium has increased significantly in the river water after the idol immersion and became came to normal after one to two month of the idol immersion; however it was below the permissible limits (Figure 13.3). Magnesium, Chromium, Cadmium, Lead and Arsenic concentration has also increased significantly in the river water ten days after the idol immersion (Figures 13.3–13.5), (Kulshrestha *et al.*,

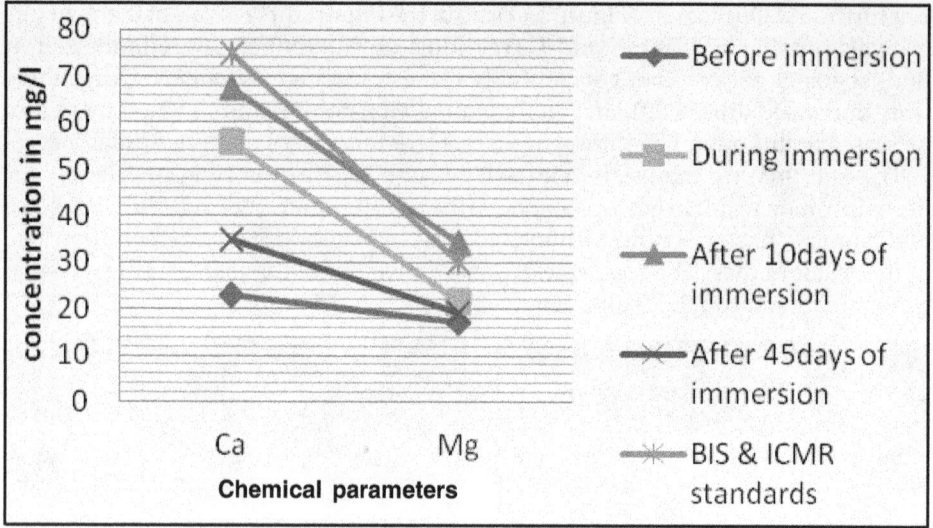

Figure 13.3: Concentration of Ca and Mg before, during, after 10 Days and after 45 Days of Immersion in Comparison to the BIS and ICMR Standards.

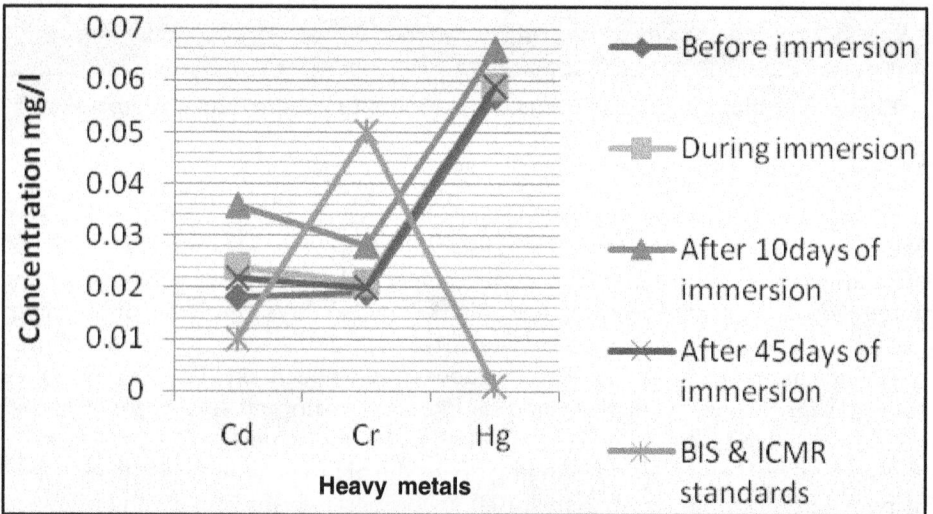

Figure 13.4: Concentration of Cd, Cr and Hg before, during, after 10 Days and after 45 Days of Immersion in Comparison to the BIS and ICMR Standards.

1988). Magnesium is non-poisons, though it increases the hardness of water. Over the year the concentration of heavy metals, especially manganese, lead and mercury has also increased considerably in the river water compared to the specifications of highest desirable limits as set by BIS and ICMR (ICMR, 1975). The concentration of cadmium, mercury and lead, the potentially obnoxious heavy metals had increased many folds in the water due to idol immersion compared to highest desirable limits of BIS and ICMR standard (Table 13.1). The heavy metal especially manganese, lead

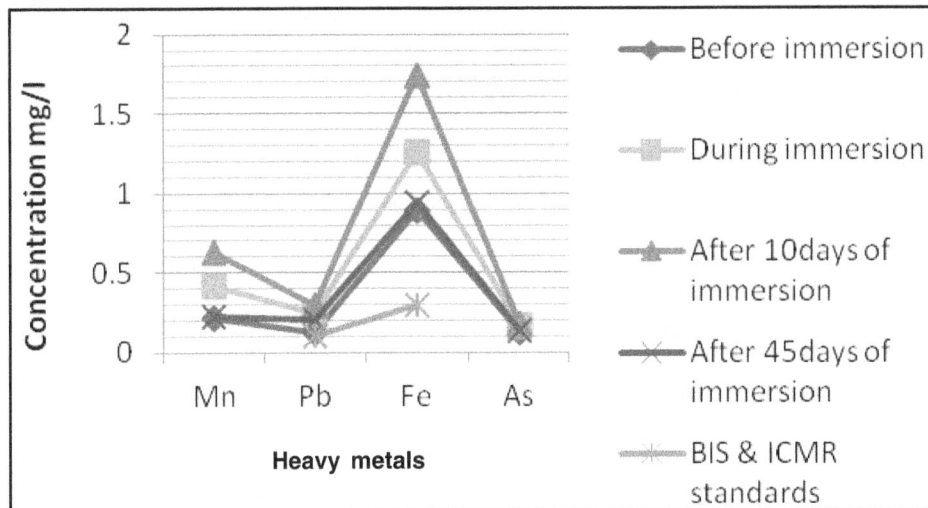

Figure 13.5: Concentration of Mn, Pb, Fe and As before, during, after 10 Days and after 45 Days of Immersion in Comparison to the BIS and ICMR Standards.

and mercury excess in water cause skin diseases (Trivedy *et al.*, 1986). The chromium concentration in the river water did not change much and was below the limits of standards.

Table 13.1: Change in Concentration (mg/l) of some Chemical Pollutants in Balighat Immersion Point of Budhabalanga River before Immersion, during Immersion and after Immersion of Idols in Month of August 2012 to October 2012.

Heavy Metals	Before Immersion	During Immersion	After 10 Days of Immersion	After 45 Days of Immersion	Standard as per BIS and ICMR
Ca	23.08	56.01	67.56	34.97	75
Mg	17.23	21.72	34.13	19.35	30
Cd	0.018	0.024	0.036	0.022	0.01
Cr	0.019	0.021	0.028	0.02	0.05
Hg	0.057	0.059	0.066	0.059	0.001
Mn	0.21	0.42	0.63	0.22	–
Pb	0.128	0.248	0.293	0.205	0.1
Fe	0.89	1.25	1.74	0.94	0.3
As	0.122	0.168	0.173	0.134	–

After the immersion of the idols, its concentration increased further and after about 45 days of immersion its concentration slightly decreases still it higher than before idol immersion (Figure 13.4). The heavy metals are known to be persistent and gradually accumulate and magnify through the process known as bioaccumulation and biomagnification, while they move up in the food chain (Vyas *et al.*, 2007). Thus

Figure 13.6: Idol Immersion Activities Kill Aquatic Fauna.

Cadmium and Mercury load may magnify in their concentrations at different tropic levels in the river ecosystem and finally reach the humans through food chain (Figure 13.6). An organic compound of Mercury, for example Methyl Mercury when it enters the human body, concentrates in the brain and destroy the brain cells, damaging the central nervous system and also cause ulceration of the digestive tracts. Therefore, it is suggested that the authorities looking into the environmental protection of the river need to take necessary steps.

Conclusion

From the mythological point of view, the water bodies are related to religious sentiments but from the scientific point of view, these water bodies like ponds, rivers and lakes are not suitable for human uses. The main cause of change in water quality in River Budhabalanga is various religious activities. The idol immersions plays an important role because the plaster of paris, clothes, iron rods, chemical colours, varnish and paints used for making the idols deteriorate water quality of river Budhabalanga. No one can stop these religious activities but awareness should be created among the people to get rid of these problems.

Suggestion

To make the idol an environmental friendly we should follow the following points:

☆ The idols should be made of traditional clay instead of baked clay.

☆ The paints of idols should be water soluble.

☆ Idols should be small as they would dissolve faster.

☆ Non-degradable chemical dyes should be banned.

☆ Stress should be given on natural colours used in food products.

Acknowledgement

The authors pay their deep sense of gratitude to the Vice-Chancellor of Fakir Mohan University. The authors also record their thanks to the person for kind co-operation in collecting the samples.

References

APHA (1995): Standard methods for examination of water and wastewater, American Publ. Heal. Assoc., Washington, D.C., 19th Edition.

Bajpai, A., Pani, S., Jain, R.K. and Mishra, S.M. (2003): Heavy metal concentration through idol immersion in a tropical lake, Eco. Env. and Cons., 8(2):157-159.

Bibicz, M. (1982): Heavy metal in the aquatic environment of some water bodies of the Lublin basin Actuatic Hydrobiologia., 24: 125-138.

Bowen, H.J.M. (1966): Trace Element in Biochemical (New Yark: Academic Press Including)

Das, K.K., Panigrahi, T. and Panda, R.B. (2012): Idol Immersion Activities cause Heavy Metal Contamination in River Budhabalanga, Balasore, Odisha, Ind. Int. J. Modern Eng. Res., 2(6): pp. 4540-4542.

Dhote, S., Varghese, B. and Mishrs, S.M. (2001): Impact of idol immersion on water quality of Twin Lakes of Bhopal, Ind. J. Env. Protect. 21: 998-1005.

ICMR, (1975): Manual of standards of quality for drinking water supplies, Special report series No.44, 2nd Edition.

Kulshrestha, S.K., George, M.P. and Khan, A.A. (1988): Preliminary Studies on the impact of certain religious activities on water quality of Upper Lake, Bhopal, National. Symp. Past present and future of Bhopal lakes, 253-257.

Leland, H.V. (1991): Transport and distribution of trace elements in a watershed ecosystem in environment. Boggess, W.R., and Wixsion, B.G. Eds. Castle House Publ., pp. 105-134.

Mukherjee, A. (2003): Religious Activities and Management of Water Bidies. Case study of idol immersion in context of Urban lakes Management, International Water History Association (3) 325.

Orissa Census (2001): District Statistics Hand Book (DSHB), Balasore (2011): Directorate of Economics and Statistics, Govt. of Orissa, Bhubaneswar, India.

Reddy, V.M. and Kumar, V.A. (2001): Effect of Ganesh Idol Immersion on some water quality parameters of Hussain Sagar, Current D.Science, 18:1412.

Trivedy, P.K. and Goel, R.K. (1986): Chemical and Biological methods water pollution studies. Karad, India, Env. Publ.

Ujjania, N.C. and Azhar, A.M. (2011): Impact of Ganesh Idol Immersion Activities on the Water Quality of Tapi River, Surat (Gujurat) India, Res. J. of Biol., 1(1): pp. 11-15.

Vyas, A., Bajpai, A., Verma, N. and Dixit, S. (2007): Heavy Metal Contamination Causes of Idol Immersion Activities in Urban Lake, Bhopal, India, J. Appl. Sci. Environ. Manage, 2(4): 37-39.

Chapter 14

Treatment of Municipal Wastewater in UASB-Reactors: An Introspection

P.K. Behera[1], S.S. Pati[2] and S.K. Sahu[3]

[1]*Central Pollution Control Board, New Delhi, India*
[2]*State Pollution Control Board, ICZMP, Odisha, India*
[3]*P.G. Department of Environmental Sciences,*
Sambalpur University, Odisha, India

ABSTRACT

The efforts in past, to improve the wastewater discharge quality, using modern sewage technology gives a ray of hope for big successes in industrialized countries. The Up-flow Anaerobic Sludge Blanket (UASB) technology was introduced in India in late eighties during the Ganga Action Plan (GAP). A set of pilot plants were installed at Kanpur initially for treatment of a mix of sewage and tannery effluent and later exclusively for sewage. UASB technology has been found to be very effective for treatment of high strength industrial effluents particularly from distilleries, pulp and paper, tanneries, and food processing industries. For high organic loads, it certainly offers advantages in terms of almost insignificant energy consumption, low operational and maintenance cost and recovery of significant amount of bio-energy. Consistent production of fairly large quantities of biogas from industrial effluents makes electricity generation for captive consumption is an attractive financial proposition. Other features of the technology, *i.e.*, lower skill requirement and sludge production; perhaps add to its attractiveness under the industrial context to a certain extent. But in contrary, can it be successful for municipal wastewater needs introspection. The major concern in today's scenario for urban bodies is alarming day by day as the discharges generated are not properly treated, thus contributing high BOD (biochemical oxygen demand) load in our river system. As reports depicted twenty-seven cities have only primary treatment facilities and forty-nine have primary and secondary treatment facilities, either defunct or not functioning properly. However, an elaborated evaluation on UASB technology is highly demanding at this juncture, which may put some rays to address the burning issue.

Keywords: UASB, Municipal wastewater treatment, Sludge blanket, Aerobic technology.

Introduction

In the past several years, the efforts to improve water quality using modern sewage technology led to big successes in industrialised countries (Switzenbaum and Grady 1986, Hulshoff *et al.*, 1997, Sasse, 1998). But, the commonly implemented treatment systems are mostly based on a high technology level which not only requires a large amount of process energy, but is also related to high investment and operation costs (Urbinati *et al.*, 2013). Plant operation furthermore requires highly qualified personnel that are very often not sufficiently available in developing countries. Speaking of these countries, a process combining a low level of mechanisation with a high purification performance is therefore highly desirable. In developing countries, pond systems are still the most widely implemented wastewater treatment process. Given advantageous geographical conditions (no need for pumps) and low land prices (depending on the desired effluent quality), they require a very low degree of mechanisation and are the most economic alternative (Perry *et al.*, 1997). With respect to the environment, they do however some major drawbacks as their land demand as well as greenhouse gas and odour emissions are considerably high (Mergaert *et al.*, 1992). In countries with a warm climate throughout the whole year, the high wastewater temperatures – which are a requirement for anaerobic degradation – allow and favour an anaerobic treatment of the entire sewage flow, not only the sludge portion (van Haandel, and Lettinga, 1994; Lettinga *et al.*, 1999, Langenhoff, and Stuckey, 2000). As ventilation or other means of aeration are not necessary for these processes, the technology can be kept considerably simpler (Dague, *et al.*, 1998). Although originally implemented for industrial rather than municipal purposes, one meanwhile widely implemented anaerobic treatment alternative in the domestic wastewater sector is the so called UASB-reactor (Up-flow Anaerobic Sludge Blanket) (Rebac *et al.*, 1996; Rodrigues *et al.*, 2001, Hassan *et al.*, 2013, Che-Jen Lin *et al.*, 2013). The process is essentially based on a special flow regime allowing the sewage to get into contact with a "sludge blanket" or "sludge bed" situated in the reactor, and a following three phase-separation of water, sludge and gas (methane) (Barbosa and Sant, 1989). Within the sludge bed, the organic matter in the sewage is reduced by bacteria. In the anaerobic milieu of the reactor, the methane is formed due to bacterial activity during the fermentation process, which can be utilised as energy source. In order to achieve the best performance of the reactor, several parameters such as COD (Chemical Oxygen Demand), required retention time and others need to be taken into consideration. The UASB-process represents one important option for sewage purification in developing countries with warm climates for sustainable operation of wastewater treatment plants (Mergaert *et al.*, 1992).

If combined with an appropriate post treatment, the effluent values reach the same level as the aerobic activated sludge process which is widely applied in industrialised countries (with temperate climate). Up-flow Anaerobic Sludge Blanket (UASB) technology was introduced in India in late eighties during the Ganga Action Plan (GAP). A set of pilot plants were installed at Kanpur initially for treatment of a mix of sewage and tannery effluent and later exclusively for sewage. This development took place when a strong need for an appropriate 'low cost' technology was felt subsequent to the experience of conventional aerobic technology based Sewage Treatment Plants (STPs) where the running costs were perceived to be unaffordable.

At that point of time, UASB which was still an evolving technology was positioned as an affordable option with potential for 'resource recovery'. It was argued that this technology will be advantageous for sewage treatment due to its following unique features:

☆ Low energy requirement

☆ Less operation and maintenance cost

☆ Lower skill requirement for operation/supervision

☆ Less sludge production, and

☆ Potential for resource recovery through generation of electricity from biogas and utilization of stabilised sludge as manure.

Based on the limited experience of the two pilots at Kanpur and the above considerations, UASB was the most preferred technology option under the Yamuna Action Plan (YAP-I) which was implemented during 1993-2002. Under this Plan 16 UASB based STPs were constructed in Haryana and UP towns with combined treatment capacity of almost 600 mld.

Subsequent to this up-scaling and wide replication, a large number of STPs based on UASB technology have been constructed in the country and considerable experience has been built-up on the technology. This chapter tries to bring out the lacunae of this technology and attempts to raise questions for its wide scale adoption specifically for sewage treatment. The findings are based on (a) a set of comprehensive case studies covering 25 STPs of 9 different technologies (b) effluent data of eight UASB based STPs and (c) a long term monitoring of the performance of pilots at Kanpur.

Advantages and Disadvantages of UASB Technology for Municipal Wastewater Treatment

Speaking of municipal wastewater treatment (MWWT) in general, the UASB technology offers a number of advantages and disadvantages in comparison with treatment alternatives such as pond systems or aerobic processes (*e.g.* activated sludge). The respective frame conditions for each specific situation – favourable or adverse in effect – do however have to be considered if a technology decision needs to be taken.

Advantages

☆ Low land demand

☆ Reduction of CH_4 emissions from uncontrolled disposal/"open" treatment (ponds) due to enclosed treatment and gas collection emissions due to low demand for foreign (fossil) energy and surplus energy production.

☆ Reduction of CO_2

☆ Low odour emissions in case of optimum operation

☆ Hygienic advantages in case of appropriate post-treatment

☆ Low degree of mechanisation

☆ Few process steps (sludge and wastewater are treated jointly)

☆ Low sludge production, high sludge quality

☆ Low demand for foreign exchange due to possible local production of construction material, plant components, spare parts

☆ Low demand for operational means, control and maintenance

☆ Correspondingly low investment and operational costs

Disadvantages

☆ Demand for know-how

☆ Insufficient standardisation and adaptation for several implementation possibilities

☆ Economically not feasible in colder climates with sewage temperature lower than 15°C

☆ Methane and odour emissions (also of end-products) in case of inappropriate plant design or operation

☆ Insufficient pathogen removal without appropriate post-treatment

☆ Sensitivity towards toxic substances

☆ Long start-up phase before steady state operation, if activated sludge is not sufficiently available

☆ Uncertainties concerning operation/maintenance due to still low local availability of know-how and process knowledge

Description of the UASB Process

In a UASB-reactor, the accumulation of influent suspended solids and bacterial activity and growth lead to the formation of a sludge blanket near the reactor bottom, where all biological processes take place. Two main features decisively influencing the treatment performance are the distribution of the wastewater in the reactor and the "3-phase-separation" of sludge; gas and water. While the sludge should remain in the reactor, the produced gas is collected before the purified water leaves the reactor. For a simple process flow scheme (Figure 14.1).

The influent point (sewage) is situated at the reactor bottom; the effluent discharge (treated wastewater) is situated in the upper part of the reactor, thus forcing the entering sewage to follow an up-flow regime and to get into contact with the sludge blanket in the reactor. Here, the organic matter in the sewage is subject to anaerobic

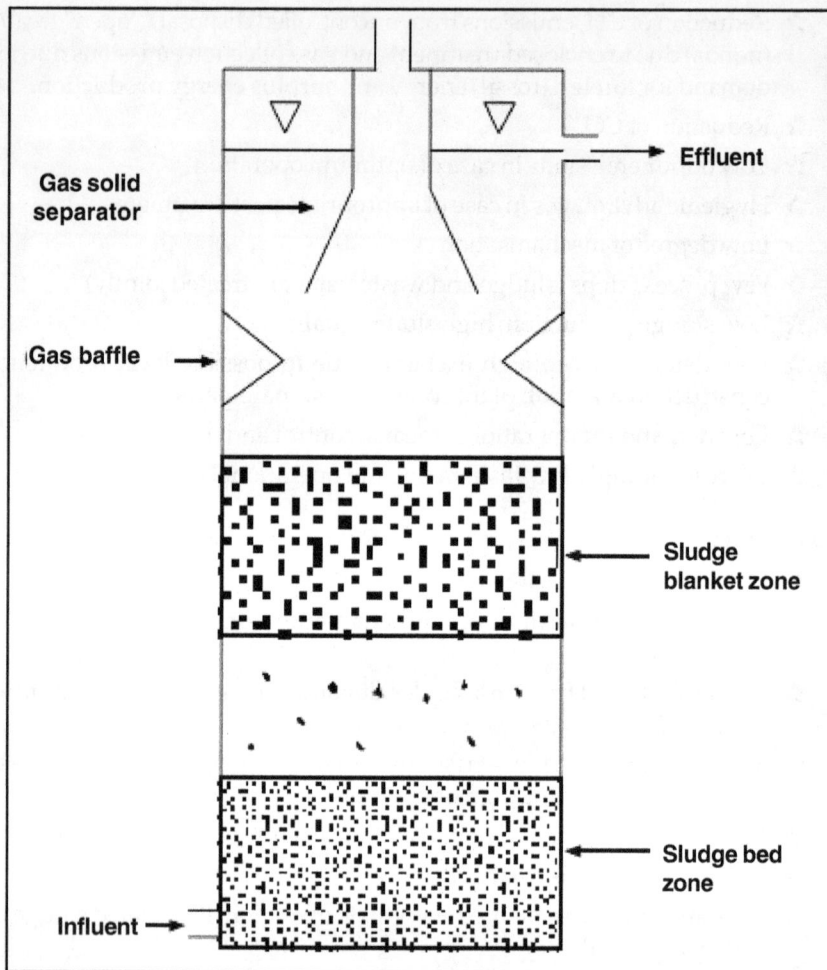

Figure 14.1: Simple Process Flow Scheme of UASB Process.

degradation by the bacteria contained in the sludge blanket, with methanogenic ("methane building") bacteria producing methane gas (CH_4) during the degradation processes. In order to prevent unwanted sludge discharge, separation devices (deflectors) are installed that prevent the further upward movement of the sludge and force it to sink back into the bed. The gas is collected in gas holders installed in the upper part of the reactor; for gas rising close to the reactor walls, an additional one may be installed.

Parameters Influencing the UASB Process

For the operation of a UASB reactor, a minimum temperature of the sewage (approx. 15°C) and an anaerobic environment are required in order to secure that the methanogenic bacteria can develop their activity. A number of additional parameters that significantly influence the process and should partly be controlled continuously are listed below.

pH

The pH-value in the digestion substrate is decisive for the activity of the very sensitive methanogenic bacteria and therefore has to be controlled continuously. The recommended range is between pH 6.3 and pH 7.8. The utilisation of hydrogen carbonate as buffer may simplify the observation of the given range.

Chemical Oxygen Demand (COD)

Basically, anaerobic sewage treatment may be applied for low as well as for high COD-concentrations. Depending on the respective local conditions, the advantages of the anaerobic process do generally only become distinct at a concentration of >250 mg COD/l and achieve an optimum at a concentration of >400 mg COD/l. An upper limit for COD concentration in the influent sewage is not known.

Temperature

The anaerobic degradation process achieves its optimum at a temperature between 35-38°C. Below this range, the digestion rate decreases by about 11 per cent for each °C temperature decrease. However, given appropriate frame conditions, anaerobic treatment has in the past years proven to have a very high potential if ambient sewage temperatures are above 20 °C. For a successful and stable microbiological degradation and the avoidance of an acidification of the process, a water temperature of at least 15°C is necessary, although bacterial activity can still be noticed at lower temperatures (10°C and less).

Wastewater Flow

The influent amount of wastewater to the plant should be considerably constant. For great flow variations, *e.g.* due to high precipitation, a sufficiently sized buffer tank should be installed prior to the UASB to guarantee an even feeding of the reactor. Flow measurements should be part of continuous process control. They are necessary for the calculation of further process, design and operational parameters (*e.g.* organic charges per volume).

Organic Acids

During anaerobic degradation in the sludge blanket, the substrate is reduced to short-chained carbon acids: butyric acid (CH_3-COOH), propionic acid (C_3H_7COOH) and finally acetic acid (C_2H_5COOH) from which a great part of the methane is formed. If the concentration of one of the carbon acids exceeds the tolerable value, a disturbance in the reduction chain and therefore problems in the fermentation process can be assumed.

Concentration of Ammonia

Ammonia (NH_3), if formed during the fermentation process, may be a cause for process inhibition. The concentrations of ammonia and ammonium (NH_4+) are in balance, the dissociation rate however depends on temperature. As opposed to the acetic acid concentration, process inhibition will be favoured by high pH-values, as these will increase the ammonia concentration. Formation of NH_3 does however only start at pH-values above 7.5. The formation of ammonia can usually be avoided by respecting the recommended pH operation range. Process inhibition can be notified only at NH_3 concentrations above 40 mg/l and temperatures above 30°C. With lower temperatures, an inhibitory effect will only occur at even higher NH_3 concentrations.

COD Charge per Volume

Measurements at UASB reactors indicate a relation between the daily COD charge per volume and the percentage of COD reduction. If a COD reduction of at least 65 per cent is required, test series proved that volumetric charges exceeding 2-3 kg COD/m^3 > d is critical. The COD-charge per volume is calculated via wastewater flow and COD concentration in the influent.

The Hydraulic Retention Time (HRT) is the time that the wastewater remains in the reactor, calculated as ratio of the reactor volume and the wastewater flow. The HRT influences the COD reduction and is an important parameter with respect to the desired degradation rate. It should not be less than 2 hours.

Up-flow Velocity

The up-flow velocity should not exceed an upper limit in order to prevent sludge wash-out, both out of the sludge blanket and out of the reactor. A minimum speed is however desired, as turbulences improve the contact between sewage and sludge. The up-flow velocity should be in the range between 0.2-1 m/h. It may be calculated as follows:

$v = Q/A$

$v = H/HRT$

v = up-flow rate; Q = flow; A = surface; H = reactor-height; HRT = hydraulic retention-time

As shown above, the up-flow velocity depends on flow and surface. In order to ensure optimum values, the reactor height should measure between 4 and 6 meters.

Optimum Wastewater Characteristics

Optimum values for the characteristics of the influent sewage are listed below in the Table 14.1.

Table 14.1: Optimum Characteristics of the Influent Sewage

Criteria	Optimum value
COD	>400 mg/l
Temperature	18-35°C
Substrate flow	Continuous flow
Nutrients	Ratio CSB : N : P : S 350 : 5 : 1 : 1
Toxic Substances/Suspended Solids	Low concentration
Micronutrients	All present

It is advantageous if the wastewater is collected in a separate sewer system so that events of high precipitation, which frequently occur in tropical climate, do not dilute the wastewater to a concentration too low for treatment (the organic loading should not be below 250 mg COD/l).

Operation and Maintenance of UASB Process

The decision whether the implementation of the UASB process may be appropriate and sustainable in a specific country and location depends on a number of factors that are too specific to be listed in conclusion. Some important "frame" data with respect to process requirements, operation and performance are however given in the followings.

Start-up Phase

During start-up, a comparably large bacterial mass for the biological degradation processes has to develop and adapt to the characteristics of the specific wastewater so that the start-up phase of anaerobic wastewater treatment plants can be rather time consuming and difficult. Municipal wastewater has however proven to be less problematic than industrial wastewater as it already contains the composition of nutrients and micronutrients required for bacterial activity and growth. Generally speaking, anaerobic sewage treatment therefore does not need inoculation to start the degradation process. The danger of overloading and acidification should however still be taken into consideration in the start-up phase.

Pre-treatment

Pre-treatment prior to the anaerobic treatment step is advisable for municipal wastewater in order to reduce the coarse and inorganic fractions (sand). Common pre-treatment steps are a screen and a grit chamber.

Degradation Performance

The digestion rate of UASB reactors with respect to its efficiency in COD reduction, is not as good as in activated sludge processes. The degradation of nutrients such as nitrogen and phosphorus are almost negligible. Therefore, if a better reduction of nutrients is required or if high standards for discharge to surface waters have to be met, the implementation of a post-treatment step following the UASB process reactors is recommended.

UASB Technology Performance

UASB technology has been found to be very effective for treatment of high strength industrial effluents particularly from distilleries, pulp and paper, tanneries, and food processing industries. For high organic loads, it certainly offers advantages in terms of almost insignificant energy consumption, low O&M cost and recovery of significant amount of bio-energy. Consistent production of fairly large quantities of biogas from industrial effluents makes electricity generation for captive consumption an attractive financial proposition. Other features of the technology *i.e.*, lower skill requirement and sludge production; perhaps add to its attractiveness under the industrial context to a certain extent. For instance, in distillery industry suitability of the technology has been amply demonstrated where due to bio-energy potential, the payback period has been found to be less than 3 years.

However, when applied for sewage treatment (where the undiluted BOD is between 200-300 mg/l), the cumulative experience has shown that these 'unique' features are not convincing for a variety of reasons. In retrospect it may be stated that for low strength wastewaters, there are more disadvantages than the upfront perceived advantages listed earlier. The issues related to effluent characteristics, requirement for secondary treatment, effluent suitability for disinfection, power generation and resource recovery are discussed below:

Effluent Characteristics

☆ UASB reactor is able to bring down BOD of sewage only to 70–100 mg/l and it perforce requires second stage aerobic treatment to enable compliance with discharge standards.

☆ The effluent from UASB is highly anoxic and it exerts a high immediate oxygen demand (IOD) on the receiving water body or land. If discharged in to a water body, it immediately sucks up the dissolved oxygen and undermines survival of aquatic life.

☆ If the raw sewage carries sulphates, it gets reduced to sulphides in the UASB reactor and upon release into an aquatic body it contributes in exerting immediate oxygen demand due to its conversion back to sulphate.

☆ While the UASB technology is perceived to require low skilled manpower and none or lower instrumentation system for operation control, its performance is characterised by frequent solids washout from the reactor. As a result the effluent BOD is found to be higher than what is normally claimed.

☆ While theoretical biogas yield is 0.35 cum/kg of COD removed, the actual yield is not more than 25-30 per cent of this value (0.08-0.1 cum/kg of COD removed). The remaining gas goes out in dissolved form with the effluent, raising its BOD and COD.

☆ The effluent has a dark brown or blackish colour which represents high concentration of dissolved and suspended humic substances in the effluent. This also leads to poor aesthetic value of the effluent.

☆ There are no reliable data correlating (a) BOD removal with biogas generation, (b) effluent BOD with COD and (c) effluent BOD with immediate oxygen demand.

☆ While the effluent BOD after final polishing unit (FPU) at various STPs is reported to be under 30 mg/l, but other independent studies carried out during similar period shows COD concentration to be above 200 mg/l.

Secondary Treatment

☆ Under YAP-I, all UASB reactors were followed by a final polishing unit (a pond) of one day retention for second stage treatment. This limited retention capacity while minimised land requirement, but from treatment point of view it at best offered only removal of solids washed out of the reactor.

☆ A retention of only 1 day does not allow growth of algal cells in the FPU as it is too short of the minimum requirement of 3 days.

☆ The FPU does not lead to re-aeration of wastewater as there is no energy input for turbulence and neither is there growth of algae which can facilitate this process naturally through the phenomenon of photosynthesis.

☆ As the settled solids are not removed regularly from the FPU (due to lack of O&M), the bottom depth for sludge storage quickly gets filled up, undermining its efficiency and leading to higher suspended solids/BOD in the final effluent.

☆ Even though a secondary treatment plant (aerobic system) will be required to bring down BOD from 70 to 30 mg/l, it would not be in any way cheaper than bringing down the raw sewage BOD from 250 to 30 mg/l, since the systems invariably need to be designed on the basis of hydraulic loading rather than the organic loading.

☆ Secondly, with input BOD less than 70 mg/l, there is not enough food for microorganisms to grow in the secondary aerobic system.

☆ Typically the power rating of an aeration system in an aerobic reactor is determined by requirements for keeping the solids in suspension and not on the basis of actual oxygen requirements. Therefore the perception of lower operating cost in the secondary stage after primary treatment in a UASB reactor is also not valid.

Unsuitability of Effluent for Chlorination

Due to anaerobic conditions, removal of total and faecal coliform in UASB is about 1-2 on log scale and it entails tertiary treatment for disinfection. However, unlike other technologies, effluent from a UASB plant cannot be readily sent for chlorination as it carries much high concentration of humic substances that lead to formation of trihalomethanes and entail higher chlorine consumption towards satisfying a part of the COD and IOD. Incidentally chlorination emerged as the only cost effective disinfection technology among a variety of options tried out under YAP-I.

Power Generation

Resource recovery in the form of bio-energy was perceived to be a major factor in favour of a UASB for sewage treatment. However, a number of limitations as listed below have been realised which prevent realisation of the claimed benefits.

☆ Among others, biogas generation is dependent on quantum of raw BOD and subject to ambient and wastewater temperature and their variations. The anaerobic bacteria culture is adversely affected with even 3-5°C fluctuation in reactor temperature. Therefore biogas production is found to go down significantly in the winter months in North India.

☆ The quantity of biogas produced in a small to medium sized UASB is not adequate enough to guarantee favourable economics of bio-energy generation.

☆ The duel fuel engines which are generally installed due to their low cost invariably require large quantity of diesel as the supplementary fuel. Apparently, the cost of diesel turns out to be not only high but disadvantageous as the electricity is made freely available to the STP operating agency. Economics of environment and resource utilisation apart, it does not make business sense for the operating agency to run the duel fuel generators on externally purchased diesel.

☆ State-of-the-art technology based gas engines are not yet made in India and the imported engines are rather expensive. Their deployment for small scale applications turns out to be unviable. Secondly, utilisation of waste heat from such cogeneration systems is not a techno-economically feasible option under the setting of an STP, which otherwise makes such systems financially attractive in colder climate countries.

☆ As the energy requirement of the UASB plant is low and the process is not vulnerable to power cuts; and energy bill of the STP is linked to the installed load any way, there in no incentive for the operating agency to generate bio-energy in-house by incurring extra expense on diesel. (These conclusions are corroborated by field observations of typical 1-2 hour operation of duel fuel engines or none at all as against the originally perceived full utilisation of biogas over 24 hour period.)

☆ Lastly, there is a risk of corrosion of the engine parts as the biogas typically contains hydrogen sulphide. The technology for desulphurisation is on one hand not widely available in India and on the other hand it entails additional recurring expenses. There have been cases of gas engines being taken off due to severe corrosion and desulphurisation plant being abandoned due to lake of required chemicals and resources.

Resource Recovery

☆ Another 'resource recovery' option through the sale of sludge and its potential to serve as a reliable and major revenue generating stream is also seem to be a failure.

☆ 'Resource recovery' through bio-energy generation and sludge, which was the guiding principle of the promoting and implementing agencies at the time of launching the UASB technology, turns out to be a myth as none of the plants have been able to contribute in any significant way towards the cost of operation and maintenance in any form.

Others

☆ Performance of the UASB based plants is, in general, adversely affected by mixing industrial effluents that contain some toxic materials or high levels of sulphate.

☆ In general, corrosion of structures in and around a UASB based plant is found to be higher compared to other technology based STPs.

Conclusion

Any wastewater treatment plant needs significant investment and efficient operation, maintenance and control, and therefore any decision to implement such a facility should be carefully considered. The adopted system must provide the best option for a low-cost, low maintenance and of effective in removing the pollutants of major concern. UASB reactor is based on that anaerobic sludge which exhibits inherently good settling properties and in which sludge is not exposed to heavy mechanical agitation. Sufficient mixing is provided by an even flow-distribution combined with a sufficiently high up-flow velocity, and by agitation that results from gas production. Biomass is retained as granular matrix or blanket, and is kept in suspension by controlling the up-flow velocity. Wastewater flows upwards through a sludge blanket located in lower part of reactor, while upper part contains three phase separation systems, which are the most characteristic feature of UASB reactor. It contributes the collection of biogas and also provides internal recycling of sludge by disengaging adherent biogas bubbles from raising the sludge particles. An advanced settling characteristic of granular sludge allows higher sludge concentrations and consequently permitted UASB reactor to achieve much higher OLRs. Granular sludge development is now observed in UASB reactors treating different types of wastewater.

But in contrary, it is evident that partial primary treatment through a UASB reactor makes the raw sewage more problematic to treat. Such systems neither deliver the required effluent quality nor produce the expected bio-energy. Considering all the pros and cons of the technology, especially the need for an elaborate secondary and tertiary treatment, the rationality for adopting a UASB for sewage (and especially diluted sewage under Indian context where the flows are intercepted in open drains) is debatable. In retrospect the less 'ambitious' conventional technologies *e.g.*, activated sludge process, trickling filter or facultative aerated lagoons would still be able to perform much better compared to the UASB. The biogas potential of sludge digesters in conventional activated sludge process plants is perceived to be more promising and consistent than the UASB reactor and therefore it is recommended that the latter option is the preferred option for sewage treatment in India at present context.

References

Barbosa, R.A. and Sant, A. G. L. (1989): Treatment of Raw Domestic Sewage in an UASB Reactor. Water Res. (G.B.), 23: 1483.

Che-Jen Lin, Zhang, P., Pongprueksa, P., James Liu, Evers, S. A. and Peter, H. (2013): Pilot- scale sequential anaerobic–aerobic biological treatment of waste streams from a paper mill. Environmental Progress and Sustainable Energy, 10: 11785.

Dague, R.R., Banik, G.C. and Ellis, G.E. (1998): Anaerobic Sequencing Batch Reactor Treatment of Dilute Wastewater at Psychrophilic Temperatures. Water Environ. Res., 70: 155.

Hassan, S. R., Zwain, M. H. and Dahla, I. (2013): Development of Anaerobic Reactor for Industrial Wastewater Treatment: An Overview, Present Stage and Future Prospects, J Adv Sci Res, 4(1): 07-12.

Hulshoff Pol, L., Euler, H., Eitner, A. and Grohganz, D. (1997): GTZ Sectorial Project: Promotion of Anaerobic Technology for the Treatment of Municipal and Industrial Sewage and Wastes. Proc. 8th Int. Conf. Anaerobic Digestion, Sendai, Jpn.; Int. Assoc. Water Qual., London.

Langenhoff, A.A.M. and Stuckey, D.C. (2000): Treatment of Dilute Wastewater Using an Anerobic Bafed Reactor: Effect of Low Temperature. Water Res.(G.B.), 34: 3867.

Lettinga, G., Rebac, S., Parshina, S., Nozhevnikova, A., van Lier, J.B. and Stams, A.J.M. (1999): High-Rate Anaerobic Treatment of Wastewater at Low Temperatures. Appl. Environ. Microbiol., 65:1696.

Mergaert, K., Vanderhaegen, B. and Verstraete, W. (1992): Applicability and Trends of Anaerobic Pre-Treatment of Municipal Wastewater. Water Res. (G.B.), 26: 1025.

Perry, R.H., Green, D.W. and Maloney, J.O. (1997): Perry's Chemical Engineer's Handbook. McGraw-Hill, New York.

Rebac, S., van Lier, J.B., Janssen, M.G.J., Dekkers, F., Swinkels, K.T.M. and Lettinga, G. (1997): High-Rate Anaerobic Treatment of Malting Wastewater in a Pilot-Scale EGSB System Under Psychrophilic Conditions. J. Chem. Technol. Biotechnol., 68: 135.

Rodrigues, A.C., Brito, A.G. and Melo, L.F. (2001): Post-treatment of a Brewery Wastewater Using a Sequencing Batch Reactor. Water Environ. Res., 73: 45.

Sasse, L. (1998): DEWATS: Decentralised Wastewater Treatment in Developing Countries. Bremen Overseas Research and Development Association (BORDA). Bremen, Germany.

Switzenbaum, M.S. and Grady, C.P.L. Jr. (1986): Anaerobic Treatment of Domestic Wastewater. J. Water Pollut. Control Fed., 58: 102.

Urbinati, E., Duda, R. M. and De Oliveira, R. A. (2013): Performance of UASB reactors in two stages under different HRT and OLR treating residual waters of swine farming., Eng. Agríc., Jaboticabal, 33(2):367-378.

Van Haandel, A.C. and Lettinga, G. (1994) Anaerobic sewage treatment. A practical guide for regions with a hot climate. John Wiley and Sons Ltd, Chichester, England.

Chapter 15

Aspects and Prospects of Traditional Knowledge, Religion and Culture in Forest Ecosystem Engineering: A Case Study

Ambarish Mukherjee

Professor, Centre for Advanced Study, Department of Botany,
Burdwan University, Burdwan – 713 104, West Bengal, India

ABSTRACT

Environmental crisis has become a global issue with rapid economic growth, advancement in science and technology. To overcome the crisis all possible scientific ways are worked out and in-depth studies on environmental concern of traditional knowledge of different indigenous communities have been presently undertaken to reveal the science cryptic in their philosophy concerning environmental perception, wisdom values and protection to the reality of environmental protection. The traditional knowledge of any ethnic group is based to a great extent on the plants occurring in his ambience. These plants get adhered to their lives and day to day activities. There is a growing recognition by the international forest policy and forest communities of the importance and relevance of Traditional knowledge about forests, and the need to consider this knowledge in the development of policies and practices that support sustainable management of forest resources. In India Traditional knowledge on the forest ecosystems is as old as ancient scriptures, bio-geographical niche, cultural history, and natural resource utilization. However, both traditional knowledge and its practice have been undergoing erosion and impoverishment. In view of this, 'in time documentation' and application of traditional knowledge in ecorestoration and environmental protection are necessary which appear certain to prove worthwhile in human welfare. The chapter applies traditional knowledge, culture and religious practices in ecosystem engineering under indigenous conditions on restoration of forests in Molandighi Beat of Durgapur Forest Range, Burdwan District of West Bengal.

Keywords: *Traditional knowledge, Indigenous, Environmental optimization, Ecosystem engineering, Forest.*

A General Comprehension

It has been realized worldwide that the indigenous societies are enormous sources of knowledge with adequate potential and relevance to sustainable use of natural resources, optimization of environment and improvisation of quality of our lives. Since this knowledge transmits orally through generations and survives in traditions of indigenous society there is an immediate need to document it on war footing; lest the uncared knowledge would be extinct. Aptly the decade beginning from 1st January 1995 was observed as the International Decade for the World's Indigenous People.

About 300 million people *i.e.* 1 in every 20 on the Earth belong to indigenous culture. Nearly half of them live in Asia and mostly in China and India. In India the ethnic communities are called "tribes". Still a large diversity and density of aboriginal human populations live today in nature and more precisely in the mega-diversity centres. In India there are nearly 68 million people of 697 indigenous communities designated as scheduled tribes. These communities belonging to these scheduled tribes still mostly live in forests sharing all moments of their life with plants, animals and the nature. Due to this intimacy the tribal communities have acquired through generations enormous experience and posses truthful knowledge about nature and biota. It is thus evident that India has enormous traditional knowledge which has been driving most of the corporate world crazy for profitable acquisition. This has given a jolt to India to arouse consciousness about her own wealth *i.e.* knowledge and resources.

The scientific world has, of late, given recognition to the indigenous traditional knowledge especially in the context of stock-taking of novel species and bioresource that are yet to be described and catalogued. The lacunae in scientific knowledge pertinent to biodiversity have been admitted and the attention is focused on the 'Linnean shortfall' (Brown and Lomolino, 1998). 'Wallacean shortfall' (Lomolino and Perault, 2004) and Centinelan extinction (Winchester and Ring, 1996). The Centinelan extinction refers to all such species which have become extinct totally being unknown to us. The 'Linnean shortfall' refers to the deficiency in the knowledge of the scientific world about the total number of species sharing the mailing address of the earth with us. The Wallacean shortfall speaks about the inadequacy of knowledge about the distribution of the species we know. To reduce these short falls utilization of the traditionally developed indigenous knowledge has been realized to be indispensible. Exploration of traditional knowledge and indigenous technical knowledge for conservation of species and their sustainable use is in progress especially in addressing the issues of conservation of forest environment and management.

The subject 'ethnobiology' has come to the limelight for its ability to give the tools to explore and document the traditional knowledge regarding taxonomy of living organisms, conservation and sustainable use of vital/biological and non-vital/material resources, meteorology, monitoring of environment, agri-horticulture, silviculture etc. aesthetics etc. Ethno-phytoresources include food, fodder, fibre, fuel, medicine, timber and pole wood, materials for thatching, musical instruments and weapons. Our interests are focused mainly on plant therapeutics, nutraceuticals, phytoprophylaxis including antioxidants, phyto-psychopathy, phyto-psychotherapy and anti-cancer plants. Biotechnologists today are in search of 'useful genes' from indigenous phytoresources to improve the quality of crops, impart in them genetic resistance to disease and pests so as to eliminate use of chemical-biocides and make the agricultural processes environment-friendly and consumer-friendly. Environmental scientists have found the traditional knowledge to be useful in ecorestoration or more contemporarily ecosystem engineering. To overcome the contemporary environmental crisis they have worked out scientific strategies for in-depth studies on environmental knowledge and concern of different communities, their religion, culture and values for application in ecosystem restoration, environmental optimization and protection.

The plant and animal resources used in the traditional customs are all of indigenous origin and outcome of the mutual relationship with the biodiversity in their ambience. As such documentation of the scientific rationale and knowledge cryptic in different types of man-plant, man-animal relationship and regular surveillance can pave the pathway for a much brighter and prosperous future.

Traditional Knowledge (TK)

A General Elucidation

The World Intellectual Property Organization (WIPO) used the term "Traditional Knowledge"- for referring to the tradition based literacy, artistic or scientific work performances, scientific inventions, discoveries, designs, marks, names and symbols, undisclosed information and all other tradition based innovations and creations resulting from intellectual activity in different scientific field. The Traditional Knowledge refers to the knowledge systems which have generally been transmitted from one generation to another by oral means only and are generally regarded as pertaining to a particular group of people or community or its territory; they are constantly evolving in response to changing environment. The message of science used to be communicated in the past through religions so that all religions are potential repositories of scientific knowledge emanating from experience and wisdom which survives through generations till today in traditional rituals, faith, taboos and cultural performances. Thus, it is a realization that the traditional knowledge and scientific rationale cryptic in different religions need to be worked out and documented immediately since there has been a progressive diminution of interests and involvements in religions. Considering the importance of TK even in modern era- the scientists have shifted their field of interest to the nature where still reside people with enormous knowledge.

In the current Intellectual Property Right (IPR) regime, documentation of TK with appropriate protection can prevent misappropriation and wrong patenting. Documentation of ethnic knowledge and wisdom can further enable the identification of promising components of biodiversity, bioresources and life sustaining systems to address issues of environment and human welfare. It is likely to help in establishing linkages with the plan and process, especially in the formation of innovative programmes and projects that can be implemented at grass root level.

TK and India

Documentation of Traditional Knowledge (TK) were carried out in India under the "Man and Biosphere" Project. Traditional Knowledge can act as a source of wisdom for decision making regarding various developmental activities. This is also considered as a valuable asset of the community, helping them to shape and control their own development.

A collaborative project 'TKDL' (Traditional Knowledge digital Library) has been undertaken in India by the Council of Scientific and Industrial Research (CSIR), Ministry of Science and Technology and Department of AYUSH, Ministry of Health and Family Welfare, and is being implemented at CSIR. An inter-disciplinary team of Traditional Medicine (Ayurveda, Unani, Siddha and Yoga) experts, patent examiners, IT experts, scientists and technical officers are involved in creation of TKDL for Indian Systems of Medicine.

The project TKDL involves documentation of the traditional knowledge available in public domain in the form of existing literature related to Ayurveda, Unani, Siddha and Yoga, in digitized format in five international languages which are English, German, French, Japanese and Spanish. Traditional Knowledge Resource Classification (TKRC), an innovative structured classification system for the purpose of systematic arrangement, dissemination and retrieval has been evolved for about 25,000 subgroups against few subgroups that was available in earlier version of the International Patent Classification (IPC), related to medicinal plants, minerals, animal resources, effects and diseases, methods of preparations, mode of administration, etc.

Presentation on Traditional Knowledge Resource Classification (TKRC) at IPC Union led to the creation of WIPO-TK Task Force consisting of USPTO, EPO, JPO, China and India by (IPC) Union for enhancing the sub-groups in IPC for classifying the TK related subject matter and considering the linking of TKRC with IPC (source: http://www.tkdl.res.in/).

Ecological Ethnicity

Ecological ethnicity is a social category that designates those people who have developed a certain respectful use of the bounty of nature and consequently a commitment to create and preserve a technology that interacts with the place and its non human collectivity in a sustainable manner. This concept includes peasants and other ecosystem people, such as fisher-folk, tribals, forest dwellers, nomadic shepherds, people marginalized by development projects and the programmes of environmental modernization (Apffel-Marglin and Parajuli, 2000). Ecological ethnicities should be theorized within the cover of ecology specialized as *moral ecology* which transcends the domains of both religion and science. The concept of ecological ethnicity adheres to the faith, attitude and practices of people who are to a great extent dependent on their immediate environment, or more precisely on its biomass and reveals the sharing of such people a great deal of environmental virtues despite differences in religious traditions.

Religion

Anthropology of Religion

The anthropology of religion involves the study of religious institutions in relation to other social institutions, and the comparison of religious beliefs and practices across cultures. Modern anthropology assumes that there is complete continuity between magical thinking and religion (Manickam, 1977) and that every religion is a cultural product, created by the human community that worships it. In virtually every major anthropological work on religion, and in most if not all introductory textbooks in cultural anthropology, the question of the truth or falsity of religious beliefs is evaded, ignored, or de-emphasized in favor of questions concerning the social, psychological, ecological, symbolic, aesthetic, and/or ethical functions and dimensions of religion. Considerations of disciplinary integrity, public welfare, and human dignity demand that religious claims be subjected to anthropological evaluation. It is precisely such areas as social, psychological, ecological, symbolic, aesthetic, and ethical functions and dimensions of religion where the anthropology of religion has made and continues to make its greatest contributions. Nevertheless, the scientific study of religion will never be fully legitimate until scientists recognize and proclaim the reality of religion (Lett, 1997).

Religion, Nature, Ecology and Environmental Studies

In the matters of relentless efforts to sustain the earth's environment as viable in optimum state for future generations, environmental studies have not so far adequately evaluated the role of different religions although ecology is deeply ingrained in it and moreover nature and its destiny, human virtues and ethics are guided and conditioned by religious beliefs and teachings. It is certain that religious tenets, views and practices can guide and mould our attitudes towards nature and her biotic as well as abiotic components and functions. Religion has been setting since time immemorial the guidelines to build up sustainable relations of our materialistic lives with the ecosystem in the best possible way and help us to reappraise our behaviour with the flora, fauna and environment and reorient us towards a successful life with optimum resources.

By the 1990s, many scholars of religion had put their sincere endeavours to generate scientific literature discussing and analyzing the envision of nature and its valuation by different religious systems of the world. A landmark setting event took place during 1996 - 1998 when a series of ten conferences on Religion and Ecology were organized at the Harvard University Center for the Study of World Religions under the esteemed leadership of Professors Mary Evelyn Tucker and John Grim of

Yale University. More than 800 international scholars, religious leaders, and environmentalists participated in the conference series. The conferences concluded at the United Nations and at the American Museum of Natural History and eventually a series of ten books (The Religions of the World and Ecology Book Series), one for each of the world's major religious traditions, *viz. Buddhism* (Tucker and Williams,1997), *Confucianism*(Tucker and Berthrong, 1998), *Hinduism* (Chapple and Tucker, 2000), *Islam* (Foltz *et al.*, 2003), *Jainism* (Chapple, 2002, De and Mukherjee,2013), *Judaism* (Tirosh-Samuelson, 2002), *Shinto* (Bernard, 2004). Other landmarks in the emerging field were the publication of the *Worldly Wonder: Religions Enter Their Ecological Phase* by Tucker *in* 2003 and the *Encyclopedia of Religion and Nature* by Taylor in 2005. Taylor also led the effort to form the International Society for the Study of Religion, Nature and Culture, which was established in 2006, and began publishing the quarterly *Journal for the Study of Religion, Nature, and Culture* since 2007.

Subsequent to the conferences, Tucker and Grim formed *The Yale Forum on Religion and Ecology* which has been instrumental in the creation of scholarship, in forming environmental policy, and in the greening of religion. In addition to their work with the Forum, Tucker and Grim's work continues in the *Journey of the Universe* film, book, and educational DVD series. It continues to be the largest international multireligious project of its kind.

The American Academy of Religion has remained active since 1991, and a number of universities, especially in North America are now offering courses on Religion and The Environment. The peer-reviewed academic journal *Worldviews: Global Religions, Culture, and Ecology* and the encyclopedia *The Spirit of Sustainability* have been putting efforts to augment the scientific message of religion for optimization of environment for peace, prosperity and improvement of qualities of life.

Environmental Concern of some Ancient Religions in India

Religion is an age-long component of Indian culture and traditions enshrining respect for nature. Every religion in India, mutually respecting each other, teaches that since God is the creator of nature its protection is a holy duty. Ancient Indian scriptures like Vedas, Upanishads, Smrities, Puranas, Mahabharata, Gita etc. teach the moral to live in harmony with nature and to respect it. Atharva-veda considers the Earth to be the mother and creations the offsprings. Water is the milk of the mother Earth. Man has no right to destroy the divine creations. In Gita Lord Krishna expresses that He exists in all living and nonliving matters which also exist in him. According to Varaha-Purana planting of trees like pomegranate, banyan, neem, wood-apple, orange, mango and ten flowering plants or creepers would lead to the Heaven. Yaur-Veda emphasizes mutual respect and kindness instead of domination and subjugation of nature by man. Many plants and animals and even patches of forests are considered sacred and protected through worship and taboos. Several animals have been enjoying protection since ancient times as 'mounts' of different Hindu Gods and Goddesses (Mukherjee, 2013).

Vedic and Upanishadic periods were followed by Buddhism and Jainism. Simplicity, nonviolence, truth, dignity, love and compassion are the basic tenets of Buddhism. Through 'Simplicity' is taught self-restraint and avoidance of overexploitation of natural resources. Jainism preaches a culture of tolerance, nonviolence, equity, universal love and compassion. Lord Mahaveera proclaimed a profound ecological truth that one who neglects or disregards the existence of the earth, air, fire, water and vegetation, himself disregards his own existence.

Distillation of science from all these religions is certain to augment conservation and optimize human living in a heavenly environment. In conformity with the cultural heritage of India, religious organizations and institutions in India have always been concerned with protection of nature and her creations for conveying the benevolence to mankind(De and Mukherjee, 2013).

Ethnobiology

The indigenous societies all over the world have traditionally developed their own knowledge about the plants and animals in their surroundings and have integrated the resources obtainable from them with their needs, cultural practices and mutually sustainable relationships. Studies on all aspects of direct man-plant and man-animal relationship compose an important subject known as ethnobiology. It is this subject which can guide us to the treasure house of bioresources, their optimum utilization and conservation under the regime of nature. The discipline became popular as the matrix of genuine academic and research activities in the second half of the twentieth century because of its potential to ensure conservation and for being synthetic and liberal in understanding the scientific rationale lying cryptic in the culture and traditions particularly of indigenous societies living in the cradle of nature, feeling and sharing moments of pleasure and sorrow with biodiversity.

The concept of sustainable use of bioresources can be comprehensively derived from the social and cultural attributes and knowledge of indigenous people of any region. Ethnobiology can resolve and assess human adaptive responses and human impact on biodiversity and *vice versa*. Once a species gets known as a resource in any cultural group the impact of this knowledge on expansion, distribution, threat to that species and in cases even its extinction, play great role (Alcorn 1984).

It has been seen that ethnic groups live in harmony with nature domesticating and selecting wild plants and animals through ages for use as resources. Necessity based folk domestication and selections have been operating on different wild species in different geographical areas to generate different types of cultivable economic plants giving us vegetables, cereals, pulses, fruits, fibres, medicines etc. Various stages of evolution of the subjects of animal husbandry and agriculture from their incipient forms can still be visualized in certain ethnic localities. The ongoing process of selection since the dawn of agriculture has led to genesis of crops with superior traits, *viz.* larger size, greater palatability, higher yield, more vitality and vigour of plants etc. The wild genotypes still used presently by indigenous societies have immense potential in improving the cultivated crops through well-planned hybridization and other gene-manipulation programmes. Ethnotaxonomic survey and evaluation of the specific and infraspecific races of natural flora and fauna co-existing with ethnic populaces are necessary for laying the foundation of bioresource management in one hand and catering their useful genes through biotechnology on the other. Efforts should thus be made on war footing to preserve as wide a range as possible of genetic resources for prevention of perilous possibilities of genetic erosion, especially in the biotic entities of evolutionary flexibility and ensure conservation even of the plants and animals the economic potential and ecological functions of which are yet to be fully known (Mukherjee 1997, 2013).

Ecosystem Engineering

Ecosystem engineering is the strategy to optimize ecosystems through creative compensations and sustenance of habitats and niches with their organisms keeping parity the laws of nature. In case of forests it has to operate biologically in the ambience of ethnic communities and in association with their livelihood and cultural activities which influence the distribution and abundance of species, resource availability, environmental resistance and biotic resilience *etc*. The work of ecorestoration is often referred to as ecosystem engineering or ecological engineering and the best known example of ecosystem engineers are the humans (Haemig, 2003)·

The term ecological engineering (the technology to restore ecosystem) was first coined by the H. T. Odum in 1962 to consider "those cases where the energy supplied by man is small relative to the natural sources but sufficient to produce large effects in the resulting patterns and processes". Ecological engineering (also called ecotechnology) presently focuses on the use of biological species (Jones *et al.*, 1994), communities and ecosystems with a reliance on self-design to encounter environmental and ecological degradation (Anon, 2005). Since man is the most important part of the ecosystem, ecotechnology is based on the self-designing capacity of nature under anthropological influences to enable the ecosystems to reach the sustainable optimum state.

Ecological engineering deals with ecosystems that are, have been, or may be in the deranged state to render services like landscaping, erosion control, renaturalization/habitat creation *etc* (Ahmed, 2004). This always aims towards stability so that the organisms including man can develop adequate resilience to appropriately encounter environmental resistance.

Need for Ecorestoration of Forests in India: A Case Study

Destruction of nature has always been a usual event in our developmental activities. Of different natural resources, forest is the most exploited natural resource. The annual loss of Indian forests was assessed to be about 1.5 million hectares (Anon, 2003). According to ISFR 2011, based on interpretation of the IRS P6 LISS-III satellite data in digital form corresponding to the period from October 2008 to March 2009, procured from National Remote Sensing Centre, Hyderabad and the Forest/TOF [Trees Outside Forests] Inventory, and socio-economic survey carried out by Forest Survey of India (FSI), the forest and tree cover of the country is 7,82,871 km^2 *i.e.* 23.81 per cent of GA (Geographical Area) and the total forest cover (including Very Dense, Moderately Dense and Open Forests) is 692,027 km^2, *i.e.* 21.05 per cent of GA (Anon, 2011). The real change in forest cover between the two assessment periods *i.e.* 2009-2011 works out to 367 km^2 which is attributable to management interventions such as harvesting of short rotational plantations, clearances on encroached areas, biotic pressures, shifting cultivation practices etc. The forests in Burdwan (Barddhaman) District of West Bengal State, according to the State Forest Report of FSI (Anon, 2011), cover only 261 km² *i.e.* about 3.72 per cent of the total geographic area *i.e.* 7024 km² in form of very dense (44 km²), moderately dense (135 km²) and open forms(82 km²). (Figure 15.1). These forests have been suffering from severe soil erosion of all types, *viz.* splash, sheet, rill, gully and toe along with attenuation of biodiversity. Under the circumstances the present authors took the initiative to apply the concept of ecosystem engineering under indigenous conditions on restoration of forests in Molandighi Beat under Durgapur Forest Range in collaboration with the Forest Department, local Forest Protection Committees and people of Santhal community.

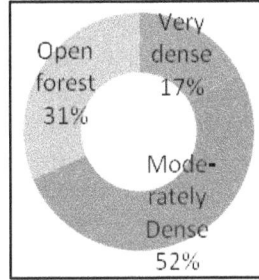

Figure 15.1: Ratio of Very Dense, Moderately Dense and Open Forests in Burdwan District, West Bengal (ISFR, 2011).

Study Site

The study site covering 54 sq Km is a natural tropical dry deciduous Sal forest under the Molandighi Beat under Durgapur Forest Range of Burdwan Forest Division. Molandighi covers six Mouzas, *viz.* Bistupur, Rakshitpur, Molandighi, Saraswatiganj, Bhagabanpur and Akandara. Forest Protection Committees are functional at Rakshitpur, Molandighi and Akandara (Anon., 2005b). There are three major seasons, *viz.* pre-monsoon (March to June), monsoon (July to October) and post-monsoon (November to February) seasons. Temperature ranges from 15°C (January) to 44°C (April). Mean annual rainfall ranges from 1000 to 1500 mm. Slope is somewhat moderate, the configuration being flat. The soils are typically lateritic, sandy-loam and severely eroded. The forest dwellers belong to Santhal community. Figure 15.2 shows the map of the district showing the study site.

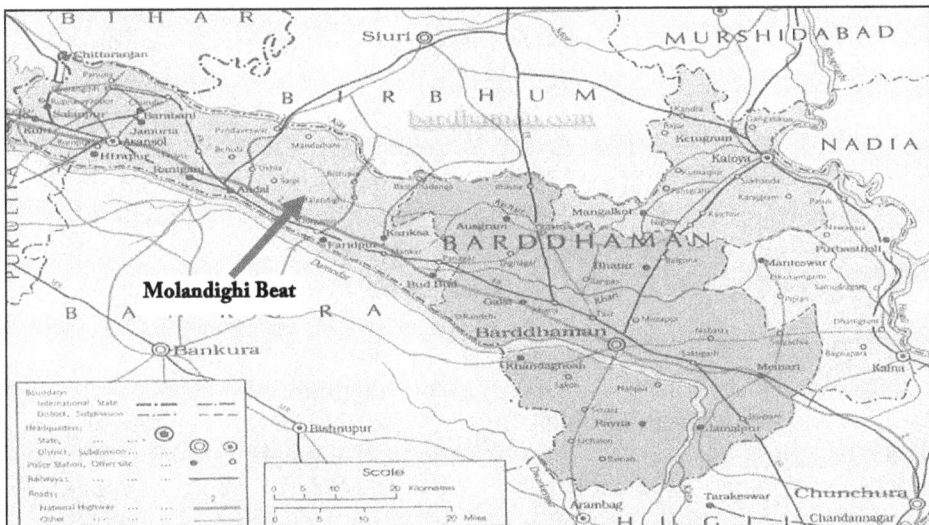

Figure 15.2: Map of the District Showing the Study Site.

Methodology

Realizing, the importance of indigenous traditional knowledge and culture in optimization of nature strategy for ecorestoration of forests in Molandighi Beat under Durgapur Range of Burdwan Forest Division was planned based in principle on the anthropological promotion of the self-designing capacity of nature. For this, the Santhal hamlets within the study site were visited in 2005 and their belief, ceremonies, festivals and traditions were thoroughly studied. With the cooperation of the Forest Department, Govt. of west Bengal and the *Naikeys* (Head priest) and *Manje Herams* (Head of the Hamlet) as many as twenty tree species which are in bondage with their culture, traditions and life were selected for planting in addition to the existing species considering that self- sustainability of a forest ecosystem depends to a great extent on the direct man-plant relationship *i.e.* the anthropological concerns of the ambient phytodiversity. Many small patches of the recently afforested areas were earmarked by the tribal priests (Naikey) as sacred (Jaher than) where trees like Sarjom (*Shorea robusta*), Matkom (*Madhuca longifolia*), Tarat (*Buchanania lanzan*), Teral (*Diospyros melanoxylon*), Pitasara (*Pterocarpus marsupium*), Rall (*Terminalia chebula*), Ull (*Mangifera indica*), Nimba (*Azadirachta indica*) etc. have been planted and the mundane boundary (Khond) for the Goddess Jaher Era prepared for worship in festivals like Sahrai, Laban, Bandna, Baha and Ashari.

Field studies were undertaken in the year 2006 prior to application of this ecorestorative strategy and again in 2009 to assess the success of the strategy. In 2006 and 2009 the constituent plant species were identified applying standard taxonomic methods and using pertinent literature (Guha Bakshi, 1984; Panigrahi and Murti, 1989; Prain, 1963; Sing, Nayar and Roy, 1994) which have been enumerated separately for the respective years according to Cronquist's (1988) system of classification.

To assess the success of this ecoengineering programme three parameters were used as indices, *viz.* Biological spectrum; ratios of tree: shrub: herb: climber and Generic Coefficient, each of which was determined initially in 2006 and eventually in 2009. The habits of the plants and their life-forms were determined according to the system of Raunkiaer (1934) as modified by Muller-Dombois and Ellenberg (1974). The biological spectrum of the area *i.e.* the percentage distribution of the constituent species in different life-forms, was determined and compared with the normal spectrum given by Raunkiaer (1934).

The generic co-efficient was determined by using the following formula given by Jaccard (1901):

Jaccard's Generic Coefficient = {(No. of genera/No. of species) × 100}.

Results and Discussions

The inventory of the angiosperms in Molandighi forest prepared in 2006 *i.e.* before ecosystem engineering, (Table 15.1) included 92 plant species under 80 genera of 41 families. Of these, 66 species were dicotyledonous and 26 were monocotyledonous (Table 15.1), their ratios at species level being1:0.39, at generic level 1:0.36 and at family level 1:0.24 (Table 15.2). Each genus on an average was allotted with 1.15 species. Generic co-efficient value (86.96) indicates that the

Table 15.1: An Enumeration of the Plants Associated with Forests of Molandighi Beat, Durgapur Forest Range, Burdwan District, West Bengal

Division: Magnoliophyta

    Class: Magnoliopsida (Dicots)

        Subclass I: Magnoliidae

            Order 7. Ranunculaceae

                (i)   Family 6. Menispermaceae

                      1.   *Tiliacora acuminata* (Lamk.) Hook. F. and Thoms. (Herb)

        Subclass II: Hamamelidae

            Order 6. Urticales

                (ii)   Family 4. Moraceae

                      2.   **Artocarpus heterophyllus* Lamk. (Tree)

                      3.   **Ficus racemosa* L. (Tree)

        Subclass III. Caryophyllidae

            Order 1. Caryophyllales

                (iii)   Family 8. Amaranthaceae

                        4.   *Aerva lanata* Juss. (Herb)

                (iv)   Family 11. Molluginaceae

                        5.   *Mollugo pentaphylla* L. (Herb)

        Subclass IV. Dilleniidae

            Order 2. Theales

                (v)   Family 4. Dipterocarpaceae

                        6.   **Shorea robusta* Gaertn. f. (Tree)

            Order 8. Capparales

                (vi)   Family 2. Capparaceae

                        7.   **Crateva religiosa* Forst. (Shrub)

            Order 12. Ebenales

                (vii)   Family 1. Sapotaceae

                      8.   **Madhuca longifolia* (J.Koenig) J.F.Macbr. (Tree)

                (viii)   Family 2. Ebenaceae

                      9.   **Diospyros melanoxylon* Roxb. (Shrub)

        Subclass V. Rosidae

            Order 3. Malvales

                (xi)   Family 2. Tiliaceae

                    10.   *Triumfetta rhomboidea* Jacq. (Herb)

                (x)   Family 3. Sterculaceae

                    11.   *Melochia corchorifolia* L. (Herb)

Contd...

Table 15.1–*Contd...*

(xi)	Family 5. Malvaceae
12.	*Abutilon indicum* (L.) Sweet (Shrub)
13.	*Sida cordata* (Burm. f.) Borssum (Shrub)
14.	*Sida cordifolia* L. (Shrub)
15.	*Urene sinuata* L. (Shrub)

Order 6. Violales

    (xii) Family 1. Flacourtiaceae

        16.   *Flacourtia indica* (Burm. f.) Merr. (Shrub)

    (xiii) Family 9. Violaceae

        17.   *Hybanthus enneaspermus* (L.) F. Muell. (Herb)

Order 2. Fabales

    (xiv) Family 1. Mimosaceae

        18.   *Acacia polyacantha* Willd. (Shrub)

        19.   **Albizzia lebbek Benth.* (Tree)

    (xv) Family 2. Caesalpiniaceae

        20.   **Cassia fistula* L. (Tree)

        21.   *Cassia occidentalis* L. (Shrub)

    (xvi) Family 3. Fabaceae

        22.   *Aeschynomene aspera* L. (Herb)

        23.   *Atylosia scarabaeoides* Benth. (Herb)

        24.   **Cajanus cajan* (L.) Huth (shrub)

        25.   *Dalbergia lanceolaria* L. f. (Tree)

        26.   *Desmodium triflorum* (L.) DC. (Herb)

        27.   **Pongamia glabra* Vent. (Tree)

        28.   **Pterocarpus marsupium* Roxb. (Tree)

        29.   *Tamarindus indicus*L. (Tree)

        30.   *Tephrosia purpurea* (L.) Pers. (Shrub)

        31.   *Vigna trilobata* (L.)Verdc. (Climber)

        32.   *Zornia diphylla* (L.) Pers. (Herb)

Order 6. Myrtales

    (xvii) Family 2. Lythraceae

        33.   **Lagerstromia speciosa* (L.) Pers. (Tree)

    (xviii) Family 5. Thymelaceae

        34.   *Cassytha filiformis* L. (Climber)

    (xix) Family 7. Myrtaceae

        35.   *Syzygium cumini* (L.) Skeels (Tree)

    (xx) Family 9. Onagraceae

        36.   *Ludwigia perennis* L. (Herb)

Contd...

Table 15.1–*Contd...*

(xxi) Family 12. Combretaceae

       37.   *Combretum roxburghii* Spreng. (Climber)

       38.   **Terminalia alata* Heyne ex Roth. (Tree)

       39.   *Terminalia arjuna* Bedd. (Tree)

       40.   *Terminalia chebula* Retz. (Tree)

       41*   *Terminalia bellirica* (Gaertn.) Roxb. (Tree)

Order 12. Euphorbiales

   (xxii) Family 4. Eupborbiaceae

       42.   *Croton bonplandianum* Baill. (Herb)

       43.   *Croton roxburghii* Balak. (Herb)

       44.   **Emblica officinalis* Gaertn. (Tree)

       45.   *Jatropha gossypifolia* L. (Shrub)

       46.   *Phyllanthus virgatus* Forst. f. (Herb)

Order 13. Rhamnales

   (xxiii) Family 1. Rhamnaceae

       47.   *Zizyphus oenoplia* Mill. (Shrub)

Order 15. Polygalales

   (xxiv) Family 5. Polygalaceae

       48.   *Polygala arvensis* Willd. (Herb)

Order 16. Sapindales

   (xxv) Family 8. Burseraceae

       49.   *Garuga pinnata* Roxb. (Tree)

   (xxvi) Family 9. Anacardiaceae

       50.   **Buchanania lanzan* Spr. (Small tree)

       51.   *Lannea coromandelica* (Houtt.) Merrill (Shrub)

       52.   **Semecarpus anacardium* (Tree)

   (xxvii) Family 13. Meliaceae

       53.   **Azadirachta indica* A. Juss. (Tree)

   (xxviii) Family 14. Rutaceae

       54.   **Aegle marmelos* (L.) Corr. (Tree)

Subclass VI. Aasteridae

   Order 1. Gentianales

     (xxix) Family 5. Apocynaceae

       55.   **Alstonia scholaris* R. Br. (Tree)

       56.   *Anodendron paniculatum* A. DC. (Climber)

       57.   *Holarrhena antidysenterica* Wall. (Tree)

Contd...

Table 15.1–*Contd...*

(xxx) Family 6. Asclepidaceae

    58.   *Hemidesmus indicus* R. Br. (Climber)

    59.   *Ichnocarpus frutescens* R. Br. (Climber)

    60.   *Tylophora indica* (Burm. f.) Merrill (Climber)

Order 2. Solanales

(xxxi) Family 3. Solanaceae

    61.   *Solanum indicum*L. (Shrub)

    62.   *Solanum surattense* Burm. f. (Herb)

(xxxii) Family 4. Convolvulaceae

    63.   *Ipomoea calycina* Clarke (Climber)

Order 3. Lamiales

(xxxiii) Family 3. Verbenaceae

    64.   *Gmelina arborea* L. (Tree)

    65.   *Lantana camara* L. (Shrub)

(xxxiv) Family 4. Lamiaceae (Labiatae)

    66.   *Anisomeles indica* (L.) O. Kunzte (Shrub)

    67.   *Hyptis suaveolens* Poit. (Shrub)

Order 6. Scrophulariales

(xxxv) Family 3. Scrophulariaceae

    68.   *Borreria articularis* (L. f.) F. N. Will. (Herb)

    69.   *Lindernia ciliata* (Colsm.) Pennell (Herb)

    70.   *Lindernia crustacea* (L.) F. Muell. (Herb)

    71.   *Scoparia dulsis* L. (Herb)

(xxxvi) Family 8. Acanthaceae

    72.   *Barleria prionitis* L. (Herb)

    73.   *Rungia pectinata* (L.) Nees (Herb)

Order 7. Campanulales

(xxxvii) Family 3. Campanulaceae

    74.   *Lobelia alsinoides* Lamk. (Herb)

Order 8. Rubiales

(xxxviii) Family 1. Rubiaceae

    75.   **Anthocephalus chinensis* (Lamk.) A. Rich. ex Walp. (Tree)

    76.   *Catunaregam spinosa* (Thunb.) Turuv. (Shrub)

    77.   *Catunaregam uliginosa* (Retz.) Sivarajan (Shrub)

    78.   *Gardenia gummifera* L. f. (Shrub)

    79.   *Meyna pubescence* (Kurz) Robyns (Shrub)

Contd...

Table 15.1–*Contd...*

Order 11. Asterales

(xxxix) Family 1. Asteraceae (Compositae) 80.*Chromolaena odorata* (L.) King and Robinson (Shrub)

      81.   *Elephantopus scaber* L. (Herb)

      82.   *Emilia sonchifolia* DC. (Herb)

      83.   *Vernonia cinerera* Less. (Herb)

Class: Liliopsida (Monocots)

    Subclass II. Arecidae

      Order 1 Arecales

        (i)   (i) Family 1. Arecaceae (Palmae)

            84.   *Phoenix acaulis* Buch. (Shrub)

(ii) Subclass III. Commelinidae

    Order 1. Commelinales

      (ii)   Family 4. Commelinaceae

         85.   *Cyanotis axillaries* Roem. and Schult. (Herb)

         86.   *Commelina bengalensis* L. (Herb)

         87.   *C.obliqua* L. (Herb)

         88.   *Floscopa scandens* Loureiro (Climber)

         89.   *Murdannia spirata* (L.) Bruckner (Herb)

      (iii) Family 5. Xyridaceae

         90.   *Xyris pauciflora* Willd. (Herb)

    Order 5. Cyperales

      (iv) Family 1. Cyperaceae

         91.   *Carex speciosa* Kunth. (Herb)

         92.   *Cyperus pangorei* Rottb. (Herb)

         93.   *Fimbristylis bisumbellata* (Forssk.) Bub. (Herb)

         94.   *Fimbristylis miliacea* (L.) Vahl (Herb)

         95.   *Fimbristylis ovata* (Burm. f.) Kern (Herb)

      (v)   Family 2. Poaceae (Gramineae)

         96.   *Aristida adscensionis* L. (Herb)

         97.   *Brachiaria ramose* (L.) Stapf. (Herb)

         98.   *Chrysopogon aciculatus* (Retz.) Trin. (Herb)

         99.   *Dactyloctenium aegypticum* (L.) Willd. (Herb)

        100. *Digitaria ciliaris* (Retz.) Koel. (Herb)

        101. *Eragrostis gangetica* Steud. (Herb)

        102. *Eragrostris tremula* Hochst. (Herb)

        103. *Eragrostis unioloides* (Retz.) Nees (Herb)

Contd...

Table 15.1–*Contd...*

	104. *Oplismenus compositus* Beauv. Bruckner (Herb)
	105. *Paspalidium flavidum* (Retz.) A. Camus (Herb)
	106. *Perotis indica* (L.) Kuntze (Herb)
Subclass V. Liliidae	
Order 1. Liliales	
(vi) Family 5 Liliaceae	
	107. *Curculigo orchioides* Gaertn. (Herb)
(vii) Family 14 Smilacaceae	
	108. *Smilax zeylanica* L. (Climber)
(viii) Family 15. Diosco-reaceae	
	109. *Dioscorea bulbifera* L. (Climber)

* Indigenous species of trees planted by the Forest Department.

community was more or less heterogeneous. Habit analysis of plants in 2006 shows that tree: shrub: herb: climber ratio was 11:22:48:11 (Figure 15.3) which speaks about an impoverished state of tree species and high proportion of shrubs and herbs. Introduction of exotic species, commercial overexploitation of forest resources, heavy grazing, collection of ground biomass (litter) *etc.* appear to be responsible for the impoverishment of the flora. These have resulted in severe soil erosion (sheet, rill, gully and toe) and loss of top soil in some places. Giving restoration the topmost priority, the Forest Department has implemented the scheme of ecosystem engineering using the species familiar to the ethnic communities and integrated with their belief, culture and traditions so that in 2009 as many as 109 species (Table 15.1) could be recorded of which 83 are dicotyledonous and 26 monocotyledonous, their ratios at specific level being 1:0.31, at generic level 1:0.29 and at familial level 1:0.18 respectively. The species quota for each genus is 1.16. Generic co-efficient was found to be 86.24 in 2009 (Table 15.2). Floristic analysis shows variation over a temporal scale of three years and interestingly, changes in values of generic co-efficient (from 86.96 to 86.24) and tree: shrub: herb: climber ratio (from 11:22:48:11 to 27:23:48:11) are likely to promote ecological welfare of the forest. Over this period of time, there has been an improvement in the status of tree species (Figure 15.3) which is a good sign of restoration.

Like a mirror, biological spectrum of a place reflects its phyto-climate through its composition of life-forms each of which is the manifestation of the sum total of adaptations to the climate in which it lives. So vegetations may be characterized and classified according to their climate-dictated life-form compositions (Oosting, 1958; Drude, 1992; Smith, 1996) which confer on them the physiognomy or the general appearance. Expression of the percentage distribution of different life-forms in the floristic composition of region is known as "the spectrum of life-forms" (Milne and Milne, 1971) or "biological spectrum" which can as well be used to calculate the stratification and layering pattern of a community (Rao, 1968; Krebs, 1994). This

Table 15.2: Taxonomic Analysis of Plants Associated with Forests of Molandighi Beat, Durgapur Forest Range, Burdwan (2006)

Angiospermic Plants	Total Number	Dicot Plants		Monocot Plants		Dicot : Monocot Ratio	Generic Index
		Total	Percentage	Total	Percentage		
Family	41	33	82.5	8	17.5	(i) 1:0.24	{(No. of genus/No. of species) × 100} = (80/92)
Genus	80	59	73.75	21	26.25	(ii) 1:0.36	
Species	92	66	71.74	26	28.26	(iii)1:0.39	× 100 = 86.96

Taxonomic Analysis of Plants Associated with Forests of Molandighi Beat, Durgapur Forest Range, Burdwan (2009)

Angiospermic Plants	Total Number	Dicot Plants		Monocot Plants		Dicot : Monocot Ratio	Generic Index
		Total	Percentage	Total	Percentage		
Family	46	39	84.78	7	15.22	(i)1:0.18	{(No. of genus/No. of species) × 100} = (94/109)
Genus	94	73	77.66	21	22.34	(ii)1:0.29	
Species	109	83	76.15	26	23.85	(iii)1:0.31	× 100 = 86.24

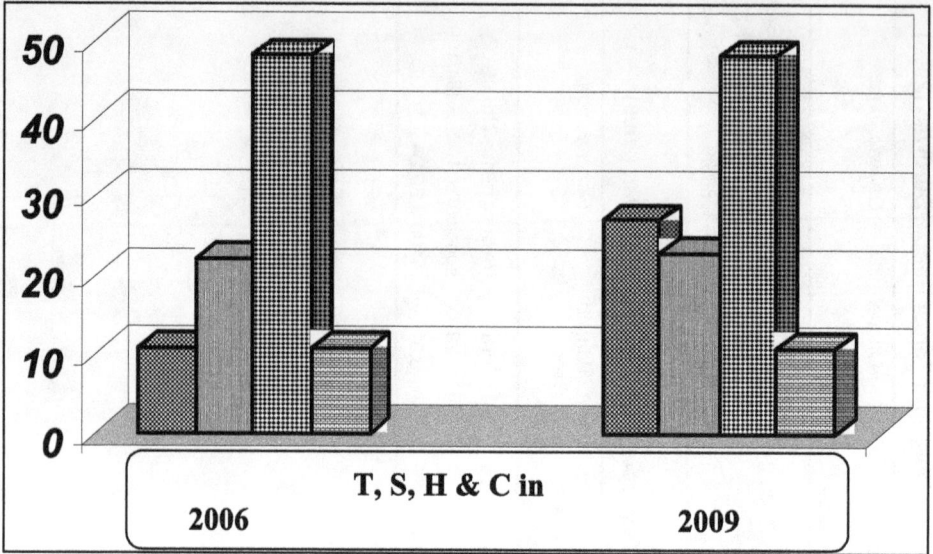

Figure15.3: Past and Present Status of Trees [T], Shrubs [S], Herbs [H] and Climbers [C] in Forests under Molandighi Beat, Durgapur Range.

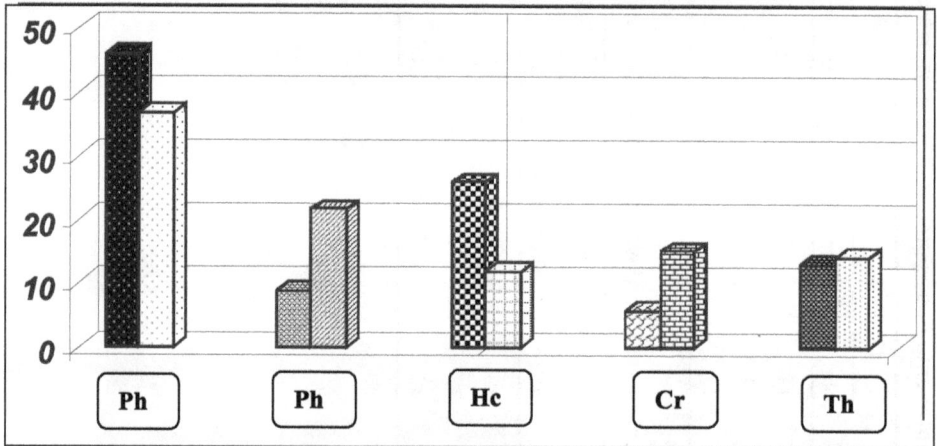

Figure 15.4: Comparison of the Biological Spectrum of Forests under Molandighi Beat in 2006 with Raunkiaer's Normal Spectrum; in each column-duo the left column shows normal value (of Raunkiaer) and the right column shows observed value.

assessment, a kind of biomonitoring, at a periodic interval is likely to set guidelines for eco-restoration and optimization. In view of this and to assess the success of the ongoing ecorestoration programme, biological spectrum of forests in Molandighi was determined.

Initially *i.e.* in 2006, out of a total of 92 species of angiosperms in the forest, phanerophytes (Ph) occupied 36.96 per cent, chamaephytes (Ch) 21.73 per cent,

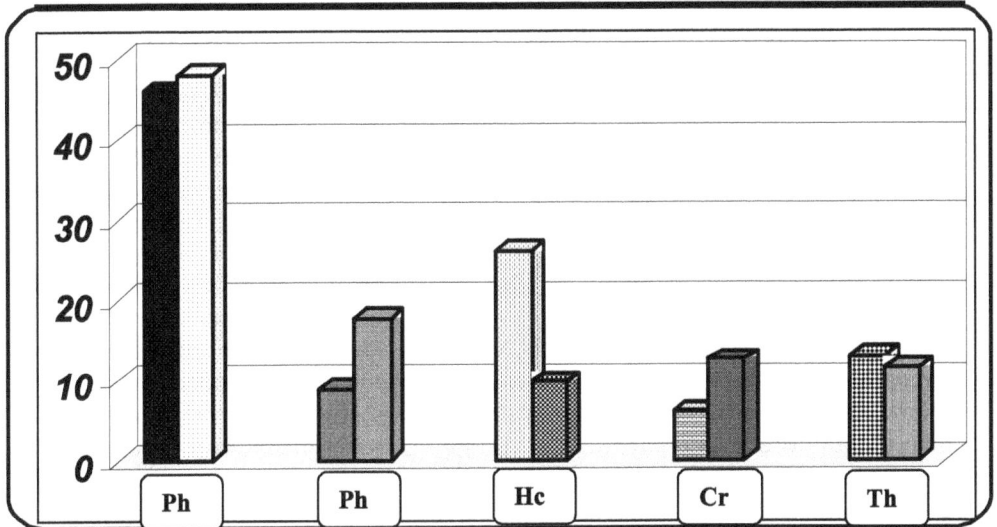

Figure 15.5: Comparison of the Biological Spectrum of Forests under Molandighi Beat in 2009 with Raunkiaer's Normal Spectrum; in each column-duo the left column shows normanl value (of Raunkiaer) and the right column shows observed value.

hemicryptophytes (Hcr) 11.96 per cent, cryptophytes (Cr) 15.22 per cent and therophytes (Th) 14.13 per cent as against the values of 46, 9, 26, 6 and 13 respectively registered in Raunkiaer's Normal Spectrum (Figure 15.4). Dominance of phanerophytes indicating the geoclimate of moist, warm, humid tropics is obvious in the area as is the case in a tropical monsoonal biome, but their percentage being less than the normal value (46 per cent). This speaks of some stress including anthropogenic ones that deter the tree vegetation and canopy structure. After the plantation of native species, the percentage of different life-forms in 2009 changed to: Ph-48.21, Ch-17.86, Hcr-9.82, Cr-12.5 and Th-11.61 (Figure 15.5). At present phanerophytes are more abundant than the normal value. Lowered proportion of hemicryptophytes at present indicates some sort of ongoing micro-environmental change to reduce stress. Proportions of chamaephytes and cryptophytes are still in higher proportions than those of Raunkiaer's Spectrum which indicate xeric trend of the community which is still likely to affect the regeneration niche (summation of all favourable conditions for birth, growth and establishment of seedlings) of trees in the forest. However the present physiognomic trend of the community is encouraging which might improve the tree cover through natural regeneration under human care and optimize the phytoclimate of the area in future. At present 52 per cent of the Indian forests do not have natural regeneration (Berwick and Saharia, 1995).

According to Jones *et al.* (1994) there are two different kinds of ecosystem engineers *viz.* allogenic- and autogenic-engineers which must co-act to restore normalcy. In the site man is the allogenic engineer regulating the restoration of the optimum environment by planting indigenous species and integrating them with his customs and traditions. The planted species, after establishment, would eventually maintain the optimum state as autogenic engineers.

Future Prospect

The prospect of the ongoing ecorestoration work at Molandighi seems to be bright since the selected species are not only indigenous but also linked intimately with the religious belief, culture and customs of the ethnic community. The nature appears to have permitted the phyto-maneuvered community to survive and proceed towards a state of dynamic equilibrium in harmony with the ethnic communities and their cultural activities. It is also encouraging to find resurrection of self respect and faith in their culture and traditions which enshrines great respect for nature and her creations. The whole hearted involvement of local Forest Protection Committees in restoration work deserves appreciation. Moreover, the Forest authorities have also started replacing the exotic tree species initially used to cover denuded forest areas with indigenous ones, creating large water reservoirs by raising rock-check dams against slope to harvest rain water and digging feeder channels to reach forest dwelling trees. Traditional practices like enrichment of soil with organic manures and phyto-mulching to stabilize soil and control erosion have been emphasized. Introduction of the nitrogen fertilizer making species like *Cajanus cajan* (L.) Huth along with Sal, the key stone species, in the revegetation programme also deserves appreciation. However, a synthetic approach at periodic intervals involving anthropology, plant taxonomy, ecology and environmental science is necessary for surveillance of the stability of the ecosystem. It is emphasized that ecorestoration is likely to be successful if there is maneuvering of creative cultural compensation for sustenance of habitats and organisms obeying the laws of nature (Figure 15.6).

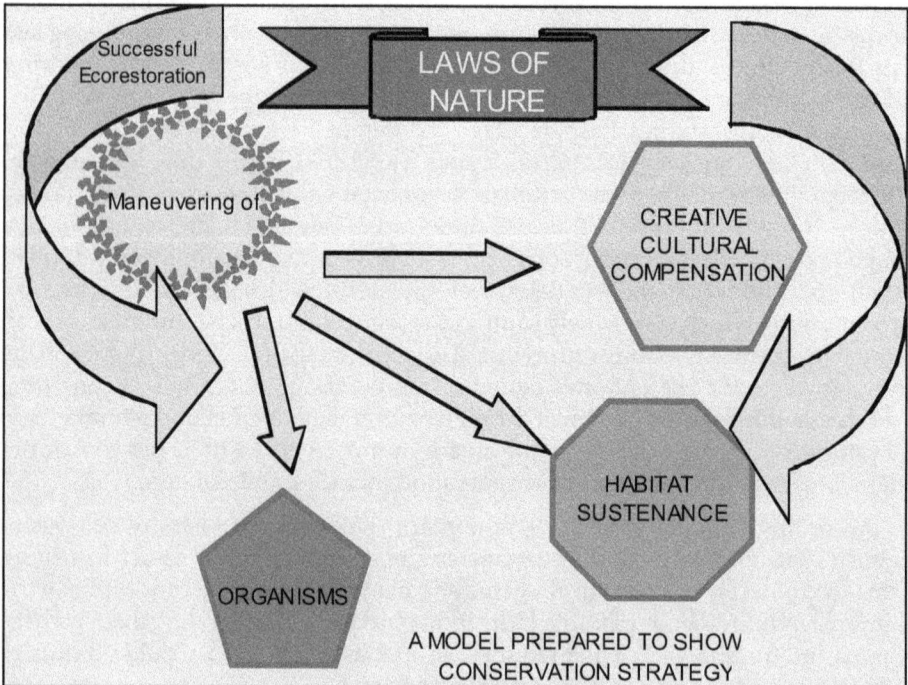

Figure 15.6: Laws of Nature.

Summing up

While summing up it may be said that the present day scientists must give priority and adequate importance to utilization of knowledge sustained through generations by indigenous people in their traditions and social life for identification and sustainable use of diverse bioresources and planning eco-engineering strategies for ecorestoration and environmental and economic welfare of mankind. Perpetuation of traditions of living in harmony with nature is certain to give mankind a brighter, peaceful and heavenly environment in all the days to come.

Acknowledgements

The author expresses is thankful to the officers and staff of The Durgapur Forest Range Office for providing all kinds of cooperation in undertaking the work. The active concern of my research students Dr. Debnath Palit, Dr. Archan Bhattacharya, Dr. Soma Chanda, Dr. Tripti Bouri, Anasua Roy, Papia Ghosh and in Animesh Maji in the field work is highly appreciated with gratitude. The author is also grateful to the forest dwellers, priests of the Santhal community and members of Forest Protection Committees for their cooperation and involvement.

References

Ahmed, M. (2004): Deforestation effects on species diversity index of terrestrial vegetation of Goalpara (Assam) and East Garo Hills District (Meghalaya), India, Plant Archives 4(1).

Alcorn, J. (1984): Huaestec Mayan. Ethnobotany. Texas, Austrin University Press.

Anon. (2003): Forest Survey of India. Ministry of Environment and Forests, Govt. of India, Dehra Dun, p. 107.

Anon. (2011): India State of Forest Report 2011. Forest Survey of India. Ministry of Environment and Forests, Govt. of India, Dehra Dun, pp.11-33 and 241-246.

Anon. (2005): Ecological Engineering: A New Paradigm for Engineers and Ecologists: 111-128, The National Academic Press, National Academy of Sciences, 500 Fifth St. N.W., Washington, D.C. 20001.

Apffel-Marglin,F. and Parajuli, P. (2000): Sacred Grove and Ecology: Ritual and Science. In: Hinduism and Ecology(eds. C.K.Chapple and M.E. Tucker) 290-316, 53-60.

Bennet, S. S. R. (1997): Name Changes in Flowering Plants of India and Adjacent Regions, Triseas Publishers, Dehra Dun, India.

Bernard, R.(ed.) (2004): Shinto and Ecology. Harvard University Press, Cambridge, MA.

Berwick, S.H. and Saharia, V. B. (1995): The Development of international principles and practices of wildlife research and management: Asian and American approaches. Oxford University Press : Delhi; New York.

Brown, J.H. and Lomolino, M.V. (1998): Biogeography (2nd ed). Sinauer, Sunderland.

Chapple, C. K. (ed.). (2002): Jainism and Ecology: Nonviolence in the Web of Life. Harvard University Press, Cambridge, MA.

Cronquist, A. (1988): Evolution and Classification of Flowering Plants. The New York Botanical Garden, Bronx, New York.

De, S. and Mukherjee, A. (2013): Study of ethnobotanical implications of places of religious activities in Kolkata: 1. Jain temples. Indian J.Applied and Pure Bio. 28(1): 5-17.

Drude, O. (1992): Pflanzengeographische Okologi–Abderhalden's Handbuch der biologischen, Arbeitistmetoden, Berlin and Vienna.

Guha Bakshi, D. N. (1984): Flora of Murshidabad District, West Bengal, India. Scientific Publishers, Jodhpur, India.

Haemig, P. D. (2003): Ecosystem Engineers: wildlife that create, modify and aintain habitats. ECOLOGY.INFO #12, © Copyright - Ecology Online Sweden.

Jaccard, P. (1901): Étude Comparative De La Distribution Florale Dans Une Portion Des Alpes Et Des Jura, Bulletin de la Société Vaudoise des Sciences Naturelles 37: 547-549.

Jones, C. G., Lawton, J. H. and Shachak, M. (1994): Organisms as ecosystem engineers, Oikos 69:373-386.

Krebs, C. J. (1994): Ecology The Experimental Analysis of Distribution and Abundance, Harper Collins, California. Lill, J. T. and Marquis, R. J. 2003. Ecosystem engineering by caterpillars increases insect herbivore diversity on white oak, Ecology, 84: 682-690.

Lett, J. (1997): Science, Religion, and Anthropology. In: Anthropology of Religion: A Handbook (ed. S. D. Glazier.) pp. 103-120. Greenwood Press, Westport, CT.

Lomolino, M.V. and Perault, D.R. (2004): Geographic gradients of deforestation and mammalian communities in a fragmented, temperate rainforest landscape. Global Ecology and Biogeography. 13: 55-64.

Manickam, T. M. (1977): Dharma According to Manu and Moses. p. 6. Dharmaram Publication, University of Michigan, USA.

Milne, L. and Milne, M. (1971): The Arena of Life The Dynamics of Ecology, Doubleday/Natural History Press, Garden City, New York, p. 240.

Mukherjee. A. (1997): Biodiversity Conservation. Journ. Asiatic Soc. 39 (2): 1-6.

Mukherjee, A. (2013): Bioresource Conservation: Traditions in India. The Ecoscan: Special issue III: 57-63.

Mukherjee, A. and Mukherjee, N. (1995): Cultural relevance of Indian Plants: An Overview. J. Indian Anthrop. Soc. 30: 67-72.

Muller-Dombois, D. and Ellenberg, H. (1974): Aims and Methods of Vegetation Ecology, John Wiely and Sons, New York.

Odum, H.T. (1962): Man and Ecosystem, Proceedings, Lockwood Conference on the Suburban Forest and Ecology, Bulletin Connecticut Agric. Station.

Oosting, H. J. (1958): The Study of Plant Communities, Freeman and Co., San Francisco.

Palit, D., Ganguly, G. and Mukherjee, A. (2002): Sci. and Cult. 68(5–6): 147–149.

Panigrahi, G. and Murti, S. K. (1989): Flora of Bilaspur District (Vols. 1 and 2), Botanical Survey of India, Calcutta.

Prain, D. (1963): (rep. ed.), Bengal Plants (Vols. I-II). Botanical Survey of India, Calcutta.

Rao, C. C. (1968.): In: Proc. Sym. Recent Advances in Tropical Ecology, Part II (Ed. R. Mishra and B. Gopal), ISTE, BHU, Varanasi.

Raunkiaer, C. (1934): The Life Forms of Plants and Statistical plant Geography, Clarendon Press, Oxford.

Sing, M. P., Nayar, M. P. and Roy, R. P. (1994): Textbook of Forest Taxonomy, Anmol Publications Pvt. Ltd., New Delhi.

Smith, R. L. (1996): Ecology and Field Biology, Harper Collins, California, pp. 598–599.

Taylor, B. (ed.).(2005): Encyclopedia of Religion and Nature (2 volumes). Continuum International Publishing Group, London.

Tirosh-Samuelson, H.(ed.). (2002): Judaism and Ecology: Created World and Revealed Word. Harvard University Press, Cambridge, MA.

Tucker, M. E. and Williams, D. R.(eds). (1997):. Buddhism and Ecology: The Interconnection of Dharma and Deeds. Harvard University Press, Cambridge, MA.

Tucker, M. E. and Berthrong, J. (eds.). (1998): Confucianism and Ecology: The Interrelation of Heaven, Earth, and Humans. Harvard University Press, Cambridge, MA.

Tucker, M. E. (2003): Worldly Wonder: Religions Enter Their Ecological Phase. Open Court, Chicago.

Winchester, N.N. and Ring, R.A. (1996): Centinelan extinctions: extirpation of Northern temperate old-growth rainforest arthropod communities. Selbyana, 17(1): 50 – 57.

Chapter 16

In vitro Solubilization of Inorganic Phosphate by Phosphate Solubilizing Microorganisms (PSM) from Rice Rhizosphere

Ranjita Panda[1], Siba P. Panda[2], C.R. Panda[3]
and R.N. Kar[3]

[1]PG Department of Environmental Sciences,
Sambalpur University, Jyotivihar, Burla – 768 019, Odisha, India
[2]Environment Cell, R&D Building, Hindalco Industries Limited,
Hirakud, Sambalpur, Odisha, India
[3]Institute of Minerals and Materials Technology (CSIR),
Bhubaneswar – 751013, Odisha, India

ABSTRACT

Ten most efficient phosphate solubilizing bacteria and four fungi (PSM) isolated from local rice rhizosphere of Sambalpur district, Odisha were studied for biochemical characteristics. PSM were grown *in vitro* for seven days on Pikovskaya's medium and following analyses were carried out *i.e.* solubilization index, pH change, phosphorus (P) solubilized and titratable acidity under *in vitro* conditions. P solubilization index of these isolates ranged from 1.22-6.68. Drop in pH of the medium ranged from 7 to 3.2 with the continuous growth of these isolates for seven days. Drop in pH of the medium ranged from 7 to 3.3 with the continuous growth of these isolates for seven days. This drop of pH was clearly indicated the production of organic acids. Fungi were found to be more active than bacteria in conversion of insoluble P to soluble P. Among bacterial strains (PSB31) showed maximum P solubilisation *i.e.* 54.76 per cent whereas, *Aspergillus niger* (PSF-1) solubilise 78 per cent of P from tricalcium phosphate.

Keywords: Phosphate solubilization, PSM, Rice.

Introduction

Phosphorus is one of the major essential macronutrients for biological growth and development. (Ehrlich, 1990). Phosphorus in soil is present in insoluble form and complexed with cations like iron, aluminium and calcium. Phosphorous is added to soil in the form of phosphatic fertilizers, part of which is utilized by plants and the remainder converted into insoluble fixed forms. Use of phosphatic fertilizers has become a costly affair for Indian farmers. Although, use of chemical fertilizers for improving soil fertility is the common approach of increasing agricultural production, a large portion of inorganic phosphate applied to soil as fertilizer is rapidly immobilized after application and becomes unavailable to plants.

Many soil microorganisms have been reported to solubilise inorganic phosphates (Chuang *et al.*, 2006; Alikhani *et al.*, 2010). In fact rock phosphate dissolution by microorganisms directly affects soil fertility (Reys *et al.*, 2002). Microorganisms substantially influence the soil productivity by solubilising this insoluble P through their metabolic processes in soil. A large number of heterotrophic and autotrophic microorganisms, such as bacteria (Louw and Webly, 1959), and fungi (Whitelaw, 2000), are reported to solubilize insoluble phosphate, *e.g.*, hydroxyapatite, tricalcium phosphate, and rock phosphate. The process of microbe mediated P solubilisation is generally due o the production of organic acid by them. (Illmer *et al.*, 2011 and Das *et al.*, 2003). The role of organic acids produced by phosphate solubilising microorganisms (PSMs) in solubilising insoluble P may be due to the lowering of pH, chelation of cation and by competing with P for adsorption sites in the soil. It has been investigated that organic acids may also form soluble complexes with metal ions associated with insoluble P (Ca, Al, and Fe) and thus P is released.

In the present study PSM from rice rhizosphere in Sambalpur district, Odisha having potential to solubilise insoluble phosphate were isolated and checked for their ability to solubilise insoluble phosphate in solid media and in liquid culture.

Materials and Methods

Isolation of Phosphate Solubilising Microorganisms (PSM)

Bacterial and fungal strains were isolated from 26 rhizosphere soil samples of rice plant from Sambalpur district, Odisha. Soil samples were collected and transferred under aseptic conditions in an ice packet and stored at 4° C in refrigerator. Ten gram (10g) of soil sample was suspended in 90 ml of sterile distilled water and 10^{-1} dilution was obtained. Serial dilutions were prepared by mixing one ml of the suspensions made into 9 ml sterile water blanks, until the 19^{-7} dilution was obtained. One ml of the appropriate dilutions (10^{-5}-10^{-7}) of the soil samples were plated on Pikovskaya's agar medium containing tricalcium phosphate as sole P source for enrichment of phosphate solubilizing bacteria and fungi. Plates were incubated at 35° C for seven days. The colonies showing clear halo zone indicating the ability to solubilise tricalcium phosphate were selected and grown on Pikovskaya agar. The purified colonies causing clear phosphate solubilising halozones were selected and purified for further study.

Identification of Fungi

For the identification of fungi a drop of Lactophenol cotton blue placed on glass slide then grown mycellium of fungi was taken on the slide and covered with cover glass and observed under microscope. Identification based upon their colony morphology, mycelial and spore characteristics (Aneja 1998).

Solubilization Index on Solid Media, Growth Condition

Qualitative estimation of phosphate solubilization of the selected fungal and bacterial isolates was conducted through plate assays using spot inoculation on Pikovskaya agar medium (Pikovskaya, 1948) supplemented with tri-calcium phosphate (TCP) at 30° C. The medium contained l^{-1} glucose, 10 g; $Ca_3(PO)_4$, 5 g; $(NH)_4SO_4$. 0.5 g; NaCl, 0.2 g; $MgSO_4.7H_2O$, 0.1 g; KCl, 0.2 g; yeast extract,4 2 42 0.5 g; $MnSO_4.7H_2O$, 0.002 g; $FeSO_4.7H_2O$, 0.002 and agar 15 g. The pH of the media was adjusted to 7.0 before autoclaving. Sterilized PVK media was poured into sterilized Petri plates after solidification of the media, a pinpoint inoculation of fungal strains was made onto the plates under aseptic conditions. They were incubated at 30°C for 7 days with continuous observation for colony diameter. The halo zone formations around the growing colony showing phosphate solubilization. Solubilization Index was measured using following formula (Edi-Premono *et al.*, 1996).

$$SI = \frac{\text{Colony diameter} + \text{Halozone diameter}}{\text{Colony diameter}}$$

Solubilisation of Phosphorous from Tricalcium Phosphate

For Quantitative estimation of phosphate solubilisation in broth culture studies, 100mL of Pikovskaya's broth medium amended with 1 per cent pulp density of tricalcium phosphate was distributed in a conical flask (250 mL) and sterilized at 120° C for 20 minutes. The pH was maintained to 7.0 before sterilization of the media. Then after sterilization five ml suspension of each bacterial and fungal culture (10^7cfu/flask) was added to the broth in triplicate. A control without any inoculation was maintained. The culture were incubated in a rotary shaker at 30°C for seven days and each day they were individually centrifuged at 15000 rpm for 20 minutes. The supernatant was collected in 100 ml volumetric flasks and the volume was made up to 100 ml with distilled water. While incase of fungi the centrifuged solution was filtered through Whatman filter paper No.42 the clear solution was collected in 100 ml volumetric flasks and the volume was made up to 100 ml. For pH of the culture was measured with a pH meter equipped with glass electrode. Dissolved phosphate concentration in the culture filtrate was determined by chlorostannous reduced molybdophosphoric acid blue method (Jackson 1987).

In order to study the titratable acidity of culture medium, three days old culture filtrates were centrifuged at 1000 rpm for 20 minutes. Five milliliter of supernatant was added with few drops of phenophthalin indicator and titrated against 0.01N NaOH. The titratable acidity was expressed as ml of 0.01N NaOH consumed per 5.0 ml of culture filtrate.

Results and Discussions

Microorganisms

Out of 85 bacterial isolates ten efficient bacteria were selected based on the zone of solubilisation in agar medium and four fungi were selected out of twenty four. These fourteen bacterial and fungal isolates were taken for further studies. The fungal strains were identified as PSF1 *(Asperillus niger)*, PSF2 *(Aspegillus fumigats)* PSF3 *(Penicillium* sp.*)* and PSF4 *(Aspergillus flavus)*.

Solubilization Index (SI)

Solublization index (SI) based on colony diameter and holozone for each PSM isolate was presented in (Table 16.1).

Table 16.1: Solubilization Index of Selected Strains of Phosphate Solubilizing Bacteria and Fungi.

Strain Code	Solubilization Index through Seven Days						
	Day 1	Day 2	Day 3	Day 4	Day 5	Day 6	Day 7
Control	0.00	0.00	0.00	0.00	0.00	0.00	0.00
PSB1	1.14	1.58	1.64	1.68	1.88	2.56	2.84
PSB3	1.22	1.45	1.34	1.76	1.88	2.04	2.11
PSB5	1.21	1.64	1.88	2.43	2.65	2.87	2.96
PSB8	1.42	1.32	1.87	1.88	1.89	2.11	2.23
PSB17	1.34	1.21	1.43	1.65	1.87	2.23	2.11
PSB18	1.11	1.34	1.43	1.56	1.84	2.22	2.26
PSB29	1.16	1.18	1.34	1.22	1.46	1.48	1.44
PSB31	1.64	1.66	2.21	3.78	3.81	3.82	3.84
PSB34	1.16	1.58	1.54	1.68	1.88	2.94	2.96
PSB47	1.22	1.46	1.48	1.95	2.23	2.84	2.87
(PSF1) *A. niger*	1.94	2.36	2.87	2.49	2.92	–	–
(PSF2) *A. fumigatus*	1.58	2.12	2.43	2.38	2.38	–	–
(PSF3) *Penicillium* sp.	1.54	1.62	2.11	2.04	2.18	–	–
(PSF4) *A.flavus*	1.34	1.68	2.23	2.42	2.43	–	–

* PSB: Phosphate solubilising Bacteria; PSF: Phosphate solubilising Fungi.

Solubilization index for fungi was determined for five days only because of fungal over growth.

Results showed that among phosphate solubilising bacteria (PSB) PSB31 was most efficient phosphate solubilizer on PVK plates with highest SI *i.e.* 3.84 after 7 days. Where as among fungi *A. niger* (PSF1) showed highest solubilizing index *i.e.*, 2.92. Studies on agar plates revealed that phosphate solubilizing microorganisms formed clear zones by solubilizing suspended tricalcium phosphate. Halozone measurements ranged from 2.4-5.7 cm for bacteria and 4.07-8.84 cm for fungi (data not shown). Generally, halozone increased with increase in colony diameter. In most of the cases it gradually increased up to seven days while in few cases it first decreased and then increased (PSB47, *A. flavus*, *A. fumigatus* and *A. niger*). Fungi produced larger halozones than bacteria. These results are in accordance to Chabot *et al.* (2010) and Nahas (1996). Similar results were found by Kucey (1983, 1987), Edi-Premono *et al.* (1996) and Kumar and Narula (2011). But solubilization index (SI) of most efficient bacteria (PSB31and PSB34) was greater than that of fungi.

The higher the value of SI the greater the activity of the tested isolate was. But with that method we could not be able to quantify the amount of phosphate solubilised. So liquid culture experiment involved evaluation of the amount of P solubilised and evolution of pH in time.

pH and Titratable Acidity

Most phosphate solubilizing microorganisms studied lowered the pH of the medium as compared to uninoculated sterile control, *i.e.*, 7.01. The pH of the broth dropped significantly as compared to the control. Fungi and bacteria both were found to be equally active in lowering the pH. The decrease in pH from 7.01(control) in the beginning to 3.36 within seven days.The highest pH decrease (3.36) was determined in PSB31 in 7th day when compared to control and the lowest pH decrease was found in PSB18 strain *i.e.*, 5.22. (Table 16.2).

In case of fungal strain the lowest pH was observed in *Aspergillus niger i.e.*, 3.21 in 7th day (Figure 16.1).

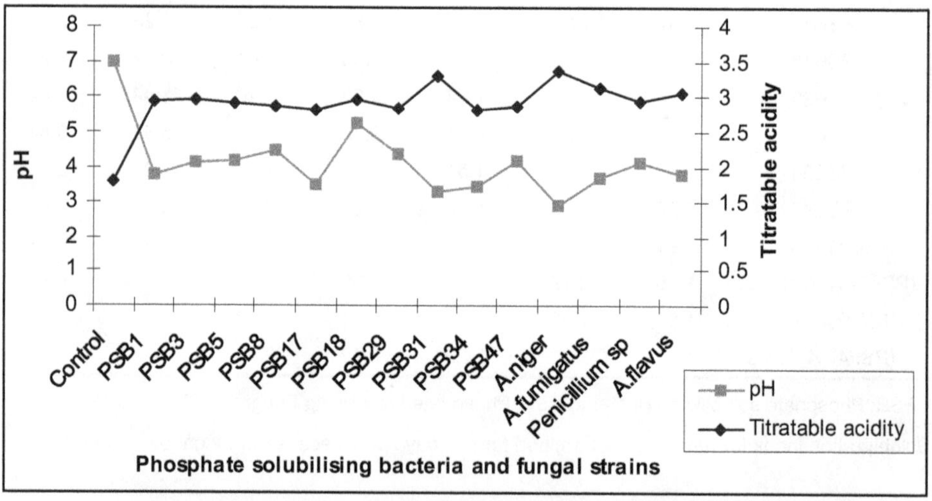

Figure 16.1: Variation of pH and Titratable Acidity by Phosphate Solubilising Bacterial and Fungal Strains after 7 Days in Culture Broth.

**Table 16.2: Drop in pH by Selected Strains of
Phosphate Solubilizing Bacteria and Fungi**

Strains	pH Drop through Seven Days						
	Day 1	Day 2	Day 3	Day 4	Day 5	Day 6	Day 7
Control	7.01	7.01	7.01	7.01	7.01	7.01	7.01
PSB1	6.54	5.64	4.91	4.11	3.96	3.88	3.78
PSB3	6.02	6.24	6.22	5.54	5.34	4.21	4.11
PSB5	6.66	5.98	5.88	4.66	4.61	4.22	4.17
PSB8	6.14	5.26	5.18	5.08	5.11	4.96	4.44
PSB17	7.01	4.86	4.21	4.16	3.89	3.63	4.48
PSB18	7.01	5.98	5.64	5.44	5.36	5.24	5.22
PSB29	7.01	5.78	5.66	5.24	5.12	4.68	4.36
PSB31	5.98	4.68	4.58	4.24	4.11	4.14	3.36
PSB34	6.11	4.22	4.17	4.08	3.96	3.84	3.44
PSB47	6.26	5.56	5.26	4.84	4.85	4.38	4.17
(PSF1) *A. niger*	5.89	4.12	3.49	3.51	3.54	3.44	3.21
(PSF2) *A. fumigatus*	5.78	4.28	3.84	3.74	3.68	4.49	4.38
(PSF3) *Penicillium* sp.	5.89	4.66	3.64	3.45	3.38	4.24	4.22
(PSF4) *A. flavus*	5.88	4.61	4.14	3.91	3.46	3.84	3.98

The titratable acidity (TA) value was increased in all the phosphate solubilising bacteria and fungi as compared to control *i.e.* 1.8. The TA value ranged from 2.81 to 3.36. The highest value was observed in *Aspergillus niger* and lowest in PSB34. From (Figure 16.1) it was observed that there was reduction in pH of the medium but an increase in titratable acidity. Decrease in pH may be explained by production of organic acids as reported by (Akintokun *et al.*, 2007). The increase in titratable acidity is responsible for the observed decrease in pH and might be due to secretions of organic acids by phosphate solubilising microorganisms (PSMs).

Phosphate Solubilisation in Broth Medium

P solubilised values in each broth sample shows that different strains solubilised the P at different rates (Table 16.3).

Phosphate solubilisation of the added tri calcium phosphate in broth was in the range of 16.44 to 54.76 per cent. Among individual culture of all bacterial strains (PSB31) solubilise highest P *i.e.* 54.76 per cent whereas the lowest P value was in PS18 strain. Among the fungi the P solubilisation was in the range of 21.67 by *Penicillium* sp. to 78.01 per cent by *Aspergillus niger*. Among all the bacterial isolates the lowest pH was observed in PSB31 where the percentage of P solubilisation was more *i.e.* 54.76 per cent. *A. niger* reduce the pH 2.89 on 7th day where the percentage of P solubilisation was 36.65 per cent. Even though maximum drop in pH was associated with higher level of P solubilisation. Overall bacterial strains solubilise greater amount of phosphorous than fungal strains. In the present study the gradual decrease of pH and the increase of P solubilisaton were observed. In the present study the negative correlation between pH and P solubilisation was observed. From the (table 3) it was observed that all the bacterial isolate increased P solubilisation with increasing period of incubation. The highest P solubilisation was observed in 7th day in case of bacterial isolates and the fungal strains.

Table 16.3: Phosphate Solubilisation as (mg per cent) during Seven Days of Incubation in PVK Broth Medium by Phosphate Solubilising Bacteria and Fungi

Strains	Per cent of P Solubilisation during Seven Days						
	Day 1	Day 2	Day 3	Day 4	Day 5	Day 6	Day 7
Control	0.12	0.15	0.16	0.16	0.17	0.18	0.18
PSB1	3.44	5.41	9.66	13.57	19.21	21.34	24.78
PSB3	4.34	7.32	8.36	10.98	14.65	18.43	22.42
PSB5	2.65	3.32	4.24	7.33	12.22	15.34	18.88
PSB8	2.85	3.78	5.66	8.87	13.32	15.11	17.43
PSB17	4.21	8.53	11.34	14.45	21.65	24.42	26.46
PSB18	2.1	3.78	4.32	8.66	12.86	15.55	16.44
PSB29	1.87	3.78	5.43	7.22	12.76	16.45	18.22
PSB31	7.53	11.32	28.67	41.65	49.98	52.91	54.76
PSB34	8.22	13.78	17.52	21.78	25.34	29.43	31.44
PSB47	2.47	6.48	10.71	11.92	16.78	18.78	19.86
A. niger (PSF1)	35.34	49.56	58.87	68.56	71.01	76.82	78.01
A. fumigatus (PSF2)	6.43	9.77	18.56	27.65	33.76	30.23	49.12
Penicillium sp. (PSF3)	4.87	7.65	14.89	18.43	21.67	20.01	19.89
A. flavus (PSF4)	5.43	9.70	17.45	21.82	24.87	22.67	21.43

P solubilisation and pH value of the culture filtrate indicated that phosphate solubility was directly correlated with the acids produced. There were reports PSMs secrete many kinds of organic acids (Rodriguez and Reynaldo, 1999). We are now investigating the analysis of organic acids in the culture medium.

Conclusion

It was concluded from the present study that PSM showed variation in their biochemical charateristics. Organic acid production was perhaps not the only possible reason for phosphosrus solubilization. Present study also showed that *Aspergillus niger* and PSB31 are the most efficient strains on the basis of their P solubilizing activity. Further research should be continued with such efficient PSM isolates. These may be used for inoculum production and their inoculation effect on the plant growth be studied *in vivo*.

References

Akintokun, A.K., Akande, G.A., Akintokun, P.O., Popoola, I.O.S. and Babarora, A.O. (2007): Solubilisation of insoluble phosphate by organic acid producing fungi isolated from Nigerian soil. International Journal of soil science 2(4) :301-307.

Alikhani, H.A., Saleh-Rastin, N., and Antoun, H. (2010): Phosphate solubilisation of rhizobia native to Iranian soils.Plant Soil 287: 35-41.

Aneja, K.R. (1998): Experiments in biology, Plant pathology, Tissue culture and Mushroom cultivation. Second edition.

Chabot, R., Antoun, H. and Cescas, M.P. (2010): Stimulation of growth of maize and lettuce by inorganic phosphorus solubilizing microorganisms. Canadian J. Microbiol., 39: 941–47.

Chuang, C.Y., Kuo, C. and Chao and Chao, W. (2006): Solubilisation of inorganic phosphates and plant growth promotion by *Aspergillus niger*. Biol. Fert. Soils, DOI: 10.1007/00374-006-0140-3

Das, K., Katiyar, K. and Goel, R. (2003): P solubilisation potential of plant growth promoting Pseudomonas mutants at low temperature. Microbiol. Res, 158: 359-362.

Edi–Premono, Moawad, M.A. and Vleck, P.L.G. (1996): Effect of phosphate solubilizing *Pseudmonas putida* on the growth of maize and its survival in the rhizosphere. Indonasian J. Crop Sci., 11: 13–23.

Ehrlich, H.L. (1990): Mikrobiologische und biochemische Verfahrenstechnik. In: Einsele A, Finn RK, Samhaber W, editors. Geomicrobiology, 2nd ed. Weinheim: VCH Verlagsgesellschaft.

Illmer, P. and Schinner, F. (2011): Solubilization of inorganic phosphate by microorganisms isolated from forest soils. Soil Biol. Biochem., 24: 389–95

Jackson, M.L. (1987): Soil Chemical Analysis. Prentice Hall Inc. Englo woodcliff, New Jersey, USA.

Kucey, R.M.N. (1987): Increased phosphorus uptake by wheat and field beans inoculated with phosphorus solubilizing *Penicillium bilaji* strain and with Vesicular Arbuscular Mycorohizal fungi. Appl. Envir. Microbiol., 53:2699–2703.

Kucey, R.M.N., Janzen, H.H. and Leggett, M.E. (1989): Microbially mediated increases in plant available phosphorus. Adv. Agron., 42:199-225.

Kumar, V. and Narula, N. (2011): Solubilization of inorganic phosphates and growth emergence of wheat as affected by *Azotobacter chroococcum* mutants. Biol. Fert. Soils, 28:301–5.

Louw, H.A. and Webly, D.M. (1959): A study of soil bacteria dissolving certain mineral phosphate fertilisers and related compounds. J. Appl. Bacteriol., 22:227-233.

Nahas, E. (1996): Factors determining rock phosphate solubilization by microorganisms isolated from soil. World J. Microbiol. Biotech., 12: 567–72.

Pikovskaya, R.I. (1948): Mobilization of phosphorus in soil in connection with the vital activity of some microbial species. Mikrobiologiya, 17:362–70.

Reyes, I., Bernier, L., Simard, R.R. and Antoun, H. (2002): Effect of nitrogen source on the solubilization of different inorganic phosphates by an isolate of *Penicillium rugulosum* and two UV induced mutants. FEMS Microbiol Ecol, 28: 281–291.

Rodriguez, H and Reynaldo, F. (1999): Phosphate solubilising bacteria and their role in plant growth promotion. Biotechnol. Adv. 17: 319-339.

Whitelaw, M.A. (2000): Growth promotion of plants inoculated with phosphate-solubilizing fungi. Adv. Agron., 69:99-151.

Chapter 17

Seasonal Variation of Soil Respiration in the Control and Dumping Sites of NACs and Municipality of Balasore District, Odisha, India

S.C. Pradhan[1] and R.C. Sahoo[2]

[1]Head, P.G. Department of Environmental Science,
F.M. University Balasore, Odisha, India
[2]Lecturer in Botany, Dr. J.N. College, Rasalpur,
Balasore, Odisha, India.

ABSTRACT

Waste materials are generated by anthropogenic activities which ultimately find their way to municipal dumping yards causing environmental pollution, more effectively soil pollution. Soil respiration is taken as an index of soil metabolism which signifies the status of soil, whether polluted or non polluted. In this work, soil respiration of municipal solid waste dumping sites of Balasore district was studied. Four dumping sites like Balasore sather, and all three NACs such as Nilgiri, Soro and Jaleswar were taken along with their control counterparts for the study. The investigation was done for three seasons *i.e.* rainy, winter and summer from July 2007 to March 2008. There was high soil respiration in the rainy season in all the sites. The rate of soil respiration gradually decreased to winter and summer. There was no significant difference between the sites. High soil respiration in rainy season was due to high moisture content of soil. There was reduced soil respiration in all the dumping sites with respect to their control counterparts, due to the presence of pollutants in the municipal solid wastes.

Keywords: Solid waste, MSW dumping, Soil respiration, Carbon dioxide evolution.

Introduction

The soil is a natural medium for plant growth which is developed by natural forces acting on natural materials. It is the natural habitat for micro-organisms, insects, earthworms and other flora and fauna. Soil metabolic activity is related to soil processes and rate of decomposition in the soil which influences soil microbial communities (Davidson and Janssen, 2006). Soil respiration is an important aspect of soil quality and an indicator of soil fertility (Staben *et al.*, 1997). As early as 1931, Smith and Humfeld noted that during decomposition of green manures, the numbers of bacterial population followed the carbon dioxide (CO_2) evolution. Gainey, 1919 reported that roughly carbon dioxide evolution from soil has been used as an indicator of the relative fertility of various soils. Soil respiration has been widely used for many years to quantify the impact of various treatment and management inputs on soil microbial activity (Haney *et al.*, 2008).

Soil Respiration is one of the largest flux pathways of carbon in terrestrial ecosystems and the source of much of the atmospheric loading of carbon dioxide (CO_2). Total soil Respiration (TSR) represents the sum of autotrophic respiration processes. A variety of interacting factors can contribute to changes in total soil respiration. (Melany *et al.*, 2004). Soil respiration consists of autotrophic root respiration and heterotrophic respiration which is associated with decomposition of litter, root and soil organic matter (SOM) (Bernhardt *et al.*, 2006 a, b). It is one of the largest fluxes in the global carbon cycle (Raich and Schlesinger, 1992).

Global modeling studies have demonstrated that even a small change in soil CO_2 emission due to global change has the potential to impact atmosphere CO_2 accumulation and global carbon budget (Cramer *et al.*, 2001). Soil processes such as decomposition are strongly influenced by soil microbial communities (Monson *et al.*, 2006). These microbial communities have been found to be quite heterogeneous establishing both temporal (Lipson and Schmidt, 2004) and Spatial variation (Fierer *et al.*, 2003) in species assemblage.

Materials and Methods

Balasore is a coastal district in the state of Odisha towards its northern side, facing the Bay of Bengal. It has one municipality, the Balasore Sather and three notified area councils (NACs) like, Nilgiri, Soro and Jaleswar. These towns have their own municipal solid waste dumping yards. Municipal solid wastes (MSW) constitute varieties of materials starting from kitchen wastes up to market wastes. Soil respiration in form of carbon dioxide evolution was measured form all the MSW dumping sites and control sites taking four replicates from each site in three seasons for one year. Control sites were taken one kilometer away from the dumping sites of each town. Soil respiration was measured by alkali absorption method of Witkamp (1966); Mishra and Dash (1982). It was expressed in terms of mg of CO_2 evolved/m²/hr.

Results and Discussions

Table 17.1 shows the soil respiration in the control and dumping sites of Balasore municipality, and other three NACs like Nilgiri, Soro and Jaleswar. Higher soil respiration was observed in the rainy season than other seasons in all the sites both in control and dumping. Jina *et al.* (2008) reported that the rate of soil respiration is highest in non-degraded sites which are 68.9 to 363.6 mg of $CO_2/m^2/hr$. soil respiration was positively correlated with soil temperature and soil moisture. They have found highest soil respiration in rainy season and lowest during winter season. In this work there was lower soil respiration in summer season which might be due to less moisture content of the soil. Wolna – Maruwka *et al.* (2007) reported that soil respiration was more in sewage sludge fertilized soil, than the control soil. Similar results were reported by Dobosz *et al.,* 2002. Deng *et al.* (2010) reported that soil respiration exhibits a clear pattern with maximum respiration rates during summer when soil temperature and moisture were high. In the present investigation, the soil respiration gradually decreased to winter and summer months. Of course there was no significant variation of soil respiration between the sites. The higher rate of soil respiration in rainy season might be due to high moisture content of the soil, which might have favoured the soil organisms for higher degradation of organic matter in the soil. Towards the winter and summer season, there was further reduction in the moisture content. This might be the cause of low rate of soil respiration these seasons.

Table 17.1: Soil Respiration of Municipalities and NACs of Balasore district of Odisha (mg of CO_2 evolved/m^2/hr).

Sites		Rainy Season July–Sept, 2007	Winter Season Oct–Dec, 2007	Summer Season Jan–Mar, 2008
Balasore Municipality	Control	0.872	0.763	0.738
	Dumping	0.861	0.709	0.708
Nilgiri NAC	Control	0.829	0.772	0.716
	Dumping	0.806	0.708	0.653
Soro NAC	Control	0.813	0.746	0.705
	Dumping	0.794	0.682	0.648
Jaleswar NAC	Control	0.851	0.765	0.712
	Dumping	0.822	0.724	0.686

Again the soil respiration was marked to be lowest in all dumping sites under investigation in comparison to control sites. This might be due to less vegetation and presence of toxic chemicals that affects the metabolism microbes present in the MSW dumping yard.

References

Bern Hardt, E.S., Barber, J.J., Pippen, J.S., Taneva, L., Andrews, J.A. and Chelesinger, W.H. (2006): Long term effects of Free air CO_2 Enrichment (FACE) on soil respiration, Biogeochemistry, 77: 91 – 116.

Bernhandt, E.S., Barber, J.J., Pipeen, J.S., Taneva, L., Andrews, J.A. and Schlcsinger, W.H. (2006): Long term Effects of Free Carbondioxide Enrichment (FACE) on Soil Respiration. Biogeo Chemistry, 77: 99 – 116.

Carner, W., Bondeau, A., Ian wood ward, F., Prentice, C., Betts, A., Brovkin, V., Cox, P.M., Fisher, V., Foley, J.A., Friend, A.D., Kucharik, C., Lomas, M.R., Raman Kutty, N., Sitch, S. and Smith, B. (2001): Global response of terrestrial ecosystem and function to CO_2 and climate change results from six dynamic global vegetation models, Glob. Change, 7(4): 357 – 373.

Davidson, E.A. and Janssen, I.A. (2006): Temperature sensitivity of soil carbon decomposition and feed back to climate change. Nature 440: 165 – 173, dou: 10.1038/Nature 04514.

Deng, Q., Zhou, G., Lice, J., Lice, S., Duan, H. and Zhang, D. (2010): Responses of Soil Respiration to Evaluate Carbon Dioxide and Nitrogen addition in Young Sub-tropical Forest Ecosystem in China. Biogeo Sciences, 7: 315 – 328.

Dobsz, K., Peterson, S.O., Kure, L.K. and Ambus, P. (2002): Evaluating effects of Sewage Sludge and Household Compost in Soil Physical, Chemical and Microbiological Properties. APPL. Soil. Ecol. 19: 237 – 2002.

Fierer, N., Schimel, J.P. and Holden, P.A. (2003): Variation in Microbial community composition through two soil depth profiles, soil, Biol, Biochem, 35: 167 – 176.

Gainey, P.L. (1919): Parallel formation of carbon dioxide, ammonia, and nitrate in soil, Soil Sci. 7: 293 – 311.

Haney, R.L., Brinton, W.H. and Evans, E. (2008): Estimating soil carbon, nitrogen and phosphorus mineralization from Short – Term carbon dioxide respiration. United

Jina, B.S., Bohra, C.P.S., Rawat, Y.S. and Bhatt, M.D. (2008): Seasonal Changes in Soil Respiration of Degraded and none-degraded sites in Oak and Pine Forest of Central Himalaya. Scientific World, 6(6): July 2008.

Lipson, D.A. and Schmidt, S.K. (2004): Seasonal Changes in an alpine soil bacterial community. Applied and environmental Microbiology. 70(5): 2862 – 2879.

Melany, C., F., Timothy, J., Fahey, Peter, M. Groffman and Patrick, J. Bohlen, (2004): Earthworm invasion, Fine root distributors and Soil respiration on North Temperate forests. Ecosystems (2004). 7 : 55 – 62 DO 1 : 10. 1007/s 10021 – 003 – 0130.

Mishra, P.C. and Dash, M. C. (1982): Diurnal variation in soil respiration in tropical pasture and forest site at Sambalpur, India, Comp. Physiol. Ecol. 7(2): 137-140.

Monson, P.K., Lipson, D.I., Burns, S.P., Turnipseed, A. A., Delany, A.C., William, M.W. and Schmidt, S.K. (2006): Winter forest soil respiration controlled by climate microbial community composition. Nature 439: 711 – 714 DOI 10.13068/Nature 04555.

Raich, J.W. and Schelesinger, W.H. (1992): Global carbon dioxide flux in soil respiration and its relationship to vegetation and climate, Tellus, 44B, 81 – 99.

States Department of Agriculture, Agriculture Research Service, 808 E. Blackland Rd, Temple, TX, 75602 USA.

Smith, N.R. and Humfeld, H. (1931): The decomposition of green manures grown on a soil and turned under compared to the decomposition of green manures added to a fallow soil J. Agr, Res, 43:715 – 731.

Staben, M.L., Bezdicek, D.F., Smith, J.L. and Fauci, M.F. (1997): Assessment of soil quality in conversion reserve program and wheat – fallow soils, soil Sci. SOC; Am, J. 61: 124 – 130.

Witcamp, M. (1966): Rate of carbon dioxide evolution from forest floor. Ecology 47: 494- 496.

Wolna Maruwka, A., Sawicka, A. and Kayza, D. (2007): Size of Selected Group of Microorganism and Soil Respiration Activity fertilized by Municipal Sewage Sludge, Polish. J. of Environ. Stud, Vol. 16, No. 1 129 – 138.

Chapter 18

Atrazine Toxicity Alters the Activities of Antioxidative Enzymes and Increases Lipid Peroxidation Level in Germinating Mung Bean Seeds

Surjendu Kumar Dey and Swarna Prasanti Pradhan

Department of Environmental Science,
Fakir Mohan University, Vyasa Vihar,
Balasore – 756 020, Odisha, India

ABSTRACT

Mung bean (*Vigna radiata* (L.) Wilczek) seeds were exposed to atrazine, a selective systemic herbicide, for 48 h and different parameters like germination percentage, soluble protein content, activities of antioxidative enzymes like superoxide dismutase (SOD), catalase (CAT) and peroxidase (POX) and lipid peroxidation level of embryonic tissues of germinating seeds were determined. Germination percentage was not found to affect by atrazine but soluble protein content of the embryonic tissues decreased with increase in the atrazine in the medium. SOD and CAT activities decreased whereas POX activity increased. Because of the alterations in the activities of these antioxidative enzymes imposition of oxidative stress in the tissues was presumed which was, in fact, substantiated by detection of higher levels of lipid peroxidation in the tissues. Further, it was also found that addition of antioxidants like ascorbate and benzoate, along with atrazine, neutralized the atrazine induced toxicity in terms of decreasing the lipid peroxidation levels in the tissues. The findings of the study indicated that atrazine imposes oxidative stress in the germinating seeds and antioxidant supplementation can reduce the atrazine toxicity.

Keywords: *Herbicide, Catalase, Superoxide dismutase, Peroxidase, Lipid peroxidation, Antioxidants.*

Introduction

Weed management and control in the agricultural field is always a challenging task not only for the farmers but also for the agronomists. Among the different approaches adopted for weed control, application of herbicides in the field has been found to give promising results. Herbicides are chemicals or any other biological derivatives that inhibit or interrupt the normal growth and development of weeds. They are used widely in agriculture, industry and urban areas for weed management. Herbicides kill plants in different ways. To be effective, an herbicide must meet several requirements. It must come in contact with the target weed, be absorbed by the weed, move to the site of action in the weed and accumulate in sufficient concentrations at the site of action to kill or suppress the target plant. The herbicide mode of action involves absorption into plant, translocation or movement in the plant, metabolism of herbicide and expression of physiological responses. Since herbicide application is cost effective and is done with minimum of labour, this has become a very popular approach for weed control and management throughout the world.

Atrazine is a selective systemic herbicide which was introduced in the market in 1958 by J.R. Geigy, which is now a part of Novartis (Hicks, 1998). It is widely used for the pre and post-emergence control of annual and broad leaved weeds and perennial grasses (Tomlin, 2000). Atrazine inhibits photosynthesis and interferes with other enzymatic processes. It is mainly absorbed through the plant roots, but can also enter through the foliage and accumulates in the apical meristems and leaves. This is mainly used in the production of maize, sugarcane, sorghum, pineapples, grasslands, conifers and also for industrial weed control (Tomlin, 2000) with its biggest market in maize production (Hicks, 1998).

The chemical properties of atrzine make it susceptible to leaching and runoff, especially during heavy rains. Atrazine has large potential to leach or to move in surface solution and a medium potential to adsorb to sediment particles. The half-life of atrazine in loamy soils ranges from 60 to 150 days. However, when the conditions in the soil are changed from aerobic to anaerobic, the rate of degradation decreases considerably and once it enters the water column, the degradation rate becomes very slow (Goolsby *et al.*, 1993). Thus, atrazine and its degradates can persist in some soils and in reservoirs. It is a very pervasive environmental contaminant and due to its persistence in soil and water, it poses a great threat for the other non-target plants and animals.

Because of continuous application of atrazine in the field, this chemical accumulates in the soil. Since it has many phytotoxic effects, it is expected to have toxic effects in the germinating seeds of non-target crop plants. Therefore, in this study attempts were made to assess the toxic effects of atrazine in the germinating mungbean seeds. Besides different parameters, attempts were made to assess the effects of atrazine on the antioxidative efficiency of germinating mungbean seeds and at the same time some approaches were adopted to neutralize the toxicity in laboratory condition.

Materials and Methods

Viable mung bean (*Vigna radiata* (L.) Wilczek) seeds were collected from local market and were selected for uniformity of size. The seeds were then surface sterilized with freshly prepared and filtered 3 per cent solution of commercial bleaching powder (calcium oxychloride, $CaOCl_2$) for 30 min followed by thorough washings with distilled water for several times for another 30 min. Then the seeds were spread in Petri dishes over filter paper, moistened with 5.0 ml solution of different concentrations (250, 500, 750, 1000, 1500 ppm) of commercially available atrazine ('Atrataf' with atrazine 50 per cent, Rallisa: A Tata Enterprise). Distilled water was taken in another Petri dish as control. In each Petri dish 25 nos. of surface sterilized seeds were spread in order to maintain uniformity in stress imposition. The seeds were allowed to germinate in dark at $30 \pm 2°C$ and after 48 h of germination, number of seeds germinated in each Petri dish were counted to determine the germination percentage. Then the seeds were collected separately for various analytical studies.

Further, to neutralize the atrazine induced toxicity, two antioxidants (*i.e.*, ascorbate and benzoate) were taken separately in two different Petri dishes along with 1000 ppm atrazine and seeds were allowed to germinate for 48 h. From these seeds, embryo portions were excised and lipid peroxidation level was determined following the method mentioned below.

For different analytical studies, the embryo portion of the germinated seed was excised carefully (taking the radicle and plumule portions, discarding the cotyledons). For extraction of soluble protein, first a cell-free preparation was made. For this, the tissues were homogenized in a mortar and pestle under ice-cold condition with sodium phosphate buffer of 50 mM, pH 7.5. The homogenates were centrifuged separately at $17,000 \times g$ for 10 min at $-4°C$. From the resultant supernatant, 1.0 ml from each sample was taken separately and to it equal volume of 20 per cent (w/v) trichloroacetic acid (TCA) was added. The samples were kept in refrigerator for overnight and then washed with 10 per cent (w/v) TCA solution. The pellets were then subsequently washed with absolute alcohol, alcohol and chloroform (in a proportion of 3:1), alcohol and ether (in a proportion of 3:1) and finally with ether. After washing, the pellets were air dried and resuspended with 0.3 N NaOH solution for 16 h at $37°C$. After that, the samples were centrifuged and supernatants were collected for soluble protein estimation after suitable dilutions. Soluble protein content was estimated following the Folin-Ciocalteu reagent method, as described by Lowry *et al.* (1951).

For enzyme extraction, cell-free extract was prepared following the procedure as mentioned above. The buffers used were sodium phosphate buffer, 50 mM and pH 7.4 for superoxide dismutase (SOD) and 50 mM, pH 7.5 for catalase (CAT) and peroxidase (POX). Superoxide dismutase (EC 1.15.1.1) activity was determined by following the method of Das *et al.* (2000) where the inhibition of superoxide driven nitrite formation from hydroxylamine hydrochloride by SOD was measured spectrophotometrically under laboratory conditions. For determination of SOD activity the formula used was Vo/V-1, where Vo is the absorbance at 543 nm of the control (without enzyme) and V is the absorbance of sample (with enzyme) at the same wavelength. Catalase (EC 1.11.1.6) activity was assayed following the method of Aebi (1983) in which the decreasing rate of H_2O_2 concentration was measured spectrophotometrically at 240 nm and CAT activity was calculated using the extinction coefficient of 40.0 mM^{-1} cm^{-1} for H_2O_2 at 240 nm. For peroxidase (EC 1.11.1.7) assay, H_2O_2 and guaiacol were taken as substrate and reduced co-substrate respectively in this study. The rate of increase in colour intensity in the assay mixture due to formation of tetraguaiacol formation was recorded at 470 nm, as described by Kar and Feierabend (1984). Peroxidase activity was calculated using the extinction coefficient of 26.6 mM^{-1} cm^{-1} due to tetraguaiacol formation.

For determination of lipid peroxidation level in the tissues malondialdehyde (MDA), a decomposition product of peroxidized polyunsaturated fatty acid components of membrane lipid, was estimated taking thiobarbituric-acid-reactive material. For this the tissues were homogenized with 5 per cent (w/v) trichloroacetic acid and the extract was directly used for MDA estimation following the method of Heath and Packer (1968). In this method, the unspecific turbidity in the reaction mixtures was corrected by subtracting the absorbance at 600 nm, and for absorbance at 532 nm originating from extract after incubation without thiobarbituric acid.

All the experiments were performed at least for three times, with three replicates in each time and the mean values with standard deviations are presented. The SOD activity is expressed in unit (U), which is defined as the amount that inhibits the superoxide driven nitrobluetetrazolium reduction by 50 per cent under the assay conditions. Activities of CAT and POX are expressed in katals (kat), *i.e.*, moles of substrate used up or product formed per second due to enzyme activities under assay conditions.

Results and Discussions

Germination is an important phase of plant growth and development during which the dormancy of the seeds breaks upon absorption of moisture and the radicle and plumule emerge out by rupturing the seed coats. During this phase, the exposure of tender embryonic tissues to different toxicants including herbicide results in different anomalies in the growing seedlings. When atrazine is used in the field against weeds it is obvious that the non-target plants including the crop plants are also exposed to it simultaneously. In this study, it has been found that the germination percentage of mung bean seeds was not affected by atrazine (Table 18.1). But soluble protein content of the embryonic tissues decreased with increase in atrazine concentrations (Table 18.1). The decrease in soluble protein content indicates the probability of decline in *de novo* enzyme contents of the tissues as a result the physiological anomalies in the tissues are also expected.

Table 18.1: Changes in the Germination Percentage of Mung Bean Seeds and Soluble Protein Content of Embryonic Tissues of Mung Bean Seeds Exposed to Atrazine for 48 h.

Atrazine (ppm)	Per cent of Seed Germination	Soluble Protein (mg/g FW)
0	100	6.5 ± 0.51
250	90 ± 3	5.9 ± 0.42
500	87 ± 3	4.45 ± 0.39
750	89 ± 2	4.1 ± 0.32
1000	91 ± 4	3.6 ± 0.21
1500	90 ± 5	3.3 ± 0.23

Oxidative stress in aerobic organisms is a regulated situation in which the reactive oxygen species are generated in large amount beyond the capacity of the cell's endogenous antioxidative defense system to scavenge them off (Halliwell and Gutteridge, 2007). The reactive oxygen species include superoxide radical ($O_2^{\cdot-}$), hydrogen peroxide (H_2O_2), hydroxyl radical ($\cdot OH$) and singlet oxygen ($^1O_2$). These species are continuously generated in aerobic cells as a consequence of normal oxygen metabolism. But their toxicity is not realized under normal situation due their efficient scavenging by antioxidative defense system which includes antioxidative enzymes like superoxide dismutase (SOD), catalase (CAT), peroxidase (POX), ascorbate peroxidase, glutathione reductase etc. and other different low molecular antioxidants like ascorbate, reduced glutathione, α-tocopherol, carotenoids, flavonoids etc. Under different developmental stages as well as under biotic and abiotic stress situations, the antioxidative defense system fails to protect against the deleterious effects of reactive oxygen species as a result of which the organism gets exposed to oxidative stress. In this study it was found that with increase in atrazine concentration in the medium there was decrease in SOD as well as CAT activities in the embryonic tissues of germinating seeds (Table 18.2). Due to decrease in the activities of these two important antioxidative enzymes, there was poor protection against $O_2^{\cdot-}$ and H_2O_2 in the cell. Even though POX activity was increased with increase in the atrazine concentration (Table 18.2), this can not be attributed to efficient H_2O_2 scavenging in the cell. This is because the elevated POX activity might be due to the release of cell wall bound peroxidases in the cell caused as a result of atrazine stress and such increase in POX activity has also been reported under various stress situations (Mittal and Dubey, 1991; Shah *et al.*, 2001; Verma and Dubey, 2003; Dey *et al.*, 2007). Thus, the alterations in the activities of SOD, CAT and POX favoured the accumulation of $O_2^{\cdot-}$ and H_2O_2 which increased the chances of imposition of oxidative stress in the tissues.

The reactive oxygen species like $O_2^{\cdot-}$ and H_2O_2 are known to react in presence of transition metal ions forming hydroxyl radical via Haber-Weiss reactions (Halliwell and Gutteridge, 2007). Among the different reactive oxygen species, hydroxyl radicals are most potentially toxic and polyunsaturated fatty acid components of membrane lipid are highly susceptible to their attack. As a result, lipid peroxidation takes place which leads to loss of membrane integrity, cellular architecture and ultimately cell

Table 18.2: Changes in the Activities of Superoxide Dismutase (SOD), Catalase (CAT) and Peroxidase (POX) in Embryonic Tissues of Germinating Mung Bean Seeds Exposed to Atrazine for 48 h.

Atrazine (ppm)	SOD Activity (U/g FW)	CAT Activity (nkat/g FW)	POX Activity (nkat/g FW)
0	1420 ± 50	8.25 ±0.61	41.0 ± 3.2
250	1002 ± 51	4.75 ± 0.21	50.5 ± 3.8
500	826 ± 24	4.25 ± 0.19	134.6 ± 11.1
750	724 ± 31	3.5 ± 0.18	500.0 ± 24.0
1000	516 ± 22	2.9 ± 0.2	673.0 ± 42.0
1500	479 ± 25	2.8 ± 0.21	704.5 ± 48.0

death. Increase in lipid peroxidation rate is therefore, considered as an indicator of free radical mediated imposition of oxidative stress in aerobic organisms (Kappus, 1985). In this study, the lipid peroxidation level in the embryonic tissue was determined by estimating the malondialdehyde (MDA) content, which is the decomposition product of the peroxidized membrane lipids, taking thiobarbituric acid as the reactive substance. The results presented in Figure 18.1 indicated that with increase in atrazine concentration in the medium there was increase in the level of lipid peroxidation in the embryonic tissues. This was a clear indication that even though there was no

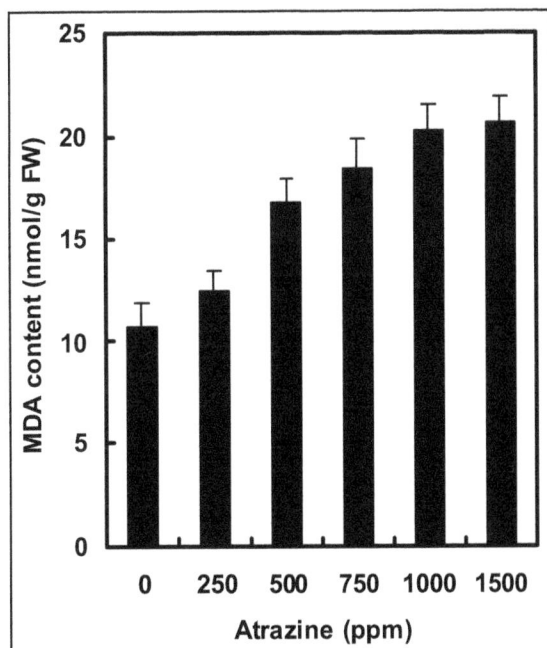

Figure 18.1: Changes in the Lipid Peroxidation Level in Embryonic Tissues of Germinating Mung Bean Seeds Exposed to Atrazine for 48 h.

inhibition in germination percentage of the seeds by atrazine, toxicity was imposed in the tissues in the form of imposition of oxidative stress. Since oxidative stress is known to exert different toxicities mainly in the forms of loss of membrane integrity, damage in DNA structure, inhibition in the activities of many enzymes involved in different metabolic pathways, damage in the natural integrity of other biological macromolecules; all these physiological anomalies are also expected to occur in the mung bean embryonic tissues because of atrazine toxicity. This was mainly because of the alterations in the activities of key antioxidative enzymes like SOD, CAT and POX in the tissues due to atrazine toxicity.

In a system if toxicity is because of imposition of oxidative stress, then addition of antioxidant in the medium is expected to alleviate the toxicity. Ascorbate and benzoate are two important antioxidants. Ascorbic acid is an important low molecular weight antioxidant of the cell which is known to scavenge the reactive oxygen species like superoxide radical, hydroxyl radical and singlet oxygen (Bodannes and Chan, 1979) where as benzoate scavenges hydroxyl radicals (Neta and Dorfman, 1968). In this study along with 1000 ppm atrazine, 100 ppm ascorbate and 10 ppm benzoate were added separately and lipid peroxidation level in the embryonic tissues of germinating mung bean seeds was assessed in order to determine the protective effect of antioxidants against the atrazine induced toxicity. The results presented in Table 18.3 shows that ascorbate decreased the atrazine induced lipid peroxidation by 12.3 per cent whereas benzoate decreased the lipid peroxidation by 39 per cent. From this it could be presumed that atrazine exerts its toxicity by imposing oxidative stress in the tissues and antioxidant supplementation may reduce the toxic effects.

Table 18.3: Effect of Antioxidants (Ascorbate and Benzoate) in Neutralizing Atrazine Induced Lipid Peroxidation in the Embryonic Tissues of Germinating Mung Bean Seeds Exposed for 48 h.

Sample	MDA (nmol/g FW)
Control	10.8 ± 1.1
Atrazine (1000 ppm)	20.3 ± 1.2
Atrazine (1000 ppm) + Ascorbate (100 mM)	17.8 ± 1.2
Atrazine (1000 ppm) + Benzoate (10 mM)	12.4 ± 1.1

Conclusion

Even though the applications of different pesticides in the crop fields are always targeted towards specific pests, the non-target organisms like the beneficial flora and fauna of the field including the crop plant itself are always exposed to these chemicals. Exposure to these chemicals is expected to cause physiological as well as morphological anomalies in plants. In this study it was found that even though there was no inhibition in the germination percentage in mung bean seeds exposed to atrazine, decrease in soluble protein content, decrease in the activities of antioxidative enzymes like SOD and CAT and increase in the activity of POX were found in the embryonic tissues of germinating seeds. Because of the alterations in the activities of antioxidative enzymes like SOD, CAT and POX imposition of oxidative stress in the tissues was presumed which was substantiated by the detection of increased lipid peroxidation in the tissues. However, further studies on the determination of low molecular weight antioxidant contents in the tissues and assaying the activities of enzymes involved in the ascorbate-glutathione path way are highly essential for such proposition. Further the toxicity was also neutralized by the addition of antioxidants like ascorbate and benzoate in the medium. The findings of this study indicate that atrazine imposes oxidative stress in the embryonic tissues of the germinating seeds and application of antioxidants may reduce the atrazine induced toxicity.

Acknowledgements

The authors are thankful to the Head, Department of Environmental Science, Fakir Mohan University, Vyasa Vihar, Balasore-756020 for providing necessary facilities to carry out this work.

References

Aebi, H.E. (1983): Catalase. In: Bergmeyer, H.U. (ed): Methods of Enzymatic Analyses, Verlag Chemie, Weinheim. 3: pp. 273-285.

Bodannes, R.S. and Chan, P.C. (1979): Ascorbic acid as a scavenger of singlet oxygen. FEBS Letter 105: 195-196.

Das, K., Samanta, L. and Chainy, G.B.N. (2000): A modified spectrophotometric assay of superoxide dismutase using nitrite formation by superoxide radicals. Ind. J. Biochem. Biophys. 37: 201-204.

Dey, S.K., Dey, J., Patra, S. and Pothal, D. (2007): Changes in the antioxidative enzyme activities and lipid peroxidation in wheat seedlings exposed to cadmium and lead stress. Braz. J. Plant Physiol. 19: 53-60.

Goolsby, D.A., Battaglin, W.A., Fallon, J.D., Aga, D.S., Kolpin, D.W. and Thurman, E.M. (1993): Persistence of herbicides in selected reservoirs in the Midwestern United States: some preliminary results. U.S. Geological Survey, Open-File Report 93-418, Denver, CO.

Halliwell, B. and Gutteridge, J.M.C. (2007): Free Radicals in Biology and Medicine. 4th ed. Oxford University Press, New York.

Heath, R.L. and Packer, L. (1968): Photooxidation in isolated chloroplasts. I. Kinetics and stoichiometry of fatty acid peroxidation. Arch. Biochem. Biophys. 125: 189-198.

Hicks, B. (1998): Generic Pesticides-The Products and Markets, Agro Reports, PJB Publications.

Kappus, H. (1985): Lipid Peroxidation: Mechanisms, analysis, enzymology and biological relevance. In: Sies, H. (ed), Oxidative Stress, pp. 273-310. Academic Press Inc., London.

Kar, M. and Feierabend, J. (1984): Metabolism of activated oxygen in detached wheat and rye leaves and its relevance to the initiation of senescence. Planta, 160: 385-391.

Lowry, O.H., Rosebrough, N.J., Farr, A.L. and Randell, R.J. (1951): Protein measurement with Folin-phenol reagent. J. Biol. Chem. 193: 265-275.

Mittal, R. and Dubey, R.S. (1991): Behaviour of peroxidases in rice: changes in enzymatic activity and isoforms in relation to salt tolerance. Plant Physiol. Biochem. 29: 31-40.

Neta, P. and Dorfman, L.M. (1968): Pulse radiolysis studies. XIII. Rate constants for the reaction of hydroxyl radicals with aromatic compounds in aqueous solutions. Adv. Chem. Ser. 81: 220-230.

Shah, K., Kumar, R.G., Verma, S. and Dubey, R.S. (2001): Effect of cadmium on lipid peroxidation, superoxide anion generation and activities of antioxidant enzymes in growing rice seedlings. Plant Sci. 161: 1135-1144.

Tomlin, C.D.S. (2000): Pesticides Manual, 12th Edn., British Crop Protection Council.

Verma, S. and Dubey, R.S. (2003): Lead toxicity induces lipid peroxidation and alters the activities of antioxidant enzymes in growing rice plants. Plant Sci. 164: 645-655.

Chapter 19

Assessing the Impact of Open Cast Mines on Surrounding Land-use Pattern: A Case Study from Keonjhar District of Odisha, India

Himansu Sekhar Patra[1] and Kabir Mohan Sethy[2]

[1]Ph.D. Scholar, [2]Associate Professor,
Department of Geography, Utkal University,
Bhubaneswar, Odisha, India

ABSTRACT

The environmental impacts of open cast mining are many and diverse. Mining causes massive damage to landscape and biological communities. Mining operations, which involve minerals extraction from the earth's crust tends to make a notable impact on the environment, landscape and biological communities of the earth. Large scale of ongoing unplanned and unscientific open cast mining in various mineral belt of state of Odisha has emerged as a major cause of concern and threat to the land-use of the region. This has got some serious implications on the vegetation pattern of this area. Periodic monitoring of environmental impacts is essential to characterize land-use and land-cover changes over time. Remote sensing technology can be effectively used as a valuable asset in completing environmental assessments at mining affected areas to generate reliable information on land use changes. The present study was undertaken to analyze the process of human-induced landscape transformation in the iron mined affected areas of Odisha, India by interpreting temporal remote sensing data and using Geographic Information System. Remote sensing and GIS techniques can be effectively used for identification of various sequential changes in land use patterns at remotely access area of mining areas. The results find out significant changes in certain land use category such as settlement, agricultural land, waste land and most importantly the forest area. This study concludes that land use changes is an important issue at open cast mining which needs to be timely addressed. The study also highlights the utility of this method to monitor the impact of large-scale mining and other extensive forms of resource exploitation such as deforestation in developing countries.

Keywords: Remote sensing, Land use, Land cover, Geographic information system.

Introduction

Minerals are critical resources in economic growth of any country. Production and utilization of minerals have often been taken as an index of development (Roy and Mishra, 2007). Minerals and metals have played a crucial role in the development and continuation of human civilization. After agriculture it is the second largest industry at all scales and regions and has played a vital role in the development of civilization from ancient times (Lodha, *et al.*, 2009). However Mining, like any other industrial activity tends to leave a strong negative impact on the environment unless it is meticulously planned and carefully executed. It is a well established fact that mining is an environmentally destructive activity. The state of Odisha is bestowed with abundant mineral resources which are found to be deposited under the large tract of forest mostly located in Keonjhar, Sundergada, Mayurbhanja, Koraput and Kalahandi district (Bhusan and Hazra, 2008). After 2000, the state government tried to turn around its development fortune by using its mineral deposits (World Bank Report, 2007). For this, the state government has leased out a large part of its mineral resources, whose exploration is found to directly affect the forest. Data of Directorate of mines, Govt. of Odisha shows that till date more than 600 mines (both functional and non functional) covering diversified minerals has been leased out in different part of state. The environmental impacts of open cast mining are many and diverse. Mining operations, which involve minerals extraction from the earth's crust tends to, make a notable impact on the environment, landscape and biological communities of the earth (Bell, *et al.*, 2001). Mining operation cause big void on earth thus degrading the land, loss of forest and topsoil (Vaghlolikar *et al.*, 2003).

Location of the mines at inaccessible location is a major challenge to assess the impact. This has lead to surfacing large scale of regulatory irregularities. A detailed understanding of the impact of the existing mining on surrounding land use/land cover (LULC) pattern is essential from regulatory point of view. Remote sensing technology gives a viable means of analyzing the changing conditions at mine sites located at inaccessible place. The present study was undertaken to analyze the process of human-induced landscape transformation at the iron ore mined affected areas of Keonjhar district of Orissa by using remote sensing and geographic information system.

Remote Sensing and its Usefulness in Mining Impact Assessment

Periodic monitoring of land use and land cover in the neighbourhood of any developmental activity is one of the most important components necessary for environmental impact assessment (Down and Stocks, 1977). Remote sensing and GIS techniques are quite useful in identification of the degraded areas cause due to mining activity (Patra and Sethy, 2013). These are important tools for studying the pattern of land scape dynamics. The changes of land cover are invariably associated with mining of natural resources. Remote sensing provides multi-spectral and multi-temporal synoptic coverage for any area of interest (Ranade, 2007). The satellite data provides a permanent and authentic record of the land-use patterns of a particular area at any given time, which can be re-used for verification and re-assessment. This can be very useful in assessment of impact of mines. As most of the mines are located at hostile terrain, remote sensing can be effectively used to collect information about the impact of the mining (Ranade, 2007).

Study Area

The present study area is located between 21°37'09"–21°40'02"N and longitudes 85°29'20"–85°31'30"E, near to Suakati town in Keonjhar district of Odisha (Figure 19.1). It is coming under Banspal block. A major iron ore deposit of state namely Gandhmardan hill is located at the centre of study area, having a reserve of 350 million tonne of iron ore. The existing iron ore mines located at this hill is one of the oldest mines of Odisha. The iron ore mining started in Gandhamardan hill range by Odisha Mining Corporation (OMC), a State Govt, owned agency in 1965 and presently it has two open cast iron mines namely Gandhamardan A and B covering around 2200 hectare. Similarly two Private owned mines are also operating at Putulpani (Talajagar) and at Urumunda village respectively covering a total area of 182.1932

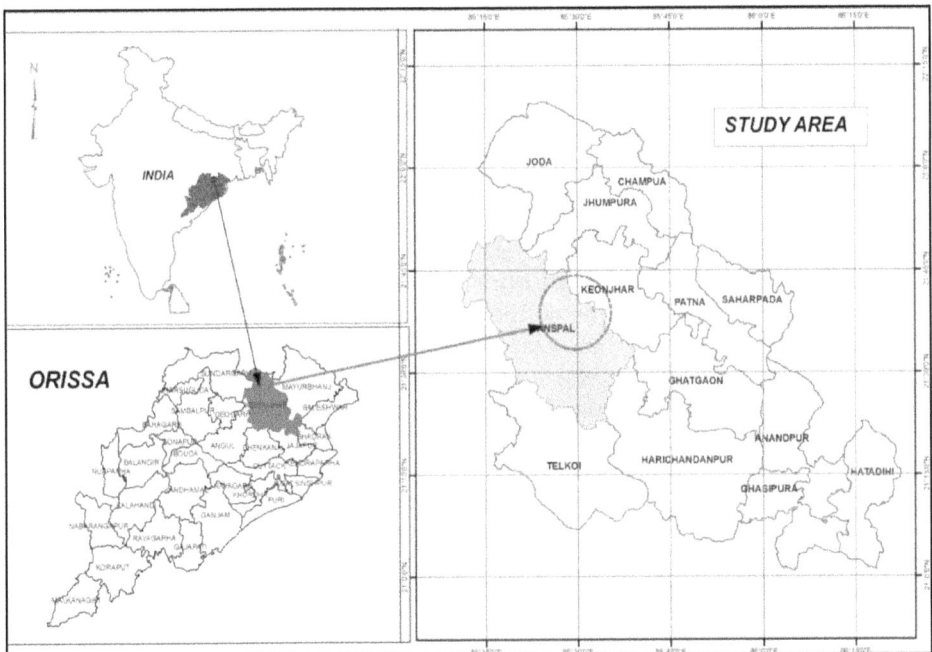

Figure 19.1: The Location of Study Area.

hectares. The core mining area as well as the surrounding area shows a good forest cover. The study area is centered around a ridge like escarpment – Gandhmardan hill which runs in North-South direction. The total top of the ridge has a peak at 1009 mRL. The study area has a dendritic pattern of drainage because of its hilly topography. The area falls under Baitarani river drainage basin.

Methodology Adopted

Survey of India toposheets (scale 1:50000) no. 73 G/6 and 73 G/10 were geo-referenced to super impose on orthorectified satellite image (IRS P-6 LISS III). Mosaicing was performed for the geo-referenced topo-sheet to form a continuous frame. A base map was prepared from the mosaic of survey of India topo-sheet comprising features such as administrative boundary, roads, river, and drainage (Figure 19.2). IRS P-6 LISS III data offers spatial resolution of 23.5m with the swath width of 141 km. The data was collected in two visible bands namely green (0.52-0.59 µ), red (0.62-0.68 µ), infrared (0.77-0.86 µ) along with new feature SWIR band (1.55-1.70 µ) with orbital receptivity period of 24 days. The standard FCC was generated by assigning blue, green and red colours to visible green, visible red and near infrared bands respectively. Image processing and rectification was done in ERDAS IMAGINE 8.4 software and spatial data was created in Arc GIS-9.1 software. Area of interest comprising the study area was then selected and extracted from the satellite image. Suitable image enhancements were then applied on the extracted area of interest. A visual interpretation followed by supervised classification was adopted to classify various land use-land cover features. Mask of mine area within 10 KM radius was superimposed on the final output to generate area statistics for different land use categories. Classification accuracy estimation was done on the supervised classified image for further rectification. Based on this, final estimation and results for a land use-land cover features existing in the study area were derived. In the present study land use status at a gap of 10 years such as 1990, 2000 and 2012 was compared to find out the impact of mining. Final output maps were prepared on 1:50,000 scale (Figures 19.3–19.5),

Results and Discussions

In the present study, LISS-III satellite imagery of 1990, 2000 and 2012 was compared qualitatively and quantitatively. This study has provided information on the land use changes during these years. Different land use categories were identified such as. Forest, Fallow land, waste land, Urban/Reclaimed area, vegetation and Water. The results of land use/land cover assessment were based on visual interpretation from satellite data. The changes occurred in the land use pattern are clearly observed in post-mining period (Figure 19.6). The study finds out that reduction in land use category such as agricultural land, dense forest, and forest plantation and water body has occurred whereas increase in the rest type of land use category has been observed.

Agriculture Land

Analysis of status of agricultural land in various periods (1990, 2000 and 2012) shows a decline in agricultural land in the surrounding 10 km radius of mining area.

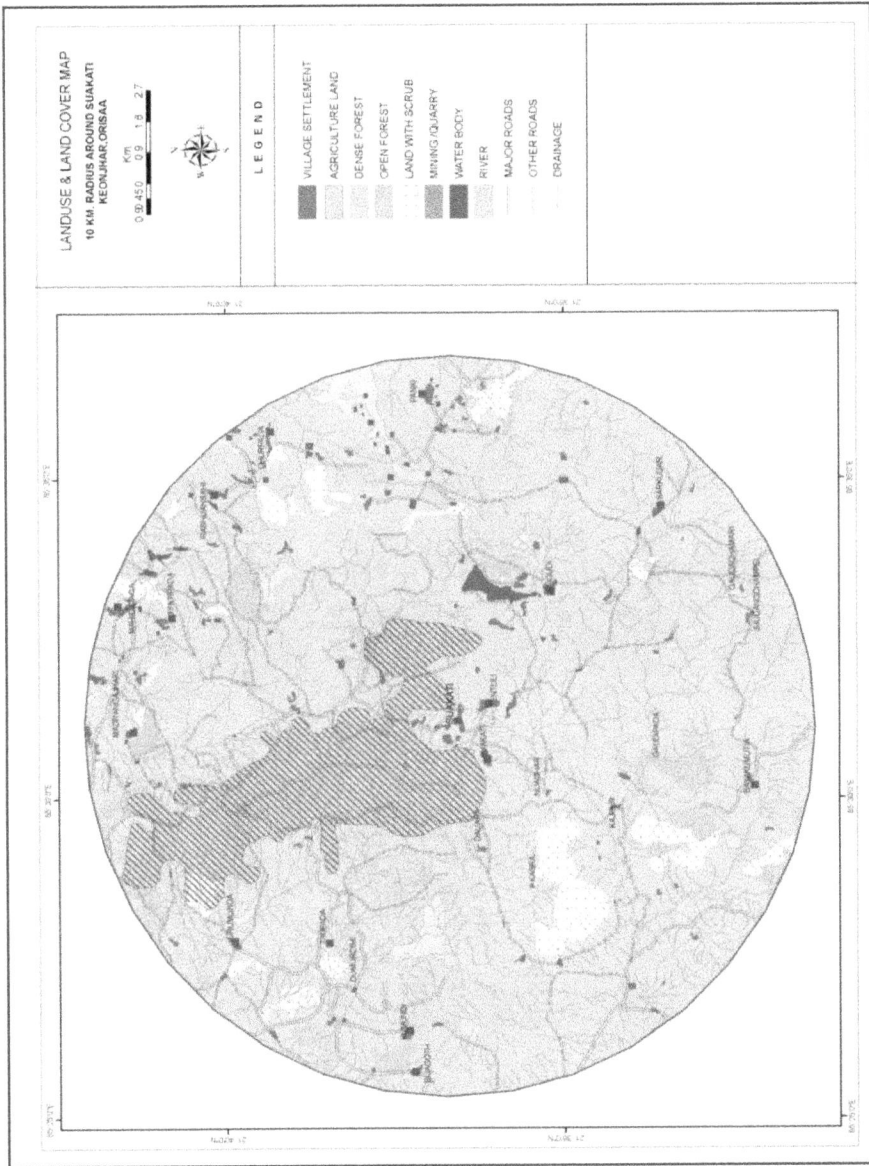

Figure 19.2: The Base Map of the Study Area.

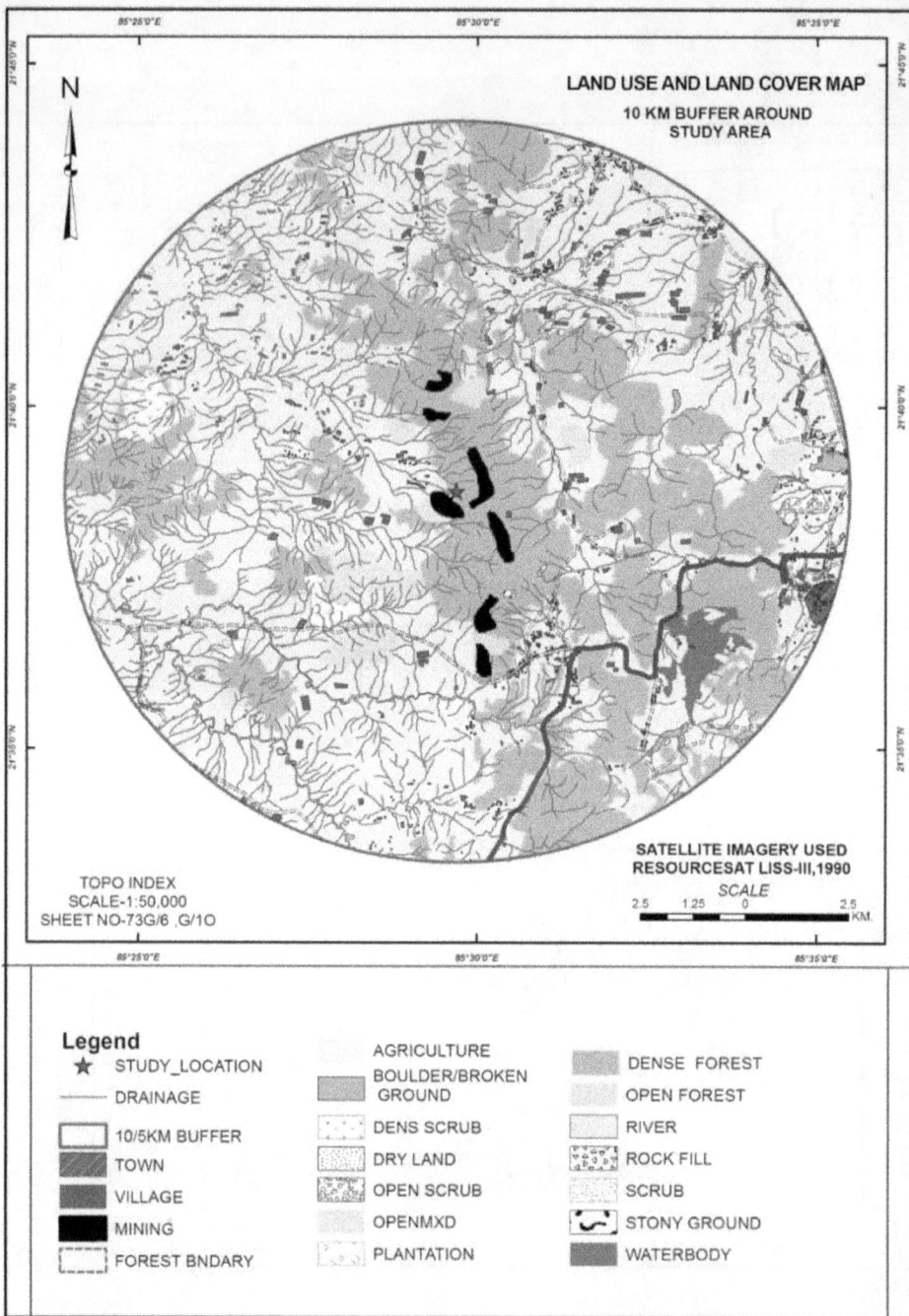

Figure 19.3: The Land use Land Cover Status of Study Area in 1990.

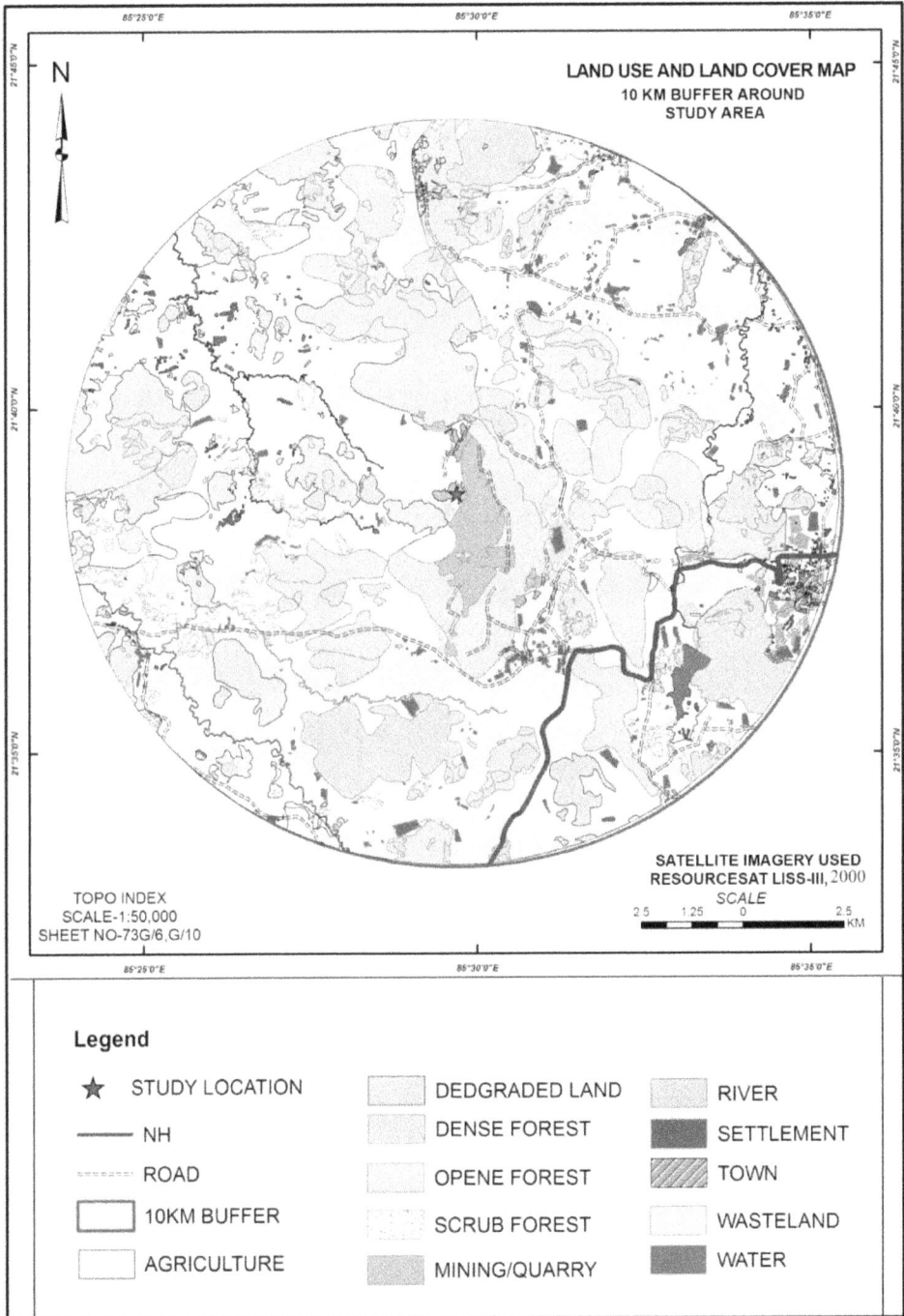

LAND USE AND LAND COVER MAP
10 KM BUFFER AROUND
STUDY AREA

SATELLITE IMAGERY USED
RESOURCESAT LISS-III, 2000

SCALE

TOPO INDEX
SCALE-1:50,000
SHEET NO-73G/6,G/10

Legend

★ STUDY LOCATION
— NH
--- ROAD
□ 10KM BUFFER
□ AGRICULTURE

DEDGRADED LAND
DENSE FOREST
OPENE FOREST
SCRUB FOREST
MINING/QUARRY

RIVER
SETTLEMENT
TOWN
WASTELAND
WATER

Figure 19.4: The Land use Land Cover Status of Study Area in 2000.

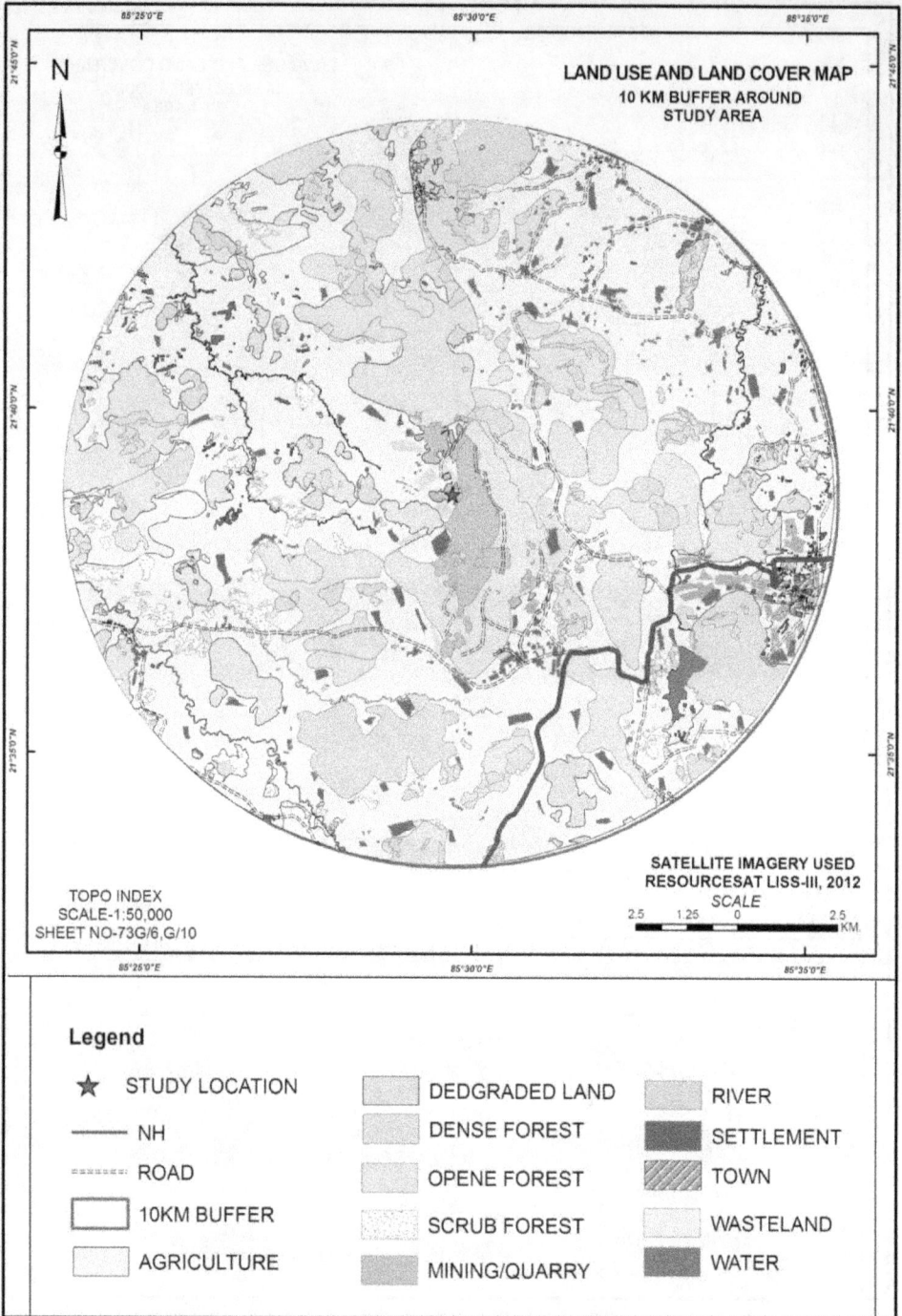

Figure 19.5: The Land use Land Cover Status of Study Area in 2012.

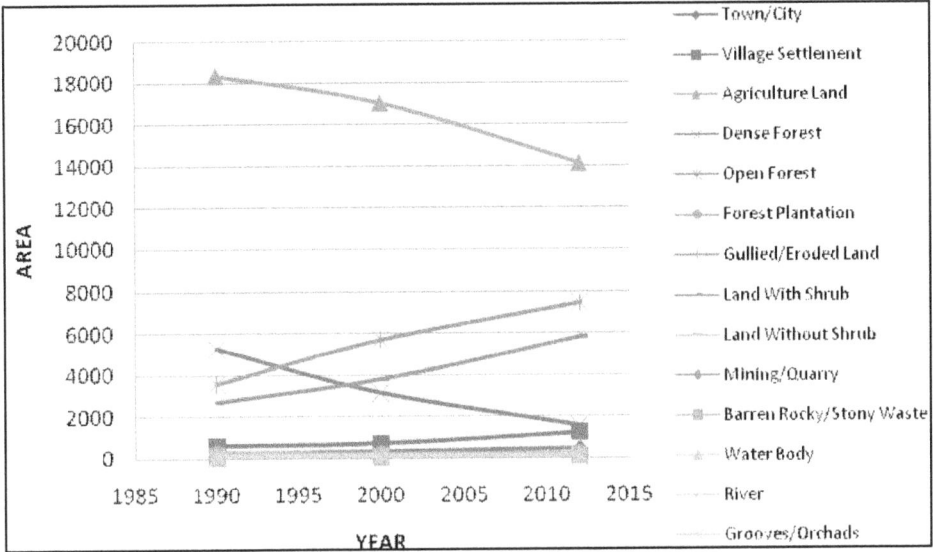

Figure 19.6: The Pattern of Changes in LULC at Study Area.

During 1990-2000, the decline was 7 per cent where as it was found at 20 per cent during 2000-2012. The analysis of maps shows that a large patch of agricultural land has been converted in to town/city, village settlement and waste land category and it has mostly occurred adjacent to mining lease area.

Forest Land

The composition of dense forest land (Area having canopy cover more than 40 per cent) in the surrounding 10 km of mine area has reduced by 68 per cent during 1990-2000, where the decline percentage was around 104 per cent during 2000-2012. It has been found out that forest coming under open category (having canopy cover less than 10 per cent) has increased significantly at the cost of dense forest. The growth of open forest was 37 per cent in the first duration where as it was 23 per cent during 2000-2012. Similarly the forest category coming under plantation has also reduced by 26 per cent during 1990-2000 and 13 per cent during 2000-2012. A comparison between maps found out that major patch of dense forests which were located adjacent to the mining site has been converted in to open forest category. This may be occurred due to anthropogenic activity like mining, road, development of settlement, setting and expansion of villages, and increase demand for fuel wood.

Waste Land

There has been a significant increase of land-scape coming under wasteland category such as mining quarry, land with shrub and land without shrub. The analysis of map shows that within the wasteland category, maximum growth has occurred in mining/quarry type (56 per cent) followed by land with shrub (28 per cent), rocky/stony waste (22 per cent), Baren land without shrub (13 per cent) and gullied and eroded land (7 per cent) during 1990-2000. However the increase pattern of waste land was found different during 2000-2012. In this period, maximum growth was observed in land with shrub (34 per cent) followed by land without shrub (33 per cent), mining/quarry type (25 per cent) and barren land (21 per cent).

Built Up Land

30 per cent increase in urban area has been observed during 1990-2000, where the land coming under rural has showed a growth of 15 per cent growth. During 2000-2012, this growth was recorded at 27 per cent and 43 per cent respectively. It has been found out that most of the development occurred in area adjacent to mining area.

Water Bodies

A growth of area coming under river and stream has been observed in both period gap while a decrease in pond and other standing water body has occurred in both the period.

The study concluded that the significant impact of mining and associated activities are large scale landscape alternation particularly land coming under dense forest, agricultural land and settlement category. Similarly rapid increase of waste land is another significant finding which needs to be timely addressed by the mining companies. Periodic mapping and monitoring the level of changes in land use land cover is quite significant in formulating strategies for reclamation during post-mining period. It has been found out that agricultural and dense forest land has been reduced drastically which is a indicator of un-sustainable practise adopted by mining companies.

Conclusion

Mining will continue to expand in upcoming days, as a number of mineral based industry are going to setup in Orissa. At this background, the present study is an attempt to find out the adverse effects of mining and associated activities on land use through RS and GIS technology. The main objective of present studies was to understand the impact of mining activity on the surrounding land use/land cover pattern (within 10 Km radius) of the iron ore mines as per the MOEF (Ministry of Environment and Forests, Govt of India) guidelines. It has found out that Remote sensing technology can play a major role in carrying out the environmental studies and the subsequent impact assessment especially for open cast mines, which are of dynamic nature. It is evident from the above discussion that the mining and associated activity in the study area has put an adverse impact on the surrounding vegetation and agricultural land. It is also advisable that the mining activities have to be strictly regulated and properly complied in-order to avoid further damage. Similarly modern technology to reduce pollution should be adopted by the mining company to minimize the externalities. This type of study by using remote sensing data needs to be carried out in a timely manner by other mining company and regulatory agencies like state pollution control board, ministry of Environment and Forest for making mining activity sustainable.

Acknowledgements

The authors would like to thank the HOD, Geography Dept, Utkal University for immense corporation in this research.

References

Bell, F.G., Bullock, S.E.T., Halbich, T.F.J. and Lindsey, P. (2001); Environmental impacts associated with an abandoned mine in the Witbank Coalfield, South Africa. International Journal of Coal Geology, 45: 195-216.

Bhushan, C. and Zeya Hazra. (2008): Rich lands poor people – Is 'sustainable' mining possible? Center for Science and Environment, New Delhi, August.

Down, C.G. and J, Stocks. (1977): The environmental impact of mining. London. Applied Science.

Patra, H.S and Sethy, K.M. (2013): Assessment of Impact of Iron ore Mining Impact by using Remote Sensing : A case study from Keonjhar District of Odisha, India. International Journal of Environmental Engineering and Management. 4 (1): pp. 17-24.

Lodha, R.M., Purohit, J.K. and Yadav, H.S. (2009): Environment and mining, A peep in to deep (reprint edtn). (book)

Ranade, P. (2007): Environmental impact assessment of land use planning around the leased limestone mine using remote sensing techniques Iran. J. Environ. Sci. Eng., l4 (1): pp.61-65.

Roy, H. and Mishra, P.K. (2007): Impact of mining activities on plant diversity with special reference to coalfield of Dhanbad district in Plant diversity and conservation. Satis Serial Publishing House, New Delhi. pp 207-212.

V, Neeraj, Moghe Kaustubh A. and Dutta, D. (2003): Undermining India: Impacts of mining on ecologically sensitive areas. Pune: Kalpavriksh.

World Bank. (2007): India, Towards Sustainable Mineral-intensive growth in Orissa – Managing Environmental and Social Impacts, Report No. 39878-IN, World Bank, May.

Chapter 20

Environmental Assessment of Marine Ornamental Fishes in the Gulf of Mannar Biosphere Reserve, Southeast Coast of India

Manish Kumar[1], B. Anjan Kumar Prusty[2]
and T.T. Ajith Kumar[1]

[1]*Centre of Advanced Study in Marine Biology, Faculty of Marine Sciences, Annamalai University, Chidambaram – 608 502, T.N., India*
[2]*Environmental Impact Assessment Division, Sálim Ali Center for Ornithology and Natural History, Anaikatti (PO), Coimbatore – 641 108, T.N., India*

ABSTRACT

Coral reef is the most fragile environment in marine ecosystems and the information on the diversity of reef dwelling ornamental fishes enrich the knowledgebase of marine biodiversity. In the present chapter, we assess the diversity of marine ornamental fishes at the Gulf of Mannar region, which is the first marine biosphere reserve in India. The samples were collected on seasonal basis (Summer, Premonsoon and Postmonsoon) from two major landing centers (Rameswaram and Tuticorin) in the Gulf of Mannar region during 2008-2009 to explore the influence of destructive fishing method, unauthorized gear application and by-catch loss of ornamental fishes in this region. In total, 32 species belonging to 11 families were collected during the survey from landing centers and local fish markets. The highest (4.1) fish diversity was recorded in Rameswaram during summer and the minimum (3.6) again in Rameswaram center during post-monsoon. The species richness was maximum (4.6) in Tuticorin during post-monsoon and the minimum (3.4) at Rameswaram in premonsoon season. The evenness was maximum (0.8) in Tuticorin during premonsoon. The association of fishes was observed by Jacard's index to understand the catching pattern of ornamental fishes. The difference in landings of ornamental fishes varies due to the frequency of gear application and the efforts made on each catch. The study reveals that the by-catch ornamental fish is high for both Rameswaram and Tuticorin. This calls for restrictions on such activities for maintaining the ecological sustainability of biosphere and the aesthetic beauty of coral paradise of India.

Keywords: *Gulf of Mannar Biosphere Reserve, Ornamental fish, Rameswaram, Southeast coast of India, Tuticorin.*

Introduction

Environmental assessments have become essential in view of the various developmental projects planned in different parts of the country. Any ecosystem is likely to get impacted because of anthropogenic activities, and marine ecosystems are no exception to this. The highest impacted stratum is the marine biodiversity including fisheries and coral reef. Coral reef is the most fragile environment in marine ecosystems and the information on the diversity of reef dwelling ornamental fishes enrich the knowledgebase of marine biodiversity. The coastal zone is being altered as fast as tropical forests, and simply knowing which ecosystems have more or less species is misleading (Ray 1988). The relative lack of knowledge concerning the loss of marine diversity is in part due to the remoteness and difficulty of monitoring marine habitats (FAO 2004). It indicates that natural population from several species may be over exploited but the impact of such activity on the reef ecosystem are still poorly understood (Job 2005). Marine ecosystems also offer various ecosystem services and their environmental and cultured values are directly linked to the human society. Most of the time, marine biodiversity alters due to the selective fishing. The increasing number of export individuals coupled with high mortality rates also influence the diversity of species in particular area.

Of the several methods of fishing, trawl fishing has both direct and indirect impacts on the marine ecosystem as well as on biodiversity, as this method of fishing collects and kills huge amount of non-target species and young ones of commercially valuable species (Knieb 1991). This method has a large impact on benthic systems, and benthic habitats not only provide shelter and refuge for juvenile fish, but the associated fauna provide food sources for a variety of important demersal fish species. Thus frequent alterations in the benthic habitats would result in decline of marine fish propagation and diversification. In general, the environmental effects of bottom trawling have been found to be more destructive in structurally complex and biodiversity-rich marine habitats such as sea grass meadows, coral reefs, sea mounts and deepwater areas (Thrush 2002). It is because of longer recovery trajectories in terms of re-colonization of the habitat by the associated fauna. The non-target species may have key roles in the marine ecosystem that fortify ecosystem processes and functioning, which in turn determines the productivity of marine biodiversity. Habitat impacts and by-catches affect stocks of commercially valuable species, the natural biodiversity and ecological services provided (Biju Kumar and Deepthi 2006).

Among the marine ornamental fishes, Pomachanthidae are one of the indicators of coral reef quality (Acero and Rivera, 1992) and found frequently in by-catches. This group comprise of the fabulous fishes species of tropical reef bottom widely spread from the Atlantic to Indian Ocean. Like other reef inhabitants they are present on attractive colouration and easy to maintain in captivity and are commercially exploited as ornamental fishes (Stratton 1994; Fenner 1996; Mcmillan *et al*. 1999; Littlewood *et al*. 2004). The diversity of marine ornamental fish was studied several times and the number of marine angel species reported only two in Lakshadweep (Murty 2002) and in the Gulf of Mannar only one species by Muralitharan (1999) and four by Venkataramani *et al*. (2004). The recent observation revealed that only two species were abundant in this region (Raja 2006). Though most of the studies carried out on marine biodiversity, but there is no composition report on marine ornamental fish of by-catches. This indicates that the organisms of hard bottom areas such as ornamental fishes etc. are in urgent need to find the effect of by-catch or discards from fishing. The present study is an attempt to find out the status of ornamental fishes in this area.

Study Area

Coral reef environment is the most complex and diversified ecosystem preceded by the mangroves. The Gulf of Mannar region is well known due to the coral beauty and it is lying in between 08°47' - 09°15' N and 78°12' - 79°14' E (Figure 20.1). The currents which makes lees disturbance to this region are "Agulhas" and "Mozambique" and Monsoon drift. Temperature is mostly not variable and the average temperature varies from 28-32 °C. The Tuticorin (station I) region experiences higher temperature in some areas due to thermal power plant discharge (Kumaraguru *et al*. 2006). Higher salinity is also observed, apparently due to the discharge of wastewater from saltpans. Rameswaram (station II) is maintaining its aesthetic beauty due to less industrialization but having many deformations due to the

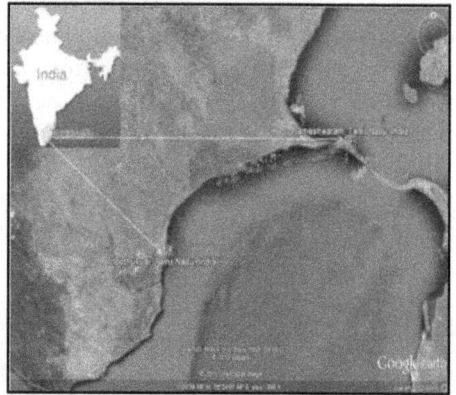

Figure 20.1: Study Area.

intrusion of people and associated activities. Available reports revealed that it comprises around 3,600 marine flora and faunal diversity. Among them the corals contributes 117, crustacean 641, fishes 441, seaweed 147, seagrass 52, molluscs 731 (Kumaraguru *et al*. 2006). 125 villages are dependent on the Gulf of Mannar for their survival.

Materials and Methods

The fish samples were collected from both the landing centers in three different seasons. Rameshwaram and Tuticorin are the two stations and summer, pre-monsoon and post-monsoon were the three seasons selected for sampling. Samplings were done directly by visiting the landing centers, fish markets and the collection centers at seasonal intervals for 12 months (Muralitharan 1999; Raja 2006). The study on biodiversity requires variety of techniques to be employed to simplify the resulting large data, involving various preprocessing of data before testing the structure. Clarke and Green (1988) and Clarke and Warwick (1994) have summarized theses steps. In the present study the diversity of ornamental fishes was statistical analyzed by some of standard methods.

These methods are used to extract the features of communities which are species independent. Compared to multivariate methods, these are obtained more easily and are also as sensitive as multivariate methods (Warwick and Clarke 1991) in terms of detecting changes. Most commonly used indices which are used in this present study are as below.

Species Diversity (Shannon-Wiener's Index)

It is the relative abundance of different species at each site or time reduced to a single index. There are many diversity indices used in ecological studies. The data were analysed using the formula proposed by Shannon-Wiener (1949). This index is denoted by H' and the value of H' is dependent upon the number of species present, their relative proportions, sample size (N) and the logarithm base.

$$H' = \sum_{i=1}^{S} Pi \log 2Pi$$

Where, $Pi = ni/N$ for the ith species, S = total number of species, (ni) = number of individuals of a species in sample, N = Total number of individuals of all species in sample.

Species Evenness (Pielou's Evenness Index)

Evenness index is also an important component of the diversity indices. This expresses how evenly the individuals are distributed among the different species. Pielou's evenness index is commonly used. It is represented by J'.

$J' = H'/H'_{max}$

Where, H' is the observed species diversity and H'_{max} is the logarithm (LN) of the total number of species (S) in the sample.

Species Richness (Margalef Index)

Margalef index is denoted by "d", which is a measure for the number of species present for a given number of individuals. The advantage of this index is that as the values can come more than 1 unlike Simpson index. It was calculated by using the following formula.

$d = (S-1)/\log N$

Where, S = total no. of species, N= total no. of individuals in the sample.

Similarity of Species Jaccard Index (CJ)

In comparing the faunastic composition of marine ornamental fish we used this similarity index which will give the idea of association of fish species with each other in coral environment. It was calculated by using the following formula.

$$CJ = J/(a + b) - J$$

Where, a is the number of species present in one population, b is the number of species present in the other population, and J is the number of species present in both populations.

Results

Survey on Ornamental By-catch Resources

Marine ornamental fishes provide aesthetic beauty to the marine ecosystems. They are very much restricted to the coral reef environment. The tiny creatures are used for decorating home and outers at the same time for mind relaxation and few in different research aspects. The hobby of keeping ornamental fish has become an industry with 10-15 per cent of annual growth. But the fast growing industry gets setback when the adverse impacts of industrialization accelerated. Research reports and articles reveal that the native of the marine, tiny, static beautiful creature and the heart of coral paradise is under threat (Venkataramani *et al*. 2004; Raja 2006). The present survey was carried out in the Gulf of Mannar region at two major fishing grounds which ranges from Ramshewarm to Tuticorin.

The traditional fishermen in this region were seen adopting the gears such as lift-nets, seines, scoop-nets and traps for collecting fishes. Further, trawlers and other mechanized boats, only few of them were using the eco-friendly type of fishing practices with fixed bamboo nets (Locally called KOODU or KOONDU or TRAP). In total, 32 species of true ornamental fishes belonging to 11 families were recorded during the survey (Table 20.1) which has come as by-catches in the landing centers with different type of fishing activities. The previous records made by Raja (2006) reported only 30 species as by-catch ornamental fishes from this region. The presence and absence of recorded fish species at different stations and seasons are denoted with (+) and (–) in Table 20.1.

The present investigation focused on the species diversity and other indices, and more on the qualitative aspects. This need to be supplemented with quantitative information on these fishes. Further studies on diversity indices will bring out a clear picture for ornamental by catch resources in the Gulf of Mannar region.

Species diversity (Shannon-Wiener's index)

The various diversity indices calculated in the two stations (St1 = Rameshwaram; St2 = Tuticorin) during three seasons (S1 = Summer; S2 = Pre-monsoon; S3 = Post-monsoon) are discussed here. During summer, the minimum diversity (3.8) was recorded in St2 and the maximum (4.1) in St1 (Figure 2). During pre-monsoon, the minimum diversity (3.7) was recorded in St1 and the maximum (3.6) in St2. During the Post-monsoon, the minimum value (3.6) was recorded in St1 and the maximum

Table 20.1: Check List of Marine Ornamental Fishes Recorded as By-catch Resources

Sl.No.	Name of Species	S 1		S 2		S 3	
		St 1	St 2	St 1	St 2	St 1	St 2
1.	Abudefduf bengalensis	+	+	+	+	+	+
2.	Acanthurus bleekeri	+	+	+	+	+	+
3.	Acanthurus dussumieri	+	+	+	+	+	+
4.	Acanthurus xanthopterus	–	+	+	+	+	+
5.	Aeoliscus strigatus	+	+	+	–	+	–
6.	Amphiprion sebae	+	–	–	–	–	–
7.	Apogon aureus	+	+	+	+	+	+
8.	Apogon fraenatus	+	+	+	–	+	+
9.	Apolemichthys xanthurus	+	+	+	–	+	+
10.	Chaetodon leucopleura	+	+	+	+	+	–
11.	Chaetodon lunula	+	+	+	+	+	–
12.	Chaetodon vagabundus	+	+	+	+	+	+
13.	Cheilinus chlorourus	+	–	–	–	+	–
14.	Coris gaimard	+	–	–	–	+	–
15.	Dascyllus trimaculatus	+	+	+	+	+	+
16.	Diagramma pictum	+	+	+	+	+	+
17.	Gnathodon speciosus	+	+	+	+	+	+
18.	Halichoeres zeylonicus	–	+	–	–	+	+
19.	Halichoeres hortulanus	+	+	–	+	–	+
20.	Heniochus acuminatus	+	+	+	+	+	+
21.	Heniochus diphreutes	+	–	+	–	+	+
22.	Odonus niger	+	–	–	–	+	+
23.	Plotosus lineatus	+	+	+	+	+	+
24.	Pomacanthus annularis	+	–	+	–	+	+
25.	Pomacanthus imperator	+	+	+	+	+	+
26.	Pomacentrus caeruleus	+	–	–	+	–	+
27.	Sargocentron rubrum	+	+	+	+	+	+
28.	Scarus ghobban	+	+	+	+	+	+
29.	Tetrosomus gibbosus	–	+	+	–	+	+
30.	Thalassoma lunare	+	+	+	+	+	+
31.	Xyrichtys pavo	+	+	–	+	+	+
32.	Zanclus cornutus	+	+	+	+	+	+

S1: Summer; S2: Premonsoon; S3: Postmonsoon

St1: Rameshwaram; St2: Tuticorin.

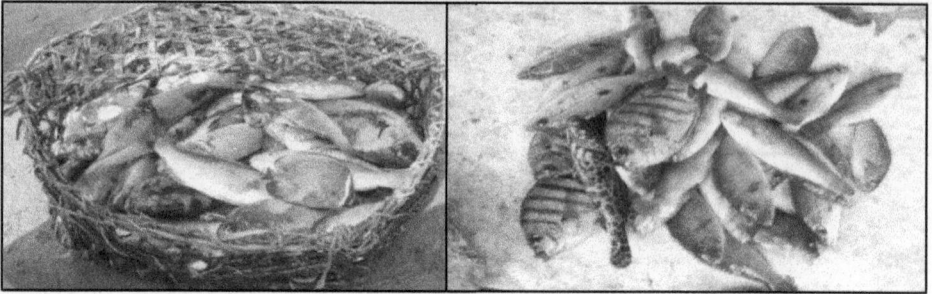

Plate 20.1: Common Landings in the Gulf of Mannar Region.

Plate 20.2: Gear Used for Capturing Ornamental Fishes.

was (3.9) in St2 (Figure 20.2). The overall diversity was maximum (4.1) in St1 was recorded during summer and the minimum (3.6) was in St1 during post-monsoon.

Species Richness (Margalef Index)

Species richness index followed a similar trend as in the case of Species diversity index. The higher values were recorded in St2 (4.0) during summer and the lower

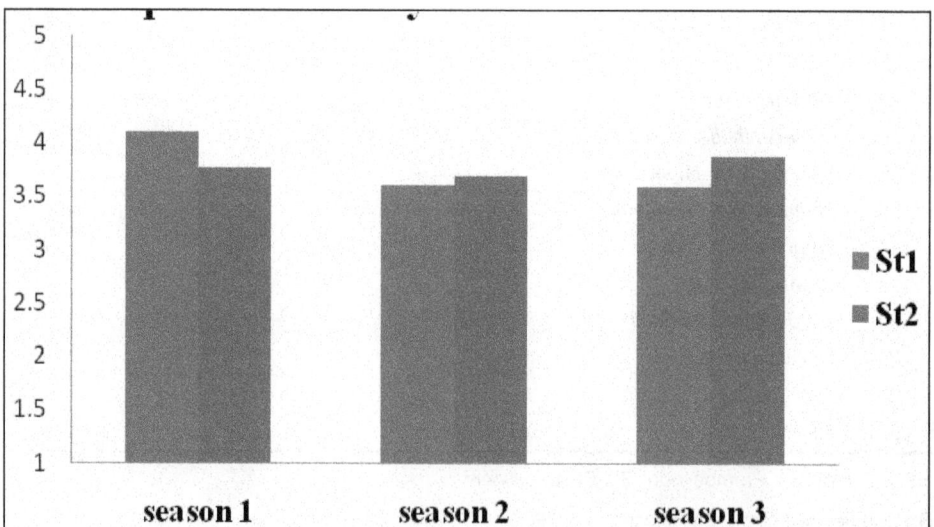

Figure 20.2: Species Diversity (H') at Two Stations in Three Seasons.

value in St1 (3.8, Figure 20.3). During pre-monsoon, the richness was lowest (3.4) in St1 and highest (4.0) in St2. During post-monsoon, the lowest value (3.8) was recorded in St1 and highest (4.6) in St2.

Species Evenness (Pielou's Evenness Index)

The species evenness index also followed a similar trend as like species diversity index. The higher values of evenness were observed in St1 (0.82) during summer and the slightly lower values in St2 (0.81, Figure 20.4). During pre-monsoon, the minimum evenness (0.78) was observed in St1 and maximum (0.84) in St2. During post-monsoon, the highest value (0.74) was recorded in St1 and the lowest (0.82) in St2.

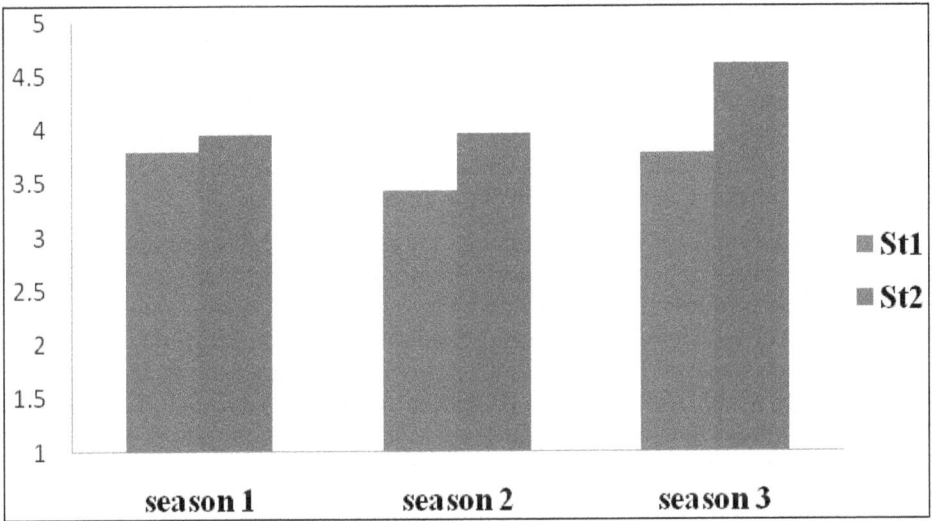

Figure 20.3: Species Richness (d) at Two Stations in Three Seasons.

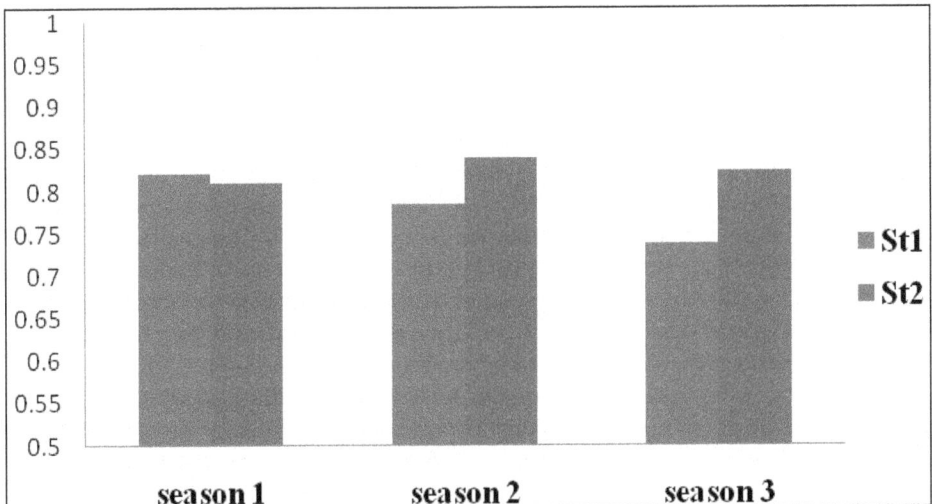

Figure 20.4: Species Evenness (J') at Two Stations in Three Seasons.

Discussions

Study on biodiversity of marine organisms is an old and well established practice in the Gulf of Mannar region, but the study on diversity of ornamental fishes were ignored most of the times. Accelerated loss of coastal and marine biodiversity components, particularly ornamentals over the last few decades has been of great concern. Environmental changes, over-exploitation, destructive fishing practices, effluent discharges and habitat loss are the major causes of species loss. Probable estimates of species diversity have been variously arrived by extrapolation of known number of species from a section of the habitat to others (Muralitharan 1999; Venkataramani *et al.* 2004; Venkatraman *et al.* 2005). Disturbance intensity is another issue that is often under-addressed or ignored in population studies of marine organisms. High levels of disturbance may cause higher rates of local population extinction and re-colonization, which may preclude the formation of significant genetic autocorrelation at local scales and instead, result in chaotic spatial genetic patterns (Johnson *et al.* 1982; Hellberg *et al.* 2002). Thus the frequent survey and study on the most admired ornamental fish group will provide clear picture on the larger scale of their population condition and health of the marine ecosystem.

The fish assemblage in any particular area represented with mixing of species from the different habitats available nearby. Sometimes these included mobile predators such as fusiliers (family: Caesonidae) which forage in the water column (column feeders), species which forage over sandy bottom (bottom feeders) such as goatfish (Mugilidae) and mojarras (Gerridae), and species which utilized the shelter provided by the boulders such as damselfish (Pomacentridae) and hawkfish (Cirrhitidae). Numerous species were observed which are more typically thought of as coral reef fish, such as angelfish (Pomacanthidae), butterflyfish (Chaetodontidae), parrotfish (Scaridae) and moorish idol (Zanclidae). Earlier Sluka and Lazarus (2003) have reported reff associated fishes in South India and have reported 94 species. Those belong to families considered reef-associated (Sluka and Lazarus, 2003), they found butterflyfish (Chaetodontidae), wrasses (Labridae) and angelfish (Pomacanthidae) in limited numbers. The present study in the Gulf of Mannar area recorded 32 species of ornamental fishes and found similar findings of Muralitharan (1999) and Raja (2006) about certain group of more valuable fishes. We observed more species diversity and richness of rabbit fishes in Tuticorin coast of Gulf of Mannar, which is similar to the findings by Venkataramani and Javahar (2004) but another study on ornamental fishes conducted by Sundaramurthy *et al.* (2000) shows that the similar species diversity and higher results in Mandapam group of islands. The available findings support the present study also but the higher diversity observed in Rameshwaram (st 1). Whereas, the previous study on species diversity of marine ornamental species (Venkataramani and Javahar, 2004) was higher as compared with the present study. This might be due to the sampling strategy followed in the present exercise (by-catch and survey of fish markets). Moreover, the present investigation was aimed at having a preliminary understanding about the ornamental fish (species composition and diversity) in view of the increasing pressure on the marine resources. In the summer season, the diversity was higher in Rameshwaram and this may be attributed to higher fishing activities and more rough oceanic currents in this region (Muralitharan

1999). On the contrary, species richness was higher in Tuticorin, which might be due to frequent trawling. The evenness was also high in Tuticorin (0.84). These results confirm the findings made by earlier researchers (Venkataramani and Javahar 2004). *Halichoeres melanurus* (17 per cent), *Pomacanthus semicerculatus* (33 per cent) and *Coris gaimard* (33 per cent) are showing minimum association (Table 20.2) with all the other species recorded but most of the recorded species were more closely associated at higher percentage levels to each other.

The habitat available for angel fishes is small and relative to the shallow reef areas to some distance of deep water habitat. The present investigation has generated preliminary information on by-catch loss of reef dwelling fishes. Further comprehensive assessments for covering different fish landing centres and involving other available fishing operations would help in preparing comprehensive database on reef dwelling fishes including ornamental fishes. This would also help in determining appropriate levels of harvest for the different fish species and corresponding families. Ecological information on reef fishes lacks clarity for Southeast coast of India, and necessitates comprehensive assessments. However, visual observations suggest that there is a number of species which are caught as by-catch in landings too, after the marine ornamental fish trading (Raja 2006). The main conservation issues related to the marine ornamental trade will be of great concern in India as destructive fishing practices, capture of species with low survival rates, high post-harvest mortality and over-exploitation (Wood 2001; Raja 2006). The present findings are in line with the finds of earlier studies on ornamental fishes from the same area. This calls for an extensive study on by-catch landings, harvesting for trade and the availability of these valuable species in various aspects for future sustainable catch.

Conservation and Management Plan Recommendations

Fisheries and aquaculture have a promising role in socio-economic development by providing nutritional security for the nation's population and contributing to the economic advancement as well as for posterity. A huge number of people engaged in the marine ornamental sector in India ranging from local fisher-folk to unskilled and skilled technocrats of different parts of the country. For sustainable fishery of these valuable fishes and maintain the ecological equilibrium, it is required that a proper sustainable harvest plan is prepared and implemented. Due to the sensitivity of coral reef environment and the livelihood of dependent populace, several strategies are made for the conservation and management of marine ornamental fishes in the Gulf of Mannar region. The strategies will cover stakeholders at various levels and are listed as below:

☆ Impact of climate change on breeding of ornamental organisms need to be studied

☆ Regulating the fishing activity in the coral region and nearby environment

☆ Trawler operation should be avoided in the reef region

☆ Permitting only eco-friendly type of cages (bamboo koods) to the coral environment for collecting ornamental fishes would aid in conservation of these precious resources

Table 20.2: Species Association (Jacquard's index) among the Recorded Ornamental Fish Species.

	Sg	Cv	Cl	Pl	Tl	Cg	Hh	Dp	Pi	Pa	Ps	Cle	Ad	Hha	Gs	Ab	As	Aa	On	Dt	Xp	Ax	Axa	Hea	Hed	Tg	Af	Hm	Pc	Zc	Sr	Abe
Sg	100																															
Cv	100	100																														
Cl	83	83	100																													
Pl	100	100	83	100																												
Tl	100	100	83	100	100																											
Cg	33	33	40	33	33	100																										
Hh	67	67	50	67	67	20	100																									
Dp	100	100	83	100	100	33	67	100																								
Pi	100	100	83	100	100	33	67	100	100																							
Pa	67	67	50	67	67	50	33	67	67	100																						
Ps	33	33	40	33	33	100	20	33	33	50	100																					
Cle	83	83	100	83	83	40	50	83	83	50	40	100																				
Ad	100	100	83	100	100	33	67	100	100	67	33	83	100																			
Hha	67	67	50	67	67	50	60	67	67	60	50	50	67	100																		
Gs	100	100	83	100	100	33	67	100	100	67	33	83	100	67	100																	
Ab	100	100	83	100	100	33	67	100	100	67	33	83	100	67	100	100																
As	67	67	80	67	67	50	33	67	67	60	50	80	67	60	67	67	100															
Aa	100	100	83	100	100	33	67	100	100	67	33	83	100	67	100	100	67	100														
On	50	50	33	50	50	67	40	50	50	75	67	33	50	75	50	50	40	50	100													
Dt	100	100	83	100	100	33	67	100	100	67	33	83	100	67	100	100	67	100	50	100												
Xp	83	83	67	83	83	40	80	83	83	50	40	67	83	80	83	83	50	83	60	83	100											
Ax	83	83	67	83	83	40	80	83	83	50	40	67	83	80	83	83	50	83	60	83	67	100										

Contd...

Table 20.2–Contd...

	Sg	Cv	Cl	Pl	Tl	Cg	Hh	Dp	Pi	Pa	Ps	Cle	Ad	Hha	Gs	Ab	As	Aa	On	Dt	Xp	Ax	Axa	Hea	Hed	Tg	Af	Hm	Pc	Zc	Sr	Abe
Axa	100	100	83	100	100	33	67	100	100	67	33	83	100	67	100	100	67	100	50	100	83	83	100									
Hea	100	100	83	100	100	33	67	100	100	67	33	83	100	67	100	100	67	100	50	100	83	83	100	100								
Hed	67	67	50	67	67	33	33	67	67	100	50	50	67	60	67	67	60	67	75	67	50	80	67	67	100							
Tg	83	83	67	83	83	40	50	83	83	80	40	67	80	80	83	80	83	83	83	83	83	83	80	80	80	100						
Af	83	83	67	83	83	40	50	83	83	80	40	67	83	80	83	80	83	83	60	83	67	100	83	83	80	100	100					
Hm	17	17	20	17	17	50	25	17	25	25	50	20	25	25	17	17	25	17	33	17	20	20	17	25	20	20	17	100				
Pc	50	50	33	50	50	25	75	50	50	40	25	33	50	40	50	50	17	50	50	60	33	33	50	40	33	33	17	50	100			
Zc	100	100	83	100	100	33	67	100	100	67	33	83	100	67	100	100	67	100	50	100	83	83	100	100	67	83	83	17	50	100		
Sr	100	100	83	100	100	33	67	100	100	67	33	83	100	67	100	100	67	100	50	100	83	83	100	100	67	83	83	17	50	100	100	
Abe	100	100	83	100	100	33	67	100	100	67	33	83	100	67	100	100	67	100	50	100	83	83	100	100	67	83	83	17	50	100	100	100

Sg: *Scarus ghobbar*; Cv: *Chaetodon vagabundus*; Cl: *Chaetodon lunula*; Pl: *Plotosus lineatus*; Tl: *Thallasoma lunare*; Cg: *Coris gaimard*; Hh: *Helichoeres hortulanus*; Dp: *Diagramma pictum*; Pi: *Pomacanthus imperator*; Pa: *Pomacanthus annularis*; Ps: *Pomacanthus semicerculatus*; Cle: *Chaetodon leucopleura*; Ad: *Acanthurus dussumieri*; Hha: *Helichoeres hartzfeldii*; Gs: *Gnathanodon speciosus*; Ab: *Acanthurus bleekeri*; As: *Aeoliscus strigatus*; Aa: *Apogon aureus*; On: *Odonus niger*; Dt: *Dascyllus trimaculatus*; Xp: *Xyrichtys pavo*; Ax: *Apolemichthys xanthurus*; Axa: *Acanthurus xanthopterus*; Hea: *Heniochus acuminatus*; Hed: *Heniochus diphreutes*; Tg: *Tetrosomus gibbosus*; Af: *Apogon fraenatus*; Hm: *Halichoeres melanurus*; Pc: *Pomacentrus caeruleus*; Zc: *Zanclus cornutus*; Sr: *Sargocentron rubrum*; Abe: *Abudefduf bengalensis*.

☆ Controlling overexploitation of ornamental resources, license system should be adopted for the sustainable harvest

☆ Provide an alternate livelihood option for the fisher-folk of this region, which will improve the socioeconomic status of the rural people

☆ Awareness programme regarding the values of coral reef ecosystem and their resources will definitely help the people to come forward for better management and to avoid destructive fishing practices

☆ Data sheet on export and import of marine ornamental resources should be maintained on behalf of the Govt. of India

☆ Continues survey on the stock assessment should be implemented through appropriate agencies

☆ A centre of excellence in ornamental aquaculture should be established in the region of Gulf of Mannar to boost further research on these precious ecological and economic resources.

Summary and Conclusion

India's vast and diverse aquatic genetic resources are essential to maintain ecological as well as socioeconomic equilibrium. In the fishery sector, ornamental fishes collected around the world to supply specimens for the international aquarium trade, which is estimated to be worth 200-330 million US$ annually (Wabnitz *et al.* 2003, Knittweis *et al.* 2009). The trading was started in India during 1969 on experimental basis with the export value of US$ 0.04 million. Now it increases to > 100 crore per year @ 20 per cent average growth rate (Mohanta and Subramaian 1999). The Indian marine ornamental fish industry is heavily dependent on the capture from the Gulf of Mannar region. Since almost the entire volume of ornamental fishes is collected from the natural aquatic resources, there is a big sustainability threat to these resources. The impetus behind the present exercise is conservation of the desirable quantities of marine ornamental fishes and its habitats. For this purpose the study was carried out in the Gulf of Mannar region on documenting the diversity and diversity of true ornamental fishes using by-catch resource data. It revealed that the higher diversity in Rameshwaram, where the density condition is better in Tuticorin. The other indices, *viz.*, species richness and evenness are higher in Tuticorin waters which may be the cause of higher landings associated with frequent boat operations.

Overharvesting, in particular sex, sizes or age classes may reduce population sizes to levels at which inbreeding and loss of genetic diversity may become inevitable. Therefore, it is required to maintain the genetic identity and integrity of the species in their natural habitat as well as genetically sustainable fishery. Further molecular studies comprising a higher number of samples and other species are still required to precisely evaluate the genetic structure of angelfishes along the Indian coast. Hence, documentation of genetic diversity is of vital significance to evolve conservation strategies with long-term impact. It is currently not clear what levels of loss/change may be regarded as acceptable. This calls for an urgent attention on formulating sound ecological and economic strategies on national level. Further, the management plan suggested through this study will help in conserving the fragile region in a sustainable way.

References

BijuKumar, A. and Deepthi, G.R. (2006): Trawling and by-catch: Implications on marine ecosystem. Current Science. 90 (7): 922-931.

Clarke, K.R. and Green, R.H. (1998): Statically design and analysis for a biological effects study. Mar. Eco. Prog. Ser., 46: 316-326.

Clarke, K.R. and Warwick, R.M. (1994).\: Similarity based testing for community pattern: the two-way layout with no replication. J. Mar. Bio. 118 (1): 167-176.

FAO, (2004): The State of World Fisheries and Aquaculture. Food and Agriculture Organization, Rome, 153pp.

Fenner, R. (1996): The French angelfish, *Pomacanthus paru*. Trop. F. Hob., 44(8): 32-38.

Hughes, R.M. and Noss, R.F. (1992): Biological diversity and biological integrity: current concerns for lakes and streams. Fisheries (Bethesda) 17: 11-19.

Job, S. (2005): Integrating marine conservation and sustainable development: Community-based aquaculture of marine aquarium fish. SPC Live Reef Fish Inf. Bul., 13: 24-29.

Knieb, R.T. (1991): Indirect effects in experimental studies of marine soft sediment communities. Am. Zool., 31: 874-885.

Knittweis, L, Kraemer, W.E., Timm, J. and Kochzius, M. (2009): Genetic structure of *Heliofungia actiniformis* (Scleractinia: Fungiidae) populations in the Indo-Malay Archipelago: Implications for live coral trade management efforts. Conservetion Genetics. 10:241-249.

Kumaraguru, A.K., Joseph, V.E., Marimuthu, N. and Wilson, J.J. (2006): Scientific information on Galf of Mannar - A Bibliography. Centre for Marine and Coastal Studies, Mdurai Kamaraj University, Tamil Nadu, India. 656pp.

Littlewood, D.T.J., McDonald, S.M., Gill, A.C. and Cribb, T.H. (2004): Molecular phylogenetics of *Chaetodon* and the Chaetodontidae (Teleostei: Perciformes) with reference to morphology. Zootaxa, 779: 1-20.

McMillan, W.O., Weigt, L.A. and Palumbi, S.R. (1999): Color pattern evolution, assortative mating, and genetic differentiation in brightly colored butterflyfishes (Chaetodontidae). Evolution, 53 (1): 247-260.

McNeeley, J.A. (1988): Economics and biological diversity. International union for conservation and of nature and natural resources, Gland, Switzerland, 236.

Mohanta, K.N. and Subramanian, S. (1999): Ornamental fish farming, a multi-million doller industry in India. Seafood Export Journal. 30(2): 29-31.

Muralitharan, J. (1999): Biodiversity of reef Ichthyofauna of Gulf of Mannar along the Southeast coast of India. Ph.D. thesis submitted to Annamalai University.172pp.

Raja, K. (2006): Trade of ornamental fishes in the Gulf of Mannar and experimental studies on clown fish *Amphiprion sebae*. Ph.D. thesis submitted to Annamalai University. 147pp.

Ray, G.C. (1988): Ecological diversity in coastal zones and oceans. In Wilson EO and Peter FM (Eds.) Biodiversity. National Academy Press, Washington, 36-50.

Shannon, C.E. and Wienner, W. (1949): The mathematical theory of communication. Univ. of Ilinois press, Urbana.

Sluka, R.D. and Lazarus, S. (2003): Community development and conservation through alternative uses of reef fish resources in South India. Center for Applied Science, Millennium Relief and Development Services Special Publication, 3:19pp.

Stratton, R.F. (1994): Practical angels. Trop. Fish Hob., 43 (1): 30-36.

Sundaramurthy, S., Shunmugaraj, T., Ramanathan, V. and Usha, T. (2000).:Annual Report to Integrated Coastal and Marine Area Management (ICMAM) Project, Directorate, Department of Ocean Development, Institute for Ocean Management, Anna University, Chennai. 84.

Thrush, S.F. and Dayton, P.K. (2002): Disturbance to marine benthic habitats by trawling and dredging: Implications for marine biodiversity. Annu. Rev. Ecol. Syst., 33: 449-473.

Venkataraman, K. and Wafer, M. (2005): Coastal and marine biodiversity of India. Indian J. Mar. Sci., 34 (1): 57-75.

Venkataramani, V.K. and Jawahar, P. (2004): Resource assessment of ornamental reef fisheries of Gulf of Mannar, Southeast coast of India. Final Report - ICAR/NATP/ CGP/Project. 66.

Venkataramani, V.K., Jawahar, P., Vaitheeswaran, T. and Santhanam, R. (2005): Marine Ornamental Fishes of Gulf of Mannar. ICAR/NATP/CGP/Publication. 115.

Warwick, R.M. and Clarke, K.R. (1991): A comparison of some methods for analyzing changes in benthic community structure. J. Mar. Bio. Assoc. UK. 71 (1): 225-244.

Wabnitz, C., Taylor, M., Green, E. and Razak, T. (2003): The global trade in marine ornamental species. From Ocean to Aquarium. UNEP-WCMC, Cambridge, UK. 1-65.

Chapter 21

Bioindicators and Biomarkers of Soil Contamination: Earthworms and Microorganisms as Models

C.S.K. Mishra[1], Snehasis Mishra[2],
D.K. Bastia[3] and Swetalina Acharya[1]

[1]Department of Zoology, Orissa University of Agriculture and Technology, College of Basic Science and Humanities, Bhubaneswar – 751 003, India
[1]School of Biotechnology, KIIT University, Bhubaneswar, India
[2]Department of Agronomy, Orissa University of Agriculture and Technology, College of Agriculture, Bhubaneswar – 751 003, India

ABSTRACT

Rapid industrialization and modern agricultural practices has been the key to economic prosperity in the developed countries and helped rapid growth of the developing nations. However, development has always been associated with environmental degradation. Toxic contaminants generated from industrial wastes and agrochemicals have been a cause of concern all over the globe. The assessment of environmental contamination requires precision in order to ensure ecosystem functioning and stability. Bioindicators and biomarkers are useful tools which reflect the abiotic or biotic stress on the environment. Biodiversity indicators, sentinel organisms, key stone and endangered species could help indicate ecotoxicity in diverse habitats. This chapter reviews the functional roles of earthworms and soil microorganisms as indicators of contamination due to pesticides, herbicides and toxic metals in various ecosystems.

Keywords: *Bioindicator, Biomarker, Ecotoxicity, Agrochemicals, Toxic metals, Earthworm, Soil microorganisms.*

Introduction

Bioindicators

Bioindicators are organisms or communities of organisms which help to evaluate a situation giving clues for the condition of an entire ecosystem. The bioindicator has particular requirements with regard to a known set of physical or chemical variables such that changes in presence/absence, numbers, morphology, physiology or behaviour of that species indicating that the given physical and chemical variables are outside their preferred limits (Gerhart, 1999). Mostly bioindicators are defined as species reacting to anthropogenic effects on the environment. In other words the bioindicator is a species or group of species that readily reflects the abiotic or biotic state of an environment. The impact of environmental changes on a habitat, community or ecosystem or is indicative of the diversity of the species within an area.

Bioindicators are useful in three situations.

1. Where the indicated environmental factor cannot be measured (Climate change).
2. Where the indicated factor is difficult to measure for example pesticides and their residues, toxic effluents etc.
3. Where the environmental factors is easy to measure but difficult to interpret (whether observed changes have ecological significance).

Different types of bioindicators can be described from different perspectives (Figure 21.1). According to the aim of bioindication three types of bioindicators can be distinguished.

1. Compliance indicators
2. Diagnostic indicators
3. Early warning indicators

Compliance indicators: For example, fish population attributes are measured at the population, community or ecosystem level and are focused on issues such as the sustainability of the population or community as a whole. Diagnostic and early warning indicators are measured on the individual or suborganisomal (biomarker) level, with early warning indicators focussing on rapid and sensitive responses to environmental change. Accumulation bioindicators (*e.g.* mussels, mosses, lichens) are distinguished from toxic effect bioindicators, with the effects being studied on different biological organization levels.

According to the different applications of bioindicators, three categories can be distinguished.

1. *Environmental indicator*: This is a species or group of species responding predictably to environmental disturbance or change (*e.g.* sentinels, detectors, exploiters, accumulators, bioassay organisms). An environmental indicator system is a set of indicators aiming at diagnosing the state of the environment for environmental policy making.

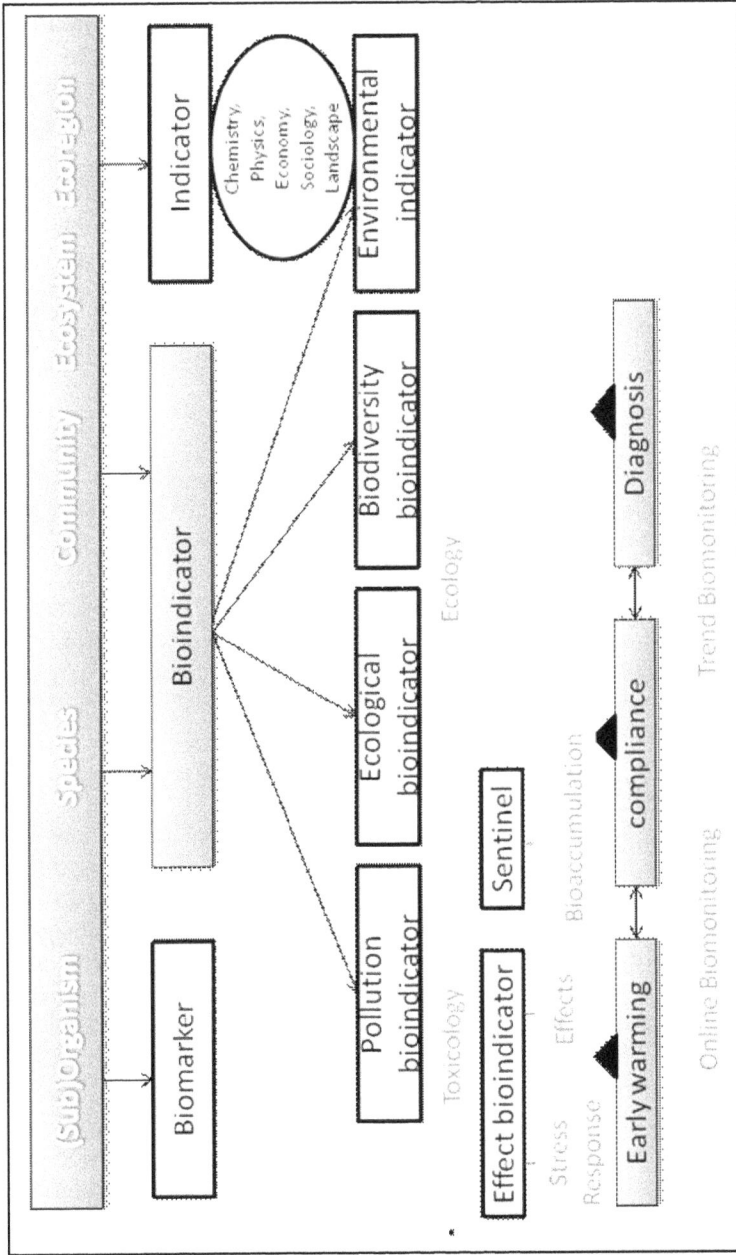

Figure 21.1: Types of Bioindicators in the Context of their Use in Biomonitoring.

2. *Ecological indicator*: This is a species that is known to be sensitive to pollution, habitat fragmentation or other stresses. The response of the indicator is representative for the community.

3. *Biodiversity indicator*: The species richness of an indicator taxon is used as indicator for species richness of a community. However the definition has been broadened to "measurable parameters of biodiversity", including *e.g.* species richness, endemism, genetic parameters, population-specific parameters and landscape parameters.

Sentinels

Bioaccumulation indicators are a special kind of indicator organisms which accumulate and concentrate pollutants from their surroundings and/or food so that an analysis of their tissues provides an estimate of the environmentally available concentrations of these pollutants. Sentinel organisms should be large in number in order to provide enough tissue for analysis and must be widely distributed to facilitate comparisons.

Key Stone Species

A keystone species is a species on which the persistence of a large number of other species in the ecosystem depends. The removal of a keystone species has significant effects on the diversity in an ecosystem and on ecosystem functions.

Endangered Species

These are the species which have decreased in distribution and abundance over a time period mainly due to anthropogenic impacts. These can be categorized as threatened to die, immensely endangered, endangered, potentially endangered species on early warning list etc.

Biological Soil Quality Indicators

The interest in soil quality management dates back to the ancient roman civilization. Over the ages use of agricultural residues, application of organic matter, rotation and tillage practices have been fundamental in maintaining soil fertility. One important discovery at the end of nineteenth century was the nitrogen fixing organisms associated with roots that opened the door for a better understanding of rhizosphere and development of soil ecology.

Soil Contamination

Soil contamination has enormously increased during the last decades due to intensive use of biocides and fertilizers during agricultural practices, industrial activity, urban waste and atmospheric deposition etc. Soil contamination generally causes decrease in soil fertility, alteration of soil structure, disturbance of the balance between soil flora and fauna, contamination of the crops, groundwater ultimately threatening the health and existence of living organisms.

The most diffusive chemicals occurring in soil are heavy metals, pesticides, petroleum hydrocarbons, polychloro biphenyl (PCBs), dibenzo-p-dioxins/

dibenzofurans (PCDD/Fs). Heavy metals from anthropogenic sources are widely spread in the environment and most of them finally reach the surface soil layers. Heavy metals can enter the soil from different sources, such as pesticides, fertilizers, organic and inorganic amedants, mining, wastes and sludge residues (Capri and Trevisan, 2002). In contrast to harmful organic compounds, heavy metals do not decompose and do not disappear from soil even if their release to the environment can be restricted (Brusseau, 1997). Therefore, the effects of heavy metal contamination on soil organisms and decomposition process persist for many years. Pesticides are widely used in agriculture for counteracting insects, fungi, rodents or other animals living in or on the crops. They are either directly applied to soil to control soil borne pests or deposited on soil as run off from foliar applications and their concentrations are high enough to affect the soil macro-organisms (Bezchlebova *et al.*, 2007). The pesticides most widely used in the past have been organochlorine pesticides, characterized by high hydrophobicity and persistence. Currently they have been replaced by less persistent compounds. Organophosphates have become the most widely used pesticides today. They are used for pest control on crops in agriculture and on livestock, for other commercial purposes, and for domestic use. Due to their water solubility, the organophosphate residues in agricultural practices are capable of infiltrating through soil into surface water. As a consequence of their wide diffusion they have been detected in food, ground and drinking water and natural surface waters (Dogheim *et al.*, 1996; Garrido *et al.*, 2000). Soil pollution by petroleum hydrocarbons usually originates from spills or leaks of storage tanks during fuel supply and discharge operations. Petroleum hydrocarbons include aliphatic and aromatic compounds; some of them are known or suspected human carcinogens, and are classified as priority pollutants. PCBs are persistent soil contaminants due to their hydrophobicity and resistance to biodegradation (Weber *et al.*, 2008). They can be released into the environment from poorly maintained hazardous waste sites that contain PCBs, illegal or improper dumping of PCB wastes, such as transformer fluids, leaks or releases from electrical transformers containing PCBs and disposal of PCB-containing consumer products into municipal or other landfills not designed to handle hazardous waste. PCBs are also currently released into the environment by municipal and industrial incinerators from the burning of organic wastes. PCDD/Fs have a high affinity to organic matter and have limited mobility unless transported in association with particulate organic matter. However these compounds are bioaccumulative, can be found in the terrestrial food chain, and have been reported to impact the biota at the higher tropic levels. Recently, the attention of the scientific community focused on emerging contaminants in the soil, such as pharmaceuticals, endocrine disruptors, personal care products, surfactants, flame retardants. They are currently not included in routine monitoring programmes, but may be candidates for future regulation depending on research on their toxicity, potential effects on the environment and occurrence in the environmental compartments.

Due to the increasing concern about chemical contamination of soil there is an increasing interest in the scientific community and international agencies for soil pollution monitoring and assessment. The traditional approach to soil pollution assessment, based on the analysis of the concentrations of pollutants in the soil and comparison with specific threshold values, does not provide indication of deleterious effects of contaminants on the biota. It neglects several essential aspects such as toxicity of chemicals not included in the selection of contaminants to be analyzed, interactive effects (synergism and antagonisms) of pollutants on biota and bioavailability. Bioavailability refers to the fraction of a contaminant that is taken up by an organism from the environmental media (*i.e.*, through both passive and active routes), and directly influences toxicity (Smith *et al.*, 2010). Bioavailability of pollutants in soil to terrestrial invertebrates and plants can be influenced by some characteristics of the soil such as pH, cation exchange capacity, and organic matter content (Bradham *et al.*, 2006; Spurgeon *et al.*, 2006; Criel *et al.*, 2008). The influence of a single factor on pollutant bioavailability is usually site-, chemical-, and soil-specific. For this reason it is difficult to model pollutant bioavailability based on total concentration and soil characteristics alone. The best integrators of these complex effects are the exposed organisms themselves.

For these reasons, new biological approaches to soil monitoring, such as the measurement of biochemical and cellular responses to pollutants (*i.e.* biomarkers) on organisms living in the soil (bioindicators) have become of major importance for the assessment of the quality of this environmental compartment (Kammenga *et al.*, 2000). Soil invertebrates may represents good sentinel organisms of soil chemical pollution because they are in direct with soil pore water or food exposure, in contrast to many vertebrates that are indirectly exposed through the food chain (Kammenga *et al.*, 2000). Among soil invertebrates earthworms are relevant organisms for soil formation and organic matter breakdown in most terrestrial environments. Because of their particular interactions with soil, earthworms are significantly affected by pollution originated on intensive use of biocides in agriculture, industrial activities, and atmospheric deposition. Hence, earthworms as been proved as valuable bioindicators of soil pollution (Lanno *et al.*, 2004).

Earthworms as Indicator Organisms

Earthworms are very important organisms for soil formation and organic matter breakdown in most terrestrial environments and traditionally have been considered to be indicators of land use impact and soil fertility. They significantly contribute to pedogenesis and affect the physical, chemical and microbiological properties of soil (Barlett *et al.*, 2010). Earthworms may increase mineralization and humification of organic matter by food consumption, respiration and gut passage (Lavelle and Spain, 2001). They may also stimulate microbial mass and activity as well as the mobilization of nutrients by increasing the surface area of organic particles and by production of casts (Emmerling and Paulsch, 2001). Moreover, their borrowing activity significantly contributes to increase water infiltration and soil aeration.

In the last several years various species of earthworms have been used as indicator organisms for assessment of the biological impact of soil pollutants (Spurgeon *et al.*, 2003). Earthworm manipulation is relatively simple which facilitates the measurement of different life cycle parameters *e.g.* growth and reproduction, as well as accumulation and excretion of pollutants and biochemical responses. Hence, these organisms have proved to be extremely suitable for soil ecotoxicological research.

Because of their interaction with soil earthworms are significantly affected by pollutants reaching the soil system. The earthworm skin is extremely permeable to water (Wallwork, 1983) and is represents a main root for contaminant uptake (Jajer *et al.*, 2003; Vijver *et al.*, 2005). Earthworms are able to accumulate various organic and inorganic contaminants (Morrison *et al.*, 2000). Laboratory investigations have demonstrated that earthworms bioaccumulate metals such as Cd, Cu, Zn and Pb. Earthworms can tolerate high tissue metal concentrations using variety of mechanism (Peijnenburg, 2002; Andre *et al.*, 2009). Bioaccumulation in earthworms can be expressed as biota to soil accumulation factor (BSAF) (Cortet *et al.*, 1999). BSAF is calculated by the formula; BSAF= metal content in earthworm/total metal content in soil. Since earthworms serves as a major food source for numerous animals such as amphibians, reptiles, birds and mammals, bioaccumulation of chemical contaminant implies the risk of transfer of pollutants to higher tropic levels (Marino *et al.*, 1992).

The importance of earthworms in testing the adverse effect of chemicals on soil organisms has been recognized by several environmental organizations all over the globe and as a result a set of standard test guidelines are now available.

Most studies on earthworm biomarkers have been conducted on *Eisenia* species while other earthworm species remain less investigated. *Eisenia fetida* is the standard testing organism used in terrestrial ecotoxicology due to its rapid life cycle and simple rearing in the laboratory. The ecotoxicological relevance of *Eisenia* species as bioindicators organism in soil monitoring based on biomarker approach has been recently questioned (Sanchez Hernandez, 2006) because they are epigeic species, forming no permanent surface burrows on the soil surface, feeding on decaying organic matter, while in most cases contaminants occur at soil depths where these earthworms are not found. *Eisenia fetida* is a north European litter-dwelling species inhabiting the soil surface, living primarily in sites rich in organic matter (Jansch *et al.*, 2005) such as compost heaps, manure piles, or sewage sludge, and thus unlikely to be present naturally in agricultural soils or contaminated landsides (Spurgeon *et al.*, 2002). The study of anecic species (*i.e. Lumbricus terrestris*), that forms temporary deep burrows and comes to the surface to feed, and endogeic species (*i.e. Aporrectodea caliginosa*), which build complex lateral burrows systems through all layers of the upper soil rarely coming to the surface, could be more ecologically relevant. In fact these species can be exposed to pollutants present not only in the soil surface but also in the soil deeper layer, providing an integrated response to soil pollution.

Biomarkers in Earthworm

Over the years the use of biomarkers in earthworms has become important for the evaluation of effects of contaminants on soil organisms. Markers such as acetylcholinesterase, metallothionein, biotransformation enzymes and antioxidant defences have been widely used (Sanchez-Hernandez, 2006; Novais *et al.*, 2011). Research is in progress in search of novel biomarkers in earthworms to monitor and asses the level of soil contamination, some of the important biomarkers which have been used recently as follows.

Metallothioneins

Metallothioneins (MTs) are low-molecular-weight cystein-rich metal-binding proteins that are involved in homeostasis of essential metals like Cu and Zn and detoxification of non essential metals such as Ag, Cd and Hg (Costello *et al.*, 2004; Amiard *et al.*, 2006). In addition to their function as metal chelators, MTs act as free radical scavengers (Min, 2007). They contain 25 per cent–30 per cent cysteine, but few aromatic or histidine residues. Vertebrate and invertebrate metallothioneins contain two unique metal-thiolate clusters determined by the presence along the sequence of the protein of metal chelating Cys-X-Cys sequences, where X can be any amino acid other than cysteine. The MT protein is dumbbell-shaped, and the polypeptide backbone is wrapped around the metal thiolate core, forming the scaffold for two domains, designated % and %, separated by a short linker region. Induction of MTs by metal exposure has been detected in a wide variety of organisms including earthworms (Stürzenbaum *et al.*, 2001). For example a significant induction of MT proteins was observed in different earthworm species such as *Lumbricus rubellus*, *Eisenia fetida*, *Eisenia andrei* exposed to cadmium (Calisi *et al.*, 2009; Demuynck *et al.*, 2006; Ndayibagira *et al.*, 2007; Brulle *et al.*, 2007), or in *Lumbricus mauritii* exposed to Pb and Zn contaminated soil (Maity *et al.*, 2011) and in *Lumbriucus terrestris* exposed to cadmium, copper and mercury (Calisi *et al.*, 2011a). It is known that earthworms share a high tolerance to heavy metal exposure (Stürzenbaum *et al.*, 1998) also thanks to the fundamental contribution of these metalbinding proteins. In *Lumbricus terrestris* Calisi *et al.* (2011b) found the major concentration of MT in the postclitellar portion of the animal body with respect to the preclitellar part. This result is in agreement with the immunohistochemical localization of MT in the intestine and chloragogenous tissue previously reported by Stürzenbaum *et al.*, in *Lumbricus rubellus* (2001).

Although the amino acid sequences of more than 50 invertebrate MT and MT-like proteins have already been determined, little is known about the biochemical properties of earthworm MTs. So far, only 5 MT genes of earthworms have been cloned from Lumbricus castaneus, *Eisenia fetida*, *Lumbricus rubellus*, and *Lumbricus terrestris* (Gruber *et al.*, 2000; Liang *et al.*, 2009) whose expression is differentially regulated by different heavy metals (Sturzenbaum *et al.*, 1998, 2001). Metallothionein induction is one of the mostly utilized biomarker in earthworms and is applied as early biomarker of exposure to heavy metals in soil monitoring.

Acetyl Cholinesterase

Acetyl cholinesterase (AChE) is a key enzyme in the nervous system, terminating nerve impulses by catalyzing the hydrolysis of neurotransmitter acetylcholine. AChE is the target site of inhibition by organophosphorus and carbamate pesticides. In particular, organophosphorus pesticides inhibit the enzyme activity by covalently phosphorylating the serine residue within the active site group. They irreversibly inhibit AChE, resulting in excessive accumulation of acetylcholine, leading to hyperactivities and consequently impairment of neural and muscle system. Acetylcholinesterase represents the main cholinesterase in earthworms (Rault *et al.*, 2007). Its activity has been identified and biochemically characterized only in a few earthworm species (Caselli *et al.*, 2006). According to Rault *et al.* (2007) and Calisi *et al.* (2011b), the highest concentration of AChE activity was found in the pre-clitellar part of the animal and suggests a main role of this enzyme in functioning of the dorsal brain localized near the prostomium. Rao *et al.* (2003) and Rao and Kavitha (2004) found a time-dependent AChE inhibition in *Eisenia fetida* exposed to chlorpyrifos and azodrin, two organophosphate pesticides, in the standardized paper contact test. Calisi *et al.* (2009) reported a significant inhibition (about 45 per cent) of AChE activity in *Eisenia fetida* after two weeks of exposure to the carbamate methiocarb added into the soil at the maximal concentrations recommended in vineyards (EEC, 2001). Moreover, Calisi *et al.* (2011b) observed that in *Lumbricus terrestris* high percentage inhibition of AChE activity by pesticide exposure was not paralleled by a corresponding high mortality value (higher than) as observed in birds and mammals (Table 21.1), where AChE inhibition higher than 50 per cent of normal is referred to be irreversible and regarded as being in the lethal range (Lionetto *et al.*, 2010).

Table 21.1: AChE Inhibition and Mortality in Earthworm and Higher Vertebrates.

AChE Inhibition (per cent)	Mortality (per cent)	Species	Ref.
70	30	Earthworm (*Lumbricus terrestris*)	Calisi *et al.*, 2011b
30	100	Mammals and birds	Walker, 1998

The lower sensitivity of animal survival to AChE inhibition compared to vertebrates was also recently documented in other earthworm species (Rault *et al.*, 2008) and suggests that the toxic action of pesticide on earthworms can involve also other molecular or cellular target beyond AChE.

Recently, the potential of some metallic ions, such as Hg^{2+}, Cd^{2+}, Cu^{2+} and Pb^{2+}, to depress the activity of AChE of fish and invertebrates, *in vitro* and or *in vivo* conditions has been demonstrated in several studies (Lionetto *et al.*, 2010). On the contrary in *Lumbricus terrestris* (Calisi *et al.*, 2011b) and *Eisenia fetida* (Calisi *et al.*, 2009) AChE activity was unaffected by copper sulphate exposure. AChE inhibition in earthworms is presently regarded as giving early warning of adverse effects of pesticides (Booth and O'Halloran, 2001), and consistently included among the batteries of biomarkers employed for early assessments of pollutant impact on wildlife in terrestrial ecosystems. However, concerning AChE in earthworms only a few pesticides in use have been tested against relatively few earthworm species both in laboratory tests and under field conditions (Rao *et al.*, 2003; Calisi *et al.*, 2009; Scott-Fordsmand and Weeks, 2000; Rao and Kavitha, 2004; Gambi *et al.*, 2007). As pointed out by Scott-Fordsmand and Weeks (2000) and by Sanchez-Hernandez (2006) the potential use of AChE in earthworms as biomarker of pesticide exposure has not been sufficiently explored.

Biotransformation Enzymes

In their habitat earthworms can be exposed to a variety of plant alkaloids, PAHs and pesticides known to be inducers of detoxification responsible enzymes. In eukaryotes, detoxification of organic compounds usually occurs in two phases. Phase I detoxification processes involve the cytochrome P450 enzyme system and results in the introduction of a functional group, such as hydroxyl or sulphonyl, to non-polar compounds. In some cases the metabolites of phase I reactions are more toxic than the parent compound. Phase II detoxification enzymes, such as glutathione S-transferase (GST), attach a large polar, watersoluble moiety to the products of phase I metabolism to promote excretion and elimination of the toxicant.

The presence of cytochrome P450 was demonstrated in Lumbrucus terrestris by Liimaitainen and Hänninen (1982), but only the occurrence of the monooxygenase activity benzoxyresorufin-Odealkylase (BenzROD) but not of other phase I enzymes was proven (Berghout *et al.*, 1991). However, induction of CYP1A in earthworms has demonstrated to be quite difficult to be measured because of interference from endogenous pigments (Liimatainen and Hänninen, 1982) and the identification of non-inducible forms of cytochrome P450 (Milligan *et al.*, 1986). Achazi *et al.* (1998), by utilizing ethoxy-, pentoxy- and benzoxyresorufin as substrates for monooxygenase activity, demonstrated pentoxy-resorufin-Odealkylase (PentROD) and BenzROD activities in *Eisenia fetida* microsomes, but exposure of the animals for up to four weeks to 100 mg fluoranthene or benzo[a]pyrene kg^{-1} soil (dry weight) did not induce significant changes in the activity of these monooxygenases. The same authors demonstrated the presence of etoxy-resorufin-Odealkylase (EROD) and PentROD activities in *Eisenia crypticus* but failed to demonstrate an induction of these activities following xenobiotic exposure. On the other hand short-term exposure to benzo[a]pyrene by feeding reduced the EROD activity significantly by 45 per cent, but did not affect PentROD activity. After long-term (8 weeks) exposure to benzo[a]pyrene in the agar–agar medium EROD activity was not changed but PentROD was decreased to zero (Achazi *et al.*, 1998).

Glutathione transferases (GSTs) form a ubiquitous superfamily of multi-functional dimeric enzymes (w50 kDa) with roles in phase-II detoxification. GSTs neutralise a broad range of xenobiotics and endogenous metabolic by-products via enzymatic glutathione conjugation, glutathione-dependent peroxidase activity or isomerisation reactions (Hayes *et al.*, 2005). Several studies have demonstrated the sensitivity of earthworm GST to metals and pesticide exposure (Aly and Schröder, 2008; Maity *et al.*, 2008; Lukkari *et al.*, 2004; Saint-Denis *et al.*, 2001; Booth *et al.*, 2000). Recently, transcriptome approaches in the earthworm *Lumbricus rubellus* highlight GSTs as responders to several classes of pollutants including inorganic (cadmium, copper), organic (fluoranthene) and agrochemicals (atrazine) (Bundy *et al.*, 2008; Owen *et al.*, 2008). LaCourse *et al.* (2009) demonstrated *Lumbricus rubellus* to possess a range of GSTs related to previously known GSTs from other taxa including nematodes and humans, with evidence of tissue-specific isoforms, activity, location, the ability to detoxify products of cellular toxicity and potential response to pollution. This study combined subproteomics, bioinformatics and biochemical assay to characterise the *Lumbricus rubellus* GST complement as pre-requisite to initialise assessment of the applicability of GST as a biomarker.

Antioxidant Enzymes

The exposure to either organic or inorganic pollutants is known to induce oxidative stress in the cells. A by-products of the metabolism of xenobiotics is the production of free radicals, on the other hand exposure to metals leads to the generation of reactive oxygen species (ROS) such as hydrogen peroxide (H_2O_2), superoxide (O_2-) and hydroxyl (OH) radicals (Dazy *et al.*, 2009). In order to scavenge ROS and avoid oxidative damage on biological macromolecules (lipids, proteins or DNA), cells protect themselves using enzymes and small molecular-weight antioxidants, such as glutathione (Valavanidis *et al.*, 2006). Superoxide dismutase, catalase, and glutathione peroxidase and glutathione reductase are important enzymatic antioxidants in the response to oxidative stress: superoxide dismutase metabolizes the superoxide anion (O_2-) into molecular oxygen and H_2O_2, which is then deactivated by catalase, thus preventing oxidative damage. The glutathione reductase enzyme also plays an important role in cellular protection by reducing glutathione in the oxidized form (GSSG) to GSH (reduced and active form). Several studies (Liu *et al.*, 2010) indicate that exposure to either organic or inorganic pollutants are able to induce a stress response in the antioxidant enzymes, suggesting their potential application as general biomarkers for assessing effects of pollutants in terrestrial ecosystems at early stages and with low concentrations. However, the dose-dependent and time-dependent response of antioxidant enzymes to pollutant exposure is sometimes complex and a better understanding of their behaviour in stress condition is needed for their application in monitoring and assessment programmes. For example Liu *et al.* (2010) demonstrated that exposure to toluene, ethylbenzene and xylene in earthworms (*Eisenia fetida*) induced a bell shaped change in superoxide dismutase and catalase activities with a tendency of inducement firstly and then inhibition with increasing concentrations of the pollutants. Moreover, Wu *et al.* (2011) found superoxide dismutase to be induced during the early period of phenanthrene exposure while with longer exposure times its activity decreased.

Cellular Biomarkers on Coelomocytes

Earthworm coelomic fluid is particularly interesting from a toxicological perspective for the development of novel cellular biomarkers. It can transport pollutants throughout the exposed organism and its cells (coelomocytes) are involved in the internal defence system (Cooper *et al.*, 2002; Reinhart and Dollahan, 2003; Engelmann *et al.*, 2004). The coelomocyte population is comprised of amoebocytes originating from mesenchymal lining of the coelom (Hamed *et al.*, 2002) and eleocytes (chloragocytes) sloughed into the coelomic fluid from the chloragogen tissue surrounding the intestine and blood vessels (Affar *et al.*, 1998). Thank to the important role played by coelomocytes in the animal physiology, any impairment of their functioning can alter the health of the entire organism. Five cell types were observed by Calisi *et al.* (2009) in *Eisenia fetida* celomic fluid, corresponding to the previously described coelomocyte cell types (Valembois *et al.*, 1985): leukocytes type I (basophilic) and II (acidophilic), granulocytes, neutrophils, and eleocytes. The most recent cellular biomarkers standardized on earthworm coelomocytes are summarized in Table 21.2.

Eleocytes are characterized by the presence of granules (chloragosomes), showing a high autofluorescence derived from riboflavin stored in (Cholewa *et al.*, 2006, Plytycz *et al.*, 2007) and from other fluorophores, putatively including lipofuscins (Cygal *et al.*, 2007, Plytycz *et al.*, 2009). Riboflavin storage was detected in all earthworm species studied, either in chloragocytes localised in chloragogen tissue of *Lumbricus* spp. and *Aporrectodea* spp. or in freely floating eleocytes (Plytycz *et al.*, 2006). This suggests that riboflavin plays an important role in immunity of lumbricid worms, as it does in vertebrates (*e.g.* Verdrengh and Tarkowski, 2005). The amount of riboflavin in eleocytes is species-specific (Plytycz *et al.*, 2006) and changes in response to environmental factors, including metal pollution, in a metal- and species-specific manner (*e.g.* Kwadrans *et al.*, 2008, Plytycz *et al.*, 2009). It was proposed as general biomarkers of exposure to environmental pollutants.

Table 21.2. Cellular Biomarkers on Earthworm Coelomocytes

Biomarker	Type	Analytical Technique	Species	Ref.
Eleocyte riboflavin concentration	General biomarker of exposure	Flow cytometry	*Dendrodrilus rubidus*	Plytycz *et al.*, 2007
Lysosomal membrane stability	General biomarker of exposure and effect	Neutral red retention assay	*Lumbricus* spp. *Eisenia* spp. *Aporrectodea caliginosa*	(for review see Sanchez Hernandez, 2006)
Granulocyte morphometric alteration	General biomarker of exposure and effect	Diff Quick® Stain	*Eisenia fetida* *Lumbricus terrestris*	Calisi *et al.*, 2009 Calisi *et al.*, 2011b
Gene expression	Specific biomarkers of exposure	Real-Time PCR	*Eisenia fetida*	Brulle *et al.*, 2010

The most investigated coelomocyte alteration is represented by lysosomal membrane stability used as an indicator of chemical exposure and associated biological effects (Svendsen *et al.*, 1996; Maboeta *et al.*, 2002; Svendsen *et al.*, 2004). Responses of the lysosomal system are generally thought to provide a first answer to pollutant exposure in a wide variety of animals including earthworms, since injurious lysosomal reactions frequently precede cell and tissue pathology (Moore *et al.*, 2006). The neutral red retention assay (NRRA) has been successfully applied for the *in vivo* evaluation of lysosomal membrane stability in earthworm coelomocytes (Svendsen *et al.*, 1996; Weeks and Svendsen, 1996; ScottFordsmand *et al.*, 1998; Svendsen *et al.*, 2004, Gastaldi *et al.*, 2007). The quantification of this biomarker is based on the time at which 50 per cent of the cells, previously incubated with neutral red, show sign of lysosomal leaking (the cytosol becoming red and the cells rounded), asevaluated by microscopic observations. Several studies have been demonstrated that lysosomal membrane destabilization is a useful predictor of adverse effect on lifecycle parameters such as survival, reproduction, and growth (Sanchez-Hetnandez 2006). Recently Calisi *et al.* (2009, 2011b) demonstrated pollutant-induced morphometric alterations in both *Eisenia fetida* and *Lumbricus terrestris* granulocytes with possible applications as sensitive, simple, and quick biomarker for monitoring and assessment applications (Calisi *et al.*, 2009; Calisi *et al.*, 2011b). Granulocyte morphometric alterations were determined by image analysis on Diff-Quick® stained cells (Calisi *et al.*, 2009; Calisi *et al.*, 2011b). The rapid alcohol-fixed Diff-Quick stain is widely utilized in clinical and veterinary applications for immediate interpretation of histological samples. It was successfully applied to earthworm coelomocyte staining (Calisi *et al.*, 2009). Granulocytes appeared as large cells with broad pseudopodial processes; they were filled with numerous acidophilic granules, presumably corresponding to the lysosomal compartment. They are the cell type mainly involved in phagocytosis (Engelmann *et al.*, 2002; Cooper and Roch, 2003). A considerable enlargement of granulocytes was observed in copper sulphate exposed earthworms with respect to control group. The enlargement was quantified by measuring the area of 2D digitalised granulocyte images. The same effect was observed also when the animals were exposed to xenobiotics, such as the pesticide carbamate methiocarb. Either copper sulphate or methiocarb exerted the same effect on granulocyte dimension. In general, cell swelling can result from the impairment of mechanisms regulating intracellular osmolarity, such as alteration in protein catabolism and/or amino acid and ion transport across cell membrane and it is often an indication of cell damage or metabolic alterations. Heavy metals and pesticides are known to interfere with a wide range of metabolic functions and membrane transport mechanisms (Lionetto *et al.*, 1998; Scott Fordsman and Weeks, 2000; Sanchez-Hernandez, 2006). This could result in an increase of intracellular osmolyte content, followed by osmotic influx of water and cellular swelling. Therefore, granulocyte enlargement could be the resulting integrated effect of the impairment of several cellular functions by different classes of toxic chemicals and can be related to manifestations of sublethal injury due to pollutant exposure. Moreover, in either copper sulphate or methiocarb exposed animals the increase in the granulocyte dimension was accompanied by cell rounding with loss of pseudopods. This effect could be ascribed to toxic chemical-induced reduction of the microfilament and microspine number. This result can be assigned at alteration on actin or tubulin cytoskeletal components by either copper or methiocarb. The cytoskeleton has been demonstrated to be an intracellular target of heavy metals and xenobiotic such as pesticides and polycyclic aromatic hydrocarbon (GomezMendikute and Cajaraville, 2003). The cytoskeleton has also been shown to play a role in cell volume regulation (Pedersen *et al.*, 2001). Therefore, a possible pollutant induced alteration of cytoskeletal components could contribute to the observed morphometric alterations of earthworm granulocytes.

A pollutant induced increase in the cell size was previously documented in the granulocytes of Mytilus galloprovincialis (Calisi *et al.*, 2008) following cadmium exposure. In earthworms granulocyte enlargement was similar in the two species investigated, suggesting the potential application of this response in several earthworm species. Due to the important immunological role of granulocytes, which mediate many of the innate immune responses in earthworms (Cooper *et al.*, 2002), the observed adverse effects of pollutants on these cells may increase the susceptibility of animals to diseases and reduce their survival ability. In fact, the immune system is extremely vulnerable to injury by chemical pollutants. Major changes in the immune system can be expressed in considerable morbidity and even mortality of the organisms involved. Therefore, early subtle alterations in some of the components of the immune system can be used as early indicators of altered organism health. Pollutant induced granulocyte enlargement in *Lumbricus terrestris* was consistent with alterations at the organism level, such as mortality and reduced reproduction rate, suggesting a possible link to organism health impairment. Compared to the other biological responses to pollutant, granulocyte enlargement showed high percentage variation, very similar to the values of specific biomarkers (such as MT induction in copper exposure and AChE inhibition in methiocarb exposure). This result pointed out the high sensitivity of the granulocyte enlargement with respect to other general standardized biomarkers, such as lysosomal membrane stability, and indicated its possible applications as a sensitive, simple, and quick general biomarker for monitoring and assessment applications (Calisi *et al.*, 2009; Calisi *et al.*, 2011b) to be included in a multibiomarker strategy. It demonstrates several of the necessary characteristics for successful application as an effective biomarker in monitoring and assessment programs. This includes an evaluation of pollutant-induced stress at the cellular level in an easy, sensitive, and inexpensive way. Moreover, it provides a sensitive generalized response to pollutants that can integrate the combined effect of multiple contaminants present in the soil.

Earthworm coelomocytes have been recently exploited for trascrittomic studies to identify genes whose expression varies during metal exposure (Brulle *et al.*, 2010). Brulle *et al.* (2008) identified and assayed (by Real-Time PCR (RTPCR)) 3 transcripts that were significantly elevated in coelomocytes when *Eisenia fetida* was exposed to a metalliferous field soil from the vicinity of a Pb-smelter. These were Cd-MT, and two hitherto unstudied earthworm immunity biomarkers (lysenin, and a transcript identified as coactosin-like protein, CLP). The lysenin is a haemolytic protein, produced in coelomocytes. CLP is a member of the ADF/cofilin group of actin-binding proteins which support the activity of the 5-lipoxygenase (5-LO), an enzyme of central importance in cellular leukotriene synthesis, which are key mediators of inflammatory disorders in vertebrates.

Genotoxicity Biomarkers

Coelomocyte are also interesting for ecotoxicological research and application being the cells of choice for the assessment of the genotoxic effect of pollutants on earthworms. Many pollutants in soil either metals or POPs can alter both the structure and integrity of DNA. Since DNA damage may result in severe consequences for individuals and species, it is considered as an important indicator to be used in the assessment of earthworm health (Reinecke and Reinecke 2004). However, so far there have been only a few studies which used earthworms for assessing the genotoxicity of field-contaminated soils (Button *et al.*, 2010; Espinosa-Reyes *et al.*, 2010; Klobuèar *et al.*, 2011; Quiao *et al.*, 2007). The single cell gel electrophoresis (or comet assay) and micronucleus test are two most extensively used methods in the detection of genotoxicity of chemicals in the environment. Compared to other assays, they are sensitive, rapid and easy to handle. Comet assay measures DNA damage in single cells, as single- and double-strand breaks, alkali-labile sites, oxidative DNA base damage (Cotelle and Ferard, 1999). The comet assay technique involves embedding cells in agarose gel on microscope slides and lysing with detergent and high salt. Slides are then soaked in an alkaline solution to allow cleavage of DNA at alkali labile sites. During electrophoresis under alkaline conditions, cells with damaged DNA display increased migration of DNA from the nucleus towards the anode. Broken DNA migrates further in the electric field, and the cell then resembles a 'comet' with a brightly fluorescent head and a tail region which increases as damage increases. The degree of migration is related to DNA damage (Lee and Steinert 2003). The Comet assay presents various advantages, because of its sensitivity for detecting low levels of DNA damage in single cells and the relative ease of application (Tice *et al.*, 2000). The Comet assay has been demonstrated to be effective in determining DNA damage levels in the coelomocytes of earthworms exposed to genotoxic compounds, both *in vivo* and *in vitro*, in several studies (Reinecke and Reinecke, 2004; Fourie *et al.*, 2007; Di Marzio 2005; Bonnard *et al.*, 2009). Dose-dependent DNA damage in earthworm coelomocytes has been demonstrated in vivo for chromium (Manerikar *et al.*, 2008), cadmium (Fourie *et al.*, 2007) nickel (Reinecke and Reinecke, 2004; Bigorgne *et al.*, 2010) and arsenic (Button *et al.*, 2010).

Besides comet assay the micronucleus test has emerged as one of the most powerful methods for assessing chromosome damage (both chromosome loss and chromosome breakage) accumulated during lifespan of the cell in vertebrates and invertebrates. A micronucleus is formed during cell division. It may arise from a whole lagging chromosome or an acentric chromosome fragment detaching from a chromosome after breakage which do not integrate in the daughter nuclei. Sforzini *et al.* (2010) provided the first step of validation of this test on earthworm (*Eisenia andrei*) cells.

Haemoglobin Oxidation

Changes in haematology are reported to be early warning signals of the toxic effects of pollutants in vertebrates (Bowerman *et al.*, 2000; Dauwe *et al.*, 2006; Rogival *et al.*, 2006), but they are poorly explored in invertebrates and for comparison in earthworms. Earthworms have a closed circulatory system. The blood contains haemoglobin which is a large extracellular hemoprotein flowing in a closed circulatory system. In spite of the fundamental role of this respiratory pigment in earthworm physiology, little is known about its sensitivity to environmental pollutants. Recently Calisi *et al.* (2011a) demonstrated heavy metal (cadmium, copper, mercury) exposure to significantly induce changes in either Hb concentration or its oxidation state in the earthworm *Lumbricus terrestris*. Exposure to heavy metals (10-5-10-3 M for Cd, 10-4-10-3 M for Hg, and 10-4-10-2 M for Cu) was found to increase blood Hb concentration. The observed effects were seen at concentrations in the order of 65 (for Cu) to 200 (for Hg) mg/l, below the LC50 value for heavy metal exposure previously observed in earthworms (Neuhauser *et al.*, 1985). Further studies are needed to demonstrate if the observed effect is due to a metal induced increased expression of Hb protein and/or to a reduced degradation of the molecule. In addition to changes in the Hb concentration, heavy metals showed a dramatic effect on the oxidation state of the respiratory pigment. A strong dose-dependent increase of blood met haemoglobin (MetHb) percentage was observed following 48 h exposure with the highest Hb oxidation sensitivity to mercury, followed by cadmium and copper. The role of trace metals in the generation of free radical mediated oxidative stress is known. This could account for the Hb oxidation observed in the earthworms during metal exposure. In addition, a direct action of metals on the earthworm haeme group cannot be excluded. In fact, copper is a known direct-acting methemoglobin producing agent in humans directly converting Fe(II) to Fe(III) in a two-stage reaction (Smith and Reed 1993, French *et al.*, 1995). Compared to other biological responses to heavy metals, such as the known metallothionein induction, MetHb increase showed a higher sensitivity. In fact, the lowest concentration able to significantly increase MetHb concentration was 10-8 M for mercury, 10-7 M for cadmium and 10-6 M for copper while 10-5M was the concentration of each metal able to significantly induce metallothionein increase in the same species and in the same exposure conditions. Moreover, it is interesting to observe that MetHb formation was very suitable for routine application in monitoring assessment in terms of measurable biological response. In fact, it showed a very high percentage variation following heavy metal exposure, being about ten fold higher compared to Mt induction. Future studies will be addressed to evaluate if the observed response is specific for heavy metal exposure or represents a biomarker of general health of earthworms in polluted sites. In any case it demonstrated to be a suitable biomarker of exposure/effect to be included in a multibiomarker strategy in earthworm in soil monitoring assessment.

Earthworm Biomarker Relevance for Soil Pollution Monitoring and Assessment

Earthworm biomarkers represent useful tools in soil monitoring and assessment as an early warning of adverse ecological effects (Sanchez-Hernandez, 2006; Rodriguez-Castellanos and Sanchez-Hernandez, 2007). As indicated by Sanchez-Hernandez (2006) four types of approaches can be performed in soil pollution monitoring : 1) biomarker analysis on native earthworm populations; 2) use of transplanted organisms in *in situ* exposure bioassays; 3) exposure of a selected earthworm population to the environmental medium (soil) in laboratory standardized conditions; 4) simulated field studies. The use of natural population offers the advantage of an ecologically more relevant approach to environmental monitoring and assessment. In addition the usefulness of native organisms arises mainly when studying pollutant long-term effects that may be emphasized in organisms from natural populations. In fact it is difficult to extrapolate effects from spiked soils to field soils, when these are already polluted for a long time. The bioavailability of pollutants in comparable soil types polluted in the field or spiked in the lab is different (Smolders *et al.*, 2003). Soil characteristics, *e.g.* pH, organic matter and clay content (Peijnenburg *et al.*, 2002) also play an important role in determining the bioavailability of pollutants. Using native earthworm populations for biomarker analysis integrates the bioavailability of pollutants, exposure pathways and temporal aspect of exposure (Spurgeon *et al.*, 2002; Sanchez-Hernandez, 2006). However, so far only few studies have explored the potentiality of the biomarker approach to native earthworms, if compared with studies on aquatic environments. For example, Laszczyca *et al.* (2004) found spatial and temporal variation of AChE and antioxidant enzymes in three natural earthworm population (*Aporrectodea caliginosa, Lumbricus terrestris* and *Eisenia fetida*) collected from meadow sites along a 32 km long transect from a Zn/Pb ore mine and a smelter metallurgic complex. Lukkari *et al.* (2004b) documented an increase in the response of three biomarkers (metallothionein, cytochrome P4501A, glutathione transferase) along a 4 km long transect from an area contaminated by a steel smelter. The response of the three biomarkers was positively correlated with decreasing distance from the steel smelter. Moreover, Svendensen *et al.* (2003) demonstrated the validity of using NRRA in biological impact assessment along gradients of contamination. The authors collected earthworms (*Lumbricus castaneus*) at the site of a large industrial plastics fire in Thetford, UK along a 200 m transect leading from the factory perimeter fence. NRRA response was positively correlated with decreasing distance from the factory. While metal residues in soil and earthworms were found to be highly elevated close to the factory perimeter and to rapidly drop to background levels within the first 50 m of the transect, the NRRA values were significantly different from the NRRA determined in control animals also in the surrounding forest along the transect. This results shows that NRRA determination represents a more sensitive indicator of pollutant exposure along a contamination gradient with respect to the analytical metal residue determination. Most of the available studies have been carried out on earthworm population inhabiting areas contaminated with heavy metals (Button *et al.*, 2010). However, in some cases the employment of native earthworms to determine soil toxicity (particularly in studies of long term exposure) is complicated by the fact that some populations appear to have developed a resistance to metals in soil (Spurgeon and Hopkin, 2000). In the case of transplanted organisms earthworms can be collected from a population at one location and translocated to the monitoring sites. This approach provides the advantage of ensuring comparable biological samples, reducing the variability of results usually encountered in field sampling programmes. Using caged organisms in biomonitoring studies, as well as in related research, makes it easier to standardize the results and to compare control organisms to animals collected from potentially polluted sites. The application of the biomarker approach to a selected earthworm population exposed to the test soil in laboratory standardized conditions offers a complementary approach to standard toxicity tests (*i.e.*, mortality and reproduction rates) to investigate the effects of contaminant toxicity on living organisms at earlier stages and lower concentrations (Lukkari *et al.*, 2004; Gastaldi *et al.*, 2007; Schreck *et al.*, 2008). Appropriate biomarkers may be applied in standardized bioassays to provide evidence of the cause-effect relationship between soil contaminants and toxic effects in the individuals. Biomarkers can give a contribution in acute bioassays as a measurement of the bioavailable and bioactive fraction of contaminants and in chronic bioassays as sublethal endpoints (Sanchez-Hernandez, 2006). For example, many studies have reported that the lysosomal destabilization linearly correlate with the bioavailable fraction of heavy metals in the soil (Sanchez-Hernandez, 2006). In addition, NRRA was demonstrated to be more sensitive to Cu exposure than reproduction rate (ScottFordsmand *et al.*, 2000). The need to develop biomarkers for earthworms to supplement standard toxicity tests has been widely discussed by Van Gestel and Weeks (2004) and Sanchez-Hernandez (2006). Simulated field studies offer the opportunity to study the effects of pollutants on the biomarker responses in the organisms under the influence of multiple environmental variables (for review see Sanchez-Hernandez, 2006). They can be carried out in microcosm or mesocosm. Soil microcosm experiments are carried out in laboratory scale, under standardized ambient conditions, while mesocosm experiments are carried out in field scale and are functionally closer to the real environmental scenarios. For the potential of the biomarker approaches to be realized in soil monitoring and assessment a crucial aspect need to be clearly addressed. Ideally, the degree to which the magnitude of the biological response relates to the dose of exposure should be known, enabling severity of the exposure to be clearly assessed. In general, the extent to which a molecular or cellular response occurs is generally related to the dose of chemical received. Nevertheless, exposure to low doses may produce no effects because of the presence of a threshold level of effect, variable for each response. In other cases where the threshold level is exceeded, protective mechanisms may mask the effects, such as the induction of metallothionein. In addition contaminated soil typically contains a complex mixture of contaminants that often interact and the organisms are exposed to multiple chemicals and multiple stresses which can confound a simple dose-response curve. It is important, however, to note that although it is useful to understand the toxic responses of contaminants and how they alter with dose and over time it is harder to quantify these responses especially in field conditions in terms of dose-response. It is well understood that no one biomarker has been validated as unique tool of detecting specific pollutant exposure and effects. The biological response of an organisms to pollutant exposure can be various because of the variety of pollutants that may be present in the environment. Thus, a suite of biomarkers is required to be effectively applicable in a biomonitoring programme. By using a suite of biomarkers future attempts should be made to try and develop a quantitative biomarker index that could simplify the complex biological alterations measured by multiple biomarkers into a single, predefined quality class.

Microbes as Indicator Organisms

Microorganisms are widely used as soil quality indicators. Soil contains a large variety of microbial taxa with a wide variety of metabolic activities (Parkinson and Coleman, 1991). Soil microbial biomass compared with that of superior organisms is a more sensitive indicator and is influenced by different ecological factors like plant diversity, soil organic matter content, and moisture and climate changes. Microorganisms play a vital role in nutrient cycling and energy flow (Li and Chen, 2004) and provide information on various soil profile disturbances. Microbial communities respond to environmental stress or ecosystem disturbance (Marinari *et al.*, 2007). It is important to know that microbial indicators have advantages and disadvantages and should be selected based on ease of measurement, reproducibility and sensitivity to variables that controls quality and soil health. In addition, many of the microbial groups are culture independent making essential the use of molecular techniques which complement the traditional culture techniques (Nielsen and Wideing, 2001). Soil microflora is the first biota that undergoes direct and indirect impacts of toxic substances introduced to soil. Due to its fast response to contaminants, ubiquity, size and recycling of elements, soil microflora is suitable to act as a biomarker reflecting the negative impact of pesticides and hence is commonly used in ecotoxicological tests to evaluate the influence of chemicals on soil system (Doelman and Vonk, 1994; Edwards *et al.*, 1996; Doran and Zeiss, 2000). Cycon and Piotrowska-Seget (2007) have worked on insecticide, herbicide and fungicide impacts on nitrogen fixing and heterotrophic bacterial count and fungal count. The results indicated that the pesticides might decrease or increase the microbial counts depending on the pesticide dose and the type of microorganisms. There was a significant increase in the total number of heterotrophic bacteria however the nitrogen fixing bacteria indicated a declean in the population with increasing dose of pesticides. Higher dose of fungicide significantly reduce the fungal counts. Sebiomo *et al.* (2011) have observed the effects of herbicides (atrazine, primeextra, paraquat and gyphosate) on soil microbial population and dehydrogenase activity and reported significant reduction in dehydrogenase activity with respect to the control and a sharp declean in the bacterial count.

Molecular Identification of Marker Microorganisms

Accurate identification of bacterial isolates is essential to identify indicator organisms. Generally microbial identification rely on phenotypic methods, however this method is time consuming and sometimes inaccurate. It is also difficult to identify a species with phenotypic variability. In this situation a molecular approach based on DNA sequence analysis can be used. Two PCR based approaches are widely used for molecular bacterial identification. The first targets group specific genes while the second approach uses PCR amplification of ribosomal genes by broad range primers for detection of bacterial strain. The 16s ribosomal RNA gene has been most suitable for use in phylogenetic studies.

Conclusion

Since the rapid population increase demands significant increase in crop production, the use of pesticides and chemical fertilizers can not be completely avoided during agricultural practices. However, the application of hyper doses of these environmental contaminants could be minimized which needs biomonitoring of soil. The soil below ground macrofauna and microorganisms could be used as useful tools for indication of the level of contamination. Remedial measures can be taken for the decontamination of soil and maintain the soil health for enhanced food production.

References

Achazi, R.K., Flenner, C., Livingstone, D.R., Peters, L.D., Schaub, K., and Scheiwe, E. (1998): Cytochrome P450 and dependent activities in unexposed and PAH-exposed terrestrial annelids. Comparative Biochemistry and Physiology Part C, 121 (1-3): pp. 339–350, ISSN 1095-6433.

Affar, E.B, Dufour, M., Poirier, G.G. and Nadeau, D. (1998): Isolation, purification and partial characterization of chloragocytes from the earthworm species *Lumbricus terrestris*. Molecular and Cellular Biochemestry, 185 (1-2): pp. 123-133, ISSN 0270-7306.

Aly, M.A. and Schröder, P. (2008): Effect of herbicides on glutathione S-transferases in the earthworm, *Eisenia fetida*. Environmental Science and Pollution Research International, 15 (2): pp. 143–149, ISSN 0944-1344.

Amiard, J.C., Amiard-Triquet, C., Barka, S., Pellerin J. and Rainbow, P.S. (2006): Metallothioneins in aquatic invertebrates: their role in metal detoxification and their use as biomarkers. Aquatic Toxicology, 76 (2): pp. 160–202, ISSN 0166-445X.

Andre, J., Charnock, J., Sturzenbaum, S.R., Kille, P., Morgan, A.J. and Hodson, M.E. (2009): Metal speciation in field populations of earthworms with multi-generational exposure to metalliferous soils: cell fractionation and high energy synchrotron analysis. Environmental Science and Technology, 43 (17): pp. 6822–6829, ISSN 0013-936X.

Barlett, M.D., Briones, M.J.I., Neilson, R., Schmidt, O., Spurgeon, D. and Creamer, R.E. (2010): A critical review of current methods in earthworm ecology: from individuals to populations. European Journal of Soil Biology, 46 (2): pp. 67-73, ISSN 1164-5563.

Berghout, A.G.R.V., Wenzel, E., Buld, J. and Netter, K.J. (1991): Isolation, partial purification, and characterisation of the cytochrome P-450-dependent monooxygenase system from the midgut of the earthworm *Lumbricus terrestris*. Comparative Biochemistry and Physiology Part C, 100 (3): pp. 389–396, ISSN 1532-0456.

Bezchlebova, J., Cernohlavkova, J., Ivana Sochova, J.L., Kobeticova, K. and Hofman, J. (2007): Effects of toxaphene on soil organisms. Ecotoxicology and Environmental Safety, 68 (3): pp. 326–334, ISSN 0147-6513.

Bigorgne, E., Cossu-Leguille, C., Bonnard, M. and Nahmani, J. (2010): Genotoxic effects of nickel, trivalent and hexavalent chromium on the *Eisenia fetida* earthworm. Chemosphere, 80 (9): pp. 1109-1112, ISSN 0045-6535.

Bonnard, M., Eom, I.C., Morel, J.L. and Vasseur, P. (2009): Genotoxic and reproductive effects of an industrially contaminated soil on the earthworm *Eisenia Fetida*. Environmental and Molecular Mutagenesis, 50 (1): pp. 60–67, ISSN 0893-6692.

Booth, L.H., Heppelthwaite, V. and Mc Glinchy, A. (2000): The effect of environmental parameters on growth, cholinesterase activity and glutathione S-transferase activity in the earthworm *Aporectodea caliginosa*. Biomarkers, 5(1): pp. 46–55, ISSN 1354-750X.

Booth, L.H. and O'Halloran, K. (2001): A comparison of biomarker responses in the earthworm *Aporrectodea caliginosa* to the organophosphorus insecticides diazinon and chlorpyrifos. Environmental Toxicology and Chemistry, 20 (11): pp. 2494–2502, ISSN 0730- 7268.

Bowerman, W.W., Stickle, J.E., Sikarskie, J.G. and Giesy, J.P. (2000): Hematology and serum chemistries of nestling bald eagles (*Haliaeetus leucocephalus*) in the lower peninsula of MI, USA. Chemosphere, 41 (10): pp. 1575–1579, ISSN 0045- 6535.

Bradham, K.D., Dayton, E.A., Basta, N.T., Schroder, J., Payton, M. and Lanno, R.P. (2006): Effect of soil properties on lead bioavailability and toxicity to earthworms. Environmental Toxicology and Chemistry, 25 (3): pp769–775, ISSN 0730-7268.

Brulle, F., Mitta, G., Leroux, R., Lemière, S., Leprêtre, A. and Vandenbulcke, F. (2007): The strong induction of metallothionein gene following cadmium exposure transiently affects the expression of many genes in *Eisenia fetida*: a trade-off mechanism?. Comparative Biochemistry and Physiology Part C, 144, (4): pp. 334–341, ISSN 1532- 0456.

Brulle, F., Cocquerelle, C., Mitta, G., Castric, V., Douay, F., Leprêtre, A. and Vandenbulcke, F. (2008): Identification and expression profile of gene transcripts differentially expressed during metallic exposure in *Eisenia fetida* coelomocytes. Developmental and Comparative Immunology, 32 (12): pp. 1441-1453, ISSN 0145-305X.

Brulle F., Morgan A.J., Cocquerelle C. and Vandelbulcke F. (2010): Transcriptomic underpinning of toxicant-mediated physiological function alterations in three terrestrial invertebrate taxa: A review. Environmental Pollution, 158(9): pp. 2793-2808, ISSN 0269-7491.

Brusseau, M.L. (1997): Transport and fate of toxicants in soils. In: Soil Ecotoxicology, J., Tarradellas, G., Bitton, and D., Rossel, (Eds.), pp. 33–53, Lewis Publishers, ISBN 978- 1566701341, New York, USA.

Bundy, J.G., Sidhu, J.K., Rana, F., Spurgeon, D.J., Svendsen, C., Wren, J.F., Stürzenbaum, S.R., Morgan, A.J. and Kille, P. (2008): 'Systems toxicology' approach identifies coordinated metabolic responses to copper in a terrestrial non-model invertebrate, the earthworm *Lumbricus rubellus*. BMC Biology, 6(6): pp. 1-25, ISSN 1741-7007.

Button, M., Jenkin, J.R.T., Bowman, K.J., Harrington, C.F., Brewer, T.S., Jones, D.D. and Watts M.J. (2010): DNA damage in earthworms from highly contaminated soils: Assessing resistance to arsenic toxicity by use of the Comet assay. Mutation Research, 696 (2): pp. 95–100, ISSN 1383-5718.

Calisi, A., Lionetto, M.G., Caricato, R., Giordano, M.E. and Schettino, T. (2008): Morphometrical alterations in Mytilus galloprovincialis granulocytes: a new potential biomarker. Environmental Toxicology and Chemistry, 27 (6): pp. 1435-1441, ISSN 0730-7268.

Calisi, A., Lionetto, M.G. and Schettino, T. (2009): Pollutant-induced alterations of granulocyte morphology in the earthworm *Eisenia fetida*. Ecotoxicology and Environmental Safety, 72 (5): pp. 1369-1377, ISSN 0147-6513.

Calisi, A., Lionetto, M.G., Sanchez-Hernandez, J.C. and Schettino, T. (2011a): Effect of heavy metal exposure on blood haemoglobin concentration and methemoglobin percentage in Lumbricus terrestris. Ecotoxicology, 20 (4): pp. 847-854, ISSN 0963-9292.

Calisi, A., Lionetto, M.G. and Schettino, T. (2011b): Biomarker response in the earthworm *Lumbricus terrestris* exposed to chemical pollutants. Science of the Total Environment, in press, Available from: <http://www.sciencedirect.com/science/article/pii/ S0048969711007145>, ISSN 0048-9697.

Capri, E. and Trevisan, M. (2002): I metalli pesanti di origine agricola nei suoli e nelle acque sotterranee, Pitagora Editrice, ISBN 9788837112622, Bologna, Italy Caselli,

F.; Gastaldi, L.; Gambi, N. and Fabbri, E. (2006). In vitro characterization of cholinesterases in the earthworm *Eisenia andrei*. Comparative Biochemistry and Physiology Part C, 143 (4): pp. 416-421, ISSN 1532-0456.

Cholewa, J., Feeney, G.P., O'Reilly, M., Sturzenbaum, S.R., Morgan, A.J. and Plytycz, B. (2006): Autofluorescence in eleocytes of some earthworm species. Folia Histochemical et Cytobiolica, Vol. 44, No. 1, (January 2006), pp. 65–71, ISSN 0239-8508 Cooper, E.L., Kauschke, E. and Cossarizza, A. (2002) Digging for innate immunity since Darwin and Metchnikoff. BioEssays, 24 (4): pp. 319–333, ISSN0265- 9247.

Cooper, E.L. and Roch, P. (2003): Earthworm immunity: a model of immune competence. Pedobiologia. 47 (5-6): pp. 1–13, ISSN 0031-4056.

Cortet, J., Gomot-De Vauflery, A., Poinsot-Balaguer, N., Gomot, L., Texier, C. and Cluzeau D. (1999): The use of invertebrate soil fauna in monitoring pollutant effects. European Journal of Soil Biology, 35 (3): pp. 115-134, ISSN1164-5563.

Costello, L.C., Guan, Z., Franklin, R.B. and Feng, P. (2004): Metallothionein can function as a chaperone for zinc uptake transport into prostate and liver mitochondria. Journal of Biological Inorganic Chemistry, 98 (4): pp. 664–666, ISSN 0949 - 8257.

Cotelle, S. and Ferard, J.F. (1999): Comet assay in genetic ecotoxicology: a review. Environmental and Molecular Mutagenesis, 34 (4): pp. 246–255, ISSN 0893-6692.

Criel, P., Lock, K., Van Eeckhout, H., Oorts, K., Smolders, E. and Janssen, C.R. (2008): Influence of soil properties on copper toxicity for two soil invertebrates. Environmental Toxicology and Chemistry, 27 (8): pp. 1748–1755, ISSN 0730-7268.

Cyco M., Kaczy ska A. and PiotrowskaSeget Z. (2005): Soil enzyme activities as indicator of soil pollution by pesticides – Pesticides, 1–2: 35–45.

Cygal, M., Lis, U., Kruk, J. and Plytycz, B. (2007): Coelomocytes and fluorophores of the earthworm *Dendrobaena veneta* raised at different ambient temperatures. Acta Biologica Cracovensia Series Zoologica, 49, pp. 5–11, ISSN 0001-530X.

Dauwe, T., Janssens, E. and Eens, M. (2006): Effects of heavy metal exposure on the condition and health of adult great tits (Parus major). Environmetal Pollution, 140 (1): pp. 71–78, ISSN 0269-7491.

Demuynck, S., Grumiaux, F., Mottier,V., Schikorski, D., Lemière, S. and Leprêtre, A. (2006): Metallothionein response following cadmium exposure in the oligochaete *Eisenia fetida*. Comparative Biochemistry and Physiology Part C, 144 (1): pp. 34–46, ISSN 1532-0456.

Di Marzio, W.D., Saenz, M.E., Lemière, S. and Vasseur, P. (2005): Improved single-cell gel electrophoresis assay for detecting DNA damage in *Eisenia fetida*. Environmental and Molecular Mutagenesis, 46 (4): pp. 246–252, ISSN 0893-6692.

Doelman P. and Vonk J.W. (1994): Soil microorganisms of global importance to consider ecotoxicology in an economical and ecological way (In: Ecotoxicology of Soil Organisms, Eds: T. La Pint, P.W. Greigh-Smith) – CRC.

Dogheim, S., Mohamed, E.Z., Alla, S.A.G., El-Saied, S., Emel, S.Y., Mohsen, A.M. and Fahmy, S.M. (1996): Monitoring of pesticide residues in human milk, soil, water, and food samples collected from Kafr El-Zayat Governorate. The Journal of AOAC International, 79 (1): pp. 111-116, ISSN 1060- 3271.

Doran J. and Zeiss, M. (2000): Soil Health and sustainability: managing the biotic component of soil quality. Applied Soil Ecology.15:3-11.

Edwards C.A., Subler S., Chen S.-K. and Bogomolov D.M. (1996):Essential criteria for selecting bioindicator species, processes, or systems to assess the environmental impact of chemicals on soil ecosystems (In: New Approaches to the Development of Bioindicator Systems for Soil Pollution, Eds: N.M. van Straalen, D.A. Krivolutskii) – Kluwer Academic Publishers, The Netherlands, pp. 1–18.

EEC (Council Regulation), (2001): Commission Directive 2001/58. Official Journal of the European Union, No. 212, pp. 24–33, ISSN 1725-2555.

Emmerling, C. and Paulsch, D. (2001): Improvement of earthworm (*Lumbricidae*) community and activity in mine soils from open-cast coal mining by the application of different organic waste materials. Pedobiologia, 45 (5): pp. 396–407, ISSN 0031-4056.

Engelmann, P., Pal, J., Berki, T., Cooper, E.L. and Nemeth, P. (2002): Earthworm leukocytes react with different mammalian antigen-specific monoclonal antibodies. Zoology, 105 (3): pp. 257–265, ISSN 0944-2006.

Engelmann, P., Molnar, L., Palinkas, L., Cooper, E.L. and Nemeth, P. (2004): Earthworm eukocyte populations specifically harbor lysosomal enzymes that may respond to bacterial challenge. Cell and Tissue Research, 316 (3): pp. 391–401, ISSN 0302-766X.

Espinosa-Reyes, G., Ilizaliturri, C.A., González-Mille, D.J., Costilla, R., Díaz-Barriga, F., Cuevas, M.D.C., Martínez, M.A. and Mejía-Saavedra, J. (2010): DNA damage in earthworms (*Eisenia* spp.) as an indicator of environmental stress in the industrial zone of Coatzacoalcos, Veracruz, Mexico. Journal of Environmental Science and Health, Part A, 45(1): pp. 49-55, ISSN 1093-4529.

Fourie, F., Reinecke, S.A. and Reinecke, A.J. (2007): The determination of earthworm species sensitivity differences to cadmium genotoxicity using the comet assay. Ecotoxicology and Environmental Safety, 67 (3): pp. 361–368, ISSN 0147-6513.

French, C.L., Yaun, S.S., Baldwin, L.A., Leonard, D.A., Zhao, X.Q. and Calabrese, E.J. (1995): Potency ranking of methemoglobin-forming agents. Journal of Applied Toxicology, 15 (3): pp. 167-174, ISSN 0260-437X.

Gambi, N., Pasteris, A. and Fabbri, E. (2007): Acetylcholinesterase activity in the earthworm *Eisenia andrei* at different conditions of carbaryl exposure. Comparative Biochemistry and Physiology Part C, 145 (4): pp. 678–685, ISSN 1532-0456.

Garrido, T., Fraile, J., Ninerola, J.M., Figueras, M., Ginebreda, A., Olivella, L. and Ginebreda, A. (2000): Survey of groundwater pesticide pollution in rural areas of Catalonia (Spain). International Journal of Environmental and Analytical Chemistry, 78 (1): pp.51- 65, ISSN 0306-7319.

Gastaldi, L., Ranzato, E., Caprì, F., Hankard, P., Pérès, G., Canesi, L., Viarengo, A. and Pons G. (2007): Application of a biomarker battery for the evaluation of the sublethal effects of pollutants in the earthworm *Eisenia andrei*. Comparative Biochemistry and Physiology, Part C, 146 (3): pp. 398–405, ISSN 1532-0456.

Gerhardt A. (ed) (1999): Biomonitoring of polluted water.- Review on Actual Topics, 301 pp. TransTech Publications, Zurich, Switzerland.

Gomez-Mendikute, A. and Cajaraville, M.P. (2003): Comparative effects of cadmium, copper, paraquat and benzo(a)pyrene on the actin cytoskeleton and production of reactive oxygen species (ROS) in mussel haemocytes. Toxicology *in vitro*, 17 (5-6): pp. 539- 546, ISSN 0887-2333.

Gruber, C., Stürzenbaum, S.R., Gehrig, P., Sack, R., Hunziker, P., Berger, B. and Dallinger, R. (2000): Isolation and characterization of a self-sufficient one-domain protein (Cd)- Metallothionein from *Eisenia fetida*. European Journal of Biochemistry, 267 (2): pp. 573–582, ISSN 0014-2956.

Hamed, S.S., Kauschke, E. and Cooper, E.L. (2002): Cytochemical properties of earthworm coelomocytes enriched by Percoll. In: A New Model for Analyzing Antimicrobial Peptides with Biomedical Applications, A., Beschin, M., Bilej, and E.L., Cooper, (Eds.), pp. 29-37, IOS Press, ISBN 1-58603-237-2, Tokyo, Japan.

Hayes, J.D., Flanagan, J.U. and Jowsey, I.R. (2005): Glutathione transferases. Annual Review of Pharmacology and Toxicology, 45, pp. 51–88., ISSN 0362-1642.

Jajer, T., Fleuren, R.H.L.J., Hogendoorn, E.A. and de Korte, G. (2003): Elucidating the Routes of Exposure for Organic Chemicals in the Earthworm, *Eisenia andrei* (Oligochaeta). Environmental Sciences and Technologies, 37 (15): pp. 3399– 3404, ISSN 0013-936X.

Jänsch, S., Amorim, M.J. and Römbke, J. (2005): Identification of the ecological requirements of important terrestrial ecotoxicological test species. Environmental Reviews, 13(2): pp. 51- 83, ISSN 1181-8700.

Kammenga, J.E., Dallinger, R., Donker, M.H., Köhler, H.R., Simonsen, V., Triebskorn, R. and Weeks, J.M. (2000): Biomarkers in terrestrial invertebrates for ecotoxicological soil risk assessment. Review of Environmental Contamination and Toxicology, 164, pp. 93-147, ISSN 0179-5953.

Klobuèar, G.I.V., Koziol, B., Markowicz, M., Kruk, J. and Plytycz, B. (2006).: Riboflavin as a source of autofluorescence in *Eisenia fetida* coelomocytes. Photochemical and Photobiology, 82 (2): pp. 570–573, ISSN 0031-8655.

Kwadrans, A., Litwa, J., Woloszczakiewicz, S., Ksiezarczyk, E., Klimek, M., Duchnowski, M., Kruk, J. and Plytycz, B. (2008): Changes in coelomocytes of the earthworm, *Dendrobaena veneta*, exposed to cadmium, copper, lead or nickel-contaminated soil. Acta Biologica Cracovensia Series Zoologica, 50, pp. 57–62., ISSN 0001-530X.

LaCourse, E.J., Riboflavin Hernandez-Viadel, M., Jefferies, J.R., Svendsen, C., Spurgeon, D.J., Barrett, J., Morgan, A.J., Kille, P. and Brophy, P.M. (2009): Glutathione transferase (GST) as a candidate molecular-based biomarker for soil toxin exposure in the earthworm *Lumbricus rubellus*. Environmental Pollution, 157 (8-9): pp. 2459–2469, ISSN 0269-7491.

Lanno, R., Wells, J., Conder, J., Bradham, K. and Basta, N. (2004): The biovailability of chemicals in soil for earthworms. Ecotoxicology and Environmental Safety. 57 (1): pp. 39-47, ISSN 0147-6513.

Lavelle, P. and Spain, A. (2001): Soil ecology, Kluwer Scientific Publications, ISBN 978-0- 7923- 7123-6, Amsterdam, Netherlands.

Li X, Chen, Z. (2004): Soil microbial biomass C and N along a climatic transect in the Mongolia steppe. Biology and Fertility of Soils. 39:344-51.

Lee, R. F. and Steinert, S. (2003): Use of the single cell gel electrophoresis/comet assay for detecting DNA damage in aquatic (marine and freshwater) animals. Mutation Research, 544 (1): pp. 43–64, ISSN 1383-5718.

Liang, S.H., Jeng, Y.P., Chiu, Y.W., Chen, J.H., Shieh, B.S., Chen, C.Y. and Chen, C.C. (2009): Cloning, expression, and characterization of cadmium-induced metallothionein-2 from the earthworms Metaphire posthuma and Polypheretima elongate. Comparative Biochemistry and Physiology, Part C, 149 (3): pp. 349–357, ISSN 1532- 0456.

Lionetto, M.G., Vilella, S., Trischitta, F., Cappello, M.S. and Schettino, T. (1998): Effects of CdCl2 on electrophysiological parameters in the intestine of teleost fish *Anguilla anguilla*. Aquatic Toxicology, 4 (3): pp. 251–264, ISSN 0166-445X.

Lionetto, M.G., Caricato, R., Calisi, A. and Schettino T. (2010): Acetylcholinesterase inhibition as a relevant biomarker in environmental biomonitoring: new insights and perspectives. In: Ecotoxicology around the globe, J.E. Visser, (Ed.), pp. 87-115, Nova Science Publishers, ISBN 9781617611261, New York, USA.

Liu, Y., Zhou, Q., Xie, X., Lin, D. and Dong, L. (2010): Oxidative stress and DNA damage in the earthworm *Eisenia fetida* induced by toluene, ethylbenzene and xylene. Ecotoxicology, 19 (8): pp.1551–1559, ISSN 0963-9292.

Lukkari, T., Taavitsainen, M., Soimasuo, M., Oikari, A. and Haimi, J. (2004a): Biomarker responses of the earthworm *Aporrectodea tuberculata* to copper and zinc exposure: differences between population with and without earlier metal exposure. Environmental Pollution, 129 (3): pp. 377–386, ISSN 0269-7491.

Lukkari, T., Taavitsainen, M., Väisänen, A. and haimi, J. (2004b): Effects of heavy metals on earthworms along contamination gradients in organic rich soil. Ecotoxicology Environmental Safety, 59 (3): pp. 340-348, ISSN 0147-6513.

Maboeta, M.S., Reinecke, S.A. and Reinecke, A.J. (2002): The relationship between lysosomal biomarker and population responses in a field population of

Microchaetus sp. (Oligochaeta) exposed to the fungicide copper oxychloride. Ecotoxicology and Environmental Safety, 52 (3): pp. 280-287, ISSN 0147-6513.

Maity, S., Roy, S., Chaudhury, S. and Bhattacharya, S. (2008): Antioxidant responses of the earthworm *Lampito mauritii* exposed to Pb and Zn contaminated soil. Environmental Pollution, 151 (1): pp. 1–7, ISSN 0269-7491.

Maity, S., Roy, S., Bhattacharya, S. and Chaudhury, S. (2011): Metallothionein responses in the earthworm *Lampito mauritii* (Kinberg) following lead and zinc exposure: A promising tool for monitoring metal contamination. European Journal of Soil Biology, 47 (1): pp. 69-71, ISSN 1164-5563.

Marino, F., Ligero, A. and Cosin, D.J.D. (1992): Heavy metals and earthworms on the border of a road next to Santiago. Soil Biology and Biochemistry, 24 (12): pp. 1705–1709, ISSN 0038-0717.

Marinari S, Liburdi, K., Masciandaro, G., Ceccanti, B. and Grego, S. (2007): Humification- mineralization pyrolytic índices and carbon fractions of soil under organic and conventional management in central Italy. Soil and Tillage Research. 92:10-7.

Milligan, L., Babish, J. and Neuhauser, F. (1986): Non-inducibility of cytochrome P450 in earthworm *Dendrobaena veneta*. Comparative Biochemistry and Physiology Part C, 85, pp. 85–87, ISSN 1532-0456.

Min K.S. (2007): The physiological significance of metallothionein in oxidative stress. Journal of the Pharmaceutical Society of Japan, 127 (4): pp. 695-702, ISSN 0031-6903.

Moore, M.N., Icarus, A.J. and McVeigh, A. (2006): Environmental prognostics: an integrated model supporting lysosomal stress responses as predictive biomarkers of animal health status. Marine Environmental Research, 61(3): pp. 278–304., ISSN 0141- 1136.

Morgan, A.J., Turner, M.P. and Morgan, J.E. (2002): Morphological plasticity in metal sequestering earthworm chloragocytes: Morphometric electron microscopy provides a biomarker of exposure in field populations. Environmental Toxicology and Chemistry, 21(3): pp. 610-618, ISSN 0730-7268.

Morgan, A.J., Stürzenbaun, S.R., Winters, C., Grime, G.W., Aziz, N.A.A. and Kille, P. (2004): Differential metallothionein expression in earthworm (*Lumbricus rubellus*) tissues. Ecotoxicology and Environmental Safety, 57(1): pp. 11-19, ISSN 0147-6513.

Morgan, J.E. and Morgan, A.J. (1999): The accumulation of metals (Cd, Cu, Pb, Zn and Ca) by two ecologically contrasting earthworm species (*Lumbricus rubellus* and *Aporrectodea caliginosa*): Implications for ecotoxicological testing. Applied Soil Ecology, 13 (1): pp. 9-20, ISSN 0929-1393.

Morrison, D.E., Robertson, B.K. and Alexander, M. (2000): Bioavailability to earthworms of aged DDT, DDE, DDD, and dieldrin in soil. Environmental Sciences and Technologies, 34(4): pp. 709–713, ISSN 0013-936X.

Ndayibagira, A., Sunahara, G.I. and Robidoux, P.Y. (2007): Rapid isocratic HPLC quantification of metallothionein-like proteins as biomarkers for cadmium exposure in the earthworm *Eisenia andrei*. Soil Biology and. Biochemistry, 39(1): pp. 194–201, ISSN 0038-0717.

Neuhauser, E.F., Loehr, R.C., Milligan, D.L. and Malecki, M.R. (1985): Toxicity of metals to the earthworm *Eisenia fetida*. Biology and Fertility of Soils. 1 (3): pp. 149–52., ISSN 0178- 2762.

Nielsen, M. and Winding, A. (2001): Microorganisms as indicators of soil health. Denmark: National Environmental Research Institute.

Novais S.C., Gomes S.I.L., Gravato C., Guilhermino L., De Coen W., Soares A.M.V.M. and Amorim M.J.B. (2011): Reproduction and biochemical responses in *Enchytraeus albidus* (Oligochaeta) to zinc or cadmium exposures. Environmental Pollution, 159 (7): pp. 1836- 1843, ISSN 0269-7491.

OECD (1984): Guidelines for testing of chemicals: earthworm acute toxicity test. No. 207, Paris, France.

OECD. (2004): Guideline for testing of chemicals: earthworms reproduction test. No. 222, Paris, France.

Owen, J., Hedley, B.A., Svendsen, C., Wren, J., Jonker, M.J., Hankard, P.K., Lister, L.J., Stürzenbaum, S.R., Morgan, A.J., Spurgeon, D.J., Blaxter, M.L. and Kille, P. (2008): Transcriptome profiling of developmental and xenobiotic responses in a keystone soil animal, the oligochaete annelid *Lumbricus rubellus*. BMC Genomics, 9: pp. 266, ISSN 1471-2164.

Parkinson, D. and Coleman, D. (1991): Microbial communities, activity and biomass. Agriculture Ecosystems and Environment. 34:3-33.

Peijnenburg, W.J.G.M. (2002): Bioavailability of metals to soil invertebrates. In: Bioavailability of Metals in Terrestrial Ecosystems: Importance of Partitioning for Bioavailability to Invertebrates, Microbes, and Plants, H.E., Allen, (Ed.), pp. 89–112, Society of Environmental Toxicology and Chemistry (SETAC), ISBN 978-1880611463, Pensacola, USA.

Plytycz, B., Homa, J., Koziol, B., Rozanowska, M. and Morgan, A.J. (2006): Riboflavin content in autofluorescent earthworm coelomocytes is species-specific. Folia Histochemica et Cytobiologica, 44 (4): pp. 275–280, ISSN 0239-8508.

Plytycz, B., Klimek, M., Homa, J., Tylko, G. and Kolaczkowska, E. (2007): Flow cytometric measurement of neutral red accumulation in earthworm coelo- mocytes: novel assay for studies on heavy metal exposure. European Journal of Soil Biology, 43: Supp. 1, pp. S116–S120, ISSN 1164-5563.

Plytycz, B., Lis-Molenda, U., Cygal, M., Kielbasa, E., Grebosz, A., Duchnowski, M., Andre, J. and Morgan, A.J. (2009): Riboflavin content of coelomocytes in earthworm (*Dendrodrilus rubidus*) field populations as a molecular biomarker of soil metal pollution. Environmental Pollution, 157 (11): pp. 3042–3050, ISSN 0269- 7491.

Quiao, M., Chen, Y., Wang, C.X., Wang, Z. and Zhu, Y.G. (2007): DNA damage and repair process in earthworm after in-vivo and in vitro exposure to soils irrigated by wastewaters. Environmental Pollution, 148 (1): pp. 141–147, ISSN 0269-7491.

Rao, J.V., Pavan, Y.S. and Madhavendra, S.S. (2003): Toxic effects of chlorpyrifos on morphology and acetylcholinesterase activity in the earthworm, *Eisenia fetida*. Ecotoxicology and Environmental Safety, 54 (3): pp. 296-301, ISSN 0147-6513.

Rao, J.V. and Kavitha, P. (2004): Toxicity of azodrin on the morphology and acetylcholinesterase activity of the earthworm *Eisenia fetida*. Environmental Research, 96 (3): pp. 323–327, ISSN 0013-9351.

Rault, M., Mazzia, C. and Capowiez, Y. (2007): Tissue distribution and characterization of cholinesterase activity in six earthworm species. Comparative Biochemistry and Physiology Part B Biochemistry and Molecular Biology, 147 (2): pp. 340-346, ISSN 1096- 4959.

Rault, M, Collange, B, Mazzia C. and Capowiez Y. (2008): Dynamics of acetylcholinesterase activity recovery in two earthworm species following exposure to ethyl-parathion. Soil Biology and Biochemistry, 40(12): pp. 3086–3091, ISSN 0038-0717.

Reinecke, S.A. and Reinecke, A.J. (2004): The Comet assay as biomarker of heavy metal genotoxicity in earthworms, Archives of Environmental Contamination and Toxicology, 46 (2): pp. 208–215, ISSN 0090-4341.

Reinhart, M. and Dollahan, N. (2003): Responses of coelomocytes from *Lumbricus terrestris* to native and non-native eukaryotic parasites. Pedobiologia, 47(5-6): pp.710–716, ISSN 0031-4056.

Rodriguez-Castellanos, L. and Sanchez-Hernandez, J.C. (2007): Earthworm biomarkers of pesticide contamination: current status and perspectives. Journal of Pesticide Science, 32(4): pp. 360-371, ISSN 1348-589X.

Rogival, D., Scheirs, J., De Coen, W., Verhagen, R. and Blust, R. (2006): Metal blood levels and haematological characteristics inwoodmice (*Apodemus sylvaticus* L.) along a metal pollution gradient. Environmental Toxicology and Chemistry, 25(1): pp. 149–157, ISSN 0730-7268.

Saint-Denis, M., Narbonne, J.F., Arnaud, C. and Ribera, D. (2001): Biochemical responses of the earthworm *Eisenia fetida* andrei exposed to contaminated artificial soil, effects of lead acetate. Soil Biology and Biochemistry, 33(3): pp. 395–404, ISSN 0038-0717.

Sanchez-Hernandez, J.C. (2006): Earthworms biomarkers in ecological risk assessment. Reviews of Environmental Contamination and Toxicology, 188, pp. 85-126, ISSN 0179- 5953.

Schreck, E., Geret, F., Gontier, L.M. and Treilhou, M. (2008): Neurotoxic effect and metabolic responses induced by a mixture of six pesticides on the earthworm *Aporrectodea caliginosa* nocturna. Chemosphere, 71(10): pp. 1832-1839, ISSN0045-6535.

Scott-Fordsmand, J.J. and Weeks, J.M. (1998): Review of selected biomarkers in earthworms. In: Advances in Earthworm Ecotoxicology, M. Holmstrup, J., Bembridge, S., Sheppard, *et al.* (Eds.), pp. 173–198, SETAC Press, ISBN 978-1880611258, Boca Raton, USA.

Scott-Fordsmand, J.J. and Weeks, J.M. (2000): Biomarkers in earthworms. Reviews of Environmental Contamination and Toxicology, 165, pp. 117-159, ISSN 0179-5953.

Scott-Fordsmand, J.J., Weeks, J.M. and Hopkin, S.P. (2000): Importance of contamination history for understanding toxicity of copper to earthworm *Eisenia fetida*(Oligochaeta: anellida), using neutral red retention assay. Environmental Toxicology and Chemistry, 19(7): pp. 1774-1780, ISSN 0730-7268.

Sforzini, S., Saggese, I., Oliveri, L., Viarengo, A. and Bolognesi, C. (2010): Use of the Comet and micronucleus assays for *in vivo* genotoxicity assessment in the coelomocytes of the earthworm *Eisenia andrei*. Comparative Biochemistry and Physiology, Part A, 157 (1): pp S13, ISSN 1095-6433.

Smith, B.A., Greenberg B. and Stephenson G.L. (2010): Comparison of biological and chemical measures of metal bioavailability in field soils: Test of a novel simulated earthworm gut extraction. Chemosphere, (6): pp. 755–766, ISSN0045-6535.

Smith, R.C. and Reed, V.D. (1993): Reversal of copper(II)-induced methemoglobin formation by thiols. Journal of Inorganic Biochemistry, 52(3): pp. 173-182, ISSN 0162-0134.

Smolders, E., McGrath, S.P., Lombi, E., Karman, C.C., Bernhard, R., Cools, R., Van Den Brande, K., Van Os B. and Walrave, N. (2003): Comparison of toxicity of zinc for microbial processes between laboratory-contamined and polluted field soils. Environmental Toxicology and Chemistry, 22(11): pp. 2592-2598, ISSN 0730-7268.

Spurgeon, D.J. and Hopkin, S.P. (2000): The development of genetically inherited resistance to zinc in laboratory-selected generations of the earthworm *Eisenia fetida*. Environmental Pollution, 109(2): pp. 193–201, ISSN 0269-7491.

Spurgeon, D.J., Svendsen, C., Hankard, P.K., Weeks, J.M., Kille, P. and Fishwick, S.K. (2002): Review of Sublethal Ecotoxicological Tests for Measuring Harm in Terrestrial Ecosystems. P5-063/Technical Report, pp 108, Environment Agency, ISBN 1 85705 682 5, Bristol, United Kingdom.

Spurgeon, D.J., Weeks, J.M. and Van Gestel, C.A.M. (2003): A summary of eleven years progress in earthworm ecotoxicology: The 7th international symposium on earthworm ecology, Cardiff, Wales, 2002. Pedobiologia, 47(5-6): pp. 588– 606, ISSN 0031- 4056.

Spurgeon, D.J., Lofts, S., Hankard, P.K., Toal, M., McLellan, D., Fishwick, S. and Svendsen, C. (2006): Effect of pH on metal speciation and resulting metal uptake and toxicity for earthworms. Environmental Toxicology and Chemistry, 25(3): pp.788–796, ISSN 0730- 7268.

Stürzenbaum, S.R., Kille, P. and Morgan, A.J. (1998): The identification, cloning and characterization of earthworm metallothionein. FEBS Letters, 431(3): pp. 437–442, ISSN 0014-5793.

Stürzenbaum, S.R, Winters, C., Galay, M., Morgan, A.J. and Kille, P. (2001): Metal ion trafficking in earthworms. Identification of a cadmium-specific metallothionein. Journal of Biological Chemistry, 276(36): pp. 34013-34018, ISSN 0021-9258.

Svendsen, C., Meharg, A.A., Freestone, P. and Weeks, J.M. (1996): Use of an earthworm lysosomal biomarker for the ecological assessment of pollution from an industrial plastics fire. Applied Soil Ecology, 3(2): pp. 99-107, ISSN 0929- 1393.

Svendsen, C., Meharg, A.A., Freestone, P. and Weeks, J.M. (2003): Use of an earthworm lysosomal biomarker for the ecological assessment of pollution from an industrial plastics fire. Applied Soil Ecology, 3(2): pp.99-107, ISSN 0929-1393.

Svendsen, C., Spurgeon, D.J., Hankard, P.K. and Weeks, J.M. (2004): A review of lysosomal membrane stability measured by neutral red retention: is it a workable earthworm biomarker? Ecotoxicology and Environmental Safety, 57(1): pp. 20–29, ISSN 0147- 6513.

Tice, R.R., Agurell, E., Anderson, D., Burlinson, B., Hartmann, A., Kobayashi, H., Miyamae, Y., Rojas, E., Ryu, J.C. and Sasaki, Y.F. (2000): Single cell gel/comet assay: guidelines for in vitro and in vivo genetic toxicology testing. Environmental and Molecular Mutagenesis, 35(3): pp. 206–221, ISSN 0893-6692.

Valavanidis, A., Vlahogianni, T., Dassenakis, M. and Scoullos, M. (2006): Molecular biomarkers of oxidative stress in aquatic organisms in relation to toxic environmental pollutants. Ecotoxicology and Environmental Safety, 64(2); pp. 178-189, ISSN 0147- 6513.

Valembois, P., Lassegues, M., Roch, P. and Vaillier, J. (1985): Scanning electron microscopic study the involvement of coelomic cells in earthworm antibacterial defence. Cell and Tissue Research, 240(2): pp. 479 484, ISSN 0302-766X.

Van Gestel, C.A.M. and Weeks, J.M. (2004): Future recommendations of the third international workshop on earthworm ecotoxicology, Aarhus, Denmark (August 2001).Ecotoxicology and Environmental Safety, 57(1): pp. 100-105, ISSN 0147-6513.

Verdrengh, M. and Tarkowski, A. (2005): Riboflavin in innate and acquired immune responses. Inflammation Research, 54(9); pp. 390–393, ISSN 1023-3830.

Vermeulen, F., Covaci, A., D'Havé, H., Van den Brink, N., Blust, R., De Coen, W. and Bervoets, L. (2010): Accumulation of background levels of persistent organochlorine and organobromine pollutants through the soil–earthworm–hedgehog food chain. Environment International, 36(7): pp. 721-727, ISSN 0160-4120.

Vijver, M.G., Vink, J.P.M., Miermans, C.J.H. and Van Gestel, C.A.M. (2003): Oral sealing using glue: a new method to distinguish between intestinal and dermal uptake of metals in earthworms. Soil Biology and Biochemistry, 35(1): pp. 125-132, ISSN 0038-0717.

Walker, C.H. (1998): Biomarker strategies to evaluate the environmental effects of chemicals. Environmental Health Perspectives, 106, Supplement 2, pp.613-520, ISSN 0091-6765.

Wallwork, J.A. (1983): Annelids: The First Coelomates. Earthworms Biology. Edward Arnold Publisher, ISBN 0713128844, London, United Kingdom.

Weber, R., Gaus, C., Tysklind, M., Johnston, P., Forter, M., Hollert, H., Heinisch, E., Holoubek, I., Lloyd-Smith, M., Masunaga, S., Moccarelli, P., Santillo, D., Seike, N., Symons, R., Torres, J.P., Verta, M., Varbelow, G., Vijgen, J, Watson, A., Costner, P., Woelz, J., Wycisk, P. and Zennegg, M. (2008): Dioxin- and POP-contaminated sites— contemporary and future relevance and challenger. Environmental and Sciences Pollution Research, 15(5): pp. 363–393, ISSN 0944-1344.

Weeks, J.M. and Svendsen, C. (1996): Neutral red retention by lysosome from earthworm (*Lumbricus rubellus*) coelomocytes: a simple biomarker of exposure to soil copper. Environmental Toxicology and Chemistry, 15(10): pp. 1801–1805, ISSN 0730-7268.

Wu, S., Wu, E., Qiu, L., Zhong, W. and Chen, J. (2011): Effects of phenanthrene on the mortality, growth, and anti-oxidant system of earthworms (*Eisenia fetida*) under laboratory conditions. Chemosphere, 83(4): pp. 429–434, ISSN0045-6535.

Chapter 22

Ascertain Genetic Diversity of *Artemisia annua* Adapted to Harsh Environment of Ladakh using RAPD and ISSR Molecular Markers

Pradeep Kumar Naik[1] and Aditya Kishore Dash[2]

[1]*Associate Professor Department of Biotechnology and Bioinformatics, Jaypee University of Information Technology, Waknaghat, Solan – 173 215, Himachal Pradesh, India*
[2]*Department of Environmental Engineering, Institute of Technical Education and Research (ITER), Siksha 'O' Anusandhan University, Bhubaneswar, Odisha, India*

ABSTRACT

Artemisia annua is an important medicinal plant valued all over the world. Twenty genotypes of *A. annua* adapted to harsh environment of Ladakh were collected from two valleys *viz.* Nubra (3,000 m) and Indus (3,500 m) and their genetic diversity was analyzed using 37 PCR markers (20 RAPDs and 17 ISSRs). RAPD analysis yielded 124 polymorphic fragments (96.9 per cent), with an average of 6.2 polymorphic fragments per primer. ISSR analysis produced 85 bands, of which 78 were polymorphic (86.1 per cent), with an average of 4.58 polymorphic fragments per primer. The primers based on (CT)n produced maximum number of bands (nine) while, (AT)n and many other motifs gave no amplification. The genetic diversity was high among the genotypes (Nei's genetic diversity = 0.336 and Shannon's information index = 0.495) as measured by combination of both RAPD and ISSR markers. The mean coefficient of gene differentiation (Gst) was 0.145, indicating 85.5 per cent of the genetic diversity resided within the genotypes.

RAPD markers were found more efficient with regard to polymorphism detection, as they detected 96.9 per cent in comparison to 86.1 per cent for ISSR markers. Genotypes collected from Ladakh region were clustered distinctly into two groups as per their sampling sites using RAPD, ISSR and in combination of RAPD+ISSR markers based on neighbour joining (NJ) method. A relatively high genetic variation was detected among the 20 genotypes, whereas the variation between the two valleys was less using AMOVA test. It was found that the genetic diversity among genotypes from Nubra valley was narrow than that of Indus valley, suggesting the importance and feasibility of introducing elite genotypes from different origins for *Artemisia* germplasm conservation and breeding programs.

Keywords: Artemisia annua, Ladakh, Genetic diversity, Molecular markers, RAPD, ISSR.

Introduction

Ladakh is a part of Indian Himalaya at an altitude of 3020-3790 m above mean sea level, is characterized by diverse and complex land formations. It has many unconquered peaks of impregnable heights, uncharted glaciers and valleys. It is located at the latitude of 31° 44′ 57″ – 32° 59′ 57″ N and longitude of 76° 46′ 29″– 77° 41′ 34″ E which covers more than 65,000 sq km area and is characterized by low annual precipitation (20-30 mm rainfall/snowfall), temperature ranges between +35° C in summer to –35°C during winter along with low relative humidity (20-40 per cent). These climatic features make this region a typical cold arid dessert. Under these unique geographical position and adverse climatic conditions many plant species were able to establish themselves and majority of such plants were nutritionally as well as medicinally potential and suitable to use as non-conventional vegetables. These plants were identified by local people through over the years of experience perhaps via trial and error method which led to the selection of plants that are having medicinal properties. Due to lack of proper records and over-exploitation of these wild medicinal plants by local people; the natural resources along with related indigenous knowledge are depleting day by day (Ferreira *et al.*, 1995; Lommen *et al.*, 2006). It is worthwhile to note that the amalgamation or maintenance of edible and medicinal wild plants resources could be beneficial to fulfill the total vegetable as well as folklore medicine requirement of the locals as well as the army deployed in these harsh environments of the country mainly due to its high perishable nature.

Like other parts of the Himalayas, these regions are considered treasure of medicinal, aromatic and other important plants including *Artemisia*. Different species of *Artemisia* have been explored from this region of India as mentioned in Table 22.1.

However, only *Artemisia annua* is so far use as the primary source of artemisinin for artemisinin-based combination therapies (ACT) and possibly antibacterial agents or as natural pesticides (Phillipson and Wright, 1991; Klayman, 1993). It thrives well in dry cold region of Ladakh having marginal rocky and sandy soils due to its aggressive fibrous root system. It is well adapted to survive at high altitudes of 2,500-6,000 m above mean sea level (MSL) and the temperature, nutrient and environmental stress that they are subjected to under the cold arid conditions. During our survey to study the distribution pattern of *A. annua* populations in Ladakh, we observed that

Table 22.1: Various Species of *Artemisia* Reported from the Ladakh Region of India.

Sl.No.	Artemisia sp.	Location (Altitude)
1.	A. brevifolia	Changthang (4,620 m), Suru valley (3,850 m)
2.	A. capilaris	Pangong Tso (3,580 m)
3.	A. dracunculus	Suru valley (3,850m),Khaltse (3,350 m)
4.	A. gmelinii	Zanskar valley (5,091 m), Indus valley (3,450 m)
5.	A. minor	Changla (4,710 m)
6.	A. macrocephala	Khaltse (3,350 m)
7.	A. moorcroftiana	North pullu (5,320 m)
8.	A. parviflora	Kargil (2,860 m)
9.	A. salsoloides	Pangong (3,750 m)
10.	A. stricta	Changthang (4,350 m), Suru valley (3,850 m)
11.	A. tournefortiana	Indus valley (3,450 m), Nubra valley (3050 m)
12.	A. wallichiana	Tsottak (4,680 m)
13.	A. annua	Indus valley (3,400 m), Nubra valley (3,050 m)
14.	A. sieversiana	Indus valley (3,400 m), Nubra valley (3,050 m)
15.	A. biennis	Changthang (3,380 m), Suru valley (3,015 m)
16.	A. desertorum	Zanskar valley (3,605 m)
17.	A. laciniata	Changthang valley (3,870 m)
18.	A. persica	Suru valley (2,980 m)
19.	A. scoparia	Changthang valley(3,810 m)
20.	A. stracheyi	Changthang valley (4,560 m)

the frequency of this important medicinal plant have considerably declined due to their unscientific exploitation, natural calamities, road construction, uprooting for fuel, overgrazing and other activities. Therefore, there is a need to conserve genetic diversity of this prized medicinal plant which may become extinct if its reckless exploitation continues. This has generated worldwide interest in studying the genetic diversity of *A. annua* populations, clonal variants, chemotypes, and in the synthesis of pure-line cultivars from the extreme environment of Ladakh. Very preliminary reports are available so far for the genetic characterization of this plant from Ladakh region and thus necessitate detail investigation. It is prerequisite towards effective utilization and protection of plant genetic resources (Weising *et al.*, 1995), identification of molecular markers linked to agronomic traits and to achieve rational conservation. Unlike the morphological and biochemical markers which may be affected by environmental factors and growth practices (Xiao *et al.*, 1996; Ovesna *et al.*, 2002), DNA markers portray genome sequence composition, thus, enabling to detect differences in the genetic information carried by the different individuals. In this study we have used random amplified polymorphic DNA (RAPDs) (Williams *et al.*, 1990) and inter simple sequence repeats (ISSR) (Zietkiewicz *et al.*, 1994) that are independent of environmental factors to determine the genetic relationships among several genotypes of *A. annua* from the trans-Himalayan (Ladakh, India) region.

Materials and Methods

Plant Materials

Twenty genotypes of *A. annua* were collected from two valleys at different altitudes; 3000m (Nubra) and 3500m (Indus) from the cold arid desert of the trans-Himalayas (Ladakh, India) (Figure 22.1). The young leaves were collected from 10 individual plants from each valley and stored in laboratory at –80 °C until further analysis. The interval between samples was 100-200m and the pair wise distance between valleys was 50–250 Km.

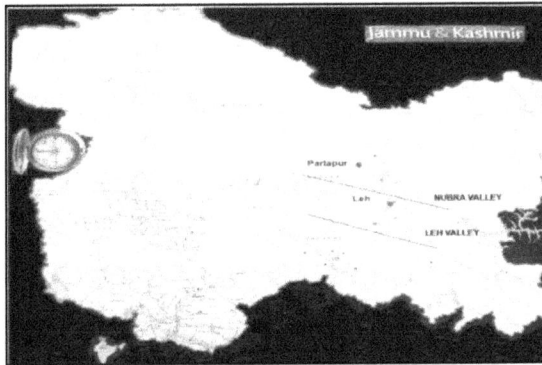

Figure 22.1: Collection Sites of 20 *Artemisia annua* Genotypes from Two Valleys (Leh and Nubra) and the Two Collection Sites (Leh and Partapur) Located in Ladakh (Jammu and Kashmir, India).

DNA Extraction and PCR Amplification

We modified the CTAB protocol (Saghi-maroof *et al.*, 1984) for the extraction of DNA using fresh plant materials: (i) incubation time of buffer and tissue mixture at 65 °C, (ii) buffer to tissue ratio and (iii) extraction with phenol:chloroform:isoamyl alcohol vs. tris saturated phenol followed by chloroform:isoamyl alcohol in extraction and purification phases. All the experiments were repeated 3-4 times to check reproducibility.

Reagent and Solution

Tris saturated phenol, phenol:chloroform:isoamyl alcohol (25:24:1), chloroform:isoamyl alcohol (24:1), 70 per cent and 80 per cent ethanol, 4M NaCl, 3M sodium acetate (pH 5.2) and TE buffer (10 mM Tris-HCl, 1 mM EDTA, pH 8.0). The extraction buffer (pH 8.0) contained 2 per cent CTAB, 100 mM Tris-HCl, 20 mM EDTA, 1.4 M NaCl, 3 per cent PVP and 0.2 per cent β-mercaptoethanol. Solutions and buffers were autoclaved at 121° C at 15 psi pressure. The stock solution 10 mg/ml of RNase A was prepared as per the user's manual (Sigma Inc).

DNA Extraction

Leaves samples were rinsed with distilled water and blotted gently with soft tissue paper. About 0.1 g of this tissue were taken and ground to fine powder using liquid nitrogen, with a precooled mortar and pestle along with 2 per cent of extraction buffer and 10 mg of PVP (Sigma). The powdered tissue was scraped into a 2.0 ml microcentrifuge tube containing preheated (65 °C) extraction buffer (1:5). After that β-mercaptoethanol was added to the final concentration of 0.2 M and mixed well. The mixture was incubated in water bath at 65 °C for 90 min and cooled for 5 min. An equal volume of chloroform:isoamyl alcohol mixture (24:1) was added to the extract and mixed by gentle inversion for 5 to 10 min to form an uniform emulsion. The mixture was centrifuged at 8000 rpm for 8 min at room temperature. Chloroform: isoamyl alcohol extraction was repeated for second time. The aqueous phase was pipetted out gently, avoiding the interface. To the above solution, 5 M NaCl (to final concentration of 2M) and 0.6 volume of isopropanol was added and incubated at room temperature for 1 h. Two volumes of 80 per cent ethanol was added and incubated again for 10 min at room temperature in order to precipitate the DNA. After incubation, the mixture was centrifuged at 10,000 rpm for 15 min. The white/translucent pellet was washed with 70 per cent ethanol, dried and resuspended in 200 µL of TE buffer.

Purification Phase

The extracted DNA was then treated with 20 µL of 10 mg/ml of RNase A and incubated at 37 °C for 60 min. After incubation with RNase, equal volume of phenol: chloroform: isoamyl alcohol (25:24:1) was added and mixed gently by inverting the microcentrifuge tube followed by centrifugation at 10,000 rpm for 5 min at room temperature. The supernatant was pipetted out into a fresh tube. The sample was then extracted with equal volume of chloroform:isoamyl alcohol (24:1), twice. The DNA was precipitated by adding 0.6 volume of isopropanol and 2.0 M NaCl. To the above, 20 µL of sodium acetate and 1 volume of 80 per cent ethanol were added, incubated for 30 min and centrifuged at 5,000 rpm for 3 min to pellet the DNA. The pellet was then washed with 70 per cent ethanol twice; air-dried and finally suspended in 40-50 µL of TE buffer. The yield of the extracted DNA and purity was checked by running the sample on 0.8 per cent agarose gel along with standard (non restriction enzyme digested) lamda DNA marker (Biogene, USA).

RAPD Analysis

Twenty random decamer primers from IDT Tech, USA (Table 22.2) were used for RAPD amplification. PCR reactions were performed in volumes of 25 µl containing 10 mM Tris- HCl (pH 9.0), 1.5 mM $MgCl_2$, 50 mM KCl, 200 µM of each dNTPs, 0.4 µM primer, 20 ng template DNA and 0.5 unit of *Taq* polymerase (Sigma-Aldrich, USA). The first cycle consisted of denaturation of template DNA at 94 °C for 4 min, primer annealing at 37 °C for 1 min, and primer extension at 72 °C for 2 min. For the next 40 cycles the period of denaturation was reduced to 1 min at 92 °C, while the primer annealing and primer extension time remained the same as in the first cycle. The last cycle consisted of only primer extension (72 °C) for 5 min.

Table 22.2: List of Primers Used for RAPD Amplification, GC Content, Total Number of Loci, the Level of Polymorphism and Resolving Power.

Primer	Primer Sequence (5' – 3')	GC (per cent)	Tm (°C)	Total Number of Loci	Number of Polymorphic Loci	Percentage of Polymorphic Loci	Total Number of Fragments Amplified	Resolving Power
S21	CAGGCCCTT C	70	36.4	6	6	100	81	4.5
S22	TGCCGAGCT G	70	40.7	7	7	100	93	5.17
S23	AGTCAGCCA C	60	34.3	7	7	100	81	4.5
S24	AATCAGCCA C	50	30.1	4	4	100	46	2.56
S25	AGGGGTCTT G	60	32.6	9	8	88.9	89	4.94
S26	GGTCCCTGA C	70	35.2	7	7	100	64	3.56
S27	GAAACGGGT G	60	33.2	6	6	100	73	4.06
S28	GTGACGTAG G	60	31.1	7	7	100	88	4.89
S29	GGGTAACGC C	70	37.4	5	5	100	72	4.0
S30	GTGATCGCA G	60	33.1	8	7	87.5	100	5.56
S31	CAATCGCCG T	60	36.7	5	5	100	75	4.17
S32	TCGGGCGATA G	60	34.0	6	5	83.3	72	4.0
S33	CAGCACCCA C	70	37.7	4	4	100	55	3.06
S34	TCTGTGCTG G	60	34.3	5	5	100	69	3.83
S35	TTCCGAACC C	60	34.2	5	5	100	65	3.61
S36	AGCCAGCGA A	60	38.3	5	5	100	58	3.22
S37	GACCGCTTG T	60	35.7	7	7	100	83	4.61
S38	AGGTGACCG T	60	36.2	6	6	100	55	3.06
S39	CAAACGTCG G	60	34.2	6	6	100	77	4.28
S40	GTTGCGATC C	60	33.5	9	7	77.8	86	4.78
	Total	–	–	124	119	96.9	1482	–

ISSR Analysis

Seventeen ISSR primers were obtained from Applied Biosciences, India (Table 22.3) and PCR amplification was performed in reaction cocktail similar to RAPD. Initial denaturation for 4 min at 94 °C was followed by next 40 cycles of denaturation at 94 °C for 45 second, 30 second at specific annealing temperature (± 5 °C of Tm), 2 min at 72 °C and a 5 min final extension step at 72 °C. PCR products were stored at 4 °C before analysis.

The amplification for each primer was performed twice independently with same procedure in order to ensure the fidelity of RAPD and ISSR markers. Amplified PCR products were electrophoresed on 1.5 per cent agarose gel at constant voltage (70 V) in 1X TAE for approximately 2 h, visualized by staining with ethidium bromide (0.5 µg ml^{-1}). A total of 2.5 µl loading buffer (6X) was added to each reaction before electrophoresis. After electrophoresis, the gels were documented on a gel documentation system (Alpha Innotech, Alphaimager, USA). Molecular sizes of amplicons were estimated using a 1 Kb DNA ladders ('Bangalore Genei, India).

Data Collection and Analysis

The banding patterns obtained from RAPD and ISSR were scored as present (1) or absent (0), each of which was treated as an independent character. Jaccard's dissimilarity coefficient (J) was calculated, subjected to cluster analysis by bootstrapping and neighbour-joining method using the program DARWIN (version 5.0.158). Statistically unbiased clustering of collected genotypes was performed using STRUCTURE (version 2.3.1) (Evann *et al.*, 2005). POPGENE software was used to calculate Nei's unbiased genetic distance among genotypes. Data for observed number of alleles (Na), effective number of alleles (Ne), Nei's genetic diversity (H), Shannon's information index (I), number of polymorphic loci (NPL) and percentage polymorphic loci (PPL) were also analyzed (Zhao *et al.*, 2006). Within species diversity (Hs) and total genetic diversity (Ht) (Nei, 1978) were calculated within the species and within two major groups (as per their collection site) using POPGENE software. The RAPD and ISSR data were subjected to a hierarchical analysis of molecular variance (ANOVA) (Excoffier *et al.*, 1992), using two hierarchical levels; among valleys and among genotypes. Correlation between both the marker types used in the study was obtained by regression analysis between similarity matrices obtained with two marker types. In this instance, the matrix regression corresponds to two independently derived dendrograms. The resolving power of the RAPD and ISSR primers was calculated according to Prevost and Wilkinson (Prevost and Wilkinson, 1999). The resolving power (Rp) of a primer is: $Rp = \Sigma\ IB$ where IB (band informativeness) takes the value of: $1-[2* (0.5-P)]$, P being the proportion of the 20 genotypes containing the band.

In order to determine the utility of each of the marker systems, diversity index (*DI*), effective multiplex ratio (*EMR*) and marker index (*MI*) were calculated according to Powell *et al.* [1996]. *DI* for genetic markers was calculated from the sum of the squares of allele frequencies: $DI_n = 1 - \Sigma\ pi^2$ (where 'pi' is the allele frequency of the *i*th allele). The arithmetic mean heterozygosity, Di_{av}, was calculated for each marker class: $Di_{av} = \Sigma\ Di_n/_n$, (where 'n' is the number of markers (loci) analyzed). The *DI* for

Table 22.3: List of Primers Used for ISSR Amplification, Sequence, GC Content, Total Number of Loci, the Level of Polymorphism, Size Range of Fragments and Resolving Power, where, (Y= C,T; R= A,G).

Primer	Primer Sequence (5'–3')	GC (per cent)	Tm (°C)	Total Number of Loci	Number of Polymorphic Loci	Percentage of Polymorphic Loci	Total Number of Fragments Amplified	Resolving Power
ISSR 1	(AG)8 T	47	47.0	4	4	100	67	3.72
ISSR 2	(GA)8 T	47	45.4	5	5	100	64	3.56
ISSR 3	(AC)8 T	47	51.4	3	2	66.7	49	2.72
ISSR 4	(TG)8 A	47	51.3	4	2	50	67	3.72
ISSR 5	(AG)8YT	47.2	49.2	6	4	66.7	87	4.83
ISSR 6	(GA)8YT	47.2	47.4	7	5	71.4	102	5.67
ISSR 7	(CT)8 RA	47.2	47.1	5	4	80	57	3.17
ISSR 8	(GT)8 YC	52.7	52.7	5	5	100	57	3.17
ISSR 9	(ACC)6	66.6	60.6	4	4	100	49	2.72
ISSR 10	CCG)6	10	76.8	4	4	100	42	2.33
ISSR 11	(GGC)6	10	77.3	11	8	72.7	97	5.39
ISSR 12	(AT)8 T	0	23.1	4	4	100	39	2.17
ISSR 13	(TA)8 RT	2.7	25.6	5	5	100	60	3.33
ISSR 14	(AT)8 YA	2.7	26.0	5	4	80	67	3.72
ISSR 15	(CT)8 T	47	45.7	4	4	100	51	2.83
ISSR 16	(TC)8 A	47	47.0	4	3	75	67	3.72
ISSR 17	(GT)8 A	47	49.4	5	5	100	65	3.61
Total		–	–	85	72	86.02	1087	–

polymorphic markers is: $(Di_{av})p = \Sigma \, Di_n/n_p$ (where 'n_p' is the number of polymorphic loci and n is the total number of loci). *EMR (E)* is the product of the fraction of polymorphic loci and the number of polymorphic loci for an individual assay. *EMR (E)* = $n_p \, (n_p/n)$. *MI* is defined as the product of the average diversity index for polymorphic bands in any assay and the *EMR* for that assay, $MI = DI_{avp} * E$.

Results and Discussions

The *Artemisia* species from high altitude (Indus and Nubra valleys) contained high amount of polysaccharides, polyphenols, essential oils and other secondary metabolites that interfere with DNA isolation. These secondary metabolites get entangled to nucleic acid during DNA isolation and hence interfere with subsequent isolation procedure. We modified the CTAB protocol of genomic DNA extraction and obtained good quality DNA from the leaf sample. The modification includes addition of 3.5 M NaCl in extraction buffer and 80 per cent ethanol with 2.0 M NaCl (final concentration) during precipitation and further purification with Tris saturated phenol during purification phase. The quality and quantity (200-400 ng) of DNA was improved significantly without contamination of polysaccharides and secondary metabolites. In the present protocol, the use of 3.5 M NaCl in the extraction buffer reduced 90 per cent of polysaccharides contamination (Danshwar and Sher-ullah, 2004) and very little or no jelly like precipitate was found during precipitation of DNA. One of the most significant steps of our protocol was the use of only Tris saturated phenol (pH 8.0), followed by chloroform: isoamyl alcohol extraction. Most of the protocols in the literature used phenol: chloroform: isoamyl alcohol (25:24:1) or chloroform: isoamyl alcohol (24:1) (Dellaporta et al., 1983; Doyle and Doyle, 1990) for removing protein. It was also observed that buffer to tissue ratio and incubation time was important factors for obtaining higher yields of DNA. In case of *A. annua* 5:1 buffer to tissue ratio and 90 min incubation at 65 °C gave best results. Figure 22.2 represents the quality of the DNAs that have been extracted.

RAPD Analysis

Twenty RAPD primers generated reproducible, informative and easily scorable RAPD profiles were preselected. These primers produced multiple band profiles (Figure 22.3) with a number of amplified DNA fragments varying from 4 to 9, with a mean of 6.2 bands per primer. All the amplified fragments varied in size from 200-1000 bp. Out of 124 amplified bands, 119 were found polymorphic (Table 22.2). The observed high proportion of polymorphic loci suggests that there is a high degree of genetic variation in the *Artemisia* genotypes. The resolving power of the 20 RAPD primers ranged from 2.56 for primer S24 to a maximum of 5.56 for primer S30. A dendrogram analysis based on bootstrapping and neighbor joining (NJ) method grouped all the 20 genotypes into two main clusters (with reference to their site of collection) (Figure 22.4a). An unbiased clustering of genotypes based on STRUCTURE program, without prior knowledge about the populations, clustered all the 20 genotypes into two major groups. Under the admixed model, STRUCTURE calculated that the estimate of likelihood of the clustering of data was greatest when K = 2, ΔK reached its maximum value when K = 2 (Figure 22.4b), suggesting that all the populations fell into one of the 2 clusters (with respect to valleys). It was found that

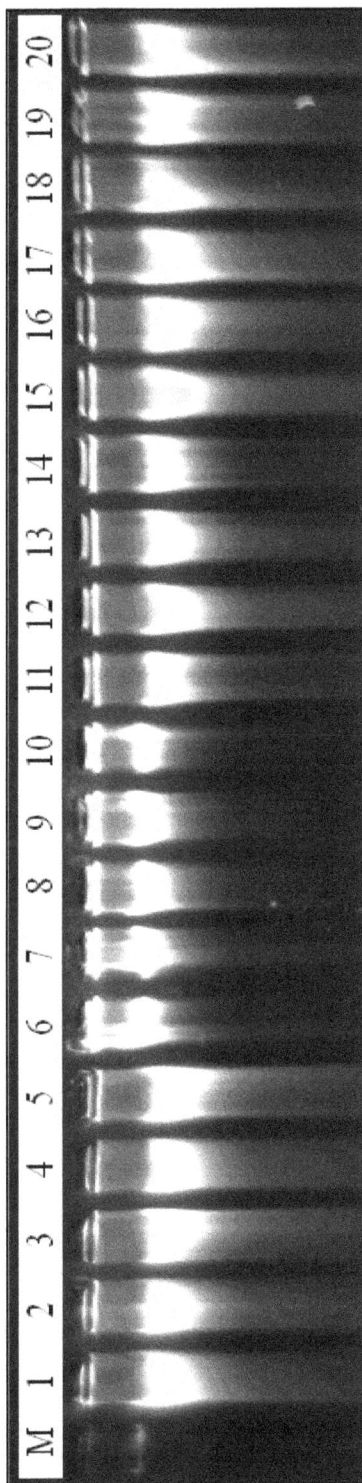

Figure 22.2: Agarose Gel Electrophoresis Showing Purified High Molecular Weight *Artemisia annua* Genomic DNA of different Species, more than 200 ng Genomic DNA from each Genotype was Electrophoresed on 0.8 per cent Agarose Gel at 65 V for 2 hr and Stain with Ethidium Nromide. Lane M, 100 ng molecular weight uncut λ DNA. Lane 1 to 10; are the genomic DNA extracted from Indus valley and Lane 11 to 20; are from Nubra valley.

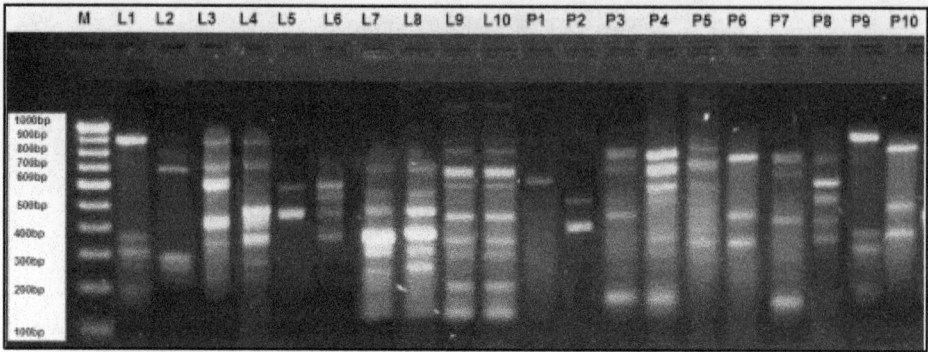

Figure 22.3: RAPD Amplification Products Obtained from the 20 Genotypes of *Artemisia annua* Studied. L1 to L10, are the genotypes collected form Indus valley and P1 to P10, are the genotypes collected from Nubra (Partapur) valley. M = the size of molecular markers in base pairs using λ DNA.

the genotypes were more likely distributed (at high probability) with respect to their geographical distribution albeit small interference (Figure 22.4c).

A relatively high genetic variation was detected among the genotypes. Genetic diversity analysis in terms of Na, Ne, H, I, Ht, Hs, and PPL for both the valleys (Indus

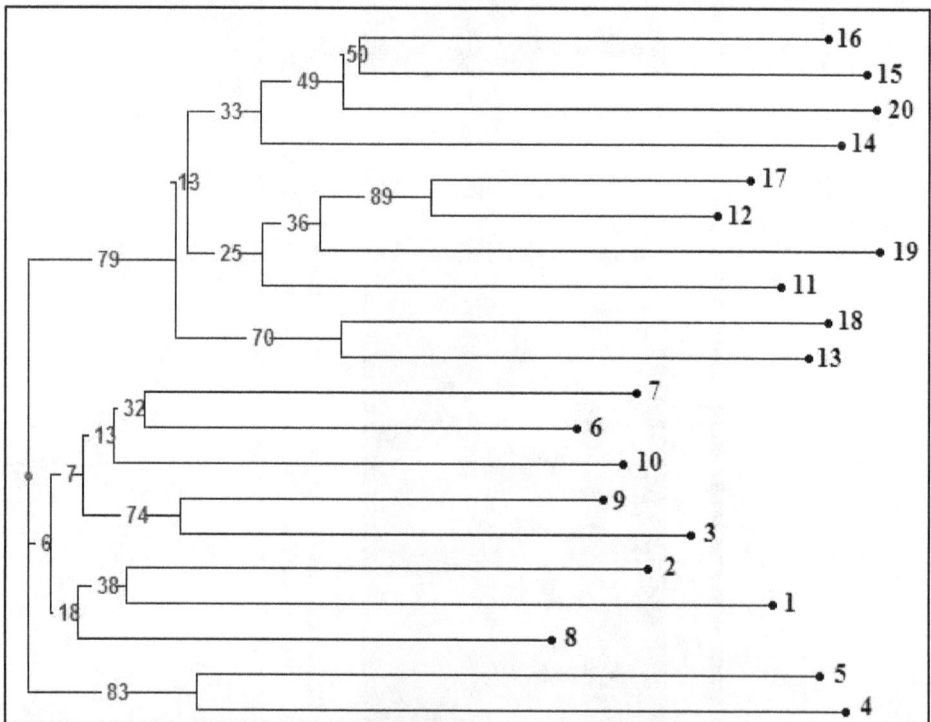

Figure 22.4(a): NJ Tree Representing Clustering of Genotypes at Populations' Level along with Supported Bootstrap Balues Based on RAPD Profiling.

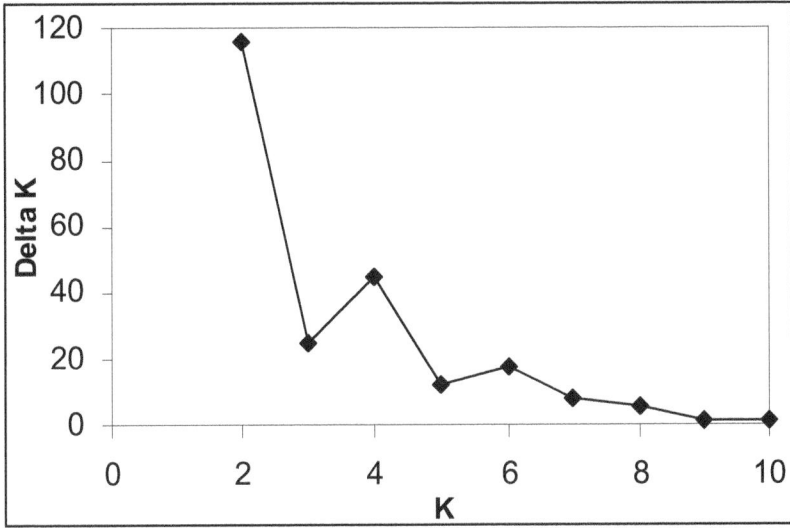

Figure 22.4(b): The Relationship between K and Δc Unbiased Clustering of Genotypes between 2 groups.

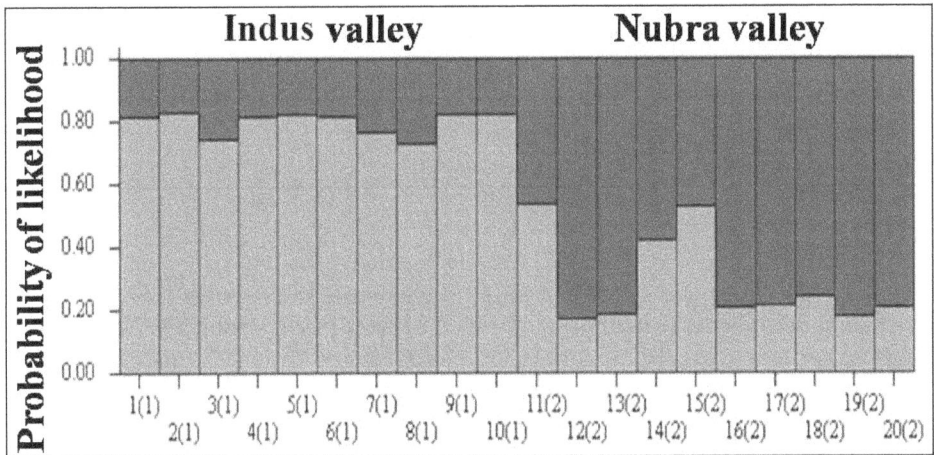

Figure 22.4(c): Statistically Unbiased Clustering of 20 Genotypes as per their Sampling Sites. The genotypes were more likely clustered corresponding to both the valleys. The value within bracket represents the different valley (1, Indus valley and 2, Nubra valley). Genotypes from both the valleys are represented with different colours: Indus valley (Green) and Nubra valley (Red).

and Nubra) revealed higher values for Indus, indicating more variability among the genotypes in comparison to Nubra valley (Table 22.4). Analysis of molecular variance among valley (10 per cent) and among genotypes within valley (90 per cent) (Table 22.5) revealed higher variations within the population. All the components of molecular variations were significant (P < 0.001). This is helpful in making strategy for germplasm collection and evaluation. The rate of gene flow estimated using Gst value was found to be 2.065. The present study and similar studies on ginger (Nayak *et al.*, 2005), *Podophyllum hexandrum* (Alam *et al.*, 2009) and *Andrographic paniculata* (Padmesh *et al.*, 1999) suggested that RAPD is more appropriate for analysis of genetic variability in closely related genotypes. It indicates that *A. annua* populations in the northwestern Himalayan region are genetically highly diverse.

Table 22.4: Summary of Genetic Variation Statistics for all Loci of RAPD among the *Artemisia annua* Genotypes with Respect to their Distributions among Two Valleys.

Valley	Sample size	Na	Ne	H	I	Ht	PPL
Indus	10	1.952	1.662	0.375	0.549	0.375	98.9
		(0.215)	(0.300)	(0.134)	(0.172)	(0.018)	
Nubra	10	1.871	1.620	0.351	0.512	0.351	94.9
		(0.337)	(0.327)		(0.162)	(0.222)	(0.026)
Mean		1.911	1.641	0.363	0.530	0.363	96.9

Na: Observed number of alleles; Ne: Effective number of alleles; H: Nei's gene diversity; I: Shannon's Information index; Ht: Total genetic diversity; PPL: Percentage of Polymorphic Loci.

Table 22.5: Summary of Analysis of Molecular Variance (AMOVA) based on RAPD Analysis. Levels of significance are based on 1000 iteration steps.

Source of Variation	Degree of Freedom	Variance Component	Percentage of Variation	P-value
Among valley	1	2.962	10	< 0.001
Among genotypes	18	24.994	90	< 0.001

ISSR Analysis

The 17 ISSR primers selected in the study generated a total of 85 ISSR bands (an average of 5 bands per primer), out of which 72 were polymorphic (86.02 per cent). Number of bands varied from 3 to 11 with sizes ranged from 200 – 1000 bp (Figure 22.5). Average number of bands and polymorphic bands per primer were 5 and 4.23 respectively, other primer amplification details are shown in Table 22.3. Amplification result of 17 primers seems to indicate that microsatellites more frequent in *Artemisia* contain the repeated di-nucleotides (AG)n, (GA)n, (TG)n, (CT)n, (AT)n, (GT)nYA, and tri-nucleotides (ACC)n, (CCG)n, (GGC)n. The number of bands produced with different repeat nucleotide were more with the (GT)n, (GA)n, (CT)n, and (AC)n primers. The primers that were based on the (GA)n, (CT)n and (GT)n motif produced more polymorphism (on average 7 bands per primer) than the primers based on any other motifs used in the present investigation. We obtained good amplification products from primers based on (CT)n and (GT)n repeats while (AT)n and some other primers gave no amplification, despite the fact that (AT)n di-nucleotide repeats are thought to be the most abundant motifs in plant species (Moreno *et al.*, 1998). Similar results were obtained in grapevine (Martin *et al.*, 2000), rice (Blair *et al.*, 1999), *Vigna* (Ajibade *et al.*, 2000) and wheat (Nagaoka and Ogihara, 1997). A possible explanation of these results is that ISSR primers based on AT motifs are self-annealing, due to sequence

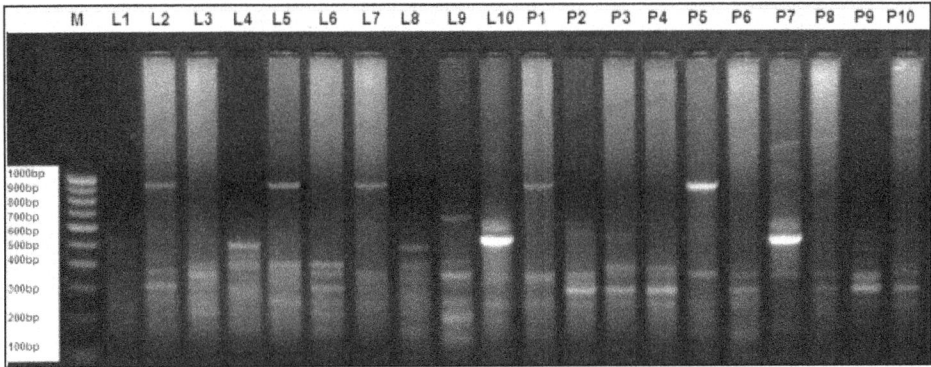

Figure 22.5: ISSR Amplification Products Obtained from the 20 Genotypes of
Artemisia annua **Studied. L1 to L10, are the genotypes collected form Indus valley**
and P1 to P10 are the genotypes collected from Nubra valley. M = the size of
molecular markers in base pairs using 100 bp DNA ladder.

complementarity, and would form dimers during PCR amplification (Blair *et al.,* 1999) or it may be due to its-non annealing with template DNA due to its low Tm. The primers with poly (GC)n and poly (GA)n motifs produced more polymorphism than any other motif. Somewhat similar result was also reported by Ajibade *et al.* (2000), where they found that the primer containing the CT repeats was one of those, which did not give interpretable phenotype when analyzed, while primers with GA and CA repeats revealed polymorphism in the genus *Vigna*. The resolving power (Rp) of the 17 ISSR primers ranged from 2.17 to 5.67 (Table 22.3).

The complete data set of 1087 bands was used for cluster analysis based on bootstrapping and NJ method. The genotypes were clustered into two clusters (with respect to their site of collection) where, cluster I represents all the genotypes from Indus valley while cluster II contains all the genotypes from Nubra valley (Figure 22.6a). The estimated likelihood of the clustering of data using STRUCTURE was found to be optimal when K = 2, ΔK reached its maximum value when K = 2 (Figure 22.6b), suggesting that all the populations fell into one of the 2 clusters (with respect to valleys). It was found that the genotypes were more likely distributed (at high probability) with respect to their geographical distribution albeit small interference (Figure 22.6c).

A relatively high genetic variation was detected among the genotypes. Genetic diversity analysis in terms of Na, Ne, H, I, Ht, Hs, and PPL for both the valleys (Indus and Nubra) revealed higher values for Indus, indicating more variability among the genotypes in comparison to Nubra valley (Table 22.6). AMOVA for among valley (11 per cent) and among genotypes within the valley (89 per cent) indicated that there are more variations within the population (Table 22.7). The estimated gene flow was 2.044.

RAPD and ISSR Combined Data for Cluster Analysis

Based on combined data set of RAPD and ISSR markers, the dendrogram obtained gave similar clustering pattern like RAPD and ISSR (Figure 22.7a). Cluster I represents

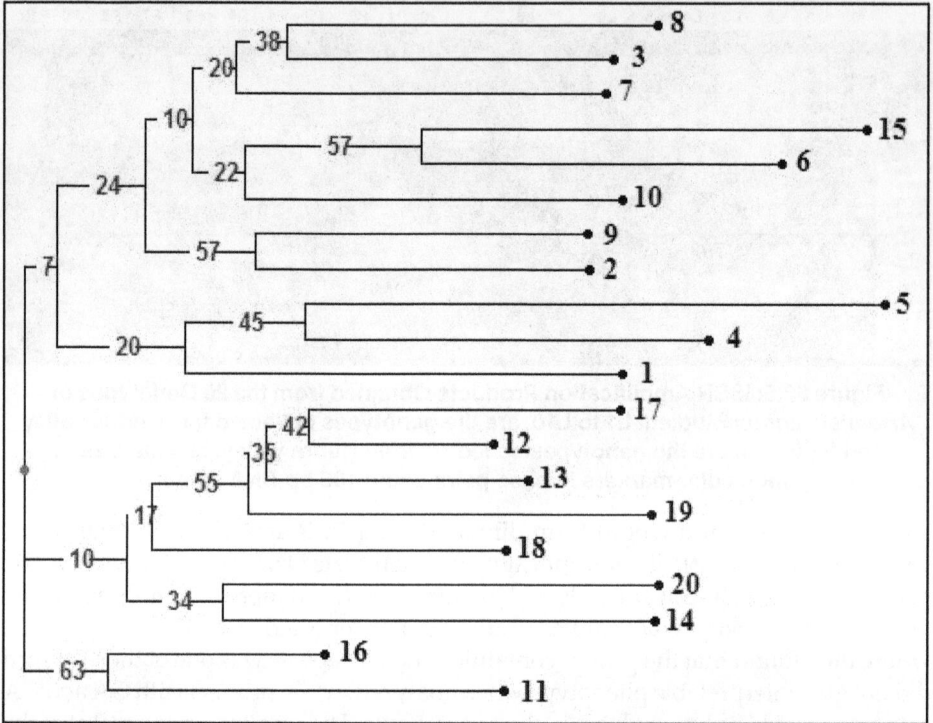

Figure 22.6(a): NJ Tree Representing Clustering of Genotypes along with Supported Bootstrap Values Based on ISSR Profiling.

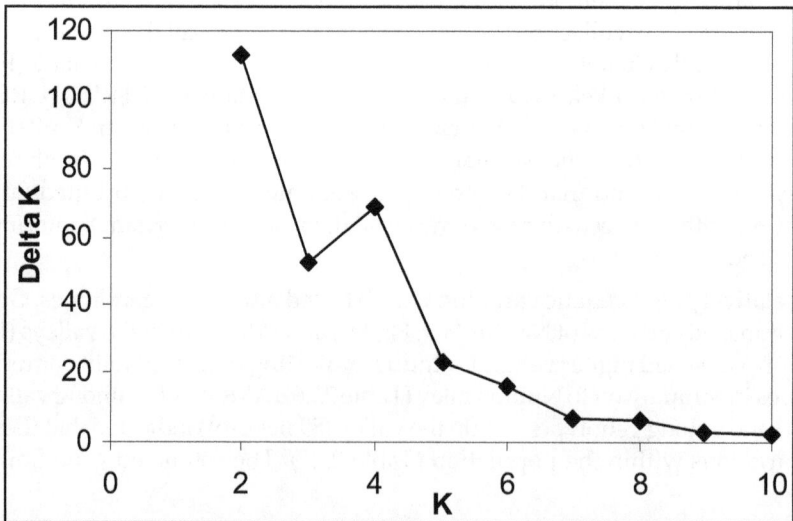

Figure 22.6(b): Statistically Unbiased Clustering of 20 Genotypes as per their Sampling Sites, the Relationship between K and ΔK.

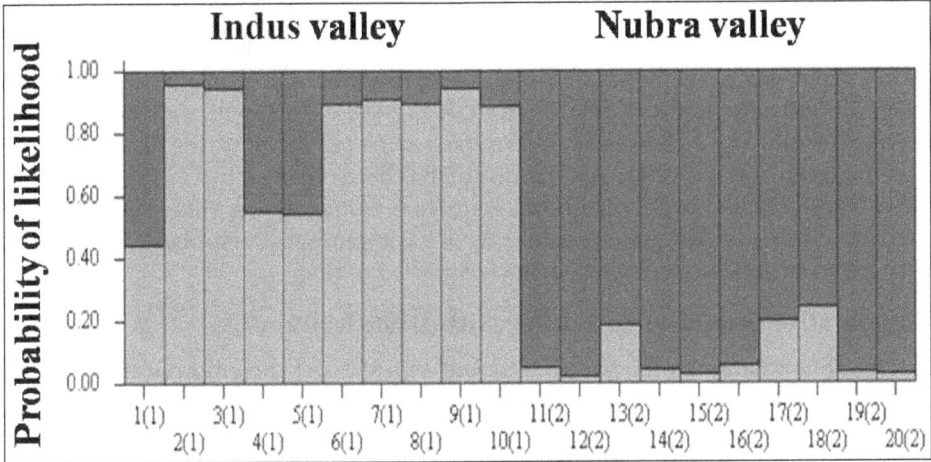

Figure 22.6(c): The Genotypes were more Likely Clustered Corresponding to both the Valleys. The value within bracket represents the different valley (1, Indus valley and 2, Nubra valley). Genotypes from both the valleys are represented with different colours: Indus valley (Green) and Nubra valley (Red).

all the genotypes from Indus valley whereas cluster II represents all the genotypes from Nubra valley. This result is also corroborative with STRUCTURE analysis (Figures 22.7b and 22.7c). Other genetic variation studies were also performed on RAPD and ISSR combined data which are represented in different Tables 22.8 and 22.9. The differences found among the dendrograms generated by RAPDs and ISSRs could be partially explained by the different number of PCR products analyzed (1482

Table 22.6: Summary of Genetic Variation Statistics for all Loci of ISSR among the *Artemisia annua* **Genotypes with Respect to their Distributions among Two Valleys.**

Valley	Sample Size	Na	Ne	H	I	Ht	PPL
Indus	10	1.823 (0.383)	1.608 (0.370)	0.337 (0.184)	0.488 (0.253)	0.337 (0.034)	87.8
Nubra	10	1.812 (0.393)	1.558 (0.358)	0.318 (0.180)	0.467 (0.250)	0.318 (0.032)	84.3
Mean		1.817	1.583	0.327	0.477	0.327	86.1

Table 22.7: Summary of Analysis of Molecular Variance (AMOVA) Based on ISSR Analysis among the Populations of *A. annua.* **Levels of significance are based on 1000 iteration steps.**

Source of Variation	Degree of Freedom	Variance Component	Percentage of Variation	P-value
Among valley	1	1.859	11	< 0.001
Among genotypes	18	15.461	89	< 0.001

P-value: Probability of Null Distribution.

for RAPDs and 1087 for ISSRs) reinforcing again the importance of the number of loci and their coverage of the overall genome, in obtaining reliable estimates of genetic relationships as observed by Loarce *et al.* (1996) in barley. Another explanation could be the low reproducibility of RAPDs (Karp *et al.*, 1997). The genetic closeness among the Indus valley and Nubra valley genotypes can be explained by the high degree of commonness in their genomes. Similar result has been obtained by Gaffor *et al.* (2001) in blackgram. In all the dendrograms, genotypes from both the valleys were found clustered distinctly. The genetic similarity of these genotypes is probably associated with their similarity in the genomic and amplified region.

Comparative Analysis of RAPD with ISSR Markers

RAPD markers were found more efficient with respect to polymorphism detection, as they detected 96.9 per cent polymorphism as compared to 86.02 per cent for ISSR markers. This is in contrast to the results obtained for several other plant species like wheat (Nagaoka and Ogihara, 1997) and *Vigna* (Ajibade *et al.*, 2000). More polymorphism in case of RAPD than ISSR markers might be due to the fact that 17 ISSR primers used in the study only amplified 1087 number of fragments (Table 22.3). While in case of RAPD, all the 20 primers which were used in the investigation

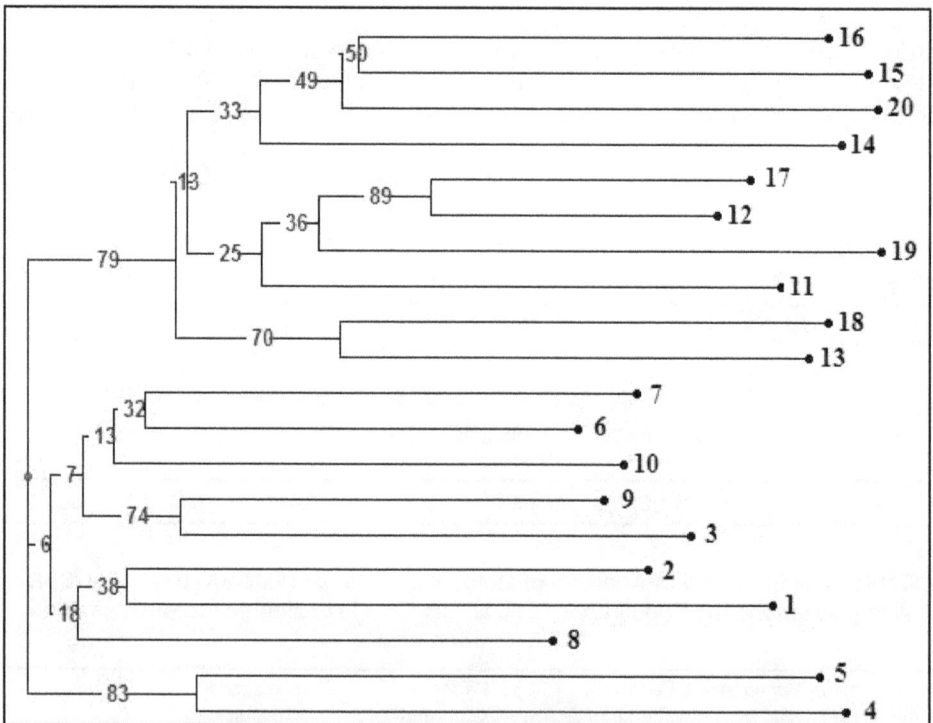

Figure 22.7(a): NJ Tree Representing Clustering of Genotypes at Populations' Level along with Supported Bootstrap Values Based on Combination of RAPD and ISSR Profiling (1482 RAPD bands+1087 ISSR bands).

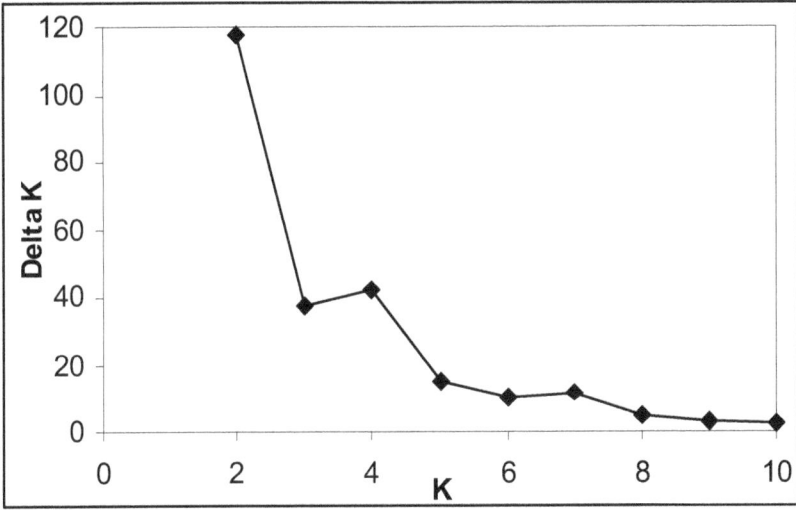

Figure 22.7(b): Statistically Unbiased Clustering of 20 Genotypes as per their Sampling Sites, Showing the Relationship between K and ΔK.

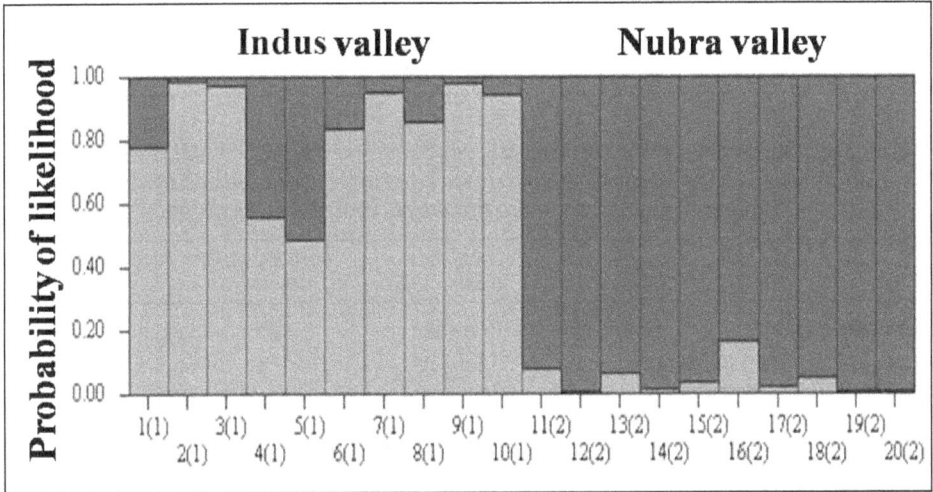

Figure 22.7(c): The Genotypes were more Likely Clustered Corresponding to both the Valleys. The value within bracket represents the different valley (1, Indus valley and 2, Nubra valley). Genotypes from both the valleys are represented with different colours: Indus valley (Green) and Nubra valley (Red).

amplified 1482 number of fragments (Table 22.2). Similar polymorphism pattern was also observed in case of *Podophyllum* (Alam *et al.*, 2009). The regression test between the Jaccard's similarity matrix resulted in low regression between RAPD and ISSR based similarities (R = 0.014), moderate for ISSR and RAPD+ISSR (R= 0.699), while it is maximum for RAPD and RAPD+ISSR based similarities (R = 0.725). This shows that RAPD data is more close to RAPD+ISSR combined data. A possible explanation for the difference in resolution of RAPDs and ISSRs is that the two-marker techniques target different portions of the genome. The diversity index, effective multiplex ratio and marker index are more for RAPD than for ISSR markers (Table 22.10). The respective values for overall genetic variability for Na, Ne, H, I, Ht, Hs, Gene flow (Nm), DI, EMR and MI across all the 20 genotypes were given in Table 22.10. Marker index of ISSR was more (0.876) in comparison to RAPD (0.851), indicating ISSR is the powerful molecular marker for genetic characterization of *A. annua* genotypes. AMOVA for among valley (21 per cent) and among genotypes (79 per cent) indicated that there are more variations within the population.

Table 22.8: Summary of Genetic Variation Statistics for the Combination of RAPD + ISSR Loci among the *Artemisia annua* Populations with Respect to their Distributions among Two Valleys.

Valley	Sample Size	Na	Ne	H	I	Ht	PPL
Indus	10	1.871 (0.347)	1.604 (0.341)	0.341 (0.169)	0.500 (0.230)	0.342 (0.028)	96.1
Nubra	10	1.861 (0.336)	1.577 (0.337)	0.332 (0.164)	0.490 (0.223)	0.332 (0.027)	85.8
Mean		1.866	1.591	0.336	0.495	0.337	91.0

Table 22.9: Summary of Nested Analysis of Molecular Variance (AMOVA) Based on Combination of RAPD and ISSR Analysis among the Populations of *A. annua*. Levels of significance are based on 1000 iteration steps.

Source of Variation	Degree of Freedom	Variance Component	Percentage of Variation	P-value
Among valley	1	6.168	21	< 0.001
Among genotypes/valley	18	23.622	79	< 0.001

P-value: Probability of null distribution.

Conclusion

In summary, the molecular analyses of both RAPD and ISSR markers were extremely useful for studying the genetic diversity of local *Artemisia* genotypes from the trans-Himalayan region of India, Ladakh. The results indicates the presence of high genetic variability, which should be exploited for the future conservation and breeding of *Artemisia* from this region. Since nosingle, or even a few plants, will represent the whole genetic variability in *A. annua*, there appears to be a need to maintain sufficiently large populations in natural habitats to conserve genetic diversity in *A. annua* to avoid genetic erosion.

Table 22.10: Overall Genetic Variability Across all the 20 Genotypes of *Artemisia annua* based Combination of RAPD and ISSR Markers.

Marker Type	Na	Ne	H	I	Ht	Hs	Gst	Nm	DI	EMR	MI
RAPD	1.984 (0.125)	1.730 (0.258)	0.407 (0.106)	0.591 (0.124)	0.407 (0.011)	0.363 (0.012)	0.108	2.065	0.817	6.2	0.851
ISSR	1.906 (0.294)	1.658 (0.325)	0.367 (0.154)	0.535 (0.207)	0.367 (0.234)	0.327 (0.021)	0.109	2.044	0.767	4.588	0.876
RAPD+ISSR	1.962 (0.192)	1.706 (0.283)	0.394 (0.125)	0.572 (0.160)	0.394 (0.016)	0.337 (0.015)	0.145	1.474	–	–	–

Nm: Estimate of gene flow from Gst; Nm: 0.25 (1-Gst)/Gst; DI: Diversity index; EMR: Effective multiplex ratio; MI: Marker Index.

References

Ajibade, S.R., Weeden, N.F. and Chite, S.M. (2000): Inter simple sequence repeat analysis of genetic relationships in the genus Vigna, Euphytica, 111: 47–55.

Alam, A., Gulati, P., Gulati, A., Mishra, G.P. and Naik, P.K. (2009): Assessment of genetic diversity among *Podophyllum hexandrum* genotypes of the North-western Himalayan region for podophyllotoxin production. Indian Journal Biotechnology, 8: 391-399.

Blair, M.W., Panaud, O. and Mccouch, S.R. (1999): Inter-simple sequence repeat (ISSR) amplification for analysis of microsatellite motif frequency and fingerprinting in rice (*Oryza sativa* L.). Theoretical and Applied Genetics, 98: 780–792.

Danshwar, P. and Sher-ullah, S.S.K. (2004): Genomic DNA extraction from Victoria amazonica. Plant Molecular Biology Reporter, 22: 195-205.

Dellaporta, S.L., Wood, J. and Hicks, J.B. (1983): A plant DNA manipreparation: version II. Plant Molecular Biology Reporter, 1: 19-21.

Doyle, J.J. and Doyle, L.J. (1990): Isolation of fresh DNA from fresh tissue. Focus, 12, 13-15.

Evann, G., Regnaut, S. and Goudet, J. (2005): Detecting the number of clusters of individuals using the software STRUCTURE: a simulation study. Molecular Ecology, 14: 2611–2620.

Excoffier, L., Smouse, P.E. and Quattro, J.M. (1992): Analyses of molecular variance inferred from metric distances among DNA haplotypes: application to human mitochondrial DNA restriction data. Genetics, 131: 479-491.

Ferreira, J.F.S., Simon, J.E. and Janick, J. (1995): Relationship of artemisinin content of tissue- cultured, greenhouse-grown, and field-grown plants of *Artemisia annua*. Planta Medica, 61: 351–355.

Gaffor, A., Sharif, A., Ahmad, Z., Zahid, M.A., Rabbani, M.A. (2001): Genetic diversity in blackgram (*Vigna mungo* L. Hepper). Field Crop Research, 69: 183-190.

Karp, A., Edwards, K.J., Bruford, M., Funk, S., Vosman, B., Morgante, M., Seberg, O., Kremer, A., Boursot, P., Arctander, P., Tautz, D. and Hewitt, G.M. (1997): Newer molecular technologies for biodiversity evaluation: Opportunities and challenge. Nature Biotechnology, 15: 625-628.

Klayman, D.L. (1993): *Artemisia annua* I: From weed to respectable antimalarial plant, In: Human medicinal agents from plants (Eds.) A D Kinghorn, M F Balandrin, (Am. Chem. Soc. Symp. Ser. Washington D.C.) 242–255.

Loarce, Y., Gallego, R., and Ferrer, E.A. (1996): Comparative analysis of genetic relationships between rye cultivars using RFLP and RAPD marker. Euphytica, 88: 107-115.

Lommen, W.J., Schenk, E., Bouwmeester, H.J. and Verstappen, F.W. (2006): Trichome dynamics and artemisinin accumulation during development and senescence of *Artemisia annua* leaves Planta Medica, 72: 336.

Martín, J.P. and Sánchez-yélamo, M.D. (2000): Genetic relationships among species of the genus Diplotaxis (Brassicaceae) using inter-simple sequence repeat markers. Theoretical and Applied Genetics, 101: 1234–1241.

Moreno, S., Martin, J.P. and Ortiz, J.M. (1998): Inter-simple sequence repeat PCR for characterization of closely related grapevine germplasm. Euphytica, 101: 117–125.

Nagaoka, T. and Ogihara, Y. (1997): Applicability of inter-simple sequence repeat polymorphisms in wheat for use as DNA markers in comparison to RFLP and RAPD markers. Theoretical and Applied Genetics, 94: 597–602.

Nayak, S., Naik, P.K., Acharya, L.K., Mukherjee, A.K., Panda, P.C. and Das, P. (2005): Assessment of genetic diversity among 16 promising cultivars of ginger using cytological and molecular markers. Science Asia, 60: 485-492.

Nei, M. (1978): Estimation of average heterozygosity and genetic distance from a small number of individual. Genetics, 89: 583-590.

Ovesna, J., Polakova, K., Lisova L. (2002): DNA analysis and their applications in plant breeding. Czech. Journal Plant Breeding, 38: 29-40.

Padmesh, P., Sabu, K.K., Seeni, S. and Pushpaangadan, S. (1999): The use of RAPD in assessing genetic variability in *Andrographis paniculata* Nees, a hepatoprotective drug. Current Science, 76: 833-835.

Phillipson, D.J. and Wright, C.W. (1991): Antiprotozoal agents from plant sources. Plant Medica, 57: 553–59.

Powell, W., Morgante, M., Andre, C., Hanafey, M., Vogel, J., Tingey, S. and Rafalski, A. (1996): The comparison of RFLP, RAPD, AFLP and SSR (microsatellite) markers for germplasm analysis. Molecular Breeding, 2: 225-238.

Prevost, A. and Wilkinson, M.J. (1999): Anew system of comparing PCR primers applied to ISSR fingerprinting of potato cultivars. Theoretical and Applied Genetics, 98: 107-112.

Saghai-maroof, M.A., Soliman, K.M., Jorgenses, R.A. and Allard, R.W. (1984): Ribosomal DNA spacer-length polymorphism in barley. Mendelian inheritance, chromosomal location and population dynamics. Proceeding of National Academy of Science, 81: 8014–19.

Weising, K., Atkinson, G. and Gardner, C. (1995): Genomic fingerprinting by microsatellite-primed PCR: a critical evaluation. PCR Methods and Application, 4: 249-255.

Williams, K., Kubelik, R., Livak, J., Rafalski, A. and Tingey, V. (1990): DNA polymorphisms amplified by arbitrary primers are useful as genetic markers. Nucleis Acid Research, 18: 6531-6535.

Xiao, J., Li, J., Yuan, L., McCouch, S.R. and Tanksley, S.D. (1996): Genetic diversity and its relationships to hybrid performance and heterosis in rice as revealed by PCR-based markers. Theoretical and Applied Genetics, 92: 637-643.

Zhao, W.G., Zhang, J.Q., Wangi, Y.H., Chen, T.T. and Yin, Y. (2006): Analysis of genetic diversity in wild populations of mulberry from western part of Northeast China determined by ISSR markers. Journal of Genetics and Molecular Biology, 7: 196-203.

Zietkiewicz, E., Rafalski, A. and Labuda, D. (1994): Genome fingerprinting by simple sequence repeat (SSR)- anchored polymerase chain reaction amplification. Genomics, 20: 176-183

Chapter 23

Monitoring Morphological Changes in a Coastal Environment: A Case Study along the Coastal Tract of Odisha with Special Reference to the Mahanadi River Mouth using Remote Sensing and GIS

G.K. Panda

Professor, Department of Geography,
Utkal University, Vani Vihar, Bhubaneswar, Odisha, India

ABSTRACT

The coastal environment of Odisha has a wide range of abiotic and biotic features arising out of the dynamic interaction of the fluvial, marine, aeolian and lacustrine processes around the river mouths and coastal tracts where the materials and processes are conspicuous. Along the Odisha coastal zone, the Mahanadi is one of the important rivers with a large catchment area, sediment and water discharge and an extensive delta with a network of distributaries. The Mahanadi river mouth is one of the major outlets of the Mahanadi delta to the sea. Large amount of sediment transport from the river to the sea through the river mouth as well as the inflow of tidal sediments and materials from the longshore drift by the waves has created an ideal depositional environment. But however the trend seems to be changing under the impact of the recent trends in the process of development along the coast along with the climate change and sea level rise. The study makes an in-depth monitoring and analysis of these changes along the coastal zone as well as around the Mahanadi river mouth region using the remote sensing data GIS. The study brings out the morphological changes over space and time in relation to the prevailing environment, shore processes and human interventions.

Keywords: Spit, Hook, Bay Mouth Bar, Barrier Beach, Island, Mud Flat, Lagoon, Mangrove.

Introduction

Seashore represents the zone between high tide and low tide water. This zone is divided in to three sections of which seaward part of the shore is the offshore zone. The intertidal zone along the shore including the beach is the shore zone. The landward part of the shore zone extending up to the backwaters and impact area of the coastal winds and tides is the onshore or backshore zone. It is the shallow zone of the continental self-adjacent to the beach. The landforms found in this region are the unique features resulting out of the dynamic interaction of the coastal agents like wind, waves, tides, littoral currents and riverine processes along the river mouths where rivers are there along the shore. Odisha's 480 km of coastline depicts a good deal of such coastal landforms and biodiversity habitats which are most prominent near the river mouths. Their geomorphologic characteristics often bear evidence about the evolution of the shoreline. These morphological features can be classified in to depositional or erosional forms.

Along the coastal tract, the river mouth areas are the most vulnerable sites for dynamic interaction of the combined effect of the fluvial, lacustrine, marine and aeolian processes. The Mahanadi is one of the major rivers of Odisha and the 6th largest river in the country in terms of its water discharge. It has produced an extensive delta at its mouth through a number of distributaries, which had created a very suitable depositional environment of riverine and marine sediments to produce a wide variety of shore and offshore landforms around its mouth with well developed tidal flats, mud swamps and mangroves.

The chapter, aims at the study of these landforms of the shore and offshore region and their changes using satellite remote sensing data. Use of these data have been found most effective through GIS in monitoring changes in the in the coastal environment and natural resources over a period of time. It can be a basis for identifying the erosional or depositional changes and degradation or deteriorations in the natural resources of the coastal region. This can further lead towards understanding its intervening causes for prediction and may help in modelling and simulating the future trends. The present paper brings out some changes along the Orissa coast using remote sensing data and GIS. The changes are related to the coastal environment with reference to the shore and offshore landforms, coastal and mangrove vegetation, prograding coastline, changing river mouths, coastal spits, hooks, sand bars and islands and nature of coastal erosion and shoreline retro-gradation.

Objective and Hypothesis

The study aims towards understanding of the morphological characteristics of the shore and offshore features along the Odisha coast and their changes over time. The study also aims towards understanding of the natural processes operating along the shore, backshore and offshore zone in the evolution of the coastal landforms. The study also attempts to focus on the changes in the landforms in the recent past comparing between the 1975 topographic maps and recent satellite imageries. This also tends to investigate into the evolution of the shoreline based on the pattern of landforms, their processes and ongoing changes. The study is based on the hypothesis that the geomorphic characteristics of the coastal landforms together with their processes and changes can form the basis for understanding the type of coastline and their nature of evolution. Thus the objectives are:

1. To study the morphological characteristics and changes in the coastal landforms along the Odisha coast with special reference to the Mahanadi River mouth region.

2. Understanding of the geomorphic and other environmental processes associated with the development of the coastal region.

3. Study the nature of the spatio-temporal changes in the morphology of the coastal landforms and their possible environmental implications.

Remote Sensing and GIS as Tools

Remote sensing is a process of acquiring information about the earth surface features and phenomena without physical contact. It is not just a data collection process; rather it includes data analysis for extracting meaningful spatial information for direct input to GIS. The advantages of remote sensing is the synoptic view it provides along with multi spectral and multi scale data with scope for monitoring changes to keep the GIS database up to date. GIS has evolved from a mapping and spatial analytical tool in spatial sciences. The shift in GIS is from the narrow focus of technology pushed application in the past to broad based approach emphasizing the total integration of data, technical and human resources within the framework of Information Technology. The remote sensing and GIS together help in the understanding the image elements like tone, texture, pattern, shadow, shape, height, site and association form the satellite images or aerial photographs. But however, the digital image interpretations are based on various computer-based algorithms that classify the pixels based on the intensity of their spectral values (Figure 23.1).

Rectification and registration followed by resampling are common operations to both GIS and remote sensing when maps or images are compared for their change detections. Because of the rapid advancement in computer hardware and software technology, there has been a great progress towards complete integration of data in GIS from remote sensing. Development of high-speed computer technology and programming has made the data structure conversion form raster to vector almost easier and faster. Thus the complete integration of GIS and remote sensing has become the most power full tool of study and analysis. This requires that both the data sets

User Interface

Image / Cartographic Database Processing

Combined Attribute / Vector / Raster Data

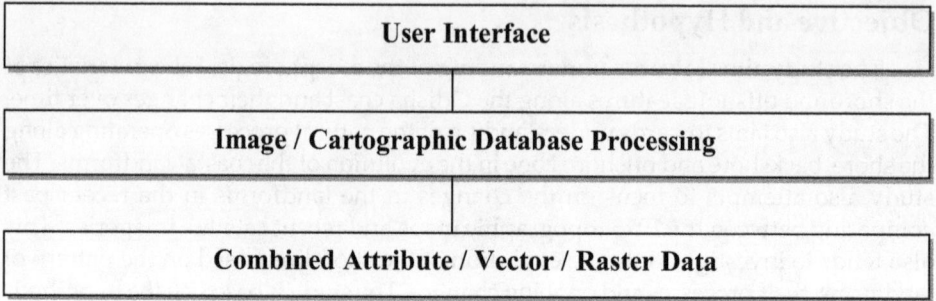

Figure 23.1: Three Stages in the Integration of Image Analysis with GIS Technology.

should be in the same geo-referencing system. This requires coordinate transformation and resampling.

The coordinate transformation can be either reclassification if the spatial dataset is to be transformed to a specific geo-referencing system, such as the UTM or registration if the coordinates of one spatial datasets are transformed to those of another spatial dataset without specific reference to a geo-referencing system. In remote sensing change detection is the process of identifying differences in the state of an object or phenomena by observing it at different times. The digital change detection in remote sensing data has been generally facilitated by use of GIS following map to map, image to image or map to image comparison, using the GIS techniques of overlay, dissolving, subtractions or band rationing.

Study Area and Methodology

Based on the above premises and scientific considerations, some case studies have been done along the Odisha coast, mapping the coastal landforms and detecting their changes in relation to their spatial dimension, land use and land cover, extension of coastal and mangrove vegetation, progradation and recession of the coast line, changing coastal morphology in relation to the emerging offshore features like bars, spits, hooks and islands, shifting river mouths and coastal lagoons. Figure 22.2 shows the flowchart of study, change detection and analysis.

The geomorphic processes along the coast have been corroborated from the existing studies and secondary literatures available on it. It is assumed that the different landforms along the coast and their characteristics are likely to reflect the processes that have produced them The study has been based on comparison of map data of Survey of India of 1975 and IRS-1C data of 1998 and 2000 and recent data of 2012 based on the visual and digital image interpretation through the following methodology (Figure 23.1).

Geomorphic Background of the Odisha Coast

Odisha has nearly 480 km of coastline extending from the Ganjam district in the south to the Baleswar district in the north spreading over six coastal districts. It comprises of the Baleswar and Bhadrak coastal plains in the north, the combined deltaic plains of Mahanadi, Brahamani, and Baitarani spreading over the districts of

```
┌─────────────────────────────┐        ┌──────────────────────────┐
│    Acquisition of Data      │        │    Source of Data        │
├─────────────────────────────┤        ├──────────────────────────┤
│ Pre-field    Visual  Inter- │        │ IRS-IA, LISS-II FCC      │
│ pretation,          Ground  │◄──────►│ Geocoded on 1:50,000     │
│ verification,  Post   Field │        │ scale                    │
│ Modification                │        │ IRS-IC LISS-III FCC      │
└─────────────────────────────┘        │ Geocoded on 1:50,000     │
              │                         │ scale                    │
              ▼                         │ SOI Toposheet on         │
┌─────────────────────────────┐        │ 1:50,000 scale           │
│    Preparation of Digital   │        └──────────────────────────┘
├─────────────────────────────┤
│ Input  &  Geo-referencing of│
│ Spatial    Data,   Topology │
│ Creation, Editing, Transfor-│
│ mation,      Extraction  of │
│ Features                    │
└─────────────────────────────┘
```

Ground Verification of Selected Sites during Field Visit

```
┌───────────────────────────┐        ┌─────────────────────────────┐
│   Analysis Functions      │        │   Analysis Themes           │
├───────────────────────────┤        ├─────────────────────────────┤
│ - Overlay of features     │        │ Land   Use/Land     Cover,  │
│ - Reclassifications &     │        │ Coastal Vegetation, Coastal │
│   Retheming               │        │ Morphology,    Shore    &   │
└───────────────────────────┘        │ Offshore Landforms          │
                                      └─────────────────────────────┘
┌───────────────────────────┐        ┌─────────────────────────────┐
│  Cartographic Presentation│        │ - Softcopy Storing          │
├───────────────────────────┤───────►│ - Hardcopy Storing          │
│ - Geographic Elements     │        └─────────────────────────────┘
│ - Cartographic Elements   │
└───────────────────────────┘
```

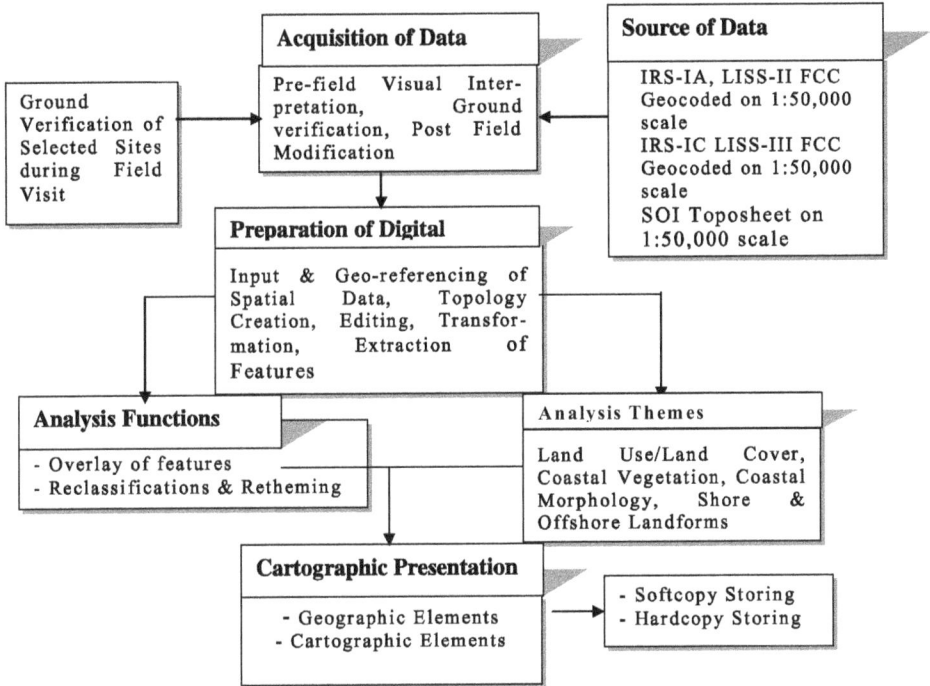

Figure 23.2: Flowchart of Study, Change Detection and Analysis.

Kendrapara, Jagatsinghpur and Puri and the narrow Ganjam plains. The Chilika Lake remains between the Puri and Ganjam plains spreading over nearly 1100 Sq Km of area along the coast. The geological characteristics of the Orissa Coastal Zone are occupied with Holocene sediments of fluvial, marine, brackish water, and lacustrine and aeolian environment. The absence of Pleistocene rock exposure has been caused due to over lying of sediments during Holocene transgression (Meijerlink, 1982). The land ward side of the shore is occupied by the occurrence of widespread palaeo-channels, natural levees, back swamps, lagoons and ancient beach ridges. The marine action is prominent in the Brahmani delta as compared to the Mahanadi delta with the predominance of mangrove swamps and mud flats.

The coastal plain consists of a narrow littoral zone adjacent to the coast extending for about 5 to 10 km inland from the shore. It is a zone of sluggish and brackish streams, parallel sand ridges, sand dunes, back swamps and mud flats. This zone is often called as the "salt tract" (Sinha, 1995). The deltaic and riverine plains of older and younger alluvium extending in width follow this zone inwards from 50 to 100 km. Most of the major rivers of Orissa originating from and beyond the Eastern ghats drain to the Bay of Bengal through this coastal tract with deltas in their mouths. The combined arcuate delta of the Mahanadi-Brahmani-Baitarani rivers extending over nearly 10,000 sq.km. area is the most prominent feature. This zone merges westward with the sub-montane tract of the Eastern Ghats. But longitudinally the coastal zone by and large consists of a sandy beach from the Andhra boarder to the mouth of the river Mahanadi with sand dunes and sand ridges of varying dimensions, the Mahanadi estuary region with the criss-crossing network of estuaries, mudflats and meandering rivers extending from the Mahanadi mouth to the Dhamra mouth of the Baitarani river and a shallow coastal zone from the Dhamra mouth to the West Bengal boarder with mud flats and absence of sand dunes. The significant river mouths along the Orissa coast are the Subarnarekha, Burhabalanga, Dhamra and Maipura of the Baitarani and Brahmani, Mahanadi, Jatadhari mouth oh Hansua river, Devi mouth, Kushabhadra mouth, Chilika lagoon mouth and the Rushikulya mouth.

The Ganjam coast is micro tidal with high wave energy. Mean wind speed is nearly 15 Km per hour and the dominant wind direction is southerly. The middle portion of the Orissa coast is deltaic with prograding coastline and high sand dunes. Mean wind velocity is nearly 10 km per hour and wind directions are northeast and southwest. There are a large number of outlets where the distributaries and creaks meet the sea across the coastline. The sand dunes become lower in altitudes towards the north of this section gradually giving up to muddy coast replacing coastal sandy beaches with tidal mudflats. The Baleswar coast is mesotidal and mean wind velocity is around 4 km per hour with low wave energy. The coast is silty and muddy. The littoral drift moving northwards also brings sediments to the coast. The coast has been endowed with rich coastal and tidal mangrove vegetation. Under these background conditions, monitoring of changes from 1975 to 2010 along the coast have shown many revelations of change in landforms, coastline and coastal mangroves.

The Mahanadi River and its Mouth Region

The Mahanadi originates from the Amarkantak hills of the Bastar Plateau of Chhattisgarh and drains into the Bay of Bengal near Paradip (Figure 23.3). Its catchment area is 141,589 sq.km of which 65,680 sq km lies inside the state. The river is 851 km. long of which 494 km lies with in Orissa. The river has a mean annual flow of 66,640 million cu.m. within the coastal plain the river has developed a network of braided streams and an extensive delta. The study area is a humid tropical delta region with monsoon climate and an average annual rainfall of 155cm of which 90 percent comes during the month of SW monsoon (*i.e.* mid June to mid October). It is one of the frequently flooded rivers of eastern India. This region comes in the mean track of the tropical cyclones, which bring huge storm tides and tidal surges contributing towards the development of erosional and depositional landforms along the mouth zone. The present study area is covered under the SoI toposheets No.73L/ 11 and 73L/15. The region has nearly 40 per cent area under mangrove vegetation, tidal swamps, mud flats and creeks where as the other 60 per cent area is predominantly agricultural.

The geology of the Mahanadi river mouth zone is totally obscured by the Holocene sediments of fluvial, marine, brackish, lacustrine and aeolian environment (Meijerlink, 1982). The tidal and aeolian sediments of the recent origin occupy the region adjacent to the shore where as the land ward part of the river is occupied by the younger alluvium, which extends into the interior part of the interfluves. It has been inferred

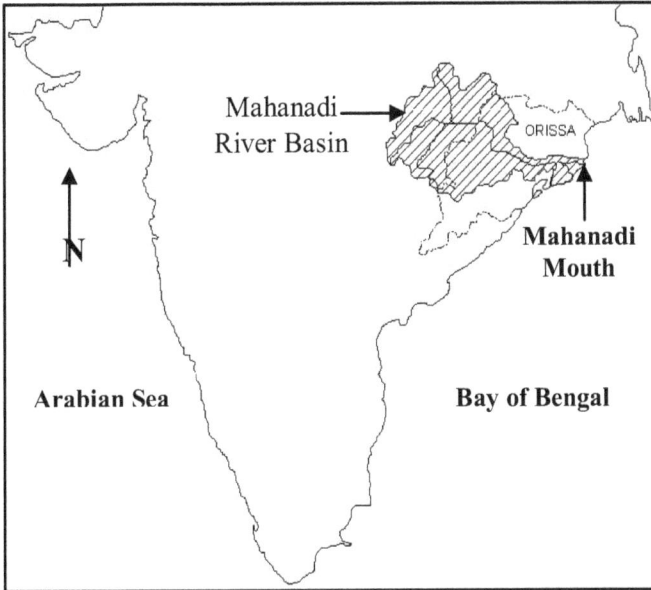

Figure 23.3: Mahanadi River Mouth along East Coast of India.

from the earlier studies that (Mahalik *et al.*, 1996) this section is the most recent evolutionary feature of the Mahanadi delta of the Holocene period. The Mahanadi arcuate delta system has been formed in a tectonic down- warp of the Gondwana graben believed to be a failed arm of the triple junction on the Eastern Indian coast passive margin (Jagannathan *et al.*, 1983). The basement floor with ridges and depressions was affected by faulting. Coastal and off-shore area record new basin development during tertiary. Coastal basins show sediment thickness of 2,400m as revealed by DSS profiling implying continued subsidence activity of growth of faults (Mohanty *et al.*, 2005). Shoreline has migrated seaward and the major delta building seems to have spanned in the late Holocene during the last 6,000 years and is prograding now in a northerly direction (Mohanty, 1993).

Morphological Characteristics and Changes

Baleswar and Bhadrak Coastal Zone

The study has revealed significant changes in the landuse and land cover, coastal vegetation and habitats along the Baleswar coast (Figures 23.4a, b).

The changes have been most conspicuous with regard to loss of coastal mangroves and it alterations to aquaculture ponds following land reclamation processes. The shallow condition of the seashore along with mesotidal regime, the sedimentation of silt, clay and mud has been responsible for a prograding muddy coast with an environment favourable for mangroves. At the backdrop of the coastline there are clear manifestation of the several ancient beach ridges in a linear pattern currently occupied with vegetation and human settlements. These are indications of

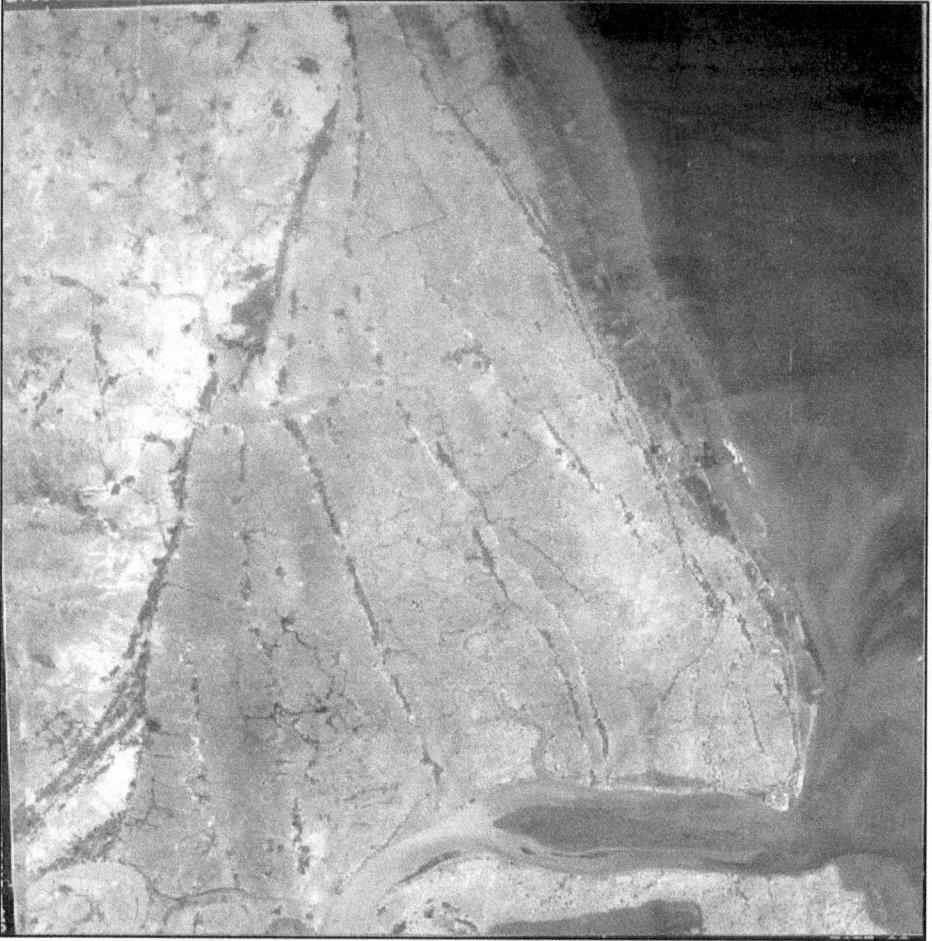

Figure 23.4(a): Baleswar Coast Showing Prograding Coastline, Degradation of Mangroves, Development of Aquaculture Ponds.

the Pleistocene coastline changes. Besides this, the human interference in the coastal ecosystem has led to a loss of coastal mangroves from 55 Sq Km in 1975 to 35 Sq Km in 2000. From the backshore zone the clearing of these forests have been replaced with aquaculture ponds over large areas.

Brahmani and Baitarani Delta Section

In this coastal zone the accretion of coarse sediments of sand and shingle in the off shore zone along with silt and clay has led to the development of many small and big islands around the Wheeler and Short's island. A number of shoals in the submerged state are clearly visible in the imagery as the emerging islands. The finer sediments are driven towards the coast by the tidal energy leaving behind the coarse sediments in the offshore zone for accretion to come up as islands.

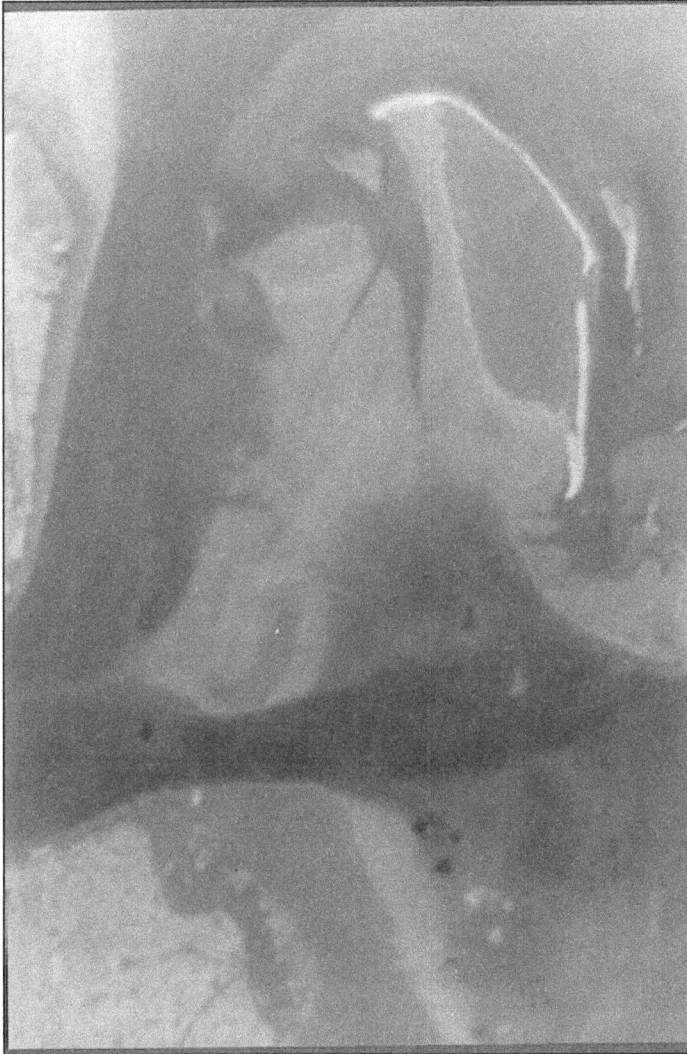

Figure 23.4(b): Dhamra River Mouth Area Showing Appearance of Offshore Islands and Shoals.

Mahanadi Delta Section

In this section of the delta there are two major distributaries *i.e.* the Mahanadi and the Devi besides many other distributaries and tidal inlets along the coast. In this coastal zone the changes are most prominent along the river mouths.

Figure 23.6 shows the mahanadi river mouth revealing erosion of the spit mangroves. Along the Mahanadi river mouth area and Hukitola Bay the changes are visible in declining forest cover. But however there are episodic manifestation of erosion and deposition along the river mouths and in its offshore bars, spits and

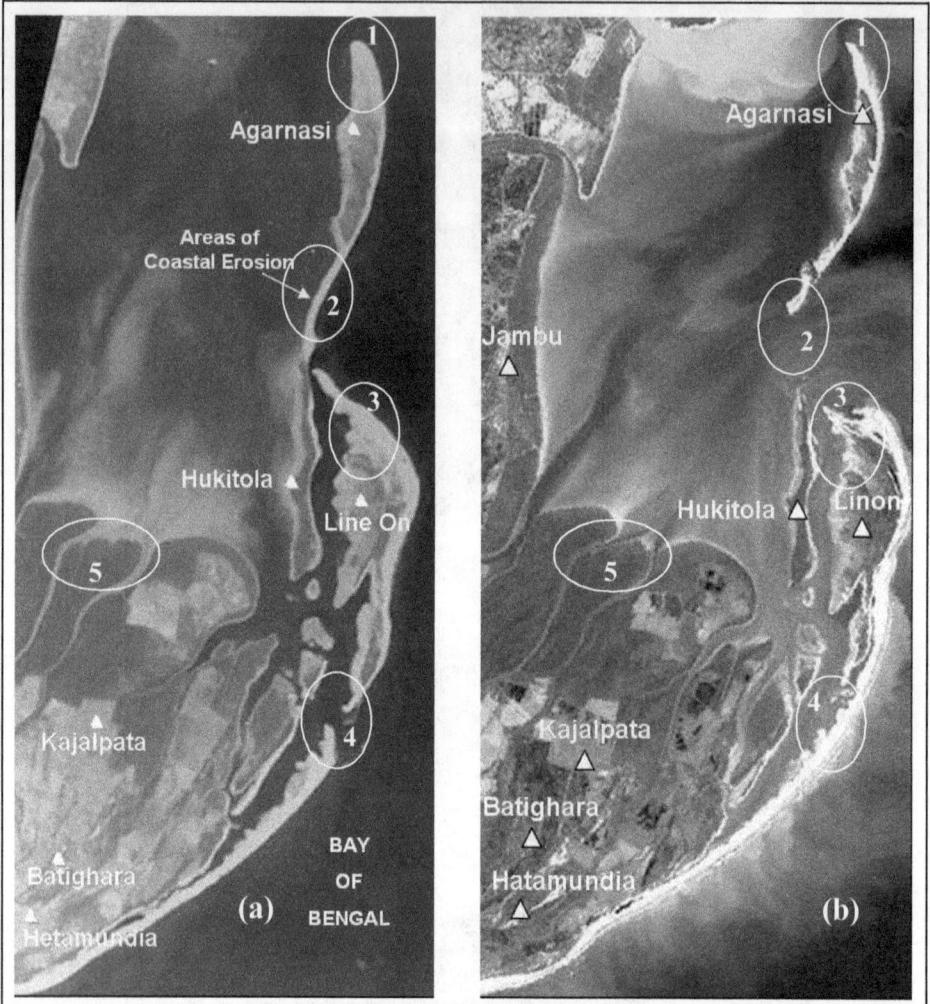

Figure 23.5: Changing Morphology of the Spits, Bay Mouth Bars and Islands around Hukitola Bay as Revealed from the (a) Satellite imageries of IRS- 1B, 1994 and (b) Landsat - 7, 2002. Areas of change are shown in circle.

hooks. There are periods of sedimentation and building up of the offshore features where as there are years of high flood discharge and severe storm surges which erode the spits and hook from the offshore position. Such changes can be marked from the river mouths of the Mahanadi, Hukitola Island and the Devi. It is also noticeable that ere has been a shifting of the river mouths northeastwards along with the prolongation of the spits from the southern side of the river mouths. This is creating a new course of the river parallel to the shore thereby reducing the flood slope and consequent delay in flooding (Figure 23.8).

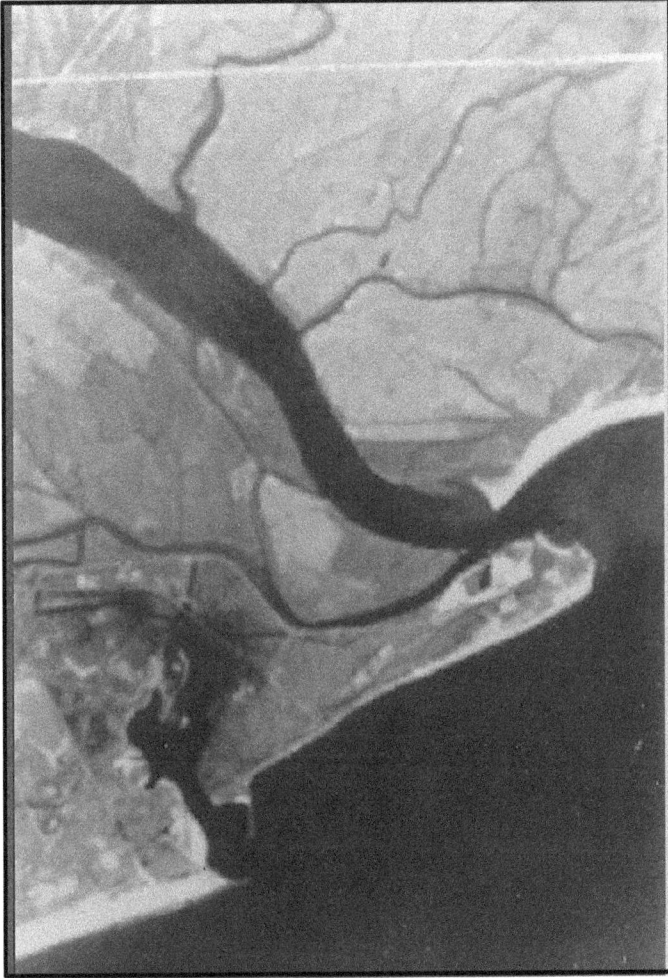

Figure 23.6: Mahanadi River Mouth Revealing Erosion of the Spit Mangroves.

(a) Offshore Bars, Cuspate Bars and Bay Mouth Bars

These are the landforms of the offshore zone, which refers to the shallow zone of the sea extending from the low tide water limit of the shore up to the continental shelf, which is mostly affected by the action of the waves, tides, littoral drift and sediment discharge from the rivers. This offshore zone corresponds with the seaward portion of the breaker zone swept by the sinusoidal wind waves, which are transformed shore ward into solitary waves. The 5 fathom and 10 fathom submarine contour along the Mahanadi mouth is about 1.5 km and 8 km from the shoreline respectively. The offshore slope of the continental shelf varies from 6' to 9' along the Orissa coast (Ahmad, 1972). Along the offshore zone of the Mahanadi river mouth, the geomorphic features are bay mouth bars, cuspate bars, spits, hooks and barrier islands enclosing shallow lagoons (Figure 23.7).

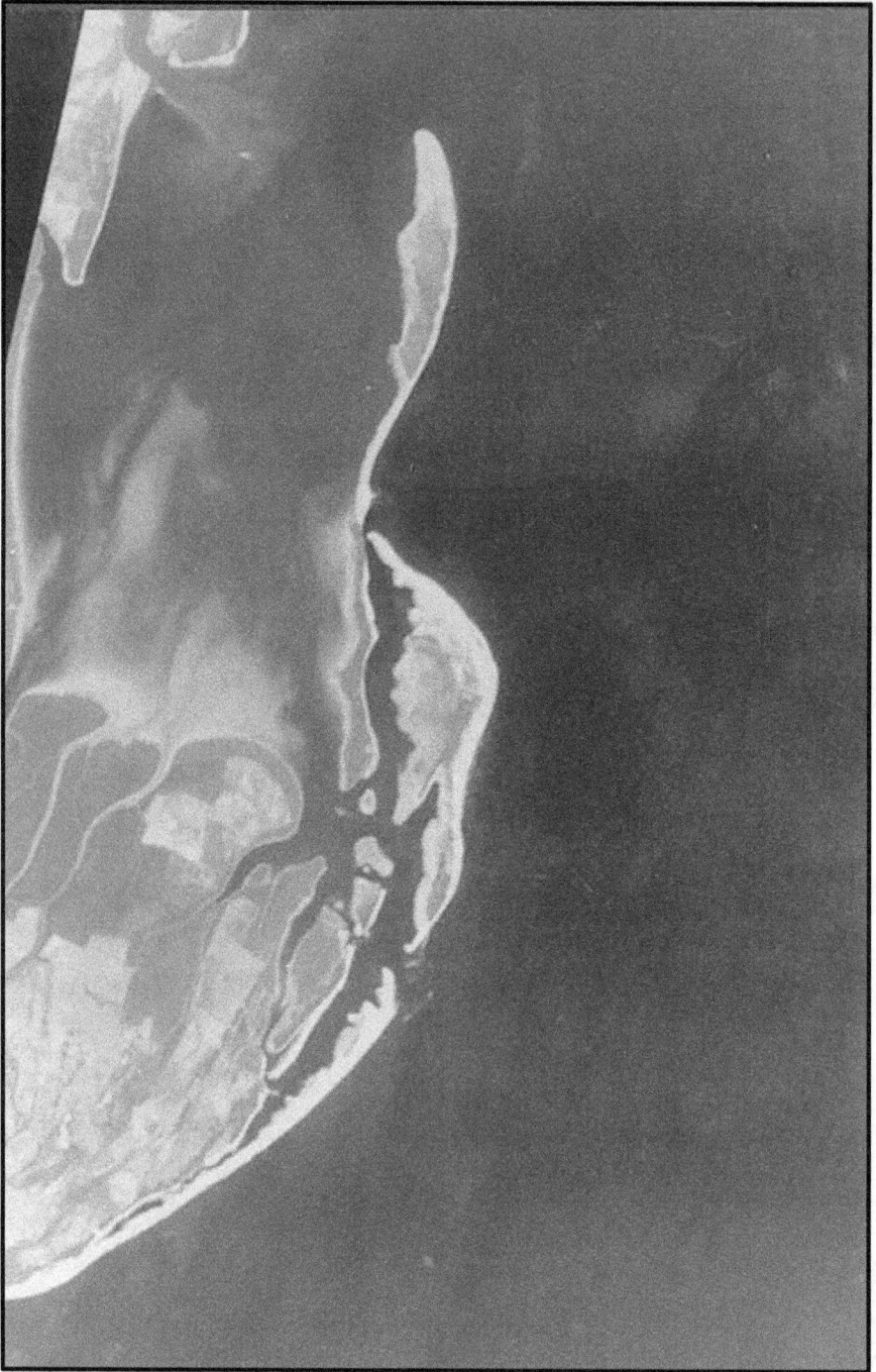

Figure 23.7: Hukitola Bay Showing Erosion of Hooks and Loss of Mangroves.

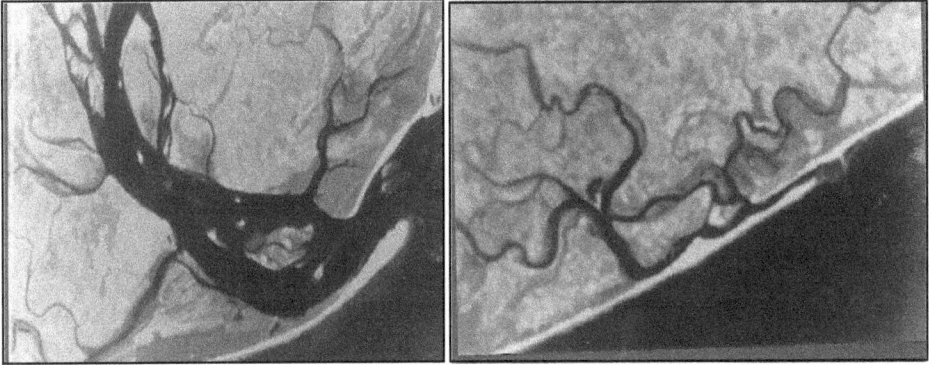

Figure 23.8(a): Devi River Mouth Area.   **Figure 23.8(b): Ersama Coast Showing the Spits.**

Offshore bars are sub-marine ridges of deposition of sediments eroded by the waves attacking the submarine floor (Johnson, 1972). These are deposited as elongated bodies of sand, shingle and clay developing sub-parallel to the shoreline and formed in an offshore position. Along the Mahanadi river mouth these features are well developed. The prolongation of the bay bar to a distance of 2 km to the south of the Mahanadi river mouth parallel to the coast has shifted the meeting point of the river to the sea to the north east. This bar has a width varying from 300 to 400 metres.

To the north of the Mahanadi mouth, the creeks which carry saline water from the sea to the river during high tides use to carry freshwater and sediments also during the flood season. The creeks are enclosed between the lobate shaped projecting landmasses *i.e.* deltaic lobes filled with tidal sediments and mangrove plants bordering the Hukitola bay. In these projecting land units deposition of sediments have created a number of bay mouth bars, cuspate bars of various shapes and sizes ranging from circular, elliptical, elongated ridge type and semi-circular and irregular land bodies in side the Hukitola bay in the lagoon zone close to the mouth and in the offshore positions. The sediments are found sorted in parallel and cross-bedded laminations.

(b) Spits, Hooks and Compound Spits

The spits, hooks and compounded spits are the most conspicuous and well developed landforms of the off-shore region near the Mahanadi river mouth unlike many other rivers of the Orissa coast. These are narrow and elongated beach features, which extend in front of the Mahanadi river mouth in the direction of the dominant wind (Figure 23.7). The spits have a northward open end and are attached to the land on the southern side. Some of the spits have curved westwards towards the main land at their extreme end, which are described as the hooked spits. These spits had extended into deeper water due to the abundant supply of sediments and elastic debris, which in turn were attacked by concentrated and refracted wave actions as well as by the storm surges.

The spit, which extends from the Mahanadi mouth near Paradip port up to its extreme limit in the northeast, spreads over a distance of about 3 km. It is an important spit, which prolongs the Mahanadi mouth northwards from its original position. The spit is nearly 0.5 km wide and 3 km long. To the north of this spit, there are two parallel set of spits which are curved in shape with bending at its apex towards the land enclosing a shallow lagoon. The outer spit is narrow at the origin but widens towards north and extends as a ridge of sand extending from the main land. But however, the plunging waves had broken the spit at its centre. The outer part of this spit to the seaside is sandy where as the inner part is silty and is occupied with mangroves.

The inner Hukitola Spit is the most elongated one and was 10 km long. Its width varies from 200m to 1.5 km, separated from the main land by a large lagoon *i.e.* Hukitola Bay extending in a northerly direction. It is of 10 km width at the base but the mouth connecting to the main sea at its extreme north is 5 km (Figure 23.7). The inward funnel shaped projection of the mouths of the spits has been the common morphological dispositions. The growth of this hooked spit enclosing the Hukitola Bay was likely to give rise to another coastal lagoon similar to that of Chilika in the near future.

(c) Beaches and Beach Ridges

These are the landforms of the shore zone between high and the low tide level often termed as the littoral zone. In this zone the main agents, which lead to morphological evolution and change are rivers, creeks, waves, tides and coastal wind. The landforms in this zone are clearly differentiated through their structure, lithological composition, nature of materials and their background conditions. The features of the backshore zone include the sand dunes and ancient beach ridges. The beach is the zone of sediments deposited by the waves along the coastal zone. The sand normally constructs the beach along with other materials derived from both local and distant sources. Near the Mahanadi mouth the riverine sediments also play a very significant part in the development of the beach. The size of the beach reflects the material in storage and is therefore a measure of the balance between the availability of sediments and wave energy. Input of sediments to the beach is derived from the local erosion, from offshore positions and from the backshore zone. To the south of the Mahanadi river mouth the beach is narrow, sandy and on the northern side it is muddy. There is a moderate beach of 10 to 15 meter between the Devi and the Mahanadi mouth. To the north of the Mahanadi, creeks are many and the beach sediments are mostly silt and clay associated with sand. Near the river mouth the sediments discharged to the sea gets sorted into course and fine fractions. While course sediments are drifted away by intense wave action, the fine sediments remain settled. Near the river mouth the beach is formed of well-sorted clean, fine to medium grained sand with inclined laminae. The composition of the sediments near the mouth of the river comprises 60 per cent sand, 26 per cent silt and 14 per cent clay. The spits are characterised by white texture of the sandy beach on the sea ward side and dark tone of the silt laden shore of the lagoon ward side on the satellite imagery (Figure 23.8).

The temporary increase in wave energy associated with frequent visit of the storm surges to this region causes erosion to the shore and offshore landforms. When the wave energy comes to the normal, offshore materials are redeposited on the whole beach and try to rebuild its former configuration. But however, currently there is erosion of the Hukitola spit on its seaward side at a faster pace as compared to the deposition on its lagoon side. There are also areas of erosion along the shore to the north of Mahanadi mouth at Satabhaya and Pentha village and south of the mouth along Ersama coast.

(d) Estuary and Creeks

Estuary is that part of the river mouth where the salt water and freshwater mixing takes place. The creeks are narrow irregular cris-crossing network of channels connecting to the sea, which carry saline water inland during the high tides. The Mahanadi also exhibits the characteristics of a barred estuary which is partially cut off from the ocean by the sand bars, spits, barrier islands, mud banks and deltaic accumulations leading to the development of the Hukitola Bay resembling a coastal lagoon (Figure 23.7). On the northern section of the Mahanadi river mouth, a large number of creeks had developed between the projecting deltaic lobes with their mouths occupied by the islands and barrier bars. Kharnasi is one of the most important creeks of the northern side. The mouths of the creeks are projected forward between the projecting lobes and cuspate forelands, which are occupied by mangrove vegetation. The mangroves create a congenial environment for trapping mud and clay and help in prograding the deltaic lobes and the mouths towards the sea. The mangrove vegetation of the deltaic lobes is well marked on the satellite imageries with dark gray tone (Figure 23.7).

(e) Islands and Bay Mouth Bars

To the north of the Mahanadi river mouth facing the Hukitola Bay there are a number of islands enclosed by the creeks. Some of these islands have grown into bay mouth bars. The causes of formation of these features are due to the obstruction of sediment load in their mouthpart. Some of these islands had started as bars but subsequently are broken by the tidal waves or storm surges and are grown over time to form the islands.

(f) Sand Dunes and Ancient Beach Ridges

These are the landforms of the backshore zone which represents the area extending landward up to which the marine influence is felt. But however, in the Mahanadi mouth region, the backshore zone extends nearly 10 km. The major features in this zone are backwaters, sand dunes, beach ridges, tidal creeks and mud swamps. The sand dunes of this region are parallel to the coastline, low in height, in the range of one meter along the shore. The tidal creeks, backwaters and mud flats occupy much of the area in the backshore zone, which gradually merges landward with the croplands and shrimp ponds. Adjacent to the river mouth there are three ridges, which are perches in a muddy sub-stratum, making the position of a pre-existing shoreline? These are berms of sand, silt and clay built by consecutive waves parallel to the coastline below the level of spring tides. On the northern side of the river the beach ridges have been washed and eroded by tidal action besides human occupancy in suitable upland sites favoring growth of settlements and agricultural activities.

The presence of multiple ridges had arisen out of the continuous swallowing of the offshore and abundant sediment supply from the Mahanadi river. These are typical sandy ridges even noticeable on the topographic maps and satellite imageries as discontinuous linear patches of human settlements surrounded by village orchards and other plantations revealed through tonal contrast. The lithology comprises course to fine sand, silt and clay. Many such parallel ridges are found in the backshore zone which reveals the old strand lines of Holocene period (Mahalik *et al.*, 1996).

(g) Tidal Inlets, Mud Flats and Mangrove Swamps

The regular astronomical tides coupled with storm tides have a profound impact over this region in producing landforms of tidal origin. The important among these are tidal inlets or creeks joining the river with the sea and even extending from the sea to the interior low lying areas of mudflats and swamps. The ill drained areas remaining in the backshore zone of the Paradip area have a number of swamps with tidal shrubs and marshy vegetation. These are revealed through their smooth texture and grey to dark grey tone due to high moisture content. The lithology of these areas comprise of greater proportion of silt and clay together comprising 30 per cent to 50 per cent. The sedimentation is from the suspended load, inter-bedded with dark grey clayey sand and silt. The tidal mud flats are distinctly visible on the imageries with light red tone associated with mangrove vegetation. The sediments are fine grained but coarse sediments are confined to the tidal channels. The tidal inlets has been favorable in propagating the tides which bring a large amount of estuarine sediments of silt and mud which are distributed in the adjoining areas to project the deltaic lobes and help in prograding the coast line (Figure 23.8).

Chilika and its South Section

The recent satellite imagery of Jan, 2004 (RESOURCE SAT) and its earlier imageries of 1980 and topomaps of 1975 and 1935 has indicated the shift of the Chilika mouth to the extreme northeast position near Arakhakuda village which has come to a closure in low tides conditions (Figure 23.8). The change in the land use and land cover in its southern part adjacent to Rambha and Khalikote is noticeable from the imageries. A significant change in the time series data is available for the weed infestation in Chilika Lake form 70 Sq Km area in 1970 to nearly 300 Sq Km in 2000. Since 2000 there has been a decrease in the freshwaterweed area after the opening of the new mouth, which has increased the salinity of the lagoon. Figure 23.9 shows the 45 km spit showing the separation of Chilika from the Sea with it's shifting mouth at the extreme northeast position.

Shore Processes and Changing Landforms

Morphological evolution of land forms and their changes depend on geologic, tectonic, geographic, climatic and environmental settings. The landforms around the Mahanadi mouth and its surrounding area bear ample evidence to the combined action of the shore processes associated with the waves, currents, tides, coastal wind and storm surges as well as river actions based on the predominance of the agents at different localities, structure of the area and the availability of the materials. The

Figure 23.9: The 45 km Spit Showing the Separation of Chilika from the Sea with it's shifting Mouth at the extreme Northeast position.

offshore features point to the dominant action of the waves followed by offshore long currents (littoral drift) and the coastal wind. The backshore features are mostly linked to the tidal actions, shore wind and flood flow in the rivers as well as the estuarine interaction of sweet and saline water and sorting of sediments. The occasional high flood flows; large tidal waves and storm surges have also a significant bearing on the morphological changes of the shore and offshore landforms. The fast protruding deltaic lobes of the study area relates to the deltaic regime, abundance of sediments from the river, tide and littoral drift.

The along shore sediment transport in Orissa compares to the highest transport rates in the world, within the order of one million cubic meters to the north at Paradip. Varadarajulu and Harikrishna (1979) studied that the predominant wave directions are from the south or southwest. The bulk of the waves attain a height of less than 1.25 m and a period of less than 7 seconds. During the storms, however, waves reach 7 to 8 m. in height as it was observed during 1999 super cyclone. The oblique waves carry sediments from the breaker zone and beach to an offshore position to develop the barrier beaches, bars and islands. The littoral drift picks up the river discharged sediments deflecting it towards the northeast. The development of the spits and bars confirm to the evidence that the littoral drift supply materials but their evolutionary process can be thought of as a result of spasmodic progradation by obliquely impinging waves at the time of major storms or depressions. The wind blown sand brings sand deposition on the barrier bars and spits. Besides deposition there has also been erosion along the shore, on the seaward side of the spits and islands (Figures 23.4 and 23.5).

Mahanadi deltaic coast is micro tidal with mean tidal range measuring 1.29 m. and the tidal cycle is semi diurnal. It is principally a wave dominated coast during the south west monsoon and during NE monsoon (Nov. to Feb.) it is mixed wave and tide dominated. The enormous quantity of sediments and water brought to the sea by the river Mahanadi has been one of the main source of supply of sediments. The northern part of Paradip is relatively meso-tidal where wave energy diminishes and the resulting high tide brings silt and mud to build the mud flats and mangrove swamps. But to the southern side of the Mahanadi river mouth, the coast is micro tidal. The wave energy is relatively more than the tidal energy, which helps in building the sandy beach. The floatation process and sorting out of heavy sand goes on for years and ultimately a high sand ridge appears near the mouth. When the river water overflowing the banks or the tidal water entering through the river mouths and creeks inundates the banks, such waters contain much sediment in suspension. It flows slowly down valley or may become fully ponded. The ponded water forms the mud flats and swamp deposits.

Summary and Conclusion

The studies indicate that there have been significant changes along the Odisha coast with respect to the coastal morphology, nature of erosion and deposition along the coast, land use and land cover. Most of these features point towards a coastline, which is developing seawards irrespective of coastal submergence, emergence and coastal erosion and deposition revealing a prograding coastline. The morphological evolution and change of the offshore and shore landforms during the last century point towards a phase of rapid sedimentation, prolongation of the deltaic lobes with the development of the tidal creeks and estuaries, development of bay mouth bars, islands, spits and hooks followed by a phase of decelerated sedimentation and accelerated erosion along the shoreline and on the seaward side of the spits and hooks. The study reveals that the late Holocene deltaic progradation under abundant sediment supply and sea ward shifting of the shoreline may be on the critical threshold of retrogression due to reduced sediment input, neotectonic sea level changes and rise of sea level besides high episodic monsoonal flood flows and cyclonic storm surges. The distinct morphological changes in the off shore landforms with increasing erosion can also be attributed to continuous coastal subsidence that might have been attenuated by possible neotectonic activities and relative sea level rise leading to shoreline retreat.

References

Aggarwal, S.N. and Lal, M. (2005): Vulnerability of Indian Coast line to Sea level Rise, Mimeographed, CAS, 111, New Delhi.

Ahmed, E. (1972): Coastal Geomorphology of India, Orient Longman, New Delhi.

Anand, S.P., Erram, V.C and Rajaram, M. (2002): Delineation of Crustal Structure of the Mahanadi Basin from Ground Magnetic Survey, Journal Geological Society of India, 60: pp 283-291, Calcutta.

Anbarasu, K. and Rajamanickam, V. (2002): Coastal Erosion around Pondicherry and Ennore along the East Coast of India, Journal of the Geological Society of India, 60: pp 519- 29528, Calcutta.

Dag, J.W., Pont, D., Hensel, P.F and Ibanez, C. (1995): Impacts of Sea level Rise on Deltas of Mexico and the Mediterranean, The Importance of Pulsating Events to Sustainability, Estuary, 18: pp-636 – 647.

Dept. of Water Resources (1986): Delta Development Plan, Govt. of Orissa, Bhubaneswar.

Dwivedi, D.N. and Sharma, U.K. (2005): Analysis of Sea Level Rice and its Impact on Coastal Wetlands of India, Proceedings of the 14th Biennial Coastal Zone Conference, New Orleans, Louisiana.

Esteves, L.S., Toldo, E. E, Dillenburg, S.R and Tomazell, L.J. (2002): Long and Short Term Coastal Erosion in South Brazil, Journal of Coastal Research, 36: Pp. 273-282, Northern Ireland.

Hemamalini, B and Rao, K.N.(2004): Coastal Erosion and Habitat Loss along the Godavari Delta Front, A fall out of Dam Construction, Current Science, 87 (9): Pp 1232-1236, IISc, Bangalore.

Jagannathan, C.R (1983): Geology of the off shore Mahanadi Basin, Petroleum Asia Journal, 4 (M): pp 101-104.

Johnson D.W. (1919): Shore Process and Shore Line Development, John Wiley and Sons Inc., New York.

Mahadevan, C. M. D., Rao, P.R. (1958): Causes of the Growth of Sand Spits North of the Godavari Confluence, Andhra Univ. Memoir of Oceanography, 2: pp. 69-74.

Mahalik, N.K., Das, C. and Maijima, W. (1996): Geomorphology and Evolution of the Mahanadi Delta, Indian Journal of Geosciences, Osaka City Univ, 39(6): Osaka, Japan.

Meijerink, A.M.J. (1982): Dynamic Geomorphology of the Mahanadi Delta, ITC Journal, Special Verstapen pp. 243-250.

Mohanty, M and Swain, M.R. (2005): Mahanadi delta, East Coast of India, an Overview on Evolution and Dynamic Processes, Mimeographed, Dept. of Geology, Utkal University, Bhubaneswar.

Mohanty, M. (1993): Coastal Processes and Management of the Mahanadi River Deltaic Complex, East coast of India, Proceedings Coastal Zone 93, ASCE, PP 75-90, New York, USA.

Nayak, N. (2002): Role of Remote Sensing to Integrated Coastal Zone Management, Mimeographed, ISRO, Ahmedabad.

Prusty, B.G., Sahoo, R.K and Mehta, S.D. (2006): Natural Causes Lead to Mass Exodus of Olive Ridley Turtles from Ekakulanasi, Orissa, India, a Need for Alternate Sites, Mimeographed.

Raman, C.V and Reddy, K.S.N.(2001): Sediment Dispersal Pattern off the Mahanadi-Nagavalli Continental Shelf, Northeast Bay of Bengal, Journal of the Geological Society of India, 58(2): pp. 123 – 133, Calcutta.

Rao, K. S, Rao, K.W., Sadakata, N., Varma, D. D and Rao, A.T.(2003): Geo-morphology and Evolution of Penner Delta, East Coast of India, Journal of the Indian Association of Sedimentologists, 22(1 and 2): pp. 171 – 181.

Rao, K.N., Sadataka, N., Hemamalini, B. and Takayasu, K.(2003): Coastal Forms and Processes of the Godavari Delta, India, Mimeographed, Dept. of Geo-engineering, Andhra Univ., Waltair

Rao, S and Vaidyanathan, R. (1979): New Coastal Landforms at the Confluence of the Godavari River, Indian Journal of Earth Sciences, 6: pp. 222-227.

Seetharamaiah, J, Bhagaban, K.V.S and Rao, K.N.(2002): Recent Morphological Changes in the Penner Estuary, East Coast of India, Journal of the Geological Society of India, 60, Calcutta.

Varadarajulu, R. and Harikrishna, M. (1979): Wave Characteristics of Paradip Port, Indian Jour. of Marine Sciences, 18: pp. 68–72, Goa.

Vinod Kumar, K and Bhattacharya, A. (2003): Geological Evolution of Mahanadi Delta Orissa Using High Resolution Satellite Data, Current Science, 85(10): I I Sc, Bangalore.

Chapter 24

Electrochemical Methods for the Treatment of High COD Loaded Industrial Effluent and their Complete Removal

S.C. Mallick[1], B.C. Tripathy[2] and N.N. Das[1]

[2]P.G. Department of Chemistry, North Orissa University,
Baripada, Orissa, India
[2]Institute of Minerals and Materials Technology (IMMT),
CSIR, Bhubaneswar – 751 013, India

ABSTRACT

Growing demand for suitable process or combination of processes for industrial water treatment, compelled to apply advance treatment technologies together to find way out to solve the mammoth problem of wastewater treatment to meet the stringent discharge standards. In the present chapter electrochemical treatment technologies such as electrocoagulation (EC) and electrooxidation (OA) processes were applied to treat a highly organic pollutant loaded industrial effluent. Electrocoagulation alone is not sufficient enough to treat the high COD loaded effluent effectively to meet the disposal standards. There are many stable organic chemicals which are very resistant to EC process and called as refractory organic compounds or recalcitrant and such compounds are effectively destroyed by electrooxidation process. Electrooxidation process alone can be able to reduce pollution loads but it is very energy intensive process to treat industrial effluents with high COD load. Combination of electrocoagulation (EC) and electrooxidation (OA) can effectively remove the pollutants with power consumption of 1 – 1.5 kWh/kg COD and where electrooxidation alone require 6-8 kWh/kg COD removal.

Keywords: Electrocoagulation, Electrooxidation, COD, Refractory organic compounds, Recalcitrant.

Introduction

Rapid industrialization, increasing population, improving standards of living and growing demands for freshwater are putting burdens on water resources. The preservation of the limited natural water supplies and, in near future, the necessity for direct recycling of water in some part of world will require improved technologies for the removal of contaminants from wastewater. There are many contaminants in wastewater vary from time to time, and they are not well characterized with respect to the chemical species for fixing suitable treatment process. Commonly, the level of organic contamination is expressed by Chemical Oxygen Demand (COD), Biochemical Oxygen Demand (BOD) or Total Organic Carbon (TOC). There several toxic contaminants which are banned by several agencies and government bodies, European Union has listed such chemicals or elements which are given the Table 24.1.

In the recent years there has been increasing interest in use of electrochemical techniques such as electro coagulation, electrochemical oxidation, electro floatation, electro decantation, non thermal plasma and wet air oxidation to treat industrial effluents and wastewater (Lawrence *et al.*, 2007). Among these techniques, electro coagulation followed by electrochemical oxidation emerges one of the promising techniques to treat industrial organic effluents (Butler *et al.*, 2011). Successful electro coagulation (EC) treatment of various industrial effluents has been reported by several researchers (Lawrence *et al.*, 2007) as it is considered to be potentially an effective tool for treatment of effluents and wastewater with high removal efficiency and applied to broad range of pollutants including organic, inorganic and pathogenic pollutants (Emamjomeh *et al.*, 2009).

Electrocoagulation involves dissolution of metal from the anode with simultaneous formation of different hydroxyl ions with polymeric forms of ions at different pH range. The reduction reaction at cathode generates hydrogen gas which helps in coagulation and flotation process. Electrocoagulation is not a new wastewater treatment technique. Treatment of wastewater by EC has been practiced since the end of 19th century with marginal success. Using electricity to treat water was first proposed in UK in 1889 and a treatment plant was built in London in 1889 for the treatment of sewage by mixing with seawater and electrolyzed (Vik *et al.*, 1983). In 1909, in the United States, J.T. Harries got patent rights for wastewater treatment by electrolysis with sacrificial aluminium and iron anodes (Vik *et al.*, 1983). A process and apparatus for electrocoagulative treatment of industrial wastewater applying electrocoagulation for the treatment of contaminated industrial water with incorporation of different tanks to manage difficulties associated with EC process is described in US patent

Table 24.1: Black List of Chemicals Substances Selected by the E.U. Group included Substances.

Chloride hydrocarbons	Aldrin, dieldrin, chlorobenzene, dichlorobenzene, chloronaphthalene, chloroprene, chloropropene, chlorotoluene, endosulfane, endrin, hexachlorobenzene, hexachlorobenzene, hexachlorocyclo-hexane, hexachloroethane, PCBs, tetrachlorobenzene, trichlorobenzene.
Chlorophenol	Monochlorophenol, 2,4-dichlorophenol, 2-amino-4-chlorophenol, Pentachlorophenol, 4-chloro-3-methylphenol, trichlorophenol.
Chloroanilines and nitrobenzenes	Monochloroanilines, 1-chloro-2,4-dinitrobenzene, dichloroaniline, 4-chloro-2-nitrobenzene, chloronitrobenzene, chloronitrotoluene, dichloronitrobenzene.
Policyclic Aromatic Hydrocarbons	Antracene, biphenyl, naphthalene, PAHs
Solvents	Benzene, carbon tetrachloride, chloroform, dichloroethane, dichloroethylene, dichloromethane, dichloropropane, dichloropropanol, dichloropropene, ethylbenzene, toluene, tetrachloroethylene, trichloroethane, trichloroethylene
Other	Benzidine, chloroacetic acid, chloroethanol, dibromomethane, dichlorobenzidine, dichloro-diisopropyl-ether, diethylamine, dimethylamine, epichlorhydrine, isopropylbenzene, tributylphosphate, trichlorotrifluoroethane, vinyl chloride, xylene.
Pesticides	Cyanide chloride, 2,4-dichlorophenoxyacetic acid and derivatives, 2,4,5-trichlorophenoxyacetic acid and derivatives, DDT, demeton, dichlorpropane, dichlorvos, dimethoate, disulfoton, phenitrothion, fenthion, linuron, malathion, MCPA, mecoprope, monolinuron, omethoate, parathion, phoxime, propanyl, pirazone, simacine, triazofos, trichlorofon, trifularin and derivatives
Inorganic substances	Arsenic and its compounds, Cadmium and its compounds, Mercury and its compounds.

20070068826A1 (Morkovsky *et al.*, 2004). Application of electrochemical methods with conventional treatments methods and their application in wastewater and industrial effluent treatment is thoroughly reviewed by the authors (Mouli *et al.*, 2004). An electrocoagulation process for removing organic and inorganic from pressurized waste fluids is described in US patent 6719894 (Gavrel *et al.*, 2004). Electrochemical methods (electrocoagulation, electrooxidation and fenton) for the treatment of organophosphorous pesticides is described Hector A. Moreno *et al.* give a detail mechanism of electrocoagulation and COD removal using aluminium and iron soluble anodes. Electrocoagulation technique was applied to remove toxic chemicals from effluents generated from pulp mill and wastewaters by using aluminium and iron (Vepsäläinen, 2012).

Materials and Methods

The studies were carried out by taking effluent from a mixed effluent stream of an industrial conglomerate of different chemical industries. There are many toxic chemicals with high salt and COD contents. The pollution load was contents of the pollutants varied frequently and estimated based on COD load of the mixed effluent stream.

The effluent samples collected for studies were alkaline of pH range 10 to 12 and containing free alkali. The pH of the effluent was adjusted using 6 N HCl and the precipitate formed was filtered through a 12 micron filter cloth. The final effluent was taken in a 10 L polypropylene electrocoagulation tank. The internal dimension of electrolytic tank was 20 cm × 20 cm × 25 cm (length × breadth × depth). Mild steel cathodes and pure aluminium anodes of 15 cm × 15 cm × 1 cm (length × breadth × thickness) were arranged in vertical disposition with anode cathode gap of 0.6 cm. The electrodes were connected in mono polar arrangement as given in the Figure 24.2. The overflow of the Electrocoagulation Cell (EC) tank was connected to a settling tank in which clear the liquid effluent continuously recirculated to EC tank through a filter press in between. Electrocogualtion was carried out by impressing DC current from 32 A and 20V DC rectifier

The final solution from EC stage was taken into a polypropylene Electrooxidation Cell of 20 cm × 18 cm × 25 cm (length × breadth × depth) for electrooxidation. Mild steel cathodes and DSA (dimensionally stable with MMO coating mesh type anode) anodes of 15 cm × 15 cm (length × breadth with 80 per cent active surface area) were used for oxidation. The electrodes were arranged at an anode cathode gap of 0.6 cm in vertical disposition with monopolar connection. The EC and EO treatment scheme is given Figure 24.1 and electrode arrangements given in Figure 24.2.

Results and Discussions

Electrocoagulation and electrooxidation processes are quite complex and may be affected by several operating parameters, such as pollutant concentrations, initial pH, applied electrical potential (voltage), COD, turbidity and above all electrode materials (Ni'am *et al.*, 2007; Zhlik *et al.*, 1996). In the present chapter, electrocoagulation and electrooxidation processes have been applied for removal of

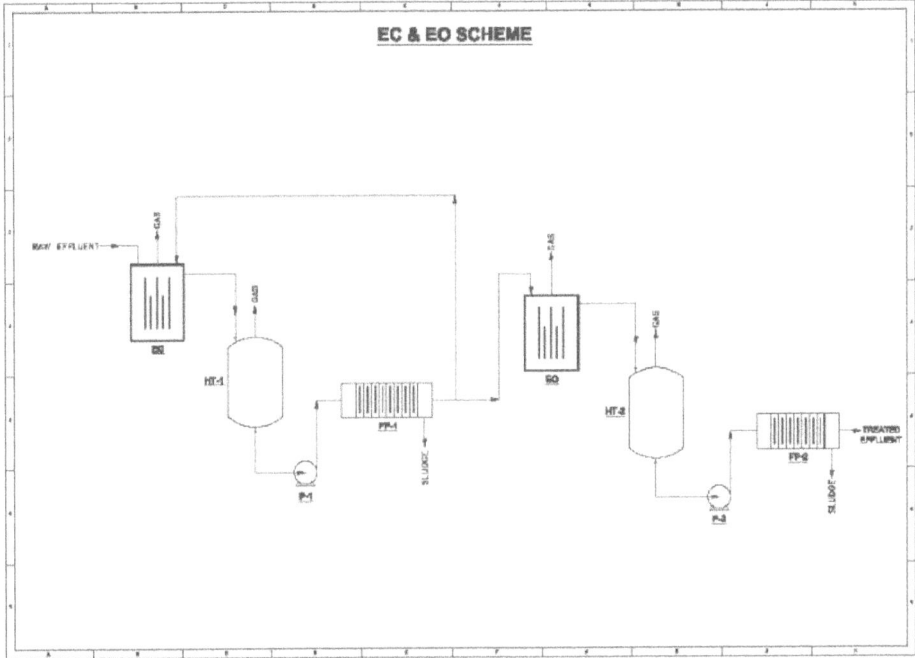

Figure 24.1: Electrocoagulation and Electrooxidation Process Scheme.

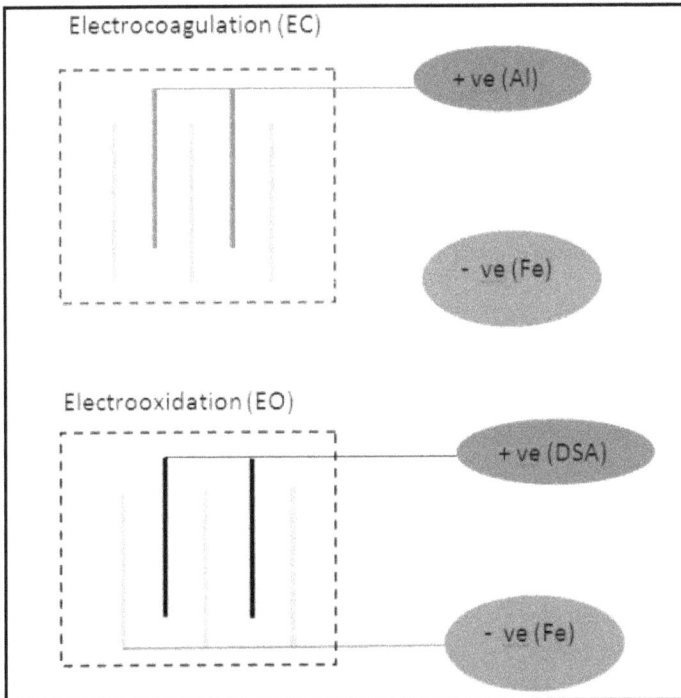

Figure 24.2: Electrode Arrangement in EC and EO Cells.

COD and colour of the complex effluent stream. There are several organic pollutants present in the effluents and wastewater streams which do not remove completely by convention chemical treatment methods cost effective and sometime not effective to remove all toxic elements. Such organic chemicals are called refractory organic material and need harsh treatment conditions such as electrooxidation (Mouli *et al.*, 2004). There are several organic chemicals which precipitate at acidic pH range. During the pH adjustment of the above effluent by 6 N hydrochloric acid, some organic chemical precipitate at lower pH (< 7) and filtered out before electrocoagulation as given in Table 24.2.

Table 24.2: Pre-treatment and pH Adjustment

Parameters	Initial	Final
pH	11.5	4.5
Free Alkali, N (NaOH equivalent)	0.2	–
Salt content, g/l (NaCl equivalent)	15	24.5
COD, mg/l	30400	27600

Chemical coagulation by alum [$Al_2(SO_4)_3.18H_2O$] is well known and which has been widely used in ages for the treatment of wastewater and effluents. In the electrocoagulation process the coagulant is generated in situ by anodic dissolution of soluble anodes such as iron and aluminium anode. The electrolytic dissolution of aluminium anode produces the monomeric cationic species, Al^{+3}, $Al(OH)_2$ at lower pH and subsequently polymerizes to $Al_n(OH)_{3n}$ (Holt *et al.*, 1999). There are other ionic species such as $Al(OH)^{+2}$, $Al_2(OH)_2^{+4}$ and $Al(OH)^-$ also generated during oxidation and take part in coagulation process. These charge species effectively remove the pollutants by adsorption to produce charge neutralization (Kobya *et al.*, 2003). During electrolysis the following anodic reactions are taken place and participate in coagulation process while at cathode reduction encountered to produce H_2 gas bubbles and helps in dispersion of coagulants. Table 24.3 shows the electrocoagulation process conditions and the electrocoagulation reasults are given in Table 24.4.

Table 24.3: Electrocoagulation Process Conditions

Parameters	Unit	Values
Anode cathode distance (ACD)	Mm	6
Current Density (cd)	A/m²	50
Cell Voltage (cv) per cell	V	0.7
No of Cells (2 anodes 3 cathodes)	No	4
Cell current	A	3.6
Anode area per volume	m²/m³	15
Duration of Coagulation	H	6
Total power consumed	W	60.5
Effluent volume	L	7

$$Al \to Al^{+3}_{(aq)} + 3e \text{ (At anode surface)}$$

$$Al^{+3} + 3H_2O \to Al(OH)_3 + 3H^+$$

$$n \, Al(OH)_3 \to Al_n(OH)_n$$

$$2H^+ + 2e \to H_2 \text{ (At cathode surface)}$$

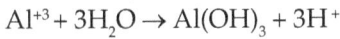

Table 24.4: Electrocoagulation Results

Parameters	Unit	Values
Colour	–	Light brown
pH	–	8.1
COD	mg/l	5450
Salt content (NaCl equivalent)	g/l	21.2
Net COD reduction	G	155
Al Consumption	Kg/kg COD	0.04
Power consumption	kWh/kg COD	0.4

Oxidation process for wastewater remediation using conventional oxidants is having their own limitation and found ineffective in many cases. The advance oxidation process involving chemical, photochemical and photocatalytic production of •OH radical act as strong oxidizing agent and can effectively oxidize organic pollutants. But, the these oxidation processes suffer due to high cost of equipments and operating cost besides generation of various other toxic by-products and low efficiency.

Electrochemical oxidation process is unique and versatile oxidation process as compared to chemical oxidation processes. There are many oxidation reactions simultaneously occur at anode during electrooxidation and systematically oxidize organic pollutants to simpler molecules (Vlyssides *et al.*, 2000). During electrooxidation process, very reactive oxygen, chlorine, hydrogen peroxide, hydroxyl radical, ozone and mediated metallic oxidants are simultaneously generated and effectively oxidize organic pollutants. There are many stable organic compounds called refractory organic pollutants which do not remove by simple oxidation processes can be oxidized effectively by electrochemical oxidation process. In electrooxidation process the selection of electrodes play an important role on effective and complete removal of pollutants. Depending upon the nature of pollutants the anode materials changes from simple steel and graphite to precious boron doped diamond (BDD) electrodes is used. The anode materials which are studied extensively for the oxidation process are Pt, PbO_2, IrO_2, TiO_2, SnO_2 and BDD besides glassy carbon, Ti/RuO_2, Ti/Pt-Ir, fiber carbon, MnO_2, Pt-carbon black, porous carbon felt, stainless steel and reticulated vitreous carbon (Lawrence *et al.*, 2007; Chen, 2004). In the present chapter, Dimensionally Stable Anode (DSA) with Mixed Metal Oxide (MMO) coating was used for oxidation. The colour of the effluent changed from dark brown of initial effluent to light brown after EC and colorless after EO (Figure 24.3). The electrooxidation process conditions are shown in Table 24.5. The final pH and COD concentration meets disposal norms set by various regulatory bodies (Table 24.6).

Table 24.5: Electrooxidation Process Conditions

Parameters	Unit	Values
No of cells (1 anode and 2 cathodes)	No	2
Anode cathode distance (ACD)	Mm	5
Current Density (cd)	A/m²	100
Cell current	A	3.6
Cell Voltage (CV) per cell	V	2.1
Anode area per volume	m²/m³	2.8
Duration of electrooxidation	H	18
Volume effluent	L	6.8

Table 24.6: Electrooxidation Results

Parameters	Unit	Values
Colour	–	Clear
pH	–	5.7
COD	mg/l	7
Salt content (NaCl equivalent)	g/l	20.9
Net COD reduction	G	37
Total power consumed	W	272.16
Power consumption	kWh/kgCOD	7.35

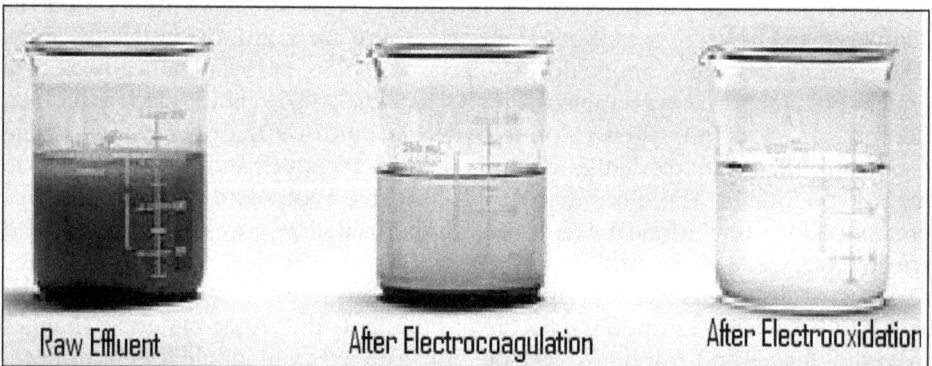

Raw Effluent   After Electrocoagulation   After Electrooxidation

Figure 24.3: Colour of Effluent at different Stages of Purification.

Conclusion

The following conclusions are drawn from the above electrochemical studies carried out on the complex industrial effluent stream:

☆ Electrochemical methods such as electrocoagulation and electrooxidation are effective treatment processes as compared to conventional wastewater or effluent treatment processes.

☆ Aluminium anode act as an effective soluble anode for EC process.

☆ No single electrochemical method can effectively used to economically treat complex and contaminant load effluents. Electrooxidation can be employed as a single treatment process to complete removal of COD. But, due to high power consumption, the electrooxidation process is alone not recommended.

☆ The final refractory organic compounds which not removed by electrocoagulation process are oxidized by electrooxidation process effectively.

☆ The total power consumption is very low as 80 per cent COD is removed by low power consuming EC process.

☆ Complex industrial effluent containing high COD can be effectively treated by applying electrochemical treatment processes systematically.

References

Butler, E., Hung, Y-T Ruth Yeh, Y-L. and Al Ahmad, M. S. (2011): Electrocoagulation in Wastewater Treatment, Water. 3: 495-525.

Chen, G. (2004): Electrochemical Technologies in Wastewater Treatment. *Separation and* Purification Technology, 38 11 – 41.

Emamjomeh, M. M. and Sivakumar, M. (2009): Review of pollutants removed by electrocoagulation and electrocoagulation/flotation processes, Journal of Environmental Management 90: 1663–1679.

Gavrel, T.G., Otto, D.W. and Vinson, I. B. (2004): Process for electro coagulating waste fluids, United States Patent 6719894.

Holt, P., Barton, G. and Mitchell, C. (1999): Electro coagulation as a wastewater treatment, The Third Annual Australian Environmental Engineering Research Event. 23-26 November Castlemaine, Victoria.

Kobya, M., Can, O. T. and Bayramoglu, M. (2003): Treatment of Textile Wastewaters by Electrocoagulation using Iron and Aluminum Electrodes. Journal of Hazardous Materials, B100: 163 – 178.

Lawrence, K., Hung, W. Y. and Shammas, N. K. (2007): Advanced Physicochemical Treatment Technologies (Handbook of Environmental Engineering, 5: (Humana Press).

Moreno, H.A.C., Cooke, D., Gomes, J.A.G., Morokovsky, P., Parga, J.R., Peterson, E. and Garcia, C (2009): Electrochemical reactions for electrocoagulation using iron electrodes. Ind. Eng. Chem. Res., 48: 2275-2282.

Mouli, P. C., Mohan, S. V. and Reddy, S. J. (2004): Electrochemical process for the remediation of wastewater and contaminated soil: emerging technology, Journal of Scientific and Industrial Research, 63: January, 11-19.

Ni'am, M. F., Othman, F., Sohaili, J. and Fauzia, Z. (2007): Removal of COD and turbidity to improve wastewater quality using electrocoagulation technique, The Malaysian Journal of Analytical Sciences, 11(1):. 198-205.

Paul, E. M. and Kaspar, D.D. (2004): Process and apparatus for electrocoagulative treatment of industrial wastewater, United States Patent 6689271.

Vlyssides, A. G., Papaioannou, D., Loizidoy, M., Karlis, P. K. and Zorpas, A. A. (2000): Testing an electrochemical method for treatment of textile dye wastewater, Waste Management 20: 569- 574.

Vik, E.A. (1983): Treatment of potable water containing humus by electrocoagulation and followed by filtration, Water supply, 1(2-3): 23.

Vepsalainen, M. (2012): Electrocoagulation in the treatment of industrial waters and wastewaters, Espoo. VTT Science 19. 96 p.

Zhlik, K. and Kalvoda, R. (1996): Electrochemistry for environmental protection, UNESCO Venice Office, Regional Office for Science and Technology for Europe (ROSTE), 1262/A Dorsoduro-Venice, Italy, 30: 123.

Chapter 25

Ambient Air Quality around OCL India Limited at Rajgangpur, Odisha, India

Sanjat K. Sahu[1], A. Mishra[2] and Aditya Kishore Dash[3]

[1] *P.G. Dept. of Environmental Sciences, Sambalpur University,*
Jyoti Vihar, Odisha
[2] *OCL India Limited, Rajgangpur, Odisha*
[3] *Environmental Engineering Department,*
Siksha 'O' Anusandhan University,
Institute of Technical Education and Research, Khandagiri,
Bhubaneswer – 751 030, Odisha, India

ABSTRACT

A work was undertaken to study the ambient air quality around OCL India Limited in Rajgangpur, Odisha. To determine the ambient air quality, parameters like suspended particulate matter (SPM), respirable particulate matter (RPM) and gaseous pollutants SO_x, NO_x and CO were monitored. Respirable dust samplers and CO monitor are used for measurement of the concentrations of all parameters mentioned above at representative assessment points situated nearer to the cement plant of OCL at four locations. Furthermore, dust and gaseous load from cement plant on the four locations near OCL were analyzed through air pollution index (API). The measured value and standards value prescribed by CPCB, OSPCB of ambient air quality were also compared and inferences were drawn accordingly.

Keywords: OCL India, SPM, RPM, SO_x, NO_x, CO, Meteorology.

Introduction

Of all the Pollution, industries find an important place. Most parts of the world are now affected in some way or other by industrial pollution. The smog, smell, and contamination of air, water or food are some of the direct effects of industrial pollution. Remote areas of the world are also affected indirectly by industrial pollution. For example, the air pollution caused by the industries can carry for many miles. The Environmental Protective Agency estimates that up to 50 per cent of the nation's pollution are caused by industry. Because of its size and scope, industrial pollution is a serious problem for the entire planet, especially in nations which are rapidly industrializing (Agrawal, 1998; Trivedi, 2000; Sharma and Kaur, 2002).

Cement industry is one of the 17 most polluting industries listed by the Central Pollution Control Board (CPCB). Cement industry is the major source of particulate matters, SO_x, NO_x and CO_2, emissions. Cement dust contains heavy metals like nickel, cobalt, lead, and chromium, pollutants hazardous to the biotic environment, with impact for vegetation, human and animal health and ecosystems (WHO, 1979; WHO, 2003).

Brief Description on OCL

OCL (Orissa Cement Limited) India Ltd, Rajgangpur, Dist Sundargarh, Orissa, formerly Orissa Cement Limited, popularly known, as "OCL" is the leading manufacturer of Cement and Refractory in the state. The Cement Factory with its captive limestone mines at Lanjiberna was established in the year 1951 and continued to progress over the years as follows:

- ☆ Diversified to refractory manufacturing in the year 1954. Current capacity of the Refractory unit is 80,000 TPA and located inside the OCL.
- ☆ Changed process of cement manufacturing from "wet to dry" in the year 1989.
- ☆ Diversified to sponge iron manufacturing in the year 2002 and integrated steel plant in the year 2005 which is located at about 8km away from the OCL.
- ☆ 1 million tone Cement grinding unit at Kapilas Road near Cuttack commissioned in July 2008.
- ☆ 1.7 million Tone Clinkerisation project at Rajgangpur is completed in March, 2009.
- ☆ 2X27 MW CPP expected to be commissioned during 2nd half of 2010-2011.
- ☆ Proposal to have more Grinding Units.

OCL is one of the largest producers of Portland Slag cement in eastern India. The total cement production capacity is 4 MTPA and Clinker capacity is 2.9 MTPA. The Company has acquired ISO 9001 for the Cement Division and the Refractory Division. It has also acquired the API certification for manufacture of Oil-Well cement being one of the only four such certified companies in India. Recently Cement Division has been certified with ISO 14001:2004 and IS 18001:2007 from BIS. OCL cement has been widely used in major constructions in eastern India such as Hirakud Dam, Second Hoogly Bridge, Mahatma Gandhi setu, Modernisation of Rourkela Steel Plant, Science city in Kolkata etc. The Cement is marketed under the brand name "Konark"and OCL is the market leader in Orissa over 5 decades. OCL also exports cement to neighbouring countries like Bangladesh and Nepal.

Manufacturing Process

The flow diagrams of the existing and proposed expanded plant are shown below and described hereunder: (OCL INTRANET-2008).

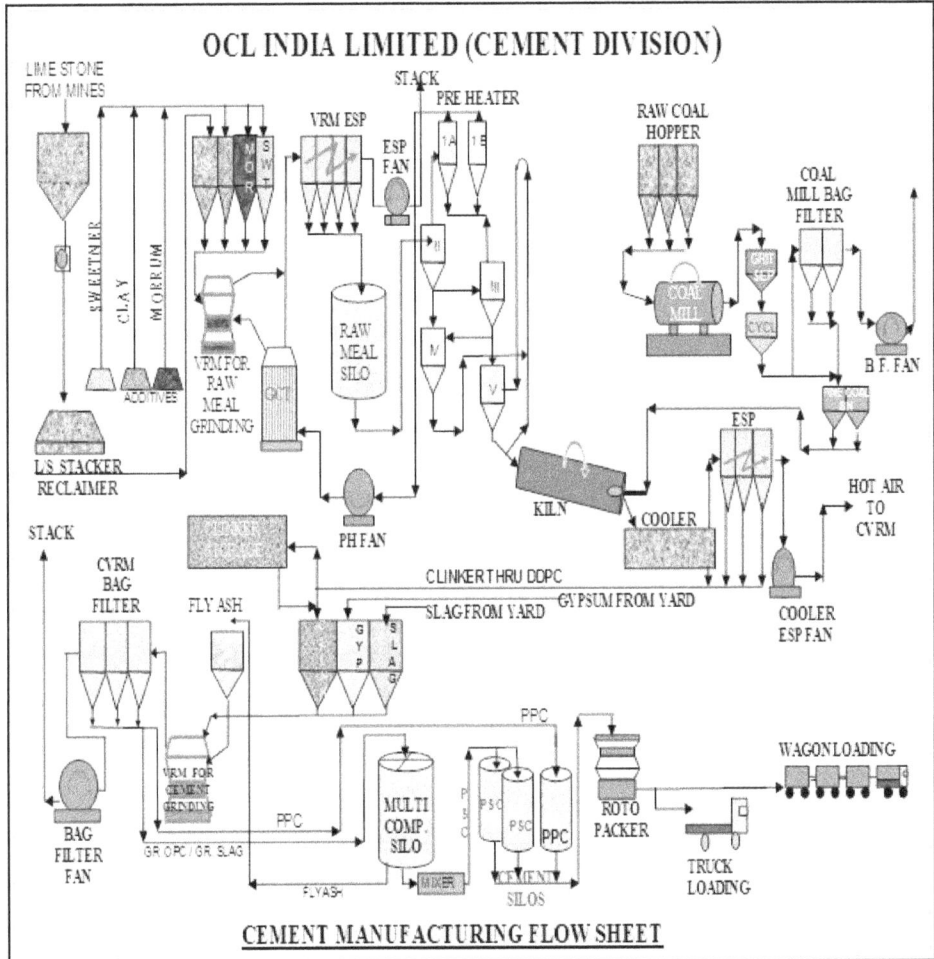

CEMENT MANUFACTURING FLOW SHEET

The manufacturing of Cement involves mining; crushing and grinding of raw materials (mostly limestone and clay); calcinations the material in rotary kiln; cooling the resulting clinker; mixing the clinker with Gypsum; and milling, storing and bagging the finished cement.

Source of Air Pollution around OCL

Various industrial installations, such as, cement manufacturing, steel and sponge iron manufacturing, engineering workshops etc. form the stationary sources of the air pollution. The automobiles such as cars, scooters, trucks, buses form the mobile sources of the air pollution. Both localized and trans-boundary sources of air pollutants are especially significant around the OCL India Ltd.

The cement manufacturing process generates lot of dust. The sources of dust emission include clinker cooler, crushers, grinders and material-handling equipments. Material-handling operations such as conveyors result in fugitive dust emission. There are so many sponge iron and mini steel plant is situated around OCL India limited, Rajgangpur which are emitting particulate emission and pollute the environment surrounding the OCL India limited.

Objectives of the Study

The main study objective was to determine the quality of ambient air on the basis of monthly measurement at four locations at the radius of 1 km and 2 km around the OCL INDIA LIMITED, Rajgangpur and to compare with the standards prescribed by CPCB. The parameters studied are: SPM, RPM, SO_2, NO_x and CO content.

Study Site and Methodology Followed

The study areas are selected around the OCL India Limited at four locations (N, S, E and W) at a distance of 1 and 2 km radius. All locations at 1 and 2 km radius from the OCL India Limited are as follows;

1 Km Radius

1. North Direction – Locogate
2. South Direction – Stores
3. East Direction - DITC
4. West Direction - Main gate

2 Km Radius

1. North Direction - Ranibandh
2. South Direction - OCL Market
3. East Direction - IT Colony
4. West Direction - Liploi

Meteorological Data

Meteorological data was taken from Weather Monitoring Station (WM 271) which is recently installed in the Cement Division of OCL India Ltd. Data on daily maximum, minimum and average temperature, relative humidity and rainfall were taken.

Period of Sampling and Parameters Studied

The period carried out for study was 8 months, *i.e.*, from October 2009 to May 2010 and measurements were taken on monthly basis with 24 – hrs sampling. All parameters of ambient air like SPM, RPM, SO_2, NO_x and CO were measured by Respirable Dust Sampler (APM 611 of Envirotech and CO monitor of NEVCO) and ambient air concentrations of suspended particulate matter above 10 microns, particulate matter less than 10 microns (PM_{10}) in µg/m³, sulphur dioxide (SO_2) in µg/m³, nitrogen dioxide (NO_2) in µg/m³, and carbon monoxide (CO) in ppm were obtained. The monitoring instruments and operation protocols of the AAQM are approved by CPCB and OSPCB (Table 25.1). These instruments are reliable and durable for 24 hours continuous field applications.

Table 25.1: Details of Air Quality Monitoring Equipments and Methods Used for Parameters Detection.

Instrument Sl.No.	Model	Make/Supplier	Parameters	Methods as per CPCB
1536-DTG-2009	APM 460BL	ENVIROTECH	SPM	Gravimetric method with RDS (Avg. Flow rate not less than 1.1 m³/min)
1537-DTG-2009	APM 460BL	ENVIROTECH	RPM	As above
1338-DTG-2009	APM 460BL	ENVIROTECH	SO_x	Improved Waste and Gaeke method
1339-DTG-2009	APM 460BL	ENVIROTECH	NO_x	Jacob and Hechheiser method
J409-M004644	–	NEVCO	CO	Gas filter correlation
39-DTA-2009	WM-271	ENVIROTECH	METEORO-LOGICAL DATA	

Air Pollution Index (API)

The overall ambient air quality of the locality was quantified and categorized in terms of Air Pollution Index (API). Pollutants responsible for affecting air quality usually pose synergetic effect. Concentration of SO_2 gas is more corrosive when combined with dust particulate. So cumulative and combined effect of pollutants in terms of Air Pollution Index (API) can be calculated with respect to specific pollutant by using the formula:

$$I = \frac{1}{2} \times 1/SA \times Ca + \{\text{Summation } n = 1 \text{ to } n \, (Cd)^2\}^{1/2}$$

Where, I = Specific pollutant Index, SA = Standard for the Pollutant, Ca = Annual average concentration, Cd = Daily concentration, n = Number of observations.

If API value is greater than 1, it is considered that concentration exceeds the limit and if it is less than 1, it is considered to be within limit.

The following formula was adopted to calculate the Air Pollution Index (API) of certain industrial locations to evaluate the index:

$$API = \frac{1}{n}\left(\frac{X_1}{a} + \frac{X_2}{b} + \frac{X_3}{c} + \frac{X_n}{y}\right) \times 100 \qquad \text{Eqn (I)}$$

where,

n = Number of pollutants studied

$X_1, X_2, X_3 \ldots$ Xn is the pollutant numbers

a, b, c, ... Y is the prescribed safer limits of their concentration.

The area under study is covering mainly three pollutants SPM, SO_2, NO_x, basing on which API equation is taken as

$$API = \frac{1}{3}\left(\frac{(SPM)}{140} + \frac{(SO_2)}{60} + \frac{(NO_x)}{60}\right) \times 100 \qquad \text{Eqn (II)}$$

Where 140 are the safer limits for SPM and 60 is the safer limit for SO_2 and NO_x as per CPCB guideline for residential localities on 24hours sampling basis. As because gaseous emissions are very less (Below Detection Level), 24 hours of sampling is undertaken for both particulate and gaseous pollutant (instead of 8 hour sampling) monitoring so that accuracy can be maintained.

Results and Discussion

The meteorological data during the period from October 2009 to May 2010 are given in Table 25.2.

Table 25.2: Meteorological Data

Date (dd/mm/yy)	Wind Direction (Deg)	Temp (Deg C)	R Humidity (per cent)	Spd (km/hr)	Rain (mm)
October - 2009					
05-10-09					
Max	330	29.2	78.4	2.7	0
Min	57	26	64.1	0.2	0
Avg	246	27.8	69.8	1.2	0
7-10-09					
Max	360	30.5	86.1	8.5	0
Min	25	24	46	0	0
Avg	163	26.7	67.6	2.1	0
12-10-09					
Max	348	32.5	87.3	3.9	0
Min	50	20.5	38.4	0	0
Avg	176.8	26.3	64.6	1.2	0
14-10-09					
Max	360	31.1	92.1	9.6	0
Min	19	21.1	38.4	0	0
Avg	152.2	26.3	64.9	2.9	0
21-10-09					
Max	342	34	84.4	4.8	0
Min	42	23.1	48.2	0	0
Avg	174.6	28	69.1	1.2	0

Contd...

Table 25.2–*Contd...*

Date (dd/mm/yy)	Wind Direction (Deg)	Temp (Deg C)	R Humidity (per cent)	Spd (km/hr)	Rain (mm)
23-10-09					
Max	360	35.4	90.3	5.8	0
Min	47	24	42.3	0	0
Avg	226.9	28.8	69.2	1.6	0
26-10-09					
Max	358	32.5	89.4	10.2	3.5
Min	23	23.3	53	0	0
Avg	199.3	26.9	76.7	2.6	1.2
28-10-09					
Max	355	30.1	93	7.9	0
Min	1	24	64.1	0	0
Avg	192.5	26.3	80	2.1	0
November - 2009					
02-11-09					
Max	180	25	94.5	8.9	7.5
Min	16	19.1	62.3	0.4	0
Avg	74.7	22.5	80	3.6	2.1
04-11-09					
Max	353	25.6	95.3	4.3	0
Min	23	20	70	0.1	0
Avg	172.1	22.1	87	1.9	0
06-11-09					
Max	359	29.5	95.3	6.3	0
Min	20	20	41.5	0.1	0
Avg	152.1	23.8	72.3	1.9	0
08-11-09					
Max	340	28.4	94.2	3.9	0
Min	9	17	38.4	0	0
Avg	173.8	22	69	1.2	0
12-11-09					
Max	360	27.3	92.2	4.8	0
Min	1	15.2	33.2	0	0
Avg	222.7	20.4	66	1.4	0
15-11-09					
Max	360	27.1	93	5.1	0
Min	1	13	29.6	0	0
Avg	179.7	19.4	65.4	1.4	0

Contd...

Table 25.2–*Contd...*

Date (dd/mm/yy)	Wind Direction (Deg)	Temp (Deg C)	R Humidity (per cent)	Spd (km/hr)	Rain (mm)
20-11-09					
Max	358	28.4	94.3	5.6	0
Min	33	12.2	33.2	0	0
Avg	175.7	19.7	66.2	1.5	0
23-11-09					
Max	341	28.6	93.5	3.4	0
Min	1	13.1	27.2	0	0
Avg	197.7	19.8	65.8	0.7	0
December - 2009					
02-12-09					
Max	358	29.4	82.2	5.1	0
Min	90	17.4	32.4	0.1	0
Avg	214.7	22.7	57.3	1.6	0
04-12-09					
Max	359	29.3	90.5	4.2	0
Min	0	14.5	26	0	0
Avg	200	20.5	63.9	1	0
07-12-09					
Max	344	29.4	91.3	5.6	0
Min	90	13.3	27.5	0	0
Avg	211.6	20.6	62.7	1.2	0
10-12-09					
Max	356	30.3	94.4	6.9	0
Min	0	16.4	36.4	0	0
Avg	190.2	22.7	67.8	1.5	0
16-12-09					
Max	353	31.1	92.5	5.6	0
Min	0	15.1	34	0	0
Avg	182.3	22.1	67.8	1.5	0
22-12-09					
Max	356	28.2	91.3	4.2	0
Min	1	13	30.2	0	0
Avg	217	19.3	64	1.3	0
24-12-09					
Max	358	27.5	88.2	5.9	0
Min	10	10.4	24.4	0	0
Avg	173.3	18.3	58.3	1.3	0

Table 25.2–*Contd...*

Date (dd/mm/yy)	Wind Direction (Deg)	Temp (Deg C)	R Humidity (per cent)	Spd (km/hr)	Rain (mm)
26-12-09					
Max	346	27.1	90.1	6	0
Min	13	11	28.1	0	0
Avg	137.8	18.8	59.3	1.9	0
January – 2010					
01-01-10					
Max	358	27.5	90.4	5.1	0
Min	0	12.2	16.4	0	0
Avg	188.9	18.8	55.8	1.2	0
05-01-10					
Max	357	27.4	89.3	4.7	0
Min	0	11.6	26.3	0	0
Avg	151.1	19	57.4	1.5	0
07-01-10					
Max	359	26.1	93.2	5.1	0
Min	1	10.4	32	0	0
Avg	213.6	17.1	63.6	1.7	0
10-01-10					
Max	354	28	82.2	3.4	0
Min	7	17.2	33.1	0	0
Avg	211.5	21.4	60	0.8	0
12-01-10					
Max	358	30.4	88.5	5.7	0
Min	6	13.3	28.1	0	0
Avg	177	21.8	58.4	1.3	0
15-01-10					
Max	330	24.6	93	6.3	0
Min	143	17	57.2	0	0
Avg	247.4	20.7	74.8	2	0
17-01-10					
Max	347	27.3	94.3	5.5	0
Min	0	14.5	33.5	0	0
Avg	241.9	20.4	67.9	1.5	0

Contd...

Table 25.2–*Contd...*

Date (dd/mm/yy)	Wind Direction (Deg)	Temp (Deg C)	R Humidity (per cent)	Spd (km/hr)	Rain (mm)
February – 2010					
01-02-10					
Max	358	30	81.3	7.5	0
Min	55	12.3	23	0	0
Avg	210.8	21.0	50.6	2.1	0.0
03-02-10					
Max	360	30.4	83.4	7.2	0
Min	2	12.1	23.2	0	0
Avg	170.0	21.7	51.0	2.2	0.0
05-02-10					
Max	354	30.3	81.3	6.6	0
Min	18	14.2	21.4	0.1	0
Avg	226.0	22.3	49.3	2.0	0.0
08-02-10					
Max	343	30	81.1	5.7	0
Min	4	14.3	24.6	0	0
Avg	166.5	22.4	48.9	2.3	0.0
22-02-10					
Max	354	30.5	81.3	4.9	0
Min	1	14	16.3	0	0
Avg	152.5	22.5	43.2	1.7	0.0
24-02-10					
Max	350	31	77.4	5.5	0
Min	4	12.1	17.1	0	0
Avg	145.9	21.7	44.0	1.8	0.0
25-02-10					
Max	360	32.1	76.3	9.1	0
Min	14	13.2	17	0	0
Avg	128.8	23.1	41.4	2.7	0.0
27-02-10					
Max	343	32.4	76.3	9.1	0
Min	4	13.2	17	0	0
Avg	127.9	22.7	44.6	2.3	0.0
March – 2010 (Data is not available due to problem in the system)					

Contd...

Table 25.2–*Contd...*

Date (dd/mm/yy)	Wind Direction (Deg)	Temp (Deg C)	R Humidity (per cent)	Spd (km/hr)	Rain (mm)
April – 2010					
01-04-10					
Max	339	44.2	29.5	22.5	0
Min	78	34.5	10	2.1	0
Avg	194.8	40.0	17.0	8.9	0.0
02-04-10					
Max	116	42.5	22.2	9.3	0
Min	34	34.4	11	2.9	0
Avg	70.2	39.2	13.8	6.2	0.0
05-04-10					
Max	272	42.2	19.2	12.2	0
Min	19	34.3	8	2.8	0
Avg	83.5	39.1	11.7	6.4	0.0
07-04-10					
Max	348	42.2	22.1	16.6	0
Min	42	33.1	12.2	0.4	0
Avg	163.8	38.2	17.3	6.9	0.0
14-04-10					
Max	358	42.2	31.1	11.2	0
Min	5	33.4	15.1	0.1	0
Avg	143.3	38.7	20.9	6.0	0.0
16-04-10					
Max	344	42.4	78.3	13.2	0.5
Min	5	24.3	18.3	8.4	0
Avg	116.6	37.5	31.4	10.7	0.1
18-04-10					
Max	321	40.2	38	8.9	0
Min	11	31.3	20	0.9	0
Avg	173.2	36.2	29.3	5.0	0.0
20-04-10					
Max	74	40.4	37	15.6	0
Min	2	33.2	21.1	7.4	0
Avg	33.3	37.4	26.3	10.4	0.0

Table 25.2–Contd...

Date (dd/mm/yy)	Wind Direction (Deg)	Temp (Deg C)	R Humidity (per cent)	Spd (km/hr)	Rain (mm)
May – 2010					
03-05-10					
Max	353	41.4	56.1	20.1	0
Min	20	27.5	17.3	2.5	0
Avg	158.3	35.1	33.3	9.3	0.0
05-05-10					
Max	320	39.2	65.2	12.2	1
Min	9	27	21.1	0.2	0
Avg	136.3	33.8	39.0	6.8	0.1
07-05-10					
Max	354	40.5	83.1	8.6	0
Min	42	26.4	23	0.1	0
Avg	155.0	32.8	51.1	3.8	0.0
10-05-10					
Max	328	43.6	68.5	11.9	0
Min	18	29.1	12	0.1	0
Avg	111.7	36.0	36.6	3.8	0.0
12-05-10					
Max	354	44.2	51.5	9.4	0
Min	12	28.1	11.1	0	0
Avg	172.0	36.3	27.2	2.4	0.0
14-05-10					
Max	316	43	50.4	13.3	0
Min	2	30.3	21.1	0	0
Avg	95.0	36.0	33.2	5.7	0.0
16-05-10					
Max	342	44.1	44.6	8.9	0
Min	18	30	13.3	0.7	0
Avg	179.5	36.4	30.4	3.8	0.0
18-05-10					
Max	359	42.4	70	25.7	0.5
Min	11	29.3	17.4	0.5	0
Avg	256.5	35.2	43.4	10.0	0.0

The windrose diagram during the month of January (Figure 25.1) and May (Figure 25.2) 2010 is given below.

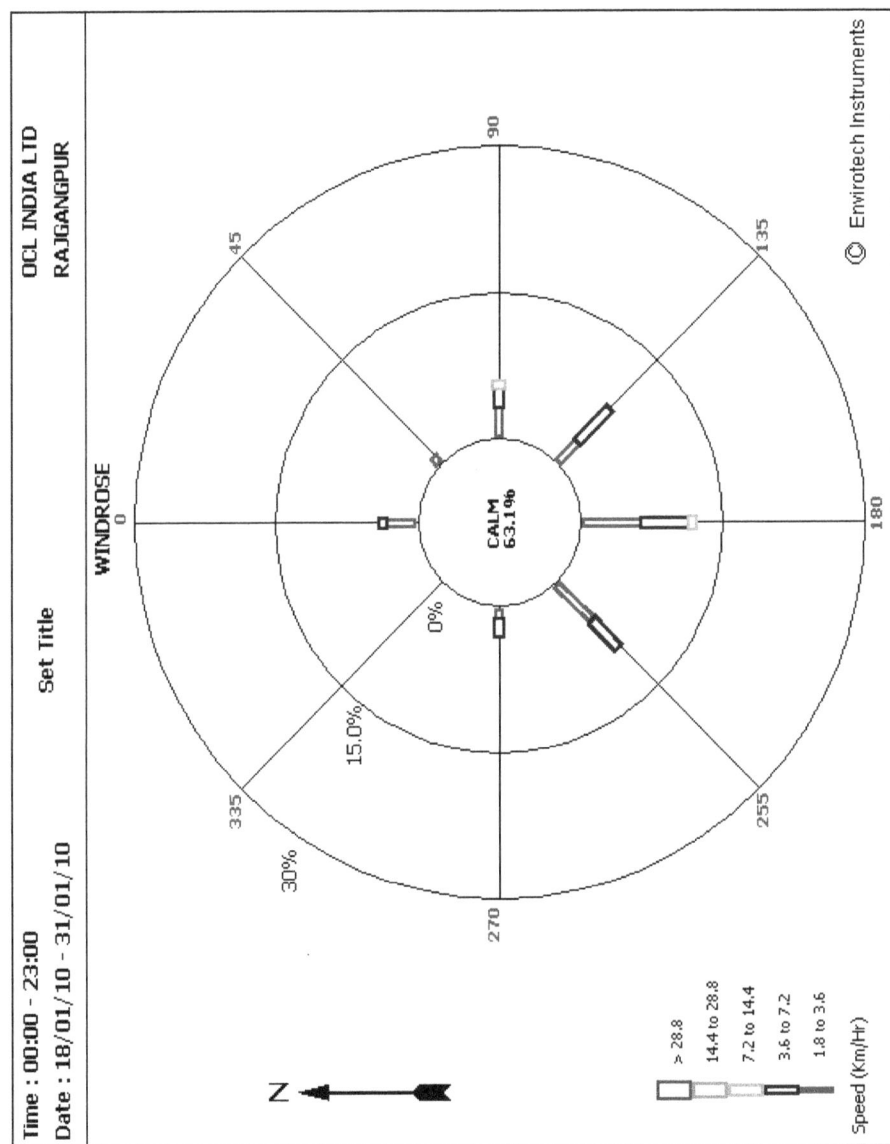

Figure 25.1: Windrose Diagram during the Month of January 2010.

472   *Advances in Environmental Sciences and Engineering*

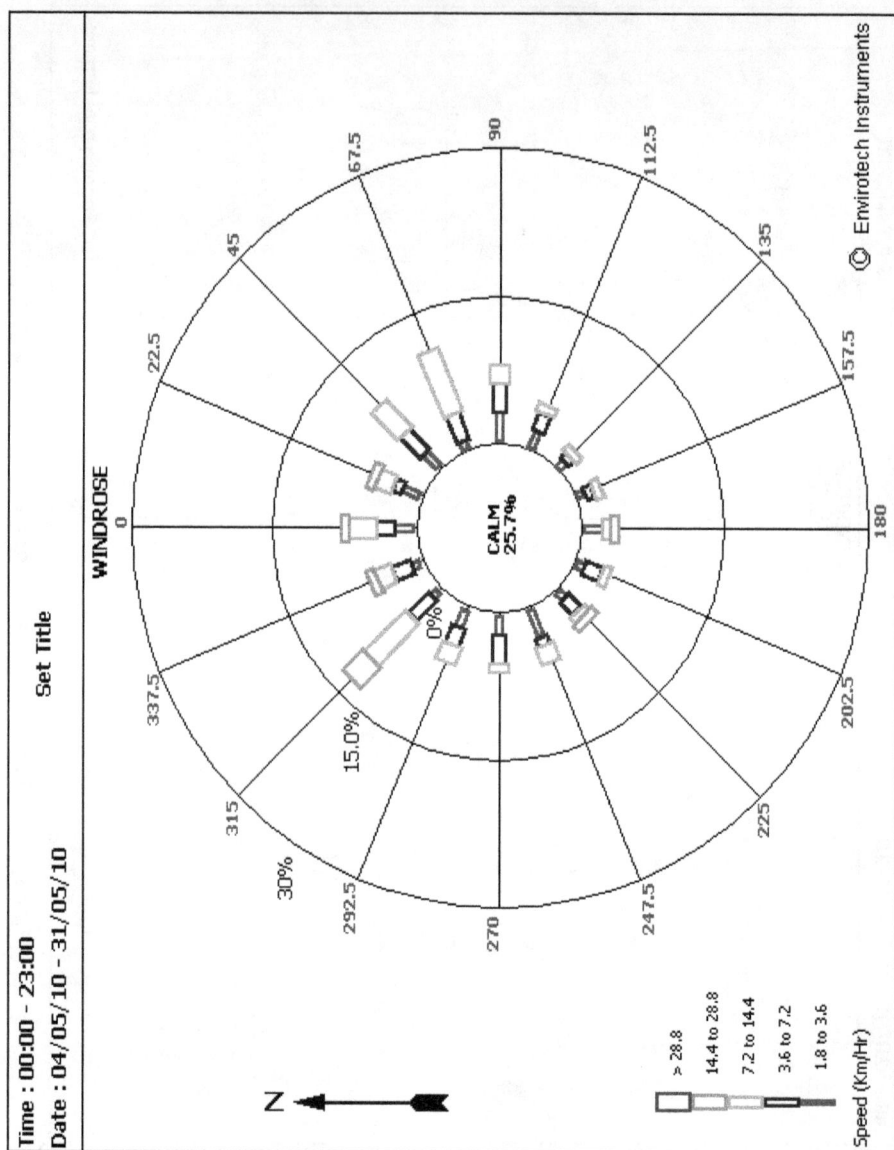

Figure 25.1: Windrose Diagram during the Month of May 2010.

The parameters like SPM, RPM, and SO_x and NO_x were measured and their valued were computed at four different locations within 1 and 2 km distance from the Cement Division of OCL India Limited. All measured values are compiled as shown in the Table 25.3 below. Table 25.4 shows Monthly variation in SPM, RPM, SO_2, NO_2 and CO at 2 km radius around OCL India Ltd.

Table 25.3: Monthly Variation of SPM, RPM, SO_2, NO_x and CO at 1 Km Radius around OCL India Ltd.

Date	Location	Sampling Time (Hrs)	Avg. Flow (m³/min)	SPM (µg/m³)	RPM (µg/m³)	SO₂ (µg/m³)	NO₂ (µg/m³)	CO PPM
OCTOBER-2009								
05.10.09	NEAR LOCOGATE	24	1.1	282.5	97.2	2.1	20.8	0.2
07.10.09	NEAR STORES	24	1.1	254.6	95.3	2.3	18.5	0.3
12.10.09	NEAR DITC	24	1.1	231.4	94.8	2.6	14.9	0.1
14.10.09	NEAR MAIN GATE	24	1.1	270.6	88.6	2.7	26.8	0.2
NOVEMBER-2009								
02.11.09	NEAR LOCOGATE	24	1.1	270.5	89.5	2.0	22.0	0.1
04.11.09	NEAR STORES	24	1.1	247.3	87.2	2.3	19.0	0.2
06.11.09	NEAR DITC	24	1.1	203.3	95.6	2.7	16.1	0.3
08.11.09	NEAR MAIN GATE	24	1.1	252.0	90.2	2.5	24.4	0.2
DECEMBER-2009								
02.12.09	NEAR LOCOGATE	24	1.1	272.7	96.8	2.1	23.2	0.1
04.12.09	NEAR STORES	24	1.1	256.8	92.5	2.5	20.2	0.1
07.12.09	NEAR DITC	24	1.1	206.1	91.1	2.5	17.3	0.2
10.12.09	NEAR MAIN GATE	24	1.1	256.8	92.5	2.7	22.6	0.1
JANUARY-2010								
01.01.10	NEAR LOCOGATE	24	1.1	293.5	93.5	2.2	17.3	0.2
05.01.10	NEAR STORES	24	1.1	309.2	98.3	2.4	32.1	0.1
07.01.10	NEAR DITC	24	1.1	274.9	98.9	2.3	12.5	0.1
12.01.10	NEAR MAIN GATE	24	1.1	259.2	98.9	2.2	26.8	0.1
FEBRUARY-2010								
01.02.10	NEAR LOCOGATE	24	1.1	276.3	93.5	3.6	24.4	0.2
03.02.10	NEAR STORES	24	1.1	270.7	95.1	3.1	31.0	0.1
05.02.10	NEAR DITC	24	1.1	292.4	97.9	3.4	19.0	0.1
08.02.10	NEAR MAIN GATE	24	1.1	296.8	96.0	4.0	23.2	0.1
MARCH-2010								
02.03.10	NEAR LOCOGATE	24	1.1	287.2	99.1	3.6	24.4	0.2
03.03.10	NEAR STORES	24	1.1	282.1	93.9	3.1	31.0	0.1
05.03.10	NEAR DITC	24	1.1	277.1	94.1	3.4	19.0	0.1
08.03.10	NEAR MAIN GATE	24	1.1	269.4	87.1	4.0	23.2	0.1

Contd...

Table 25.3–*Contd...*

Date	Location	Sampling Time (Hrs)	Avg. Flow (m³/min)	SPM (µg/m³)	RPM (µg/m³)	SO₂ (µg/m³)	NO₂ (µg/m³)	CO PPM
APRIL-2010								
01.04.10	NEAR LOCOGATE	24	1.1	286.6	98.6	3.8	26.8	0.1
02.04.10	NEAR STORES	24	1.1	281.2	93.0	2.9	32.7	0.1
05.04.10	NEAR DITC	24	1.1	276.5	93.4	3.5	20.8	0.2
07.04.10	NEAR MAIN GATE	24	1.1	269.2	86.5	4.2	19.0	0.1
MAY-2010								
03.05.10	NEAR LOCOGATE	24	1.1	284.7	96.7	4.0	25.0	0.1
05.05.10	NEAR STORES	24	1.1	280.5	92.4	3.1	30.4	0.2
07.05.10	NEAR DITC	24	1.1	279.0	96.1	3.2	19.0	0.1
10.05.10	NEAR MAIN GATE	24	1.1	268.9	86.5	3.7	20.8	0.1

Table 25.4: Monthly Variation in SPM, RPM, SO₂, NO₂ and CO at 2 Km Radius around OCL India Ltd.

Date	Location	Sampling Time (Hrs)	Avg. Flow (m³/min)	SPM (µg/m³)	RPM (µg/m³)	SO₂ (µg/m³)	NO₂ (µg/m³)	CO PPM
OCTOBER-2009								
21.10.09	RANIBANDH	24	1.1	238.8	114.3	5.2	41.1	0.4
23.10.09	OCL MARKET	24	1.1	189.9	94.7	4.6	20.2	0.2
26.10.09	IT COLONY	24	1.1	114.6	62.3	3.5	17.3	0.2
28.10.09	LIPLOI	24	1.1	213.2	106.7	3.9	22.6	0.3
NOVEMBER-2009								
12.11.09	RANIBANDH	24	1.1	226.9	108.0	5.0	48.8	0.5
15.11.09	OCL MARKET	24	1.1	200.8	88.4	4.7	21.4	0.3
20.11.09	IT COLONY	24	1.1	124.4	69.4	2.5	31.5	0.3
23.11.09	LIPLOI	24	1.1	225.3	108.6	3.4	14.9	0.4
DECEMBER-2009								
16.12.09	RANIBANDH	24	1.1	230.1	110.5	5.3	54.8	0.6
22.12.09	OCL MARKET	24	1.1	213.4	98.5	3.8	31.0	0.2
24.12.09	IT COLONY	24	1.1	111.7	63.3	3.1	32.7	0.1
26.12.09	LIPLOI	24	1.1	207.0	106.1	3.2	26.8	0.3
JANUARY-2010								
10.01.10	RANIBANDH	24	1.1	244.6	122.5	5.1	46.4	0.4
12.01.10	OCL MARKET	24	1.1	207.1	93.4	4.1	22.6	0.1
15.01.10	IT COLONY	24	1.1	106.7	62.5	3.7	28.6	0.1
17.01.10	LIPLOI	24	1.1	200.7	101.0	4.1	30.4	0.2

Contd...

Table 25.4–*Contd...*

Date	Location	Sampling Time (Hrs)	Avg. Flow (m³/min)	SPM (µg/m³)	RPM (µg/m³)	SO₂ (µg/m³)	NO₂ (µg/m³)	CO PPM
FEBRUARY-2010								
22.02.10	RANIBANDH	24	1.1	235.0	114.3	4.8	53.0	0.3
24.02.10	OCL MARKET	24	1.1	189.9	94.7	4.4	29.2	0.2
25.02.10	IT COLONY	24	1.1	202.3	62.3	3.0	23.2	0.3
27.02.10	LIPLOI	24	1.1	232.1	100.4	2.8	34.5	0.2
MARCH-2010								
14.03.10	RANIBANDH	24	1.1	244.6	120.6	5.3	31.0	0.2
16.03.10	OCL MARKET	24	1.1	207.1	94.7	3.9	15.5	0.1
18.03.10	IT COLONY	24	1.1	190.2	70.1	3.0	37.5	0.1
22.03.10	LIPLOI	24	1.1	244.3	96.0	3.8	20.8	0.3
APRIL-2010								
14.04.10	RANIBANDH	24	1.1	242.7	101.6	4.2	42.9	0.1
16.04.10	OCL MARKET	24	1.1	207.7	85.9	4.3	13.1	0.2
18.04.10	IT COLONY	24	1.1	124.4	75.9	4.1	20.8	0.2
20.04.10	LIPLOI	24	1.1	238.6	99.7	2.1	32.7	0.2
MAY-2010								
12.05.10	RANIBANDH	24	1.1	253.0	116.2	5.3	28.6	0.2
14.05.10	OCL MARKET	24	1.1	219.7	99.7	4.6	17.9	0.1
16.05.10	IT COLONY	24	1.1	119.3	70.1	3.3	40.5	0.1
18.05.10	LIPLOI	24	1.1	249.9	106.7	2.5	36.3	0.1

Figures 25.3 a and b shows the variation in SPE, RPM, SO_x, NO_x and CO concentration at 1 and 2 km radius around OCL India Ltd.

Table 25.5 shows the ambient air quality standard as per CPCB. From the above results it is evident that the SPM, RPM, SO_x, NO_x and CO value of 2 km radius are less than 1 km radius which indicates that the effect of pollutants are less with greater distance from the source of pollution.

The values of SPM are found to be maximum in south direction *i.e.* at stores in the month of January with respective wind direction. However, SPM values are almost same in all the directions. It is observed that measured values of all the pollutants are within the limits but somewhere the RPM values are nearer to the new limits prescribed by CPCB which is not by OCL but by other mini sponge iron industries near by the OCL India Limited. SO_x, NO_x and CO values are found to be very less than the prescribed limit of new notification by CPCB at both the areas within 1 and 2 km radius.

The results of Study of Air Pollution Index (API) reveals that ambient air quality around OCL India Limited are coming 94,84,79 and 79 at 1 km radius and 94,72,57

Table 25.5: Ambient Air Quality Standards.

Pollutant	Time Weighted Average	Concentration in Ambient air				
		OLD			As per New Notification on 26.11.2009	
		Industrial Area	Residential, Rural and Other	Sensitive Area	Industrial, Residential, Rural and other Area	Ecologically Sensitive Area (notified by Central Government)
Sulpher Dioxide (SO_2) µg/m³	Annual Avg.	80	60	15	50	20
	24 Hrs.	120	80	30	80	80
Oxides of Nitrogen (NO_2) µg/m³	Annual Avg.	80	60	15	40	30
	24 Hrs.	120	80	30	80	80
Suspended Particulate Matter (SPM) µg/m³	Annual Avg.	360	140	70	60 (PM_{10})	30 (PM_{10})
	24 Hrs.	500	200	100	100 (PM_{10})	80 (PM_{10})
Respirable Particulate Matter (RPM) µg/m³	Annual Avg.	120	60	50	40 ($PM_{2.5}$)	40 ($PM_{2.5}$)
	24 Hrs.	150	100	75	60 ($PM_{2.5}$)	60 ($PM_{2.5}$)
Lead (Pb)	Annual Avg.	1.0	0.75	0.5	0.5	0.5
	24 Hrs.	1.5	1.0	0.75	1.0	1.0
Carbon Monoxide (CO) mg/m³	Annual Avg.	5.0	1.0	1.0	2.0	2.0
	1 Hr.	10	2.0	2.0	4.0	4.0

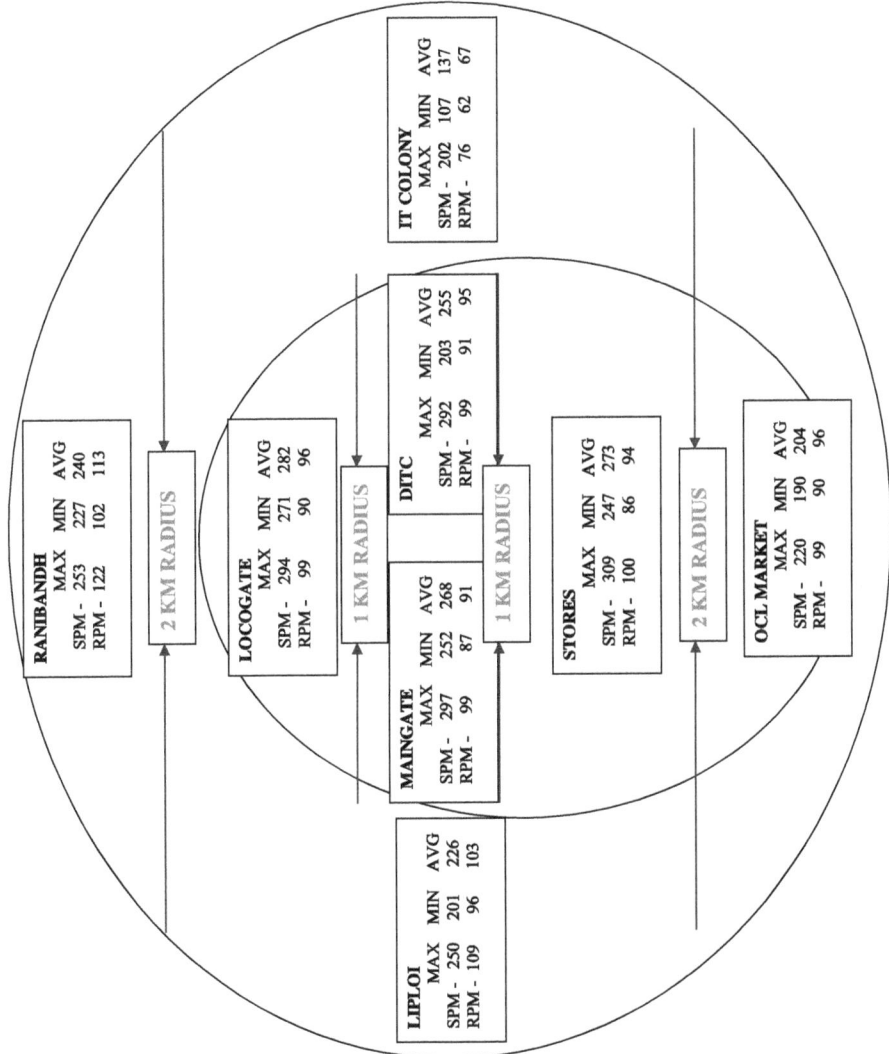

Figure 25.3a: Variation in SPE, RPM Concentration at 1 and 2 km Radius around OCL India Ltd.

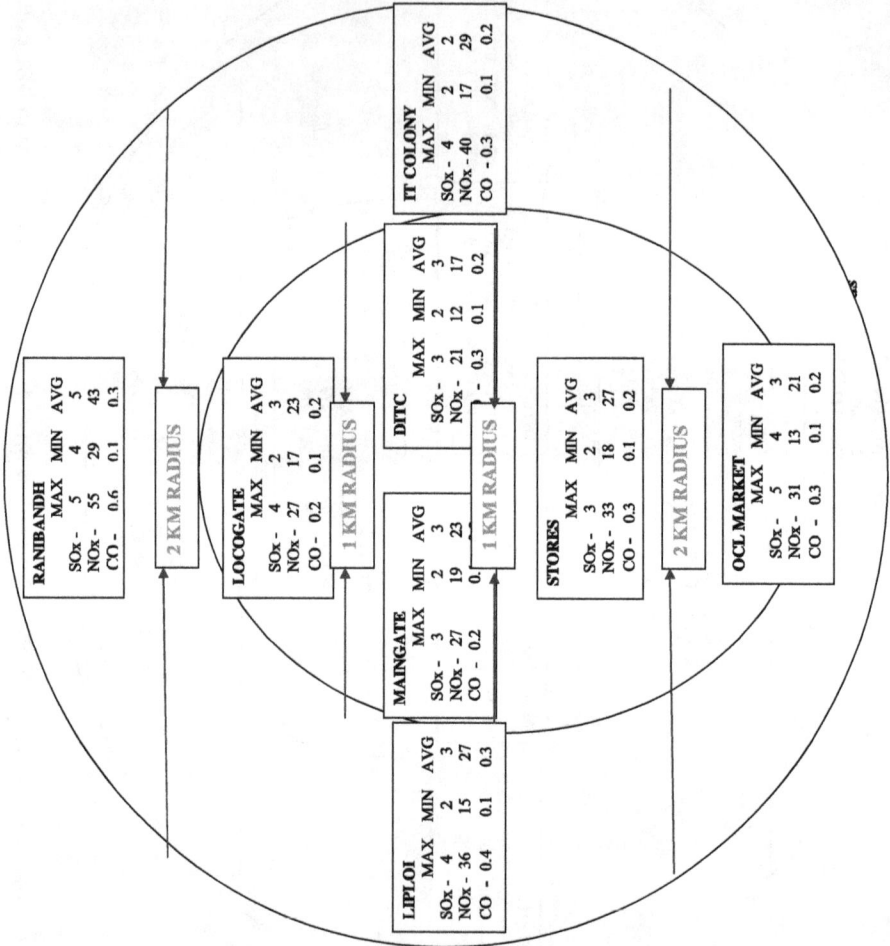

Figure 25.3b: Variation in SO$_x$, NO$_x$ and CO Concentration at 1 and 2 km Radius around OCL India Ltd.

and 83 at 1 km radius which are under safer zone (API Value < 100) with respect to dust load or overall pollution load.

Conclusion

From the above observation it is evident that most of the readings are within permissible value. Values within tolerance level signify a positive approach towards sustainable development. Ambient air pollution study aimed at the impact of industrial pollution on surrounding with special reference to human health (Cohen *et al.*, 1972; Dockery *et al.*, 1994; Ferin and Leach,1973; Kinney and Ozkaynak,1991). The observed values showed that the pollution load and emission levels are well within acceptable limit. Industry under study, that is, OCL India Limited, has no significant adverse effect upon the environment with reference to air pollution. This may be due to new technological measures implemented by OCL India Limited in clinkerisation plant as well as some modifications made in other old plant to reduce stack and fugitive dust emission besides adequate pollution control measures.

References

Agrawal, S. K. (1998): Air Pollution, APH Publishers, New Delhi, India.

Cohen, A.A., Bromberg, S., Buechley, R.W., Heiderscheit, L.T. and Shy, C.M. (1972): .Asthma and air pollution from a coal-fuelled power plant. Am. J. Public Health, 62:1181-1188.

Dockery, D.W., Pope, C.A., Xu, X., Spengler, J.D., Ware, J.H., Fay, M.E., Ferris, B.G.Jr. and Spezer, F.E. (1994): An association between air pollution and mortality in six US cities. The New Eng. J. Med. 329(24): 1753-1759.

Ferin, J. and Leach, L.J. (1973): The effect of SO_2 on lung clearance of TiO_2 particles in rats. Am. Ind. Hyg. Assoc. J., 34: 260-263.

Gauderman, W.J., Mcconnell, R., Gilliland, F., London, S., Thomas, D., Avol, E., Vora, H., Berhane, K., Rappaport, E.B., Lurmann, F., Margolis, H.G. and Peters, J. (2000): Association between air pollution and lung function growth in Southern California children. Am. J. of Resp. and Crit. Care Med., 162: 1383-1390.

Kinney, P.L. and Ozkaynak, H. (1991): Associations of daily mortality and air pollution in Los Angeles Country. Environ. Res., 54: 99-120.

Sharma, B. K. and Kaur, S. (2002): Air pollution, Krishna Publishers, Meerut.

Trivedi, P. R. (2000): Air Pollution, APH Publishers, New Delhi.

WHO (World Health Organization). (1979): Environmental Health Criteria 8. Sulfur Oxides and Suspended Particulates Matter. Geneva, Switzerland.

WHO (World Health Organization). (2003): Report on a World Health Organization Working Group on Health aspects of air pollution with particulate matter, ozone and nitrogen dioxide. (online) http://www.euro.who.int/document/e79097.pdf (26 June 2003).

Chapter 26

Changes in Physico-chemical Characteristics of Soil Following Application of Rice Mill Wastewater under Field Irrigation Conditions

Abanti Pradhan[1], Sanjat K. Sahu[2] and Aditya Kishore Dash[1]

[1]*Environmental Engineering Department,*
Siksha 'O' Anusandhan University Institute of Technical Education and
Research, Khandagiri, Bhubaneswer – 751 030, Odisha, India
[2]*P. G. Department of Environmental Sciences (Auto.),*
Sambalpur University, Jyoti Vihar, Burla – 768 019, Odisha, India

ABSTRACT

A work was undertaken to study the physico-chemical characteristics of soil irrigated with rice mill wastewater during winter crop period under field irrigation conditions. Prior to onset of the field experiment, the physico-chemical characteristics of the rice mill wastewater were measured. The wastewater revealed an alkaline pH (8.0) with low concentration of DO (0.9 mg/l), moderate concentration of COD (630 mg/l), TDS (670 mg/l) and high concentration of total suspended solids (530 mg/l) and BOD (450 mg/l). Moreover the wastewater was rich in sodium (235 mg/l), total phenols (35/l) as well as silica (58 mg/l). An upland, uncontaminated field of 30 m × 30 m area was selected near the rice mill factory site. Eight equal plots of 5m × 5m size were prepared and the distance between individual plots were kept 10 m apart to maintain hydrological isolation of plots and drains. On the basis of randomized block design method, four plots were identified as control plots and four plots as experimental plots. Four random soil samples of 400 g each were collected from control and experimental plots soon after the preparation of the plots (*i.e.* prior to irrigation of normal water/ rice mill wastewater). After collection of the soil samples, the control plots were irrigated with normal canal water of river the Mahanadi, whereas the experimental plots were irrigated with wastewater from the rice mill. Rice saplings (Jaya-T$_{90}$ variety) of 20 days old were planted in all the plots. The experiment was carried out during the month of winter (December-February) when prevailing atmospheric temperature was between 10-30°C. After experiment was over (*i.e.* at the time of harvest after 105 days) four random soil samples from each plot were again collected at 0-10 depth for physico-chemical analysis. The textural parameters like clay and silt were found to be increased remarkably in the experimental plots over control plots. The course sand, on the other hand, showed a reverse trend when the soil was irrigated with rice mill wastewater The values of chemical parameters like conductivity, OM, OC, N, C/N ratio, Na$^+$, SO$^{2-}_4$, PO$_4^{2-}$, Cl$^-$, phenols, SAR, CEC, ESP and per cent Na showed two to five fold increase in experimental plots as compared to the control plots, whereas the parameters like Ca^{++}, Mg^{++} and K$^+$ were 1.57-2.2 fold less in rice mill wastewater irrigated plots as compared to the control plots. In the present study the SAR and ESP of normal water irrigated plots was 2.09 and 17.02 (meq/100gm) respectively, whereas it was respectively 16.27 and 58.88 (meq/100gm) in effluent irrigated plots. The soil of the experimental plots showed the characteristics of saline/sodic soil. The increase in alkalinity of the soil may be attributed to formation of sodium phenolate, ammonium phenolate, sodium silicate, potassium silicate and other silicate in soil. These salts increase pH of the soil. From the above findings, it seems that wastewater in its present form is unsuitable for irrigation. So, it may be diluted to a suitable proportion before irrigation and for this pot experiment with different concentrations of wastewater be carried out to find out the percentage of dilution.

Keywords: *Rice mill wastewater, Soil physico-chemical characteristics, Soil quality.*

Introduction

India has 16 per cent of the World's population and 4 per cent of the World's water resources. Demands on water resources for domestic industrial and agricultural purposes are increasing greatly day by day in India. According to a recent estimate, the per capita availability of water, which was in the order of 5000 cubic meter per year at the time of independence, has drastically come down to 2000 cubic meter per year at present. Because of growing water problem, wastewater reclamation and reuse is one of the best alternatives for compensating water shortages. Irrigation of agricultural lands with industrial effluent have become a common practice in arid and semi arid regions, where it is readily available and inexpensive potion to freshwater (Angelakis and Bontoux, 2001; Oved *et al.*, 2001).

The agricultural use of wastewater serves many objectives, such as promoting sustainable agriculture, recycling scarce water resource and maintaining environmental quality. Also irrigating with wastewater may reduce purification levels and fertilization costs because soil and crops serve as a bio-filter and wastewater contain nutrients. Nutrient concentration in wastewater and feasibility for irrigation use has been evaluated for several industries by different workers (Soderquist and

Graham, 1972; Srivastava and Sahai, 1987; Narasimharao and Narasimharao, 1992; Cegarra *et al.*, 1997; Liu *et al.*, 1998; Paredes *et al.*, 1999). By contrast no report is available on the potentiality of rice mill wastewater for crop irrigation although rice mill industries in comparison to other industries are much more old and history of rice mill industries are associated with the history of human civilization.

The purpose of the present chapter was to investigate the feasibility of rice mill wastewater for crop irrigation through analysis of physico-chemical characteristics of soil under field conditions.

Table 26.1: Physico-chemical Characteristics of the Wastewater of a Rice Mill at Sambalpur, Orissa with their Maximum Permissible Limits as Recommended by Indian Standard Institution.

Parameters	Range	Mean±SD	ISI Limit for Discharge of Industrial Effluents	
			On Land for Irrigation (ISI, 1977)	Into Inland Surface Waters (ISI, 1974)
Colour		Brown	–	–
Odour		Unpleasant	–	–
Temperature (° C)	35.0-48.0	38.0±5.09	–	40
Conductivity (m mho cm^{-1})	0.46-0.86	0.66± 0.15	–	–
pH	7.2 - 8.8	8.0±0.54	5.5-9.0	5.5-9
Total solids (mg l^{-1})	998.1-1459.1	1200.0±189.48	–	–
TSS (mg l^{-1})	432.5-576.0	530.0±53.00	100	100
TDS (mg l^{-1})	522.1-883.1	670.0±149.2	2100	2100
Dissolved oxygen (mg l^{-1})	0.2-1.6	0.9±0.52	–	–
BOD at 20 °C (mg l^{-1})	312.1-540.1	450.0±76. 61	100	30
COD (mg l^{-1})	400.2-892.1	630.0±183.03	–	–
Total Alkalinity (mg l^{-1})	180.7-340.1	272.0±58.29	–	–
Total hardness (mg l^{-1})	98.3-256.4	182.0±59.84	–	–
Ca hardness (mg l^{-1})	38.4-98.3	78.0±22.22	–	–
Mg Hardness (mg l^{-1})	14.1-24.3	21.0±3.68	–	–
Chloride (mg l^{-1})	95.1-170.3	140.0±28.06	600	1000
Sulphate (mg l^{-1})	28.4-70.1	40.0±15.66	1000	1000
Phosphate (mg l^{-1})	10.1-35.2	21.0±11.11	–	–
Nitrate (mg l^{-1})	0.3-0.8	0.5±0.15	–	–
Sodium (mg l^{-1})	213.4-263.7	235.0±20.34	60 per cent	–
Potassium (mg l^{-1})	14.1-32.1	20.0±7.12	–	–
Phenols (mg l^{-1})	13.3-50.4	35.0±13.98	–	1.0
SiO$_2$ (mg l^{-1})	35.4-75.1	58.0±15.5	–	–

Materials and Methods

The wastewater of a nearby rice mill having milling capacity 10 MT/day has been used for the experiment. The physico-chemical characteristics of the wastewater (Table 26.1) were analyzed following the procedures recommended by APHA (1989) and values have been given in Table 26.1.

Experimental Design for Field Study

The field experiment was conducted from IVth week of January to IInd week of May during the winter crop period. An upland, uncontaminated field of 30 m × 30 m area was selected near the rice mill factory site. Eight equal plots of 5m × 5m size were prepared and the distance between individual plots were kept 10 m apart to maintain hydrological isolation of plots and drains. On the basis of randomized block design method, four plots were identified as control plots and four plots as experimental plots. Four random soil samples of 400 g each were collected from control and experimental plots soon after the preparation of the plots (*i.e.* prior to irrigation of normal water/rice mill wastewater). After collection of the soil samples, the control plots were irrigated with normal canal water of river the Mahanadi, whereas the experimental plots were irrigated with wastewater from the rice mill. Rice saplings (Jaya-T$_{90}$ variety) of 20 days old were planted in all the plots. The experiment was carried out during the month of winter (December-February) when prevailing atmospheric temperature was between 10- 30°C. After experiment was over (*i.e.* at the time of harvest after 105 days) four random soil samples from each plot were again collected at 0-10 depth. The soil samples were brought to the laboratory in polyethylene packets and made air dried for 7 days, and then passed through 2 mm sieve before analysis.

Soil Physico-chemical Analysis

pH

A soil suspension was prepared with distilled water in 1:5 ratio and pH of the unfiltered soil suspension was measured by a digital pH meter (Systronics make -335 model).

Electrical Conductivity

Like pH here also a soil suspension was prepared with distilled water in 1:5 ratio and electrical conductivity was measured by a conductivity meter (Systronics make -304 model). The electrical conductivity was expressed in terms of m mho/cm at 25°C.

Organic Matter and Organic Carbon

Walkley and Black's (1934) rapid titration method was followed to determine the organic carbon content of the soil. This method consist of oxidizing the soil sample with a mixture of potassium dichromate and H_2SO_4 diluting the suspension with water and back titrating the excess dichromate with ferrous ammonium sulphate solution. Percent organic matter (OM) was calculated as per cent OM = 1.724 × per cent C.

Nitrogen

Nitrogen content of soil was estimated by per sulphate oxidation method of Raveh and Avnimlech (1979). Soil sample was first autoclaved with potassium per sulphate and water (at 1.5 atm.) for 2 hrs, and then mixed with Deverda's alloy it was left overnight for conversion of all nitrogen into ammonia, and finally it was determined calorimetrically by indophenol's blue method and the value was expressed in g per cent.

Na⁺ and K⁺

The exchangeable Na⁺ and K⁺ in soil sample were determined by a Flame photometer using ammonium acetate extract directly and was expressed in meq/100g soil.

Ca²⁺ and Mg²⁺

The exchangeable Ca²⁺ and Mg²⁺ was determined by complexometric titration using EDTA. The metal ion indicators used for the titrations are Murexide (ammonium purpurate) for Ca and Erichrome Black T (Solochrome black) for Ca and Mg and were expressed in meq/100 g soil.

Sulphate

Soil suspension (1:5) was prepared and filtered as described earlier. The suitable aliquot of the sample was taken in a conical flask and to it 5 ml of conditioning reagent (75 mg NaCl + 20 ml Conc. HCl + 100 ml 95 per cent ethyl alcohol + 50 ml glycerol) was added. The mixture was stirred and during stirring a spoonful of BaCl₂ was added. After 4 minute the optical density of the resultant mixture was read in a spectrophotometer at 420 nm. The concentration of sulphate was read from a standard graph and the value was expressed in mg/100 gm soil.

Phosphate

Available phosphorous of soil was determined by using ammonium molybdate solution. The soil sample was treated with 0.002N H₂SO₄ and filtered. To the clear filtrate (coloured filtrate to be decolourised with activated charcoal) ammonium molybdate and SnCl₂ were added and the resulting blue colour was measured in a spectrophotometer at 690 nm. The value was expressed as mg/100g soil.

Chloride

Soil suspension (1:5) was stirred for one hour at regular interval and then it was filtered. Chloride of the suspension was measured by titration against silver nitrate solution using potassium chromate as indicator and was expressed in mg/100 g soil.

Polyphenols

Soil suspension 1:5 was stirred for one hour at regular interval and then it was filtered. A suitable aliquot of water sample was taken in a conical flask and to it 3 ml of sodium carbonate and 1 ml of Folin-ciocalteau reagent was added and allow to stand for 1 hr at room temperature (25 °C). After incubation the optical density of the colour developed was read at 750 nm against blank (distilled water). The concentration of polyphenols was estimated in mg/l using vanillic acid as standard.

Sodium Absorption Ratio

Sodium absorption ratio (SAR) was calculated using the formula

$$SAR = \frac{Na^+}{\sqrt{\dfrac{Ca^{2+} + Mg^{2+}}{2}}}$$

Cation Exchange Capacity

The cation exchange capacity (CEC) and exchangeable bases (Na$^+$, K$^+$, Ca^{2+} and Mg^{2+}) of soil samples were determined by saturating the exchange complex with ammonium ions (NH$_4^+$) from ammonium acetate (pH), washing with ethanol to remove ammonium from the soil solution, distillating the ammonium from soil suspension in the presence of NaCl and base into Boric acid and then titrating the ammonium with standardized acid. CEC was calculated by using the formula.

$$CEC = \frac{ml \; HCl \times NHCl \times 100}{Wt. \, of \, soil \, used}$$

Exchangeable Sodium Percentage (ESP)

ESP is calculated from the measured exchangeable sodium and cation exchange capacity being expression.

$$ESP = \frac{ex. \, Na^+ \, meq/100g \times 100}{CEC \, meq/100g}$$

Sodium Percentage

Na per cent was calculated using the formula

$$\% \, Sodium = \frac{Na^+}{Ca^{2+} + Mg^{2+} + K^+ + Na^+} \times 100$$

Results

Physico-chemical Characteristics of Soil

The textural and chemical characteristics of the soil prior to irrigation and after irrigation with normal water (control plots) and rice mill wastewater (experimental plots) for 105 days under field conditions are given in Table 26.2.

From the Table 26.2 it is evident that the difference between values of all the textural and chemical parameters of the soil before and after irrigation with normal water was insignificant. In contrast to this significant differences were observed

Table 26.2: Textural and Chemical Characteristics of Soil Prior to Irrigation and after Irrigation with Normal Water and Rice Mill Wastewater for 105 Days under Field Conditions.

Sl.No.	Parameter Study	Soil Characteristics Before Irrigation (A)	Soil Characteristics After Irrigation with Normal Water for 105 Days (B)	Calculated 't' between A vs. B	Soil Characteristics After Irrigation with Rice Mill Wastewater for 105 Days (C)	Calculated 't' between B vs. C
			Textural composition			
1.	Clay per cent	7.25±0.26	7.7±0.48	2.33 NS	10.3±0.88	7.33*
2.	Silt per cent	12.65±0.87	13.62±0.84	2.26 NS	22.1±1.07	17.62*
3.	Find sand per cent	26.79±1.57	28.76±1.73	2.38 NS	30.0±1.37	1.59 NS
4.	Coarse sand per cent	53.31±3.82	49.92±3.74	1.79 NS	37.6±1.38	8.73*
			Chemical characteristics			
1.	pH	6.79±0.04	6.82±0.04	1.49 NS	7.8±0.03	55.42*
2.	EC m mho/cm	7.0±0.34	7.4±0.4	2.15 NS	21.35±2.0	13.95*
3.	OM (g per cent)	3.9±0.31	4.13±0.32	1.45 NS	10.0±0.66	22.63*
4.	OC (g per cent)	2.3±0.18	2.4±0.18	1.11 NS	5.8±0.48	18.75*
5.	N(g per cent)	0.2±0.02	0.21±0.02	0.99 NS	0.25±0.02	3.99*
6.	C/N	11.5±1.07	12.0±1.09	0.925 NS	23.2±1.71	15.61*
7.	Ca^{++} meq/100g	2.71±0.17	2.76±0.16	0.60 NS	1.25±0.11	21.99*
8.	Na^{+} meq/100g	3.18±0.26	3.23±0.21	0.423 NS	18.55±1.36	31.48*
9.	Mg^{++} meq/100g	1.98±0.18	1.97±0.23	0.058 NS	1.215±0.12	8.23*
10.	K^{+} meq/100g	0.55±0.05	0.54±0.05	0.399 NS	0.35±0.04	8.39*
11.	SO_4^{2-} mg/100g	2.29±0.16	2.45±0.14	2.12 NS	6.14±0.39	25.18*
12.	PO_4^{2-} mg/100g	0.02±0.002	0.022±0.002	1.99 NS	0.049±0.003	21.17*
13.	Cl^{-} meq/100g	0.63±0.06	0.69±0.06	1.99 NS	9.78±0.6	42.63*
14.	Phenol mg/100g	32.32±2.79	35.88±3.98	2.07 NS	52.59±3.7	8.69*
15.	SAR	2.07±0.1	2.09±0.09	0.42 NS	16.7±0.27	145.17*
16.	CEC meq/100g	17.98±1.12	18.85±1.09	1.57 NS	33.8±2.44	15.82*
17.	ESP meq/100g	17.68±1.21	17.13±1.27	0.88 NS	54.88±3.26	30.51*
18.	Per cent Na	37.77±2.79	37.98±2.08	0.17 NS	86.22±3.11	36.46*

* $P < 0.05$; NS: Not Significant.

between values of textural and chemical parameters of soil irrigated with normal water and rice mill wastewater. The textural parameters like clay and silt were found to be increased remarkably in the experimental plots over control plots. The course sand, on the other hand, showed a reverse trend when the soil was irrigated with rice mill wastewater (Table 26.2). The values of chemical parameters like conductivity, OM, OC, N, C/N ratio, Na^+, SO_4^{2-}, PO_4^{2-}, Cl^-, phenols, SAR, CEC, ESP and per cent Na showed two to five fold increase in experimental plots as compared to the control plots, whereas the parameters like Ca^{++}, Mg^{++} and K^+ were 1.57-2.2 fold less in rice mill wastewater irrigated plots as compared to the control plots. The statistical analysis of the parameters by student 't' test between control and experimental plots also showed significant difference between them ($t \geq 3.99$, $p < 0.05$) (Table 26.2).

Discussion

Physico-chemical Characteristics of Soil

pH

The pH of soil is a measures of intensity of acidity and alkalinity. It is an important parameter because it effects many chemical reaction and biological system function only at relatively narrow pH ranges (Bear, 1976). It provides various clues about other soil properties. The soil pH greatly affects the solubility of minerals (Coleman and Mehlich, 1957). The soil pH can also influence plant growth. The optimum pH ranges from cultivation of different crop have been given in Table 26.3.

Table 26.3: Optimum pH Range of Soil for Cultivation of Various Crops.

Crops	Optimum pH Range	Crops	Optimum pH Range
Maize, wheat	5.5-7.5	Brinjal	5.5-6
Bean, soybean	6-7.5	Tomato	5.5-7.5
Groundnut, green gram	5.5-7.5	Cotton	5.5-7.5
Pea, tobacco, mustard,	6-7.5	Rice	5.0-6.5
Onion, carrot, cucumber	5.5-7.0	Raddish	5.0-7.0
Sugarcane	6.0-8.0	Potato	4.8-5.6
Beat, coffee	6.0-7.5	Coconut	6.0-8.0

From the table it is evident that soil pH less than 5 and greater 7.5 is not good for most of the vegetation (Brady, 1996). In the present study the irrigation of rice mill wastewater had a significant effect on soil pH under field conditions. The increase in the pH may be attributed to higher content of sodium in the effluent.

Electrical Conductivity (EC)

The electrical conductivity is directly related to the cations present in the soil solution. The more is the absorption of cation in the soil fraction, the more will be their availability in the soil solution (Brady, 1990). The significant increase in electrical conductivity was observed in the present study with increase in the concentration of the wastewater. This may be due to higher quantity of soluble nutrient released by the effluent to soil. Liu *et al.* (1998) also suggested that the inorganic constituent could raise the electrical conductivity of soil. The United State Salinity Laboratory, based upon the conductivity of soil has developed a broad generalization of plant susceptibility to soluble salts, which is given in Table 26.4 (Richards, 1947).

Table 26.4: Relation between Electrical Conductivity of Soil and Plant Growth

Conductivity (m mho/cm)	Plant Growth Condition
<2	Salinity affect mostly negligible
>2 to <4	Yield of very sensitive crops may be restricted
>4 to <8	Citrus, bean, yield of many crops restricted
>8 to <16	Only tolerant crops yield satisfactory (wheat, grape, olive)
>16	Only very important crops yield satisfactory (Barley, Sugar beet)

Organic Matter and Organic Carbon

Soil organic matter is a direct source of plant nutrient and it affects the cation exchange capacity (Tan and Dowling, 1994). The soil is classified according to the amount of organic compound and it has been given in Table 26.5.

Table 26.5: Classification of Soil According to Organic Matter Content of Soil

Organic Matter (per cent)	Soil Type
0-1	Very low humic
1-2	Low humic
2-4	Medium humic
4-8	High humic
8-20	Very high humic

As per the above classification the soil receiving 100 per cent rice mill wastewater comes under very high humic type. The higher content of organic matter in wastewater treated soil may be attributed to continuous input of organic suspended solids. Besides this, the alkaline nature of the wastewater has resulted in reduction of the decomposition of organic matter owing to adverse effect on microorganisms (Robson and Abbott, 1989).

Nitrogen

The total nitrogen concentration of soil ranges from <0.2 g per cent in sub -soil to >2.5 g per cent in peats. In the present study the nitrogen content was 0.21 g per cent in soil irrigated with normal water and 0.25 g per cent in soil irrigated with rice mill wastewater. This may be attributed to nitrogen mineralization due to continuous input of alkaline wastewater.

C/N Ratio

Mineralization of organic matter and immobilization by microorganism occurs simultaneously in soil. Therefore C/N ratio is a suitable measure of decomposition status of soil. Because of the accumulation of organic matter with relatively reduction in nitrogen mineralization a higher values of C/N ratio was noticed in soil irrigated with rice mill wastewater in the present study. This is in accordance with the findings of Odel *et al.* (1984).

CEC and Exchangeable Bases

One of the important chemical functions of soil is the exchange of cation. The ability of soil or sediment to exchange cations is expressed as Cation Exchange Capacity (CEC), the quantity of which can be exchanged per 100g of soil. Both the organic and mineral fractions of soil exchange cation. Clay minerals exchange cations because of the presence of negatively charged sites on the minerals, resulting from the substitution of an atom of lower oxidation number, whereas the organic matter exchanges cation because of the presence of carboxylase groups. The four cations namely Na^+, K^+, Ca^{++} and Mg^{++} constitute about 99 per cent of exchangeable bases in soil. The other 1 per cent is also important because it includes micronutrient cations such as Fe^{++}, Zn^{++}, Cu^{++}, Co^{++} and Mn^{++}. Trace amounts of these ions are adequate to meet the needs of plant (Troeh and Thompson, 1993). Irrigation of rice mills wastewater in the present study enhanced the cation exchange capacity and this may be attributed to higher amount of organic carbon as well as sodium in the effluent.

Sulphate

Sulphur occurs in numerous forms in soil, *i.e.* sulphites, sulphates, sulphides and in organic compounds. However, plant absorbs sulphur mostly a sulphates. Field soils frequently have 5-50 mg/100g of sulphates (Tabatabai, 1986). In our work, the sulphate contains of soil irrigated with normal water and rice mill wastewater was found to be varied between 2.45 and 6.14 mg/100g soil.

Chloride

Chloride ion usually has no effect on the physical properties of soil. Since chloride of alkali and alkaline earth groups are readily soluble in water, chloride hazards have not entered in the water classification system. It our study the chloride content of soils irrigated with rice mill wastewater were more than the soil irrigated with normal water. This might be due to the fact that rice mill wastewater contained varying quantity of chloride containing dissolve salts. (Brady, 1990).

Phosphate

Phosphorus is the second keen plant nutrient next to nitrogen. Plant use 1/10 phosphorous as they do of nitrogen, yet adequate availability of phosphorous to plant is a wide spread problem because in most soil, it is not readily available for plant use (Brady, 1990). The total phosphorous in an average arable soil is a approximately 0.1 per cent by weight of which only an infinitesimal part is available to the plant as orthophosphate ions ($H_2PO_4^-$) and (HPO_4^{2-}) in the soil solutions (Bear, 1964). Soil pH influences the availability of phosphorous; the optimum phosphorous availability in arable soil is near pH 5 (Troeh and Thompson, 1993). In the present work the average phosphate content of soil (mg/100g) irrigated with normal water was 0.022 and soil irrigated with rice mill effluent was 0.049.

Phenols

Two general groups of compound collectively make up humus, the humic group and non-humic groups. The humic substances make up about 60-80 per cent of the soil organic matter. They are comprised of most complex materials including polyphenols, which are resistant to microbial attack. Polyphenols are high molecular compound interact with nitrogen containing amino compound and give rise to a significant component of resistant humus. The formation of polymers is encouraged by the presence of the colloidal clay. In the present study phenol content of normal water irrigated soil was 35.88 mg/100g and soil irrigated with rice mill effluent was 52.59 mg/100g.

Exchangeable Sodium Percentage (ESP) and Sodium Absorption Ratio (SAR)

Several parameters are commonly used to characterize the sodium status of the soil. The Exchangeable Sodium Percentage (ESP) identified the degree to which the exchange complex is saturated with sodium. The ESP is complemented by a second more easily measured characteristic, the Sodium Absorption Ratio (SAR), which gives information on the comparative concentrations of Na^+, Ca^{2+}, Mg^{2+} in soil solution. As a thumb rule, the ESP is about 15 per cent to 30 per cent larger than SAR. Using electrical conductivity, ESP, SAR and soil pH soil are classified as Saline, Saline Sodic and Sodic (Table 26.6).

Table 26.6: Classification of Soil on the Basis of ESP, Conductivity and pH

Classification	Conductivity (m mhos/cm)	Soil pH	ESP	Soil Physical Condition
Saline	>4.0	<8.5	<15	Normal
Sodic	<4.0	>8.5	>15	Poor
Saline/Sodic	>4.0	<8.5	>15	Normal

In the present study the SAR and ESP of normal water irrigated plots was 2.09 and 17.13 (meq/100gm) respectively, whereas it was 16.7 and 54.88 (meq/100gm) in wastewater irrigated plots. The soil of the experimental plots showed the characteristics of saline/sodic soil. The increase in alkalinity of the soil may be attributed to formation of sodium phenolate, ammonium phenolate, sodium silicate, potassium silicate and other silicate in soil. These salts add basic pH to the soil.

The feasibility of industrial wastewater for irrigation depends on the several parameters like pH, EC, total solids (both suspended and dissolved) and total polyphenols. In light soil the parameters like organic solids increase the water holding capacity, silt and clay content, CEC and organic matter content. In heavy soil, organic solids, clog, capillary pores mainly in the upper soil layer and bring about a decrease in the rate of infiltration. Organic solids also clog capillaries deeper in the soil profile, where under aerobic condition decomposition of organic matter proceeds at a very low rate and makes the soil unfit for crop production. Further the enrichment of polyphenol and silica may cause disruption in the population of micro flora, leading to inhibition of soil microbial activities (Perez *et al.*, 1992). This might be the reason for inhibitory activities of microorganisms in rice mill wastewater irrigated soil.

Conclusion

From the above findings it seems that wastewater in its present form is unsuitable for irrigation. So, it may be diluted to a suitable proportion before irrigation and for this pot experiment with different concentrations of wastewater be carried out to find out the percentage of dilution.

References

Angelakis, A.N. and Bontoux, L.(2001): Wastewater reclamation and reuse in Eureau countries. Water Policy, 3: 47-49.

APHA (1989): Standard methods for the examination of water and wastewater. 18th ed., American public health association, Washington, D.C.

Bear, F.E. (1964): Chemistry of the soil 2nd ed. Oxford and IBH Publ. Co. Pvt. Ltd., New Delhi.

Brady, N.C. (1990): The nature and properties of Soils. Xth ed. Macmillan Publ. Company, USA.

Cegarra, L. Paredes, C., Roig, A., Bernal, M.P. and Carcia, D. (1997): Use of olive mill wastewater compost for crop production. Int. Biodeter. Biodegra., 38: 193-203.

Coleman, N.T. and Mehlich, A. (1957): The chemistry of Soil pH, In: Soil, The Year book of Agriculture,1957, pp 72-79, USDA, Washington.

Liu, F., Mitchell, C.C., Odum. J.W., Hill, D.T. and Rochester, E.W. (1998): Effect of swine lagoon effluent application on chemical properties of loamy sand. Biores. Tech., 63: 65-73.

Narasimharao, P. and Narasimharao, Y (1992): Quality of effluent water discharged from paperboard industry and its effect on alluvial soil and crops. Ind. J. Agr. Sc. 62(1): 9-12.

Odell, R.T., Metsted, S.W. and Walker, V.M. (1984): Changes in organic carbon and nitrogen of morrow plots under different treatments 1904-1973. Soil Sci. Am. J., 137: 160-171.

Oved,T., Shaviv, A., Goldrath, T., Mandelbaum, R.T. and Minz, D. (2001): Influence of effluent irrigation on Community composition and function of ammonia-oxidising bacteria in soil. Applied Environmental Microbiology, 67: 3426-3433.

Paredes, C., Cegarra, J., Roig, A., Sanchez-Monedera, M.A. and Bernal, M.P. (1999): Characterization of olive mill wastewater (alpechin) and its sludge for agricultural purposes. Biores. Tech., 67: 111-115.

Perez, J., De La Rubia, T., Moreno, J. and Martizez, J. (1992): Phenolic content and antibacterial activity of olive mill wastewaters. Environ. Toxicol. Chem., II: 489-495.

Richards, L.A. (1968): Diagnosis and improvement of saline and alkali soils. (River side, California Vs. Salinity Laboratory), Oxford and IBH Publ., Co., New Delhi.

Reveh, A. and Avnimlech, Y. (1979): Total nitrogen analysis in water, soil and plant material with persulphate oxidation. Plant Res., 13: 911-912.

Robson, A.D. and Abbott, L.K. (1989): The effect of soil acidity on microbial activity in soils. In: Soil acidity and plant growth, A.D. Robson (Ed.), Academic Press, Australia, pp. 139-165.

Soderquist,M.R. and Graham,J.L.(1972): Fruit, vegetable and grain processing waste, Journal of Water Pollution control Federation, 49: 1118-1123.

Srivastava, N and Sahai, R (1987): Effect of distillery wastes on the performance of *Cicer arietinum* L. Environ. Poll., 43:91-102.

Tabatabai,M.A. (1986): Sulfur in Agriculture, vol. 27, American society of Agronomy publ., USA

Tan, K.H. and Dowling, P.S. (1984): Effect of organic matter in CEC due to permanent and variable changes in selected temperate region soils. Geoderma, 32: 89-101.

Troeh, F.R. and Thompson, L.M. (1993): Soil and Soil fertility. Vth ed, Oxford University Press, New York.

Walkley, A. and Black, I.A. (1934): An examination of the Degtiareff method for determing soil organic matter and a proposed modification of the chromic acid titration method. Soil Sci., 37:29-38.

Chapter 27

Study of Effluent Treatment at Denzong Breweries Pvt. Ltd., Khurda, Odisha, India

Sushree Sasmita[1], Anand Mohan Roy[2]
and Upendra Nath Dash[1]

[1]*Department of Environmental Engineering,*
Siksha 'O' Anusandhan University, Institute of Technical Education and
Research, Bhubaneswar – 751 030, Odisha, India
[2]*Denzong Breweries Pvt. Ltd., Khurda, Odisha, India*

ABSTRACT

In this chapter, treatment of brewery wastewater in Denzong Breweries at Khurda, Odisha was investigated. The plant dealing with the production of beer and the effluent treatment plant would be having the specific reactors to treat the effluent properly so that the final outlet is free from various harmful contaminants and traces of organics and organic compounds. The effluent treatment plant has various reactors in which the effluent is treated in various stages as it progresses through during treatment. The performance efficiency of the effluent treatment plant was estimated with regard to the Up flow Anaerobic Sludge Blanket (UASB) by monitoring the parameters like Biochemical Oxygen Demand (BOD), and Chemical Oxygen Demand (COD). Other parameters like Volatile Fatty Acids (VFA), and Alkalinity in UASB were also analyzed. The other reactors and storage tanks in the effluent treatment plant were also monitored with parameters like BOD, COD and pH.

Keywords: *Brewery wastewater, Performance, BOD, COD, UASB reactor.*

Introduction

Water is essential to all known forms of life. Water pollution means contamination of water due to introduction of some external materials. Water may be polluted either due to introduction of some external materials. It may be polluted from natural sources or human activities. The pollution of water probably causes illness to human beings. Wastewater treatment is mandatory for all the industries which produce them during the processes in industrial operations.

As per the standards, the industries are required to monitor the effluents which are produced and only after treatment should discharge them into the environment. In wastewater treatment the parameters usually analyzed are Dissolved Oxygen (DO), Biochemical Oxygen Demand (BOD), Chemical Oxygen Demand (COD), pH, Volatile fatty acids (VFA), and Alkalinity.

Belay and Sahile (2013) studied the impact of Dashen brewery effluent on the bacteriological and some physicochemical qualities of Shinta River, Gondar town by analyzing temperature, total suspended solids (TSS), total dissolved solids (TDS) and Biochemical Oxygen Demand (BOD).

Nyilimbabazi, Banadda, Nhapi and GarbaWali (2011) studied the Characterization of Brewery Wastewater for Reuse in Kigali, Rwanda by analyzing Temperature (T), Total Dissolved Solids (TDS), Total Suspended Solids (TSS), Turbidity, Ammonium Nitrogen (NH_4-N), Total Nitrogen (TN), Total Phosphorus (TP), Chemical Oxygen Demand (COD), Biological Oxygen Demand (BOD), Electro-Conductivity (EC), Salinity, equivalent OH^- and Residual Chloride. Noorjahan and S.Jamuna (2012) studied physico-chemical characterisation of Brewery Effluent and its degradation using Native Fungus - *Aspergillus niger*, Aquatic Plant - Water Hyacinth- *Eichhornia* SP and Green Mussel –Pernaviridis by analyzing pH, EC, TSS, TDS, BOD, COD, Total Hardness (TH), Chloride, Copper and Zinc of untreated brewery effluent. Parawira, Kudita, Nyandoroh and Zvauya (2004) studied the industrial anaerobic treatment of opaque beer brewery wastewater in a tropical climate using a full-scale UASB reactor seeded with activated sludge by analyzing the performance of the UASB reactor during anaerobic digestion of opaque beer brewery wastewater in terms of treatment efficiency by monitoring the Chemical Oxygen Demand (COD) and total and settleable solids reduction. Bedu-Addo and Akanwarewiak (2012) determined the pollution loads of Brewery X and the impacts of the pollutant on the Sisai River and analyzed the BOD, COD, and TSS. Inyang, Bassey and Inyang (2012) studied the characterization of Brewery Effluent Fluid analyzed Biochemical Oxygen Demand (BOD) and Chemical Oxygen Demand (COD). Igboanugo and Chiejine (2011) studied the Mathematical Analysis of Brewery Effluent Distribution in Ikpoba River in Benin City, Nigeria, analyzed the concentration of effluent parameters namely; BOD, COD, DO, and pH. Oktem, and Tufekci (2006) studied the treatment of brewery wastewater by pilot scale upflow anaerobic sludge blanket reactor and analyzed the removal efficiency of COD and methane yield. Mata,Melo, Simoes and Caetano (2012) carried out the parametric study of a brewery effluent treatment by microalgae the potential of using microalgae Scenedesmus obliquus for a brewery wastewater treatment and biomass production by analyzing the Chemical Oxygen Demand (COD), Total Nitrogen (TN) and Total Carbon (TC) was followed in time, and the influence of light exposure, light intensity and culture aeration were studied. Sudarjanto, Sharma, Gutierrez and Yuan (2010) carried out a laboratory assessment of the impact of brewery wastewater discharge on sulphide and methane production in a sewer studied the impact of brewery wastewater discharge on sulphide and methane production in a sewer. Ekhaise and Anyasi (2005) studied the influence of breweries and effluent discharge on the microbiological and physicochemical quality of Ikpoba River, Nigeria estimated the bacteriological and physicochemical qualities of the Ikpoba River, Benin city was investigated to assessed the extent of pollution of the water due to effluent discharge from the two brewery industries in Benin City. Bello-Osagie, and Omoruyi (2012) studied the effect of brewery effluent on the bacteriological and physicochemical properties of Ikpoba river, Edo state, Nigeria, investigated the impact of brewery effluent on the bacteriological and physiochemical properties of Ikpoba River, Edo state, Nigeria.

Methodology

Effluent Treatment Plant

The effluent being discharged from various sections of the brewery will be brought to the effluent treatment plant. The brewery effluent mainly consists of beer and wort residues, last running, wastewater containing trub, wastewater containing yeast, wastewater from CIP plant, waste caustic from CIP plant, waste acid from CIP plant, spent caustic solution from the filtration section.

Wastewater may contain kieselguhr, alkaline cleaning water, and warm dirty wastewater and in particular waste caustic from the bottle washer containing insoluble substances such as shredded papers from labels, slurries etc. Other substances like soluble substances comprising of adhesives, caustic, metal salts and belt lubricants, traces of oil and fat from plant lubricants and residual beer from returned bottles and waste beer from the bottle filler are also present.

Basis of Design

The Wastewater Treatment Plant is designed to treat the wastewater from a brewery plant. Input flow rate of wastewater is 400 m^3/day. Figure 27.1 shows the effluent treatment plant.

Characteristics of wastewater input is given in Table 27.1.

Characteristics of Treated Wastewater

After the sand filter

BOD: <30 ppm, COD :< 250 ppm, Total SS: 35 ppm, pH: 6-8

Sludge

Anaerobic: 0.05 to 0.1 KG/kg of COD, destroyed on dry basis

Aerobic: 0.1- 0.4 KG/kg of BOD, destroyed on dry basis

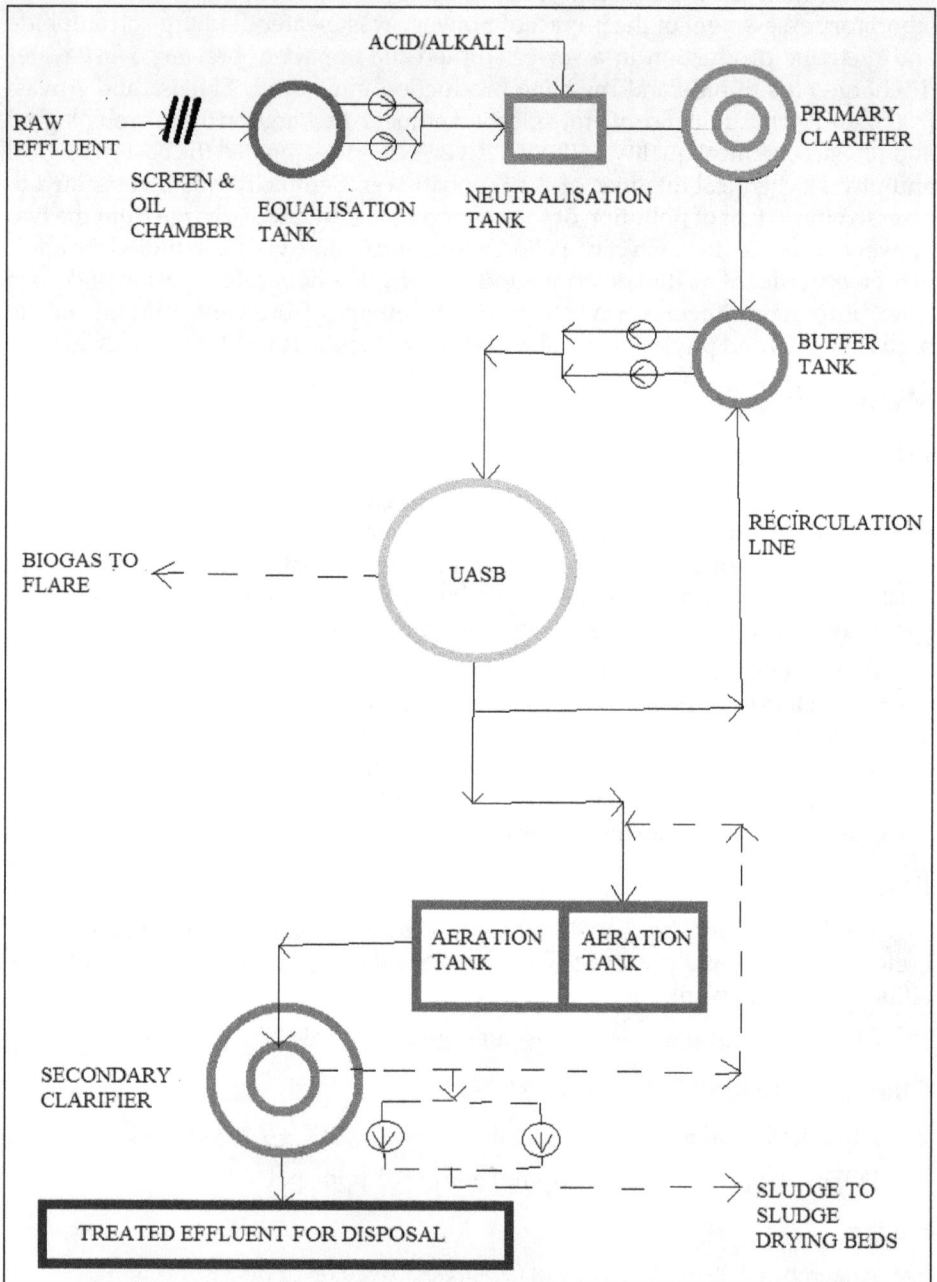

Figure 27.1: Effluent Treatment Plant.

Table 27.1: Wastewater Input Characteristics

Description	Minimum	Maximum	Design Value	Unit
BOD	1000	2500	3000	Ppm
COD	2000	3500	4000	Ppm
BOD:COD ratio	0.6	0.72	0.6	–
SS	500	1500	500	Ppm
N content	10	30	30	Ppm
P load	5	20	14	Ppm
pH	4	11	4-11	–
Temperature	25	35	35	°C
Chlorides	50	100	–	ppm

Process Description

Wastewater treatment scheme broadly consists of screening and oil removal followed by equalization of the wastewater and then by biological treatment system. Biological treatment system consists of anaerobic system followed by aerobic system.

Preliminary System

The wastewater from brewery plant is received in the screen chamber. The screen chamber is equipped with one number of manually raked inclined bar screen. The main purpose of the screen chamber is to remove large floating objects like cork, pieces of broken glasses, gunny bags, rags, etc. which may hamper function of the pump.

Screened wastewater is then taken to the oil removal tank. In the oil removal tank sufficient detention time is provided to separate free-floating oil from the wastewater. The floating oil is removed with slotted pipe skimming mechanism. From oil removal tank the wastewater comes to the Equalization tank by gravity.

The main function of the Equalization tank is to provide sufficient detention period so that the wastewater streams are homogenized properly. Equalization tank will help equalize the wastewater, both in terms of quality and quantity. Equalization tank will also ensure supply of wastewater at uniform rate to the treatment units with minimum variation in the quality, which is necessary for any biological treatment for achieving optimum performance.

The Equalization tank is provided with aeration grid through, which compressed air is supplied to the contents of the Equalization tank. Agitation created by compressed air will keep the suspended solids in suspended solids in suspension and will also help in mixing the effluent streams. Oxygen present in compressed air will avoid septic conditions and, hence, will minimize odour problem. The Equalization tank is provided with a level switch to avoid dry running of pumps. The equalized wastewater then enters into Neutralization tank. If necessary, pH correction will be carried out in the Neutralization tank. Acid and alkali dosing tank with dosing pumps are provided for this purpose.

The neutralized water is further subjected to chemical precipitation treatment. Coagulants in the form of alum, ferric chloride and/or polyelectrolyte will be added to the wastewater, which alter physical state of dissolved and suspended solids and facilitate their removal by sedimentation. The flocs formed due to coagulation of the effluent will be settled in a pre-clarifier. The settled sludge from the pre-clarifier is taken to the sludge pit. The equalized and neutralized wastewater from the pre-clarifier will be taken to the biological treatment system. The wastewater will first enter the Primary Aeration tank, which will be the first unit of the system.

Biological Treatment

The primary treated wastewater is then taken into the downstream buffer tank. Buffer tank serves as pH adjustment tank wherein the bicarbonate and alkalinity produced in the UASB reactor is used to control small variations in pH. The wastewater is then pumped to a UASB (Up flow Anaerobic Sludge Blanket) Reactor for degradation of organic matter by anaerobic bacteria where major portion of BOD/COD is reduced. The UASB reactor works on principle of degradation of organic matter within sludge bed situated at the bottom of the reactor. The anaerobic bacteria present in the sludge bed come in intimate contact with the organics, which are reduced to first low molecular fatty acids.

These acids are then converted to simpler end products such as CO_2, methane gas, water molecules and more bacteria. The gas produced gets attached to small sludge particles and tends to rise through the sludge bed. The three phase separator placed at the top of the reactor helps in detaching the sludge particles from gas bubbles. The sludge particles being heavier than water settle back to the sludge bed and the biogas leaves the system. The gas produced can be flared off.

The partially treated wastewater then undergoes anaerobic treatment for further reduction of BOD/COD through an Activated Sludge Process (ASP). The Activated Sludge Process comprises of an Aeration Tank followed by a clarifier. The supernatant from the clarifier is taken to a Chlorine Contact Tank where chlorine gas solution would be dosed. Residual COD/BOD in the form of oxidizable substrate matter can be oxidized with a strong oxidizing agent like chlorine.

Chlorine acts on the organic matter through formation of hypochlorous acid molecule. Chlorine gas is first dissolved in water and the solution prepared is then thoroughly mixed with the effluent ensuring good dispersion. The biogas generation in anaerobic reactor is flared of and reused in the boiler for domestic purposes. The end products in this reactor are fatty acids, biogas like methane, carbon dioxide and hydrogen. Figure 27.2 shows the flowchart of biogas reaction.

Fatty Acids Usually Comprise of Acetic, Propionic, Butyric, Valeric Acids or their Combinations

In the aeration tank, residual organic matter will be removed with the help of aerobic bacterial culture. The oxygen required for microbial activity is supplied with introducing atmospheric oxygen into the contents of the aeration tank by splashing. The aerators help to maintain completely mixed conditions. The Aeration tank is provided with nutrient dosing system to dose necessary nutrients so as to maintain

Complex polymers

(polysaccharides, lipids, proteins)

Hydrolysis by

microbial enzymes

Monomers

(sugars, fatty acids, amino acids)

Fermentation   Fermentation

Acetate $H_2 + CO_2$

Methanogenesis

$CH_4 + CO_2$ CH_4

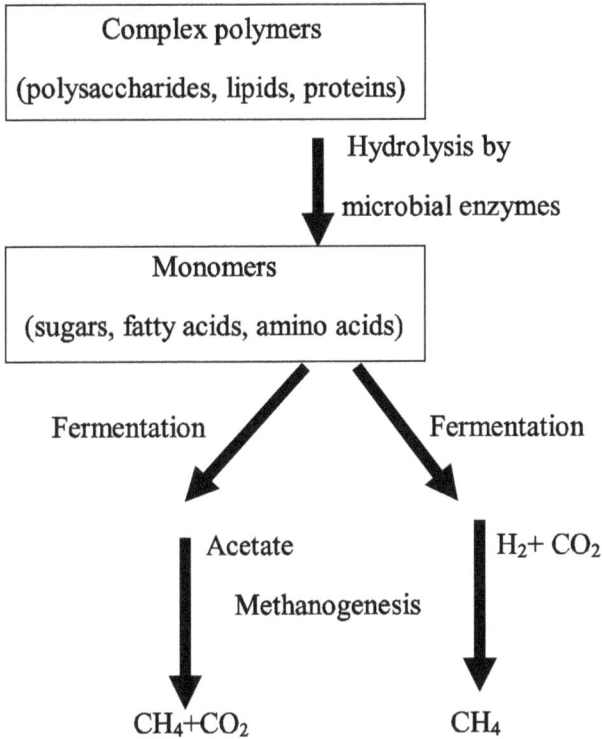

Figure 27.2: Flowchart of Biogas Reaction.

desired level of BOD : N : P. Aeration Tank is equipped with sludge recirculation system to maintain desired level of Mixed Liquor Suspended Solids (MLSS). Aerated Mixed Liquor is then taken to clarifier for separation of biomass.

In the clarifier the biomass settles at the bottom under gravity whereas treated effluent will overflow into the peripheral outlet launder on the top. Sludge settled at the bottom is then collected in the central sludge pocket with the help of centrally driven rotating scrapper mechanism. A part of the sludge from the central sludge pocket is to be pumped back to the inlet of the aeration tank to maintain desired level of MLSS, whereas excess sludge will be disposed on sludge drying beds for atmospheric drying. The overflow from the anaerobic reactor enters the aeration tank, which has activated sludge process.

Results and Discussion

The data were collected by taking samples from the various reactors in the effluent treatment plant of the brewery. The various parameters were analyzed and tabulated on various days with respect to the concerned reactors from where the samples were collected from the treatment plant. The observed values of pH, COD and BOD for different reactors were tabulated in Tables 27.2 to 27.6 from day 1 to day 5.

Table 27.2: pH, COD (ppm) and BOD (ppm) Data of different Reactors in Day 1

Day 1	pH	C.O.D	B.O.D
Equalization Tank	7.6	4120	1900
Neutralization Tank	7.5	4120	1900
Primary Clarifier	7.5	4120	1900
Buffer Tank	7.5	4120	1900
UASB	7.5	380	110
Secondary Clarifier	8.1	120	28

A perusal of Table 27.2 shows that in day 1 the pH values are the same for the Neutralization tank, Primary clarifier, Buffer tank and UASB but different in Equalization tank and Secondary clarifier. In latter two reactors, the medium becomes alkaline and more alkaline in Secondary clarifier. The input water in Equalization tank is basic in nature, to which HCl is added to decrease the value which is maintained at pH 7.5 in Neutralization tank. If the solution is acidic in Equalization tank, then $CaCO_3$ or $NaCO_3$ can be added to make it alkaline. The higher pH value in Secondary clarifier is due to the occurrence of citric acid cycle by aerobic bacteria present in Aeration tank placed before the Secondary clarifier.

It is observed from Table 27.2 that the values of COD and BOD remain constant in day 1 for the first four reactors and high, whereas these values are very low in the latter two reactors (*e.g.*, UASB and Secondary clarifier). The lower values of COD and BOD in UASB and Secondary clarifier are due to the decomposition of the organic substances (viz, starch, glucose, etc.) by anaerobic bacteria which consume dissolved oxygen in UASB reactor.

Table 27.3: pH, COD (ppm) and BOD (ppm) Data of different Reactors in Day 2

Day 2	pH	C.O.D	B.O.D
Equalization Tank	7.5	3900	1800
Neutralization Tank	7.4	3900	1800
Primary Clarifier	7.4	3900	1800
Buffer Tank	7.4	3900	1800
UASB	7.4	360	105
Secondary Clarifier	7.96	115	25

In day 3, the values of pH, COD and BOD follow the similar trend as in day 1 and day 2. More basic the medium, lower the COD and BOD values in Secondary clarifier. But this is not the case in UASB having lower values of COD and BOD with the same pH value as in other reactors. This might be due to the consumption of DO by aerobic bacteria in the same pH value.

Table 27.4: pH, COD (ppm) and BOD (ppm) Data of different Reactors in Day 3

Day 3	pH	C.O.D	B.O.D
Equalization Tank	7.6	3500	1760
Neutralization Tank	7.5	3500	1760
Primary Clarifier	7.5	3500	1760
Buffer Tank	7.5	3500	1760
UASB	7.5	370	110
Secondary Clarifier	7.95	120	27

Table 27.5: pH, COD (ppm) and BOD (ppm) Data of different Reactors in Day 4

Day 4	pH	C.O.D	B.O.D
Equalization Tank	7.4	4000	1820
Neutralization Tank	7.4	4000	1820
Primary Clarifier	7.4	4000	1820
Buffer Tank	7.4	4000	1820
UASB	7.4	340	120
Secondary Clarifier	7.9	110	28

Table 27.6: pH, COD (ppm) and BOD (ppm) Data of different Reactors in Day 5

Day 5	pH	C.O.D	B.O.D
Equalization Tank	7.6	3850	1750
Neutralization Tank	7.6	3850	1750
Primary Clarifier	7.6	3850	1750
Buffer Tank	7.6	3850	1750
UASB	7.6	350	108
Secondary Clarifier	8.0	105	24

In day 4 and day 5, the values of pH, COD and BOD are in the same order in all the reactors expecting the USAB and Secondary clarifier. But, the pH value in the USAB is equal to that of the other reactors.

The estimated DO values in Aeration Tank show irregular variation from day 1 to day 5 due to the variation of population of aerobic bacteria. Table 27.8 shows the estimation of DO (ppm) in Aeration tank for five days. Table 27.7 shows estimation of volatile fatty acids and alkalinity in UASB for five days. From the above tables it is observed that the variaion of pH is almost similar from day 1 to day 5 in all the reactors except in Secondary clarifier. The values of COD from day 1 to day 5 for the individual reactors show that COD varies in the similar fashion in four reactors while in different ways in UASB and Secondary clarifier.

Table 27.7: Estimation of Volatile Fatty Acids and Alkalinity in UASB for Five Days

UASB	Day 1	Day 2	Day 3	Day 4	Day 5
Volatile fatty acids (ppm)	540	540	530	540	545
Alkalinity (ppm)	1170	1130	1100	1110	1120

Table 27.8: Estimation of DO (ppm) in Aeration Tank for Five Days.

Aeration Tank	Day 1	Day 2	Day 3	Day 4	Day 5
Dissolved Oxygen	2.0	2.1	2.0	2.1	2.1

As observed, the variation of BOD is almost similar in Equalization tank, Neutralization tank, Primary clarifier and Buffer tank, but different in USAB and Secondary clarifier.

Estimation of the Efficiency of the UASB Reactor

1. Efficiency of UASB for

COD= [(UASB inlet-UASB outlet)/UASB inlet] × 100

2. Efficiency of UASB for

BOD= [(UASB inlet-UASB outlet)/UASB inlet] × 100

1) For Table 27.2, *i.e.*, for day 1,

☆ The efficiency of UASB for COD= [(4120 – 380)/4120] × 100=90.77 per cent

☆ The efficiency of UASB for BOD= [(1900 – 110)/1900] × 100=94.2 per cent

2) For Table 27.3, *i.e.*, for day 2,

☆ The efficiency of UASB for COD= [(3900 – 360)/3900] × 100=90.76 per cent

☆ The efficiency of UASB for BOD= [(1800 – 105)/1800] × 100=94.16 per cent

3) For Table 27.4, *i.e.*, for day 3,

☆ The efficiency of UASB for COD= [(3500 – 370)/3500] × 100=89.42 per cent

☆ The efficiency of UASB for BOD= [(1760 – 110)/1760] × 100=94.31 per cent

4) For Table 27.5, *i.e.*, for day 4,

☆ The efficiency of UASB for COD= [(4000 – 340)/4000] × 100=91.5 per cent

☆ The efficiency of UASB for BOD= [(1820 – 120)/1820] × 100=93.40 per cent

5) For Table 27.6, *i.e.*, for day 5,

☆ The efficiency of UASB for COD= [(3850 – 350)/3800] × 100=90.90 per cent

☆ The efficiency of UASB for BOD= [(1750 – 108)/1750] × 100=93.82 per cent

The overall efficiency of the UASB for COD and BOD was calculated.

1. The value was estimated by finding the sum of the COD of UASB and getting average of it.

   Average for COD= (90.77+90.76+89.42+91.5+90.90)/5=90.67 per cent ≈91 per cent

2. The value was estimated by finding the sum of the BOD of UASB and getting average of it.

   Average for BOD= (94.21+94.16+94.31+93.40+93.82)/5=93.98 per cent ≈94 per cent

Estimation of the Condition of UASB

It is estimated calculating the ratio of volatile fatty acids by alkalinity for all days. For day 1, VFA/Alk=550/1170=0.47

For day 2, VFA/Alk=540/1130=0.47

For day 3, VFA/Alk=530/1100=0.48

For day 4, VFA/Alk=540/1110=0.48

For day 5, VFA/Alk=540/1120=0.48

The average ratio of volatile fatty acid to alkalinity is found out by getting the sum of ratio for five days and getting an average of it.

Average = (0.47+0.47+0.48+0.48+0.48)/5=0.476 ≈0.48

- For a good effluent plant the ratio of VFA/Alk should be in between 0.1 to 0.7.

Conclusion

Water is the essential component on which life depends. It is not only required by human beings, but also is a necessary and a factor on which plants, animals depend for survival. Water is present in rivers, seas, oceans, ponds and is precipitated as rainfall by undergoing hydrological cycle. But, due to anthropogenic causes and water pollution from industrial effluents and waste discharges, which are discharged into rivers and lakes would hamper the normal quality of water is degrading with time. The most alarming situation is the depleting groundwater condition on earth which is falling with time. Another factor which the industries need to implement is the maintenance and proper establishment of effluent treatment plant. The industries need to reuse and recycle wastewater which would happen only after proper treatment of wastewater. They would further develop technologies for zero discharge for complete recycling of wastewater.

The effluent treatment at "Denzong Brewery" was studied and the performance efficiency of the treatment plant was estimated. The observed parameters of the outlet were compared with standards laid down by the Pollution Control Board and found to be within limits. The final outlet water which is recovered from the secondary clarifier is used in gardening. Hence, it can be concluded that the effluent treatment in the plant is done properly.

References

Bedu-Addo, K. and Akanwarewiak, W. G.(2012): Determination of the pollution loads of Brewery X and the impacts of the pollutant on the Sisai River", Journal of Public Health and Epidemiology. 4(10): 316-319.

Belay, A. and Sahile, S. (2013): The Effects of Dashen Brewery Wastewater Treatment Effluent on the Bacteriological and Physicochemical Quality of Shinta River in Gondar, North West Ethiopia, World Environment, 3(1): 29-36.

Bello-Osagie, I. O. and Omoruyi, I. M. (2012): Effect of Brewery Effluent on the Bacteriological and physicochemical properties of Ikpoba River, Edo State, Nigeria, Journal of Applied Technology in Environmental Sanitation, 2: 197-220.

Ekhaise, F. O. and Anyasi, C. C. (2005): Influence of breweries effluent discharge on the microbiological and physicochemical quality of Ikpoba River,Nigeria, African Journal of Biotechnology, 4 (10): 1062-1065.

Igboanugo, A. C. and Chiejin, C. M. (2011): A Mathematical Analysis of Brewery Effluent Distribution in Ikpoba River in Benin City, Nigeria, ARPN Journal of Engineering and Applied Sciences, September,9(6).

Inyang, U. E., Bassey, E. N. and Inyang, J. D. (2012): Characterization of Brewery Effluent Fluid, Journal of Engineering and Applied Sciences, 4.

Mata, T. M., Melo, A. C., Sim es, M. and Caetano, N. S. (2012): Parametric study of a brewery effluent treatment by microalgae Scenedesmus obliquus, Bioresource Technology, 107: 151–158.

Noorjahan, C. M. and Jamuna, S. (2012): Physico-Chemical Characterisation of Brewery Effluent and Its Degradation using Native Fungus - Aspergillus Niger, Aquatic Plant - Water Hyacinth- Eichhornia SP and Green Mussel –Pernaviridis, Journal of Environment and Earth Science, 2 (4).

Nyilimbabazi, N., Banadda, N., Nhapi, I. and GarbaWali, U. (2011): Characterization of Brewery Wastewater for Reuse in Kigali, Rwanda, The Open Environmental Engineering Journal, 4: 89-96.

Oktem, Y. and Tufekci, N. (2006): Treatment of brewery wastewater by pilot scale upflow anaerobic sludge blanket reactor in mesophilic phase, Journal of Scientific and Industrial Research, 65:248-251.

Parawira, W., Kudita, I., Nyandoroh, M. G. and Zvauya, R. (2004): A study of industrial anaerobic treatment of opaque beer brewery wastewater in a tropical climate using a full-scale UASB reactor seeded with activated sludge, DOI:10.1016/j.procbio.01.036, 593-599.

Sudarjanto, G., Sharma, R. K., Gutierrez, O. and Yuan, Z. (2010): A laboratory assessment of the impact of brewery wastewater discharge on sulfide and methane production in a sewer, 6th International Conference on Sewer Processes and Networks, Nov. 7(10): 98-120.

Chapter 28

Traffic Noise Models:
A Comparative Case Study

[1]Bijay Kumar Swain and [2]Shreerup Goswami

*[1]Department of Geology, [2]Head, Department of Geology,
Ravenshaw University, Cuttack – 753 003, Odisha, India*

ABSTRACT

A number of traffic noise models have been developed from noise descriptors to predict noise along the city roads. Experimental data are gathered for five sites of Balasore, India (Fakir Mohan Golei square, Station square, ITI square, Hemkapada square and Padhuanpada square) in the present study. Different traffic noise models are applied to predict noise level along the studied squares. A systematic comparison of predicted and monitored data is made, in order to examine the perfection of the applied models in five different studied sites.

Keywords: Road traffic noise, Noise prediction models, Equivalent noise levels, Noise variables.

Introduction

One of the most invasive types of noise pollution is traffic noise and has become an issue of immediate concern of public health and authorities. The sources of traffic noise are primarily vehicle engines, exhaust systems, tyre-pavement interaction and aerodynamic friction (Steele, 2001; Cirianni and Leonardi, 2012). The road traffic noise, the most deadly sources of environmental pollution prompted the researchers to design models to predict noise level. A number of models have been developed from noise descriptors such as the traffic flow, speed of vehicles, volume of the traffic and sound emission level using regression analysis of experimental data (Stefano and Morri, 2001; Golmohammadi *et al.*, 2007; Guarnaccia *et al.*, 2011; Kumar *et al.*,

2011). Several models for predicting noise levels generated by urban road traffic under interrupted flow conditions were developed (Griffths and Langdon, 1968; Lyons, 1973; Radwan and Oldham, 1987). Tansatcha *et al.*, 2005 have proposed motor traffic noise model based on the perpendicular propagation analysis technique (direction perpendicular to the centerline of motorways carriageway), which is performing well in a statistical goodness of fit test against the field data. A statistical model of road traffic noise in an urban setting, which is based on the fact that percentage of heavy vehicles plays an important role over road traffic noise emission, was developed by Calixto *et al.* (2003). Subsequently, some developed models for prediction of road traffic noise were suggested (Lam *et al.*, 2009; Steele, 2001; Stefano, 2001; Li, 2002; To *et al.*, 2002; Parida, 2003) and are reviewed (Steele, 2001; Golmohammadi *et al.*, 2007). Models are generally used to predict sound pressure levels specified in terms of L_{eq}, L_{10}, L_{50}, L_{90} etc. (Golmohammadi *et al.*, 2007). In these models, vehicles are mostly classified in two groups *i.e.*, heavy and light vehicles and sometimes in three groups, light, medium and heavy vehicles.

The chapter study is an attempt of comparison of simulated and experimental data monitored in Fakir Mohan Golei square, Station square, ITI square, Hemkapada square and Padhuanpada square of Balasore with an objective of identifying the best predicted traffic noise model(s). It is possible to assimilate reliability analysis of predicted noise level by using advanced mathematical techniques.

Review of some Traffic Noise Models

In this section, some of the most used TNMs (Traffic Noise Models) are briefly described. In all the formulas, L_{eq} is the equivalent noise level, 'Q' is the vehicles flow, 'P' is the percentage of heavy vehicles and 'd' is the distance of source to receiver.

Burgess, 1977

$$L_{eq} = 55.5 + 10.2 \log (Q) + 0.3P - 19.3 \log (d)$$

This model is used first time in Sydney in Australia, which is one of the first models for equivalent noise level (L_{eq}) applied in Australia.

Griffith and Langdon, 1968

$$L_{eq} = L_{50} + 0.018 (L_{10} - L_{90})^2$$

Where the statistical percentage indicator are evaluated with the following formulae:

$$L_{10} = 61 + 8.4 \log (Q) + 0.15P - 11.5 \log (d)$$

$$L_{50} = 44.8 + 10.8 \log (Q) + 0.12P - 9.6 \log (d)$$

$$L_{90} = 39.1 + 10.5 \log (Q) + 0.06P - 9.3 \log (d)$$

CSTB, 1991

$$L_{eq} = 0.65 L_{50} + 28.8 \text{ [dBA]}$$

The value of L_{50} is calculated taking in to account only the equivalent vehicular flows (Q_{eq}) and is given by:

$L_{50} = 11.9 \log Q + 31.4$ [dBA]

for urban road and highway with vehicular flows lower than 1000 vehicles/hour;

$L_{50} = 15.5 \log Q - 10 \log L + 36$ [dBA]

for urban road with elevated buildings near the carriageway edge, with L, the width (in meters) of the road near the measurement point.

Materials and Methods

The present chapter, a continuation of our previous study (Swain and Goswami, 2013), was conducted in five squares (Fakir Mohan Golei square, Station square, ITI square, Hemkapada square and Padhuanpada square) of Balasore (Figure 28.1) during March-June, 2012. Balasore is located at 21°18'North Latitude and 86°34' East Longitude. The noise levels were measured following standard procedure using calibrated sound level (dB) meter (Goswami, 2009; 2011; Goswami and Swain, 2011; 2012; 2013; Goswami *et al.*, 2011; Mohapatra and Goswami, 2012a, b; Swain *et al.*, 2011; 2012; Swain and

Figure 28.1: Map of India Showing Location of Balasore Town (study area).

Goswami, 2012; 2013; Pradhan *et al.*, 2012). Altogether 25 measuring points were selected along the road sides above five selected sites of Balasore. Each point was selected along the main street. In this study, the A-weighted continuous equivalent sound level values (L_{eq}), L_{max}, L_{min} and statistical levels of L_{10} (peak noise), L_{50} and L_{90} (background noise) were manually measured at each site separately. The L_{10}, L_{50} and L_{90} were obtained by manual recording of noise level in every 10 second interval. Measurement time for each sample was in the range of 20 minute to 1 hour depending upon the traffic flow. The number of light and heavy vehicles was counted. Distance of receiver point to the nearest edge of road was measured by tape measure. Noise measurements were at a distance of three meters from the nearest road band (Hendriks, 1998; Alimohammadi *et al.*, 2005). Equivalent noise level (L_{eq}) was calculated by using the formula of Robinson, 1971 [$L_{eq} = L_{50} + (L_{10} - L_{90})^2/56$]. Thus, experimental data (L_{eq}) are gathered for five investigated sites (Fakir Mohan Golei square, Station square, ITI square, Hemkapada square and Padhuanpada square) of Balasore in the present study. Predicted Leq are deduced by applying a number of already discussed Traffic Noise Models (Burgess, 1977, Griffith and Langdon, 1968, CSTB, 1991). A methodical comparison of simulated and experimental data is made, in order to test the perfection and behaviour of these models in five different studied sites in an Indian road condition. The measurements were taken on various days of the week, except Sunday. All the experimental data have been collected in absence of rain, with a wind speed below 5 m/s and relative humidity below 80 per cent (maximum value).

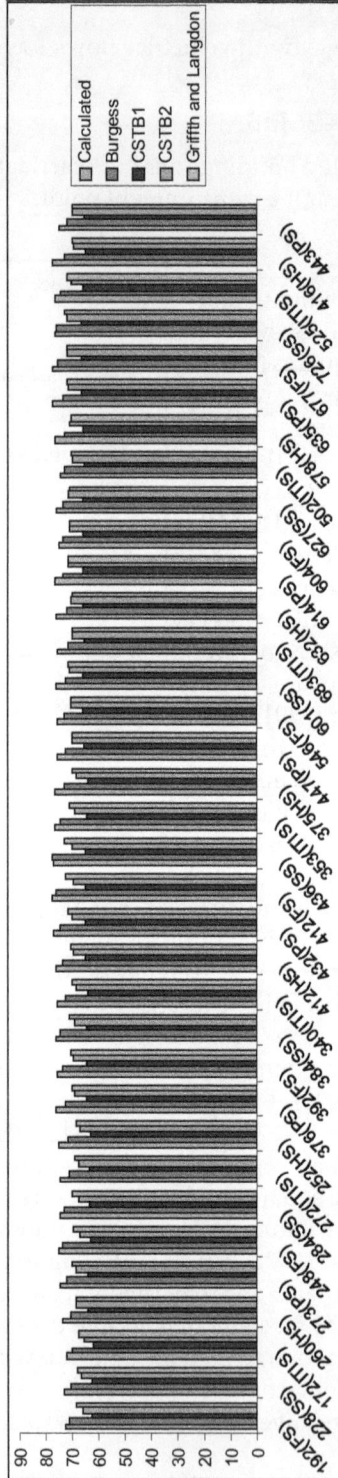

Figure 28.2: Calculated and Predicted L_{eq} Plotted Versus Hourly Vehicles Flow for different Measurements in both Sites, where X-axis represents hourly vehicle flow and Y-axis represents L_{eq} (dBA). Here, FS, SS, IT, IS, HS and PS stand for Fakir Mohan Golei square, Station square, ITI square, Hemkapada square and Padhuanpada square, respectively.

Results and Discussions

A quantitative comparison between TNMs (Traffic Noise Models) and experimental data is made to examine the perfection of established models. In Figure 28.2, the comparison of monitored and predicted data is presented against hourly vehicle flow. The figure depicts that the slope of models is not regular due to different percentage of heavy vehicles in each measurement points. In Figures 28.3–28.7 the comparison in five different sites are plotted versus time of measurement. Here, the x-axis is not on scale. The minimum noise levels (Lmin) measured at Fakir Mohan Golei square, Station square, ITI square, Hemkapada square and Padhuanpada square were 53.1, 53.4, 52.6, 51.2 and 52.1 dB, respectively. Similarly, measured Lmax values at Fakir Mohan Golei square, Station square, ITI square, Hemkapada square and Padhuanpada square were 102.8, 101.4, 93.5, 93.7 and 96.5 respectively. The average noise levels at morning and evening time at Fakir Mohan Golei square, Station square, ITI square, Hemkapada square and Padhuanpada square were 74.8± 6.3 and 75.1± 5.4; 75.1± 5.7 and 75.6± 5.8; 73.6± 5.5 and 74.2 ±4.5; 73.1± 5.8 and 73.7± 6.6; 74.2± 5.1 and 74.9± 6.7, respectively. Prevailing high noise levels in Balasore town are revealed from these data and are more than day time permissible limit of road traffic noise

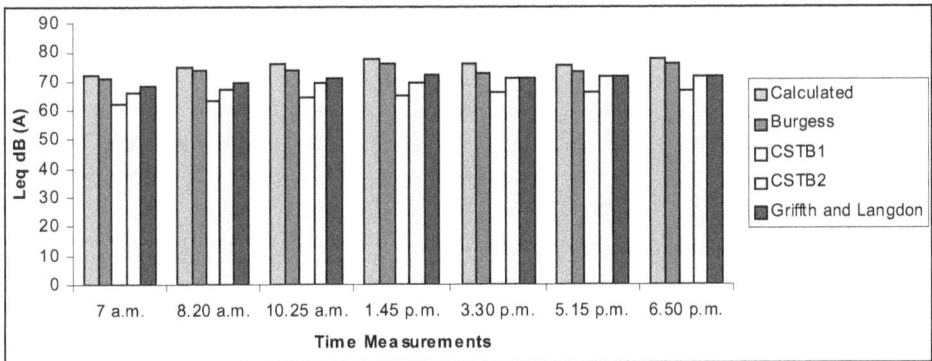

Figure 28.3: Calculated and Predicted L_{eq} Plotted Versus Time of Measurements at Fakir Mohan Golei Square.

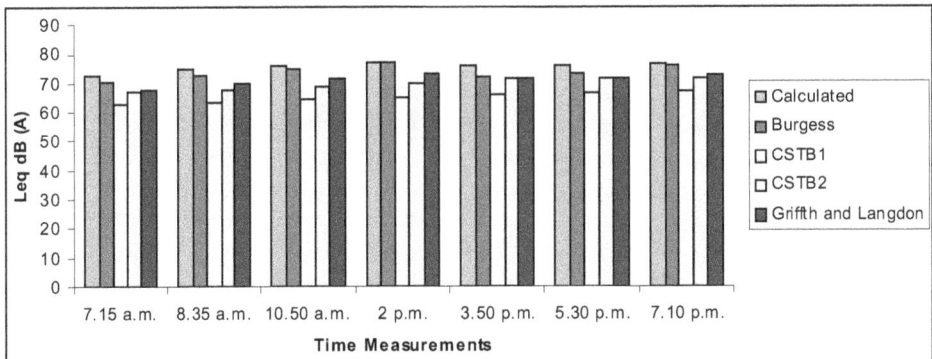

Figure 28.4: Calculated and Predicted L_{eq} Plotted Versus Time of Measurements at Station Square.

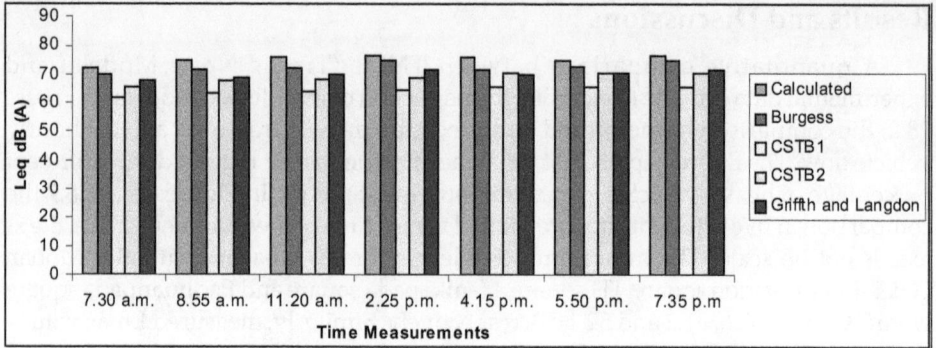

Figure 28.5: Calculated and Predicted L$_{eq}$ Plotted Versus Time of Measurements at ITI Square.

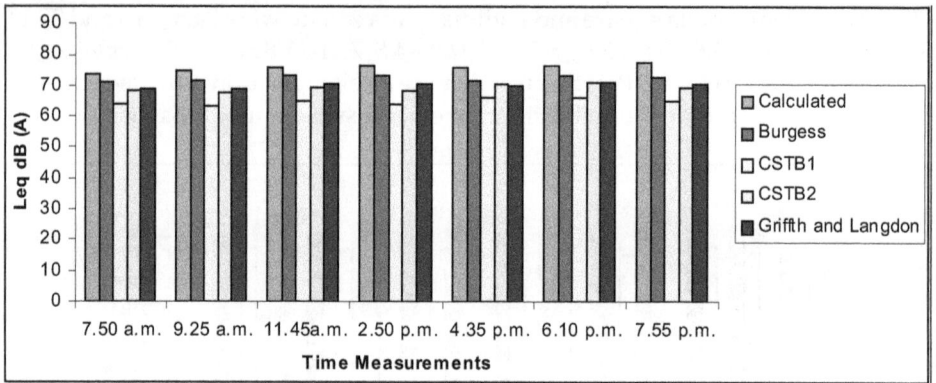

Figure 28.6: Calculated and Predicted L$_{eq}$ Plotted Versus Time of Measurements at Hemkapada Square.

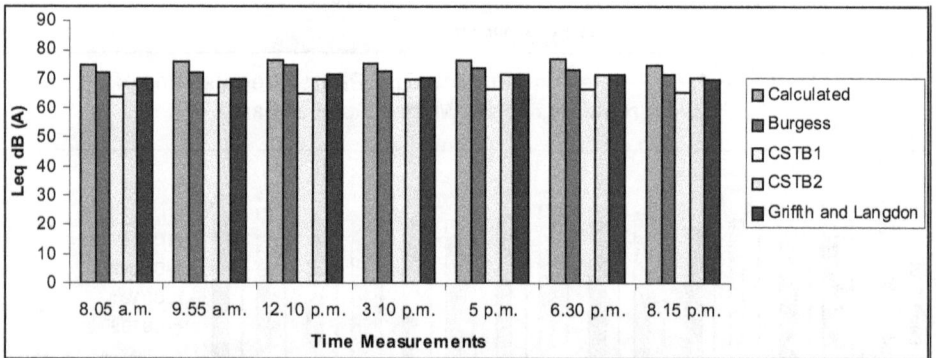

Figure 28.7: Calculated and Predicted L$_{eq}$ Plotted Versus Time of Measurements at Padhuanpada Square.

(70dB) (WHO, 1999). The calculated χ^2 values for Fakir Mohan Golei square, Station square, ITI square; Hemkapada square and Padhuanpada square at 6 degrees of freedom are 0.04, 1.769, 0.536, 0.326; 0.039, 1.698, 0.483, 0.256; 0.093, 1.816, 0.577, 0.387; 0.16, 1.873, 0.611, 0.445; and 0.123, 1.723, 0.497 and 0.383 for different models

(Griffith and Langdon, Burgess, CSTB1 and CSTB 2) in different time intervals. But $\chi^2_{tabulated}$ at 5 per cent and 1 per cent level of significance at 6 degrees of freedom are 12.592 and 16.812, respectively. Since our χ^2 value in different time interval is too small than tabulated χ^2 value, therefore $L_{calculated}$ and $L_{observed}$ values are in good agreement at 6 degrees of freedom and at 5 per cent and 1 per cent significance levels.

The comparison of the predicted and measured data of L_{eq} demonstrates that all four models are reliable to predict traffic noise in Indian conditions. However, amongst five models applied in the present study, Burgess Traffic Noise Model is most reliable, followed by Griffth and Langdon, CSTB2 and CSTB1. Such noise measurements and prediction could be helpful in understanding the problem of noise pollution along the roadways. As unplanned roadways are passing through heart of most of the cities; by-pass road, over-birdges/flyovers should be constructed to avoid agonizing road traffic.

Acknowledgements

The authors are thankful to esteemed Prof. P.C. Mishra and Prof. Aditya K. Dash for encouraging us to write this article.

References

Alimohammadi, I., Nassiri, P., Behzad, M. and Hosseini, M.R. (2005): Reliability analysis of traffic noise estimation in highways of Tehran by Monte Carlo simulation method. Iran J. Environ. Health. Sci. Eng. 2 (4): 229-236.

Burgess, M. A. (1977): Noise prediction for urban traffic conditions. Related to measurement in Sydney Metropolitan Area. Applied Acoustics 10: 1-7.

Calixto, A., Diniz, F. B. and Zannin, P. H. T. (2003): The statistical modeling of road traffic noise in an urban setting. Cities 20 (1): 23-29.

Cirianni, F. and Leonardi G. (2012): Environmental modeling for traffic noise in urban area. American Journal of Environmental Science 8 (4): 345-351.

Centre Scientifique et Technique du Batiment. (1991): Etude théorique et expérimentale de la propagation acoustique, Revue d'Acoustique, 70.

Golmohammadi, R., Abbaspour, M., Nassiri, P. and Mahjub, H. (2007): Road Traffic Noise Model. J. Res. Health Sci. 7 (1): 13-17.

Goswami, S. (2009): Road traffic noise: A case study of Balasore town, Orissa, India. Int. J. Environ. Res. 3 (2): 309-316.

Goswami, S. (2011): Soundscape of Bhadrak Town, India: An Analysis from road traffic noise perspective. Asi. J. Wat. Environ. Poll. 8 (4): 85-91.

Goswami, S., Nayak, S., Pradhan, A. and Dey, S. K. (2011): A study of traffic noise of two campuses of University, Balasore, India. J. Environ. Bio. 32 (1): 105-109.

Goswami, S. and Swain, B. K. (2011): Soundscape of Balasore City, India: A study on urban noise and community response. J. Acoust. Soc. Ind. 38 (2): 59-71.

Goswami, S. and Swain, B. K. (2012): Preliminary information on noise pollution in commercial banks of Balasore, India. J. Environ. Bio. 33 (6): 999-1002.

Goswami, S. and Swain, B.K. (2013): Soundscape of Baripada, India; An appraisal and evaluation from urban noise perspective. The Ecoscan special issue 3: 29-34.

Griffiths, I. D. and Langdon, F. J. (1968): Subjective response to road traffic noise. J. Sound and Vibra. 8: 16-32.

Guarnaccia, C., Lenza, T. L. L., Mastorakis, N. E. and Quartieri, J. (2011): A comparision between traffic noise experimental data and predictive models results. Recent Res. Mech. 5 (4): 379-386.

Hendriks, R. (1998): Technical noise supplement. Environmental Program on behalf of Environmental Engineering-Noise, Air Quality and Hazardour waste management office, Department of Transporation, Carlifornia www.dot.ca.gov; accessed on 7th January, 2013.

Kumar, K., Katiyar, V. K., Parida, M. and Rawat, K. (2011): Mathematical modeling of road traffic noise prediction. Int. J. of Appl. Math. and Mech. 7 (4): 21-28.

Lam, K. C., Chan, P. K., Chan, T. C., Au, W. H. and Hui, W. C. (2009): Annoyance response to mixed transportation noise in Hong Kong. App. Acoust. 70: 1-10.

Li, B., Shu, T., Dawson, R.W. and Lam, K. (2002): A GIS based road traffic noise predication model. Applied Acoustics 63: 679-691.

Lyons, R. H. (1973): Propagation of environmental noise. Science. 179: 1083-1090.

Mohapatra, H. and Goswami, S. (2012b): Assessment and analysis of noise levels in and around Ib River coalfield, Orissa, India. J. Environ. Bio. 33 (3): 649-655.

Mohapatra, H. and Goswami, S. (2012a): Assessment of Noise levels in various residential, commercial and industrial places in and around Belpahar and Brajrajnagar, Orissa, India. Asian J. Wat. Environ. Poll. 9 (3): 73-78.

Pardia, M., Jain, S. and Mittal, N. (2003): Modelling of metropolitan traffic noise in Dehli. ITPI J. 21(1): 5-12.

Pradhan, A., Swain B.K., Goswami, S. (2012): Measurements and model calibration of traffic noise pollution of an industrial and intermediate city of India. The Ecoscan, special 1: 377-386.

Radwan, M. M. and Oldham, D. J. (1987): The prediction of noise from urban traffic under   interrupted flow conditions. Applied Acoustics 21 (2): 163-185.

Robinson, D. W. (1971): Towards a unified system of noise assessment. J. Sou. Vibra. 14 (3): 279-288.

Steele, C. M. (2001): A critical review of some traffic noise prediction models. Applied Acoustics 2: 271-87.

Stefano, R. D. and Morri, B. (2001): A statistical model for predicting road traffic noise on poisson type traffic flow. Noise control Eng J. 49(3): 137-43.

Swain, B. K., Goswami, S., and Panda, S. (2012): Road Traffic Noise Assessment and Modeling in Bhubaneswar, Capital of Odisha State, India: A Comparative and Comprehensive Monitoring Study. International Journal of Earth Sciences and Engineering 5 (5): 1358-1370.

Swain, B. K., and Goswami, S. (2013): Integration and comparison of assessment and modeling of road traffic noise in Baripada town, India. International Journal of Energy and Environment 4 (2): 303-310.

Swain, B. K., Panda, S. and Goswami, S. (2012): Dynamics of road traffic noise in Bhadrak city, India. J. Environ. Bio. 32 (6): 1087-1092.

Swain, B. K., Goswami, S. and Tripathy, J. K. (2011): Stone Crushers Induced Noise at and around Mitrapur, Balasore, India. Anwesa 6: 12-16.

Tansatcha, M., Pamanikabud, P., Brown, A. L. and Affum, J. K. (2005): Motorway noise modeling based on perpendicular propagation analysis of traffic noise. Applied Acoustics 66 (10): 1135-1150.

To, W. M., Rodney, C. W. I., Lam, G. C. K. and Yau, C. T. H. (2002): A multiple regression model for urban traffic noise in Hong Kong. J. Acoust. Soc. Am. 112 (2): 551-556.

W. H. O. (1999): Guideline values. In: Guidelines for Community Noise (Eds. B. Berglund, T. Lindvall, D.H. Schwela). World Health Organisation. Geneva.

Chapter 29

Ambient Air Study Near the Industrial Complex 'Nisha' at Angul, Odisha, India

Subhasish Parida, Mira Das and Aditya Kishore Dash

S'O'A University, Department of Environmental Engineering, ITER, Jagamohan Nagar, Jagamara, Bhubaneswar – 30, Odisha, India

ABSTRACT

Air pollution has its peculiarities due to its transboundary dispersion of pollutants over the entire world. To arrest deterioration of air quality, government of India has enacted the Air (Prevention and Control of Pollution) Act, 1981 with subsequent amendments. This act prescribes various duties and functions of the Central Pollution Control Board (CPCB) at the apex level and the State Pollution Control Boards (SPCBs) at the state level. To meet the regulatory requirements, it is very much important to monitor the ambient air of an area, particularly near the industrial belts, which are considered as the major point sources of air pollution. The present chapter was undertaken to assess the ambient air quality status near the industrial complex 'Nisha' at Angul, Odisha. Ambient air parameters like SO_2, NO_x, PM_{10} and $PM_{2.5}$ were monitored at a network of four ambient air monitoring stations like Basudevpur, Goalbandh, Nisha, Paranga. Guidelines as per CPCB were followed during the monitoring, sampling and analysis of ambient air parameters. In general it was found that, in all the four monitoring stations, ambient air parameters are highest during the month of January and February and lowest during the month of August. The higher values during the month of January and February were might be due to the fact that, most of these times the wind speed remains calm and dispersion of air pollutants is very less. But during August, the air pollutants get diluted with the rain water and come to the earth surface for which the average monthly values are lowest in this month. Air Quality Index Value for Basudevpur (41 µg/m³), Goalbandh (40 µg/m³), Nisha

(39 µg/m³) and Paranga (36 µg/m³) shows that, all the four monitoring stations are coming under light air pollution category. Further, it was observed that, all the four ambient air quality parameters (SO_2, NO_x, PM_{10} and $PM_{2.5}$) are within the prescribed standard of CPCB. In the present study area, humidity is generally high during monsoon and post monsoon period and remain between 50 to 98 per cent. The relative humidity was comparatively low during summer and annual average varies between 25 to 85 per cent. During winter, the mean daily minimum temperatures reaches up to 12.9°C and mean daily maximum temperature reaches up to 26.7°C. The normal average rain fall of the district was 1421.1mm. The rainfall increased from South-Western part to the North-Eastern part of the study area. The predominant wind directions were from South- West (SW), North - West (NW) and West (W) and during most of the period, the wind speed remains within the range of 12.0 to 21.6 km/hr. The annual average wind speed was 15.6 km/hr. Wind velocity in general was low to moderate with little increase in speed during summer and monsoon season.

Keywords: Ambient air monitoring, Air quality index, SO_2, NO_x, PM_{10}, $PM_{2.5}$.

Introduction

Developed and developing economy and globalization have resulted in migration of fast changing energy intensive life style, mechanization and automation as a consequence of scientific advances including those of newer branches of science and engineering. Various contaminants continuously enter the atmosphere through both natural and manmade activities. Such substances, which interact with the environment causes toxicity, diseases, aesthetic distress which have been labeled as pollutants. The air we breathe is a mixture of gases and small solid and liquid particles. Air pollutants are added in the atmosphere from variety of sources that change the composition of atmosphere and affect the biotic environment. Air pollution occurs when the air contains substances in quantities that could harm the comfort or health of humans and animals, or could damage plants and materials. These substances are called air pollutants and can be either particles, liquids or gaseous in nature (Alias, *et al.*, 2007). Continuous mixing, transformation and trans-boundary transportation of air pollutants make air quality of a locality unpredictable. Particulate matter and gaseous emissions from industries and auto exhausts are responsible for rising discomfort, increasing airway diseases and deterioration of artistic and cultural patrimony in urban centers. The concentration of air pollutants depend not only on the quantities that are emitted from air pollution sources but also on the ability of the atmosphere to either absorb or disperse these emissions. The air pollution concentration vary spatially and temporarily causing the air pollution pattern to change with different locations and time due to changes in meteorological and topographical conditions.

It is important to study the ambient air quality of an area in order to know the existing ambient air status of the area and to protect the human health and ecosystem of the area (Simkhada *et al.*, 2005; Giri, *et al.*, 2006; Horaginamani and Ravichandram, 2010; Sathe, 2012; Balashanmugam *et al.*, 2012; Vedamadavan and Sarithabanuraman,

2012; Sharma *et al.*, 2012). The ambient air quality monitoring network involves measurement of a number of air pollutants at number of locations in the country so as to meet objectives of the monitoring. Any air quality monitoring network thus involves selection of pollutants, selection of locations, frequency, duration of sampling, sampling techniques, infrastructural facilities, man power and operation and maintenance costs. The network design also depends upon the type of pollutants in the atmosphere through various common sources, called common urban air pollutants, such as Sulphur Dioxide (SO_2), Oxides of Nitrogen (NO_x), Carbon Monoxide (CO), PM_{10}, $PM_{2.5}$, etc. The areas to be chosen primarily should be such areas which represent high traffic density, industrial growth, human population and its distribution, emission source, public complaints if any and the land use pattern etc. Generally the basis of a monitoring network design is the pollution source and the pollutant present.

In most of the European countries, industrialization and high volumes of traffic mean that anthropogenic sources predominate, especially in urban areas, and sources of anthropogenic particles are similar throughout Europe. The most significant of these are traffic, power plants, combustion sources (industrial and residential), industrial fugitive dust, loading/unloading of goods, mining activities, human-started forest fires and, in some local cases, non-combustion sources such as building construction and quarrying. In UK, emissions from traffic are a major contributor of harmful pollutants such as nitrogen oxides (NO_x) and particulate matter (PM) (Lima *et al.*, 2005). Potential health hazards due to particulate air pollution are a significant concern in both urban and rural areas in the United States. Several air sheds are currently classified by United States Environmental Protection Agency (USEPA) as non attainment areas for airborne particulate matter with an aerodynamic diameter of less than 10 μm (PM_{10}). Non-attainment areas are identified based on National Ambient Air Quality Standards (NAAQS) set by the Clean Air Act Amendments (CAAA) of 1990. Lombardia is a densely inhabited and industrialized region located in Po Valley (Northern Italy) which is often affected by high ozone levels during summer months and by elevated PM_{10} and NO_x concentrations during the cold season. The major urbanized and industrial conglomeration within the region is the "Milanoe-Comoe-Sempione critical area", for which local authorities have designed, during last few years, different emission abatement strategies (use of cleaner fuels for domestic heating, limitation of the circulation for non-catalyzed vehicles during colder seasons, etc.) (Silibello, *et al.*, 2008). Kathmandu Metropolitan City has taken the initiative to introduce electric vehicles in Kathmandu to control air pollution (Shrestha, *et al.*, 2002). Various other countries throughout the world are also making strategies to deal with the air pollution problem especially due to sulphur dioxide, nitrogen oxides and PM_{10} and $PM_{2.5}$.

In India, pollution has become a great topic of debate at all levels and especially the air pollution because of the enhanced anthropogenic activities such as burning fossil fuels, *i.e.* natural gas, coal and oil-to power industrial processes and motor vehicles. Burning of these fissile fuels releases carbon dioxide (CO_2), carbon monoxide (CO), nitrogen oxides (NO_x), sulphur dioxide (SO_2) and tiny solid particles-including lead from gasoline additives-called particulates (Goyal and Sidhartha, 2003). In India, outdoor air pollution is restricted mostly to urban areas, where automobiles are the major contributors, and to a few other areas with a concentration of industries and thermal power plants. Apart from rapid industrialization, urbanization has also resulted in the emergence of industrial centers without a corresponding growth in civic amenities and pollution control mechanisms. In India, coal consumption contributes 64 per cent of total SO_2 emissions followed by oil products 29 per cent, biomass 4.5 per cent and non-energy consumption 2.5 per cent (Garg *et al.*, 2001). Power and transport sector emissions equally dominate NO_x emissions contributing nearly 30 per cent each (Garg *et al.*, 2001). Kaushik, *et al.*, 2007 on their study on the rural-industrial site at Satna shows significant different from urban, urban-industrial, rural, rural-remote and rural-urban influenced sites.

In the present work, an attempt has been made to study the ambient air quality near the industrial complex 'Nisha' at Angul, Odisha India.

Study Area

Angul district is located in the centre of the state of Odisha and lies between 20° 31' N and 21° 40' N latitude and 84° 15' E and 85° 23' E longitude. The altitude is between 564 and 1187 meters. The district has a total area of 6232 km^2. It is bounded by Dhenkanal and Cuttack district in East, Deogarh, Keonjhar and Sundargarh district in North, Sambalpur and Sonepur in West and Boudh and Nayagarh in the South side. The district is abundant with natural resources, which ultimately help the district to contribute maximum amount of revenues to the state government. Even though Angul district blessed with rich natural resources, it is one of the hottest districts in India where maximum temperature goes up to 50°C during summer. A study jointly conducted by Indian Institute of Technology, Delhi and Central Pollution Control Board (CPCB) reveals that Angul district is among the top 10 most polluted Indian cities where the pollution level reached to a 'very alarming' level. The location map of Angul is given in Figure 29.1.

Monitoring Sites

The main concern of the present study is to monitor the concentration of ambient air parameters like Sulphur Dioxide (SO_2), Oxides of Nitrogen (NO_x), PM_{10} and $PM_{2.5}$ by taking readings at four different stations. These two critical gaseous pollutants and the particulate pollutants are in abundance in Angul environment since it is an industrial area. The monitoring stations were so chosen that, there can be adequate safety measures as well as reduced interference of the local public with the devices used for monitoring. The monitoring stations chosen are as follows:

1. Basudevpur village
2. Goalbandh
3. Nisha
4. Paranga

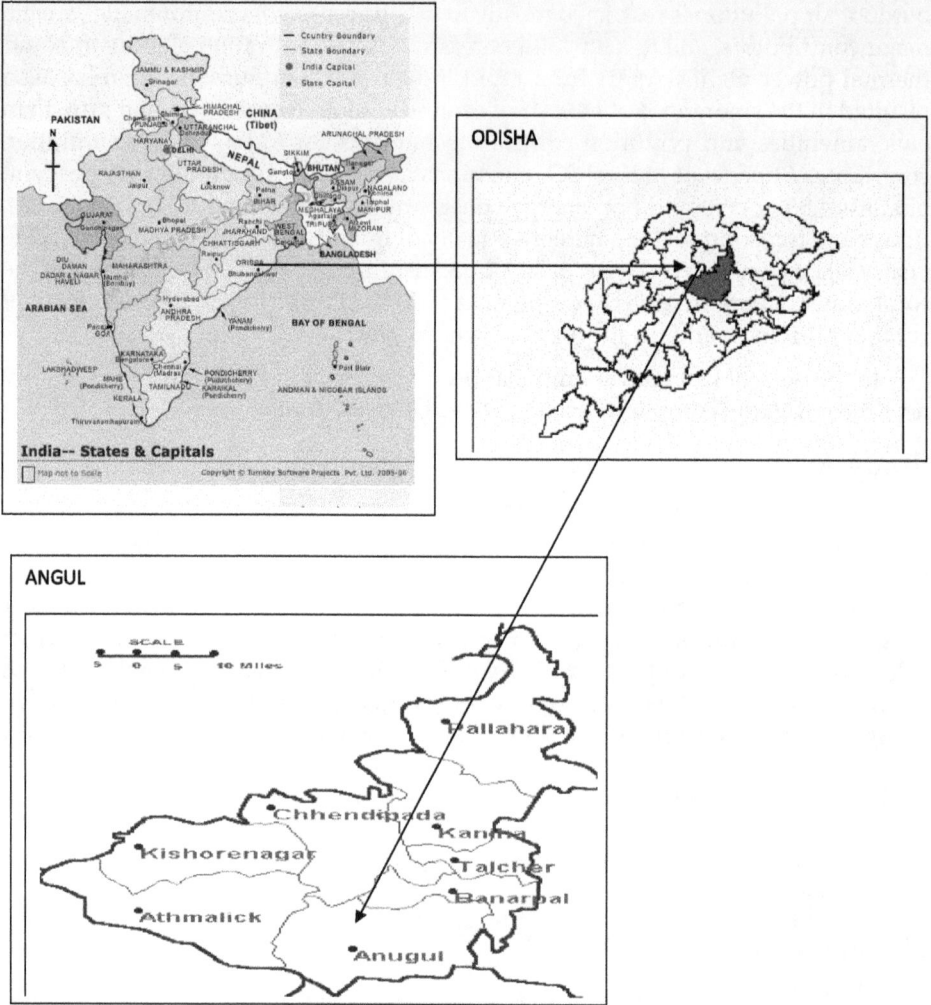

Figure 29.1: Location Map of Angul District.

Materials and Methods

Monitoring Criteria

The ambient air parameters monitored were gaseous pollutants like SO_2 and NO_x and particulate matters like $PM_{2.5}$, PM_{10}. The monitoring was carried out twice in a week during the study period. PM_{10} and $PM_{2.5}$ were monitored on 24-hrly basis at uniform interval while SO_2 and NO_x were monitored on 4 hourly basis. The monitoring stations chosen are free from any interference from the surrounding living stock. Sampling is usually done with 3 m height as per the standard. It was ensured that, filter is parallel to the ground. To obtain a representative sample, the sample was not being placed under a tree, near a wall or other obstructions that would prevent free air flow from the ambient atmosphere. During inclement weather (including high winds), the entire setup was moved to a protected location. Before the new filter was installed, loose particles from the inside surfaces of the sampler were removed and the surfaces around the filter holder were cleaned with a clean cloth.

Methodology Adopted for Air Quality Survey

The sampling and analysis of ambient air quality parameters was carried out as per the guidelines of CPCB. Brief of the testing procedures used are given in Table 29.1.

Table 29.1: Testing and Monitoring Procedures

Sl.No.	Parameter	Code of Practice	Sampler	Instruments for Analysis	Methodology Adopted
1.	Particulate Matter ($< 10\mu$)	IS: 5182 (Part IV)	RDS with cyclone separator	Balance Oven Desiccator	Gravimetric Method
2.	Particulate Matter ($< 2.5\mu$)	–	Fine particle sampler	Balance	Gravimetric Method
3.	Sulphur Dioxide (SO_2)	– do –	RDS with gaseous attachment	Colorimeter	Improved West and Gaeke method
4.	Nitrogen Oxide (NO_x)	– do –		Colorimeter	Jacob and Hochheiser Method

Sampling and Analytical Techniques

Respirable Dust Sampler (RDS), APM-460 and fine particle sampler "Envirotech" instruments as shown in Figure 29.2 were used for sampling.

Climate and Meteorology of the Study Area

Meteorology deals with atmosphere and characteristics of the weather. In the weather, dispersion and transport of pollutants which enter the atmosphere from a source, is affected by a number of atmospheric parameters. The basic micro-meteorological parameters are rainfall and precipitation, high and low pressure, heat, solar radiation, temperature, environmental lapse rate and atmospheric stability, wind speed, wind direction and mixing height, moisture and relative humidity. The extreme air pollution event, *i.e.* the maximum air pollution concentration, is governed by many complex and interrelated factors, which include the source emissions and the cumulative effect of typically complex climatological conditions such as low surface wind speed, mixing height, temperature inversion, anticyclone conditions etc. These interrelated meteorological factors exert a large influence on the pollutions level resulting in extreme air pollution events (Sfetsos, *et al.*, 2006). An air pollution problem involves three parts: the source, the movement of the pollutant and the recipient. All meteorological phenomena are a result of interaction of the elemental properties of the atmosphere, heat, pressure, wind and moisture.

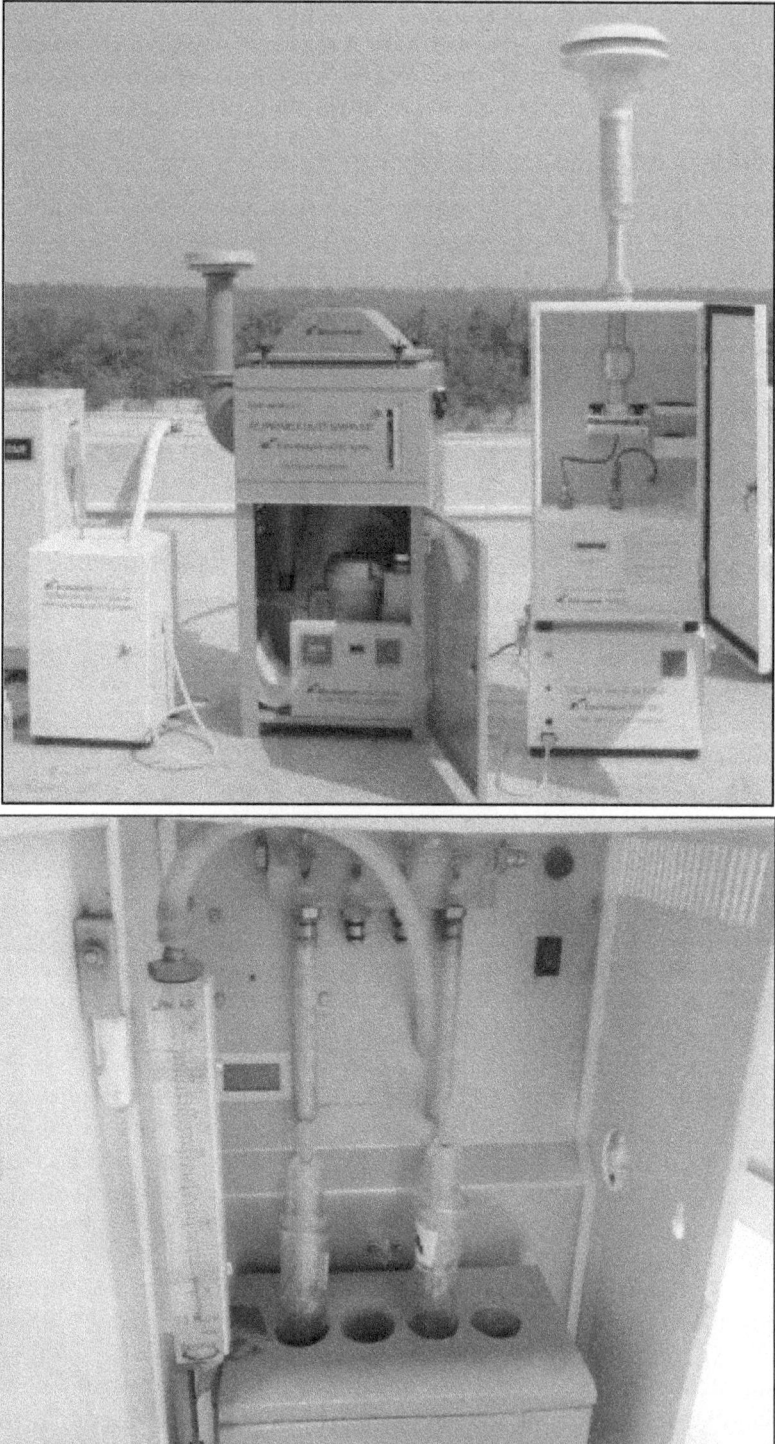

Figure 29.2: Respirable Dust Sampler with Gaseous Attachment.

Climate

The climate of the study area is generally dry and arid except in monsoon season. It is influenced by prevalence of dry air of the continental type. It is characterized by extreme conditions, summers being intensely warm and winter with cold. The climate of this region resembles with that of Deccan plateau. Climatically, the area experiences four distinct seasons as mentioned below:

☆ Monsoon: from mid-June to September.

☆ Post Monsoon: from October to November.

☆ Winter: December to February.

☆ Summer: March to mid-June.

Weather

In rainy season when South-West monsoon causes precipitation the weather remains generally cloudy. The rains in the monsoon are also associated with thunder. Thunderstorms with heavy precipitation also occur in the summer months and October, particularly in the afternoon. Foggy weather is generally observed during winter season. In other months sky remains clear.

Cloudiness

As per Indian Meteorological Department (IMD) observation, usually pre-monsoon and post-monsoon months are clear to lightly cloudy and winter seasons are usually bright. Moderate to heavy clouds are commonly observed during monsoon seasons.

Special Weather Phenomena

The district is affected by storms and depressions in the monsoon season, resulting wind speed increases and widespread heavy rain. Thunderstorm, mainly in the afternoon occurs in the summer months and also during the month of October. Rain during monsoon season is generally associated with thunder. Occasional fog occurs in the cold season.

Wind

It is generally observed that, the wind speed in the area is light to moderate except in the early monsoon period when it is generally strong. Higher speed wind blows during latter part of summer or rainy season in the direction of South-West or North-East. Wind blow with slow or moderate speed in rest part of the year. In winter, the wind blow either from West or North. Frequent variation in wind speed takes place only during summer. During most of the period the wind speed remains within the range of 12.0 to 21.6 km/hr. The annual average wind speed is 15.6 km/hr. The predominant wind directions are SW, NW and W, as in Figures 29.3–9.7.

Wind velocity in general is low to moderate with some increase in speed in summer and monsoon season. Wind mostly blows from southwest and northeast direction during monsoon period. In the cold season winds are mainly from west or

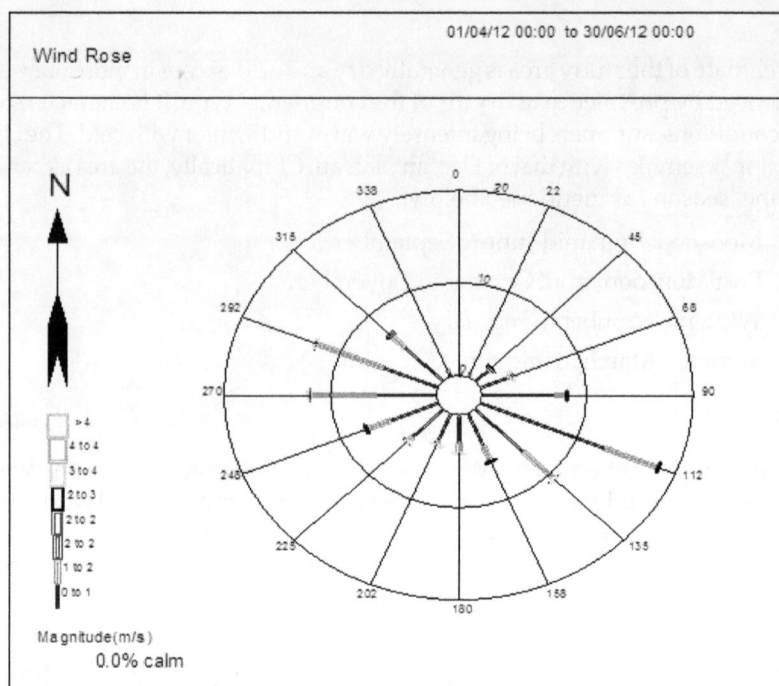

Figure 29.3: Wind Rose Diagram for the Month of April - June, 2012.

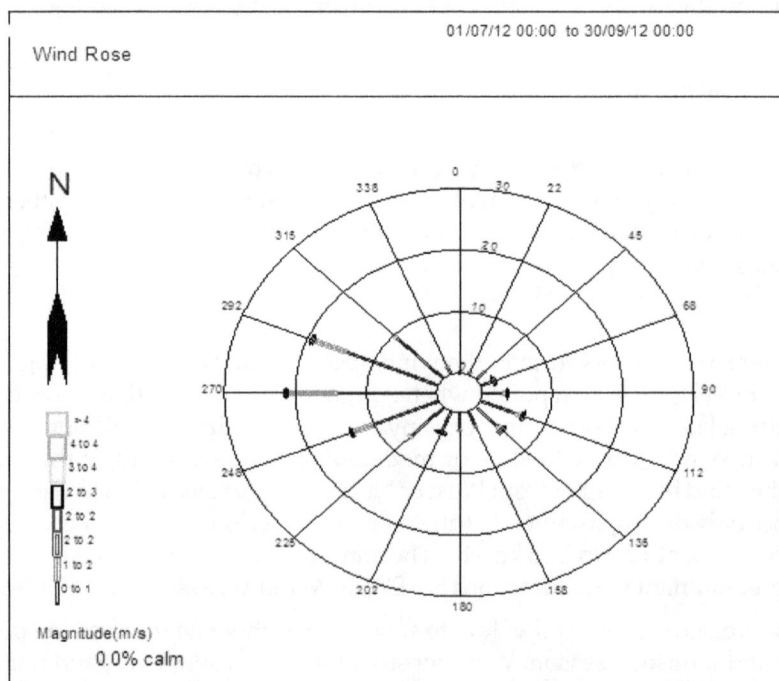

Figure 29.4: Wind Rose Diagram for the Month of July- September, 2012.

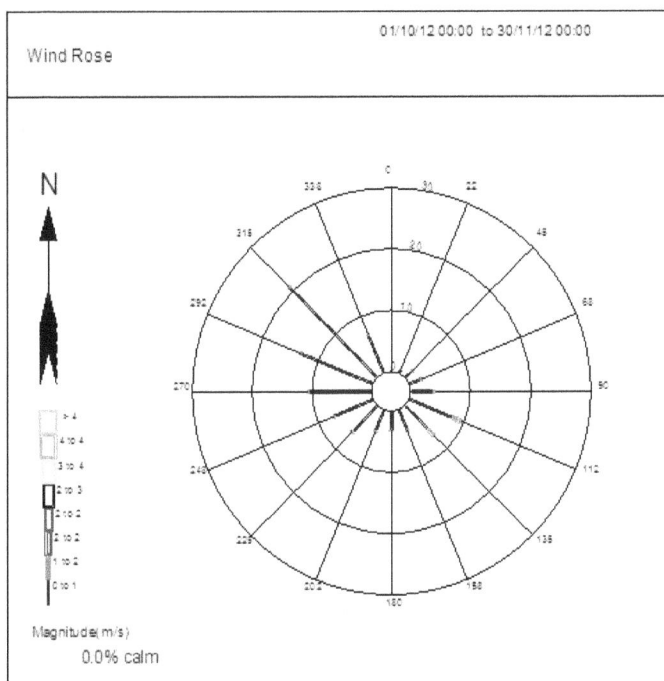

Figure 29.5: Wind Rose Diagram for the Months of October and November, 2012.

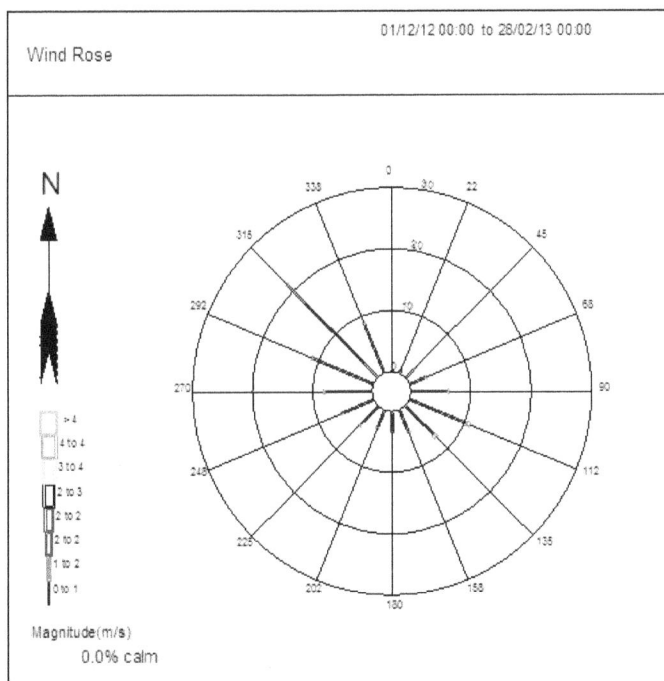

Figure 29.6: Wind Rose Diagram for the Months of December 2012 to January 2013.

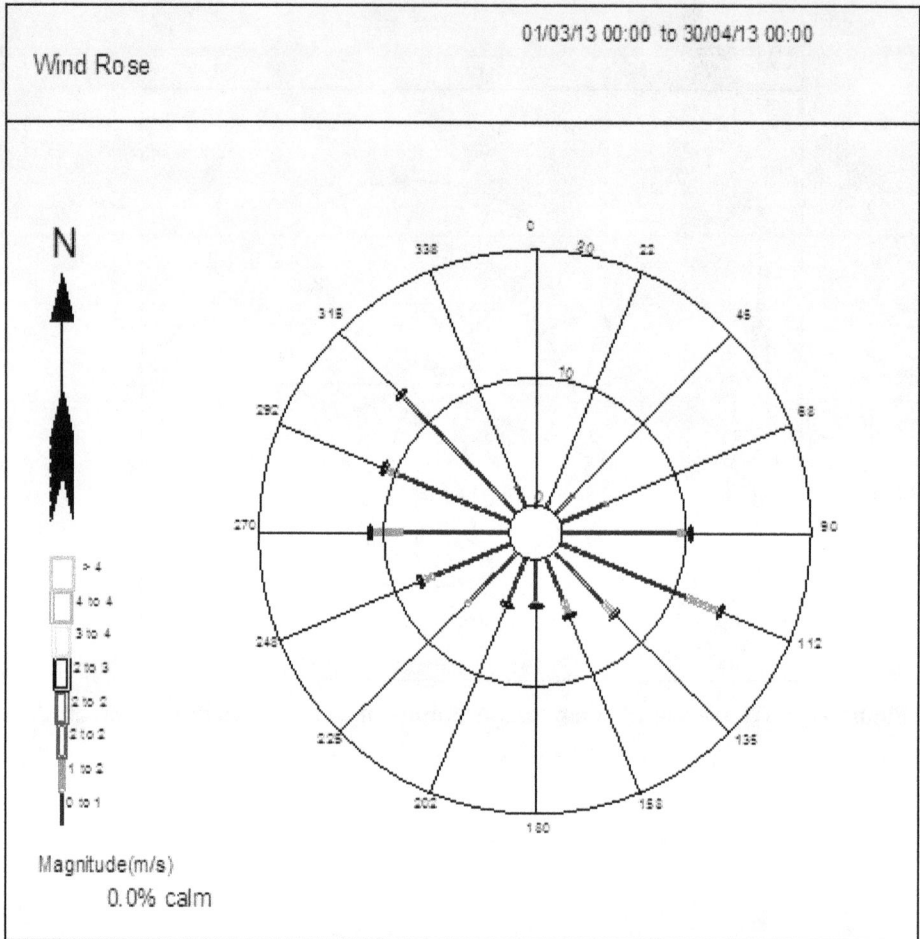

Figure 29.7: Wind Rose Diagram for the Months of February to April, 2013.

north. In the summer months, wind blows from variable directions. The mean annual wind speed during the study period was 6 km/hr.

Rainfall

The study area has monsoon type climate with rain fall predominantly during the months of June to September. Indian Meteorological Department (IMD) is having one observatory at Angul while the Govt. of Orissa maintains rain gauge stations at all block head quarters. The normal rainfall of the district is 1421.1mm. The rainfall has increased from south-western part to the north-eastern part of the study area.

Temperature

The cold season starts usually during November and continues till the end of February, December is the coldest month with mean daily maximum temperature at

26.7°C and the mean daily minimum temperature at 12.9°C. Both day and night temperature increases rapidly from March and by May the mean daily maximum temperature reaches to 39.9°C, while the mean minimum temperature is 26.3°C.

Humidity

The humidity in the study area is generally high in monsoon and post monsoon months and remains between 50 to 98 per cent. The relative humidity is comparatively low in summer months and ranges from 25 to 85 per cent.

In general the climate of the study area is of subtropical type, and is characterized by an oppressive hot summer, a mild winter and well-distributed rainfall during the south-western monsoon season.

Results and Discussions

Basudevpur

Sulphur dioxide gas is mostly produced by the combustion of fossil fuels. Sources include industrial activities such as flaring at oil and gas facilities and diesel power generation, commercial and home heating and vehicle emissions. In the present study, near Basudevpur concentration of SO_2 (Figure 29.8) was found to be highest in the month of January ($47\,\mu g/m^3$) and lowest ($2\,\mu g/m^3$) during the most of August. The average value is high in the February ($15\,\mu g/m^3$) whereas lowest in the month of August ($3\,\mu g/m^3$).

Nitrogen oxides are mostly formed when fuel is burned at high temperatures, as in a combustion process. NO_x represents the sum of the various nitrogen gases found in the air out of which, Nitric Oxide (NO) and Nitrogen Dioxide (NO_2) are the dominant forms. Near Basudevpur, the concentration of nitrogen oxides (Figure 29.9) was found to be highest during the month of February ($33\,\mu g/m^3$) and lowest ($4.5\,\mu g/m^3$) during

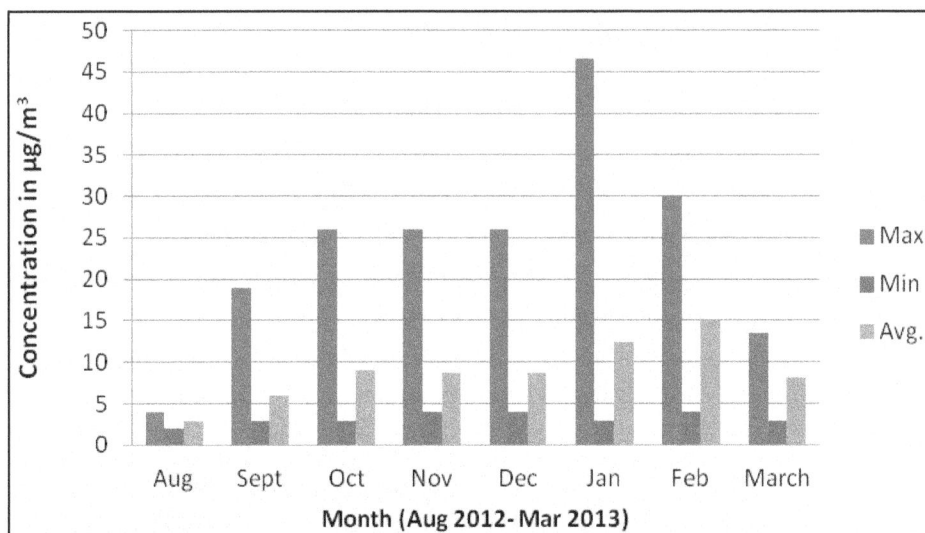

Figure 29.8: Concentration of SO_2 at Basudevpur.

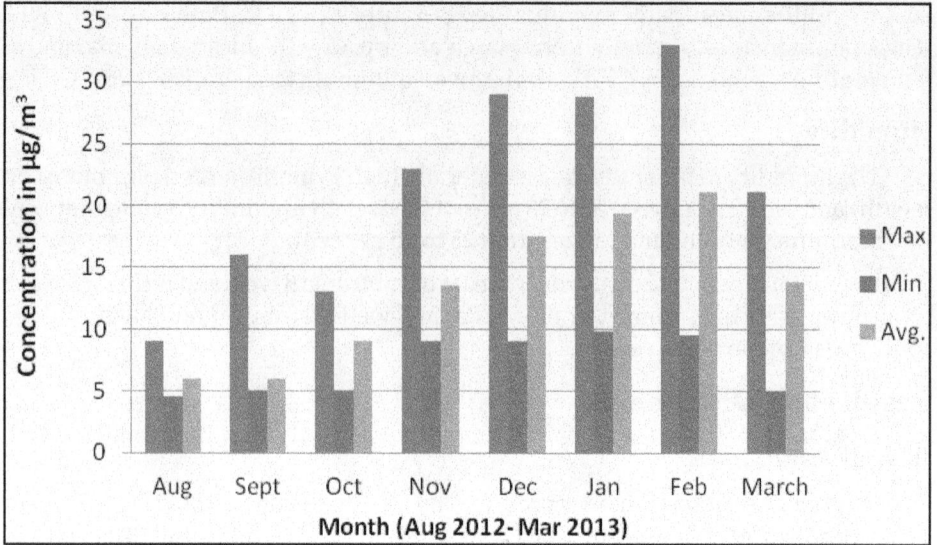

Figure 29.9: Concentration of NO_x at Basudevpur.

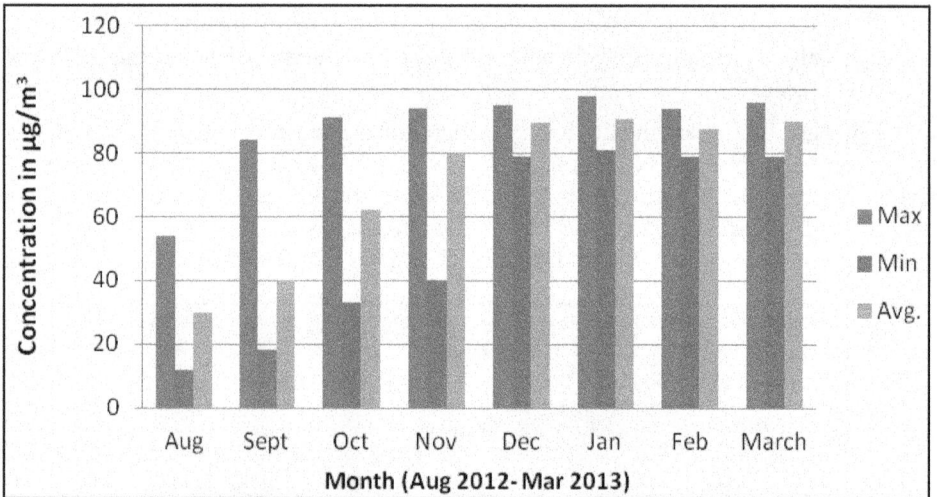

Figure 29.10: Concentration of PM_{10} at Basudevpur.

August. The average NO_x value was highest in the month of January (21 µg/m³) whereas lowest both during August and September (6 µg/m³).

Particulate Matter is the general term used for a heterogeneous mixture of solid particles and liquid droplets found in the air, including dust, dirt, soot, smoke and liquid droplets. It is a unique pollutant, reflecting the fact that it has both natural and anthropogenic sources. PM_{10} refers to the particulate matter of size 10 µm or less in diameter. At the Basudevpur, during the study period, concentration of PM_{10} (Figure 29.10) was found to be highest in the month of January (98 µg/m³) and lowest in the month of August (12 µg/m³). The average value was high during January (91 µg/m³) whereas lowest (30 µg/m³) in August.

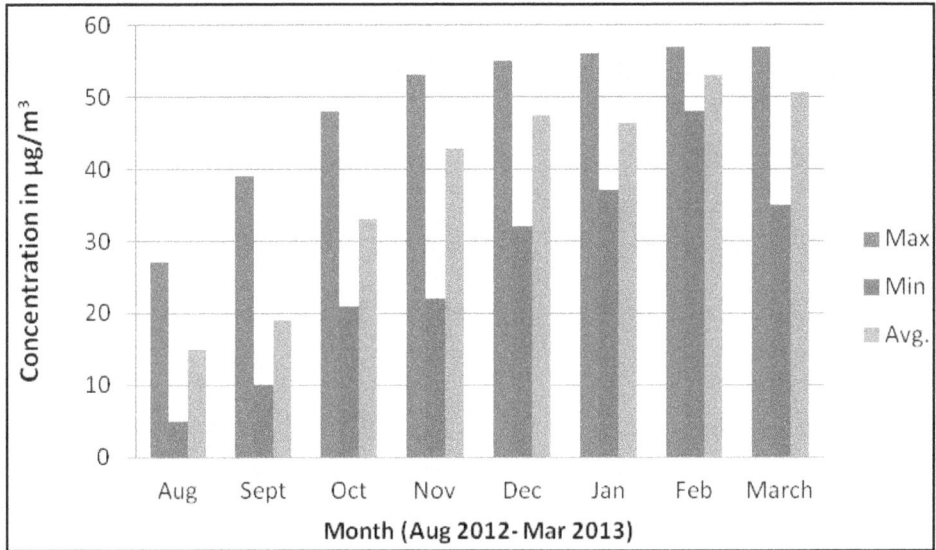

Figure 29.11: Concentration of PM$_{2.5}$ at Basudevpur.

PM$_{2.5}$ refers to the particulate matter that are of 2.5 μm or less in diameter. These are under the category of Respirable Particulate Matter. At Basudevpur, the concentration of PM$_{2.5}$ (Figure 29.11) was highest in the month of February and March *i.e.* 57 μg/m^3 and lowest in the month of August *i.e.* 5 μg/m^3. The average value is also high in February *i.e.* 53 μg/m^3 whereas lowest (15 μg/m^3) in the month of August.

Golabandha

At Golabandha, SO$_2$ was maximum in the month of January *i.e.* 36 μg/m^3 and minimum during August (2 μg/m^3). The average value was also high in January *i.e.* 13 μg/m^3 whereas lowest in both August and September (3 μg/m^3) (Figure 29.12).

NO$_x$ at the Golabandha was maximum in the month of February *i.e.* 35 μg/m^3 and minimum value was also in month October *i.e.* 4.5 μg/m^3. The average value is high in February *i.e.* 13.3 μg/m^3 whereas lowest in the month of October (4.5 μg/m^3) (Figure 29.13). The average NO$_x$ was maximum in February (18 μg/m^3) and minimum during August (6 μg/m^3).

Maximum concentration of PM$_{10}$ was during the month of December- March *i.e.* 96 μg/m^3 and minimum value for the month of September *i.e.* 20 μg/m^3. The average value was highest in March *i.e.* 92.3 μg/m^3 whereas lowest in September (39 μg/m^3) (Figure 29.14).

At the Golabandha, PM$_{2.5}$ was maximum in the month of February *i.e.* 59 μg/m^3 and minimum value for the month September *i.e.* 8 μg/m^3. The average value was highest in the February month *i.e.* 52 μg/m^3 whereas lowest (16 μg/m^3) in August (Figure 29.15).

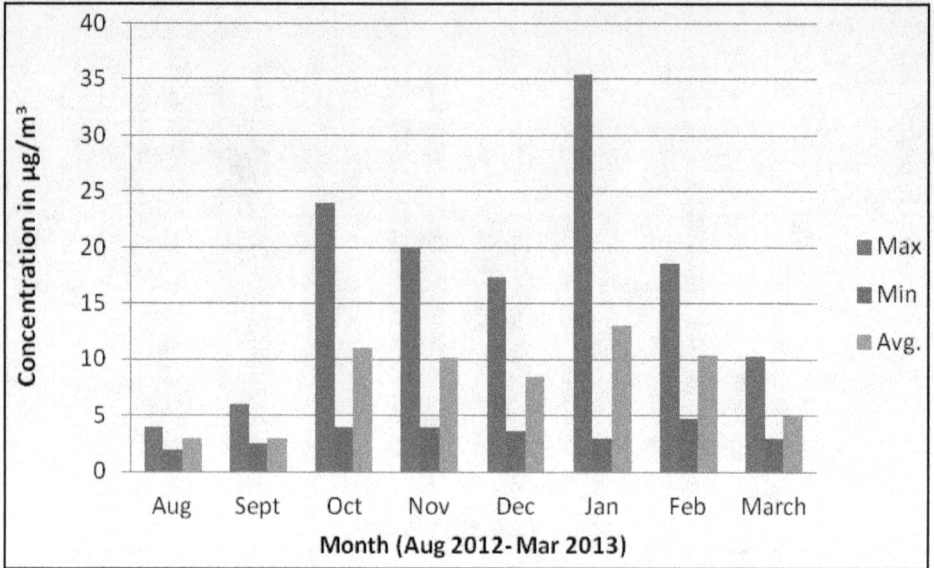

Figure 29.12: Concentration of SO$_2$ at Goalbandh.

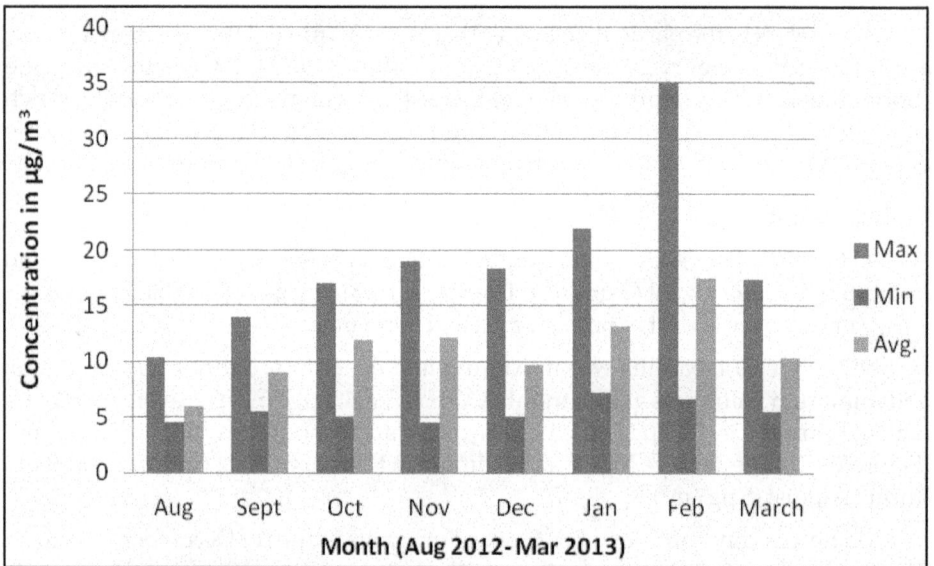

Figure 29.13: Concentration of NO$_x$ at Goalbandh.

Nisha

At Nisha, the maximum concentration of SO$_2$ was found to be 36 µg/m^3, which was in the month January and the minimum concentration was in August (2 µg/m^3). The average value was highest both in January and February *i.e.* 15 µg/m^3 whereas lowest (3 µg/m^3) in August (Figure 29.16).

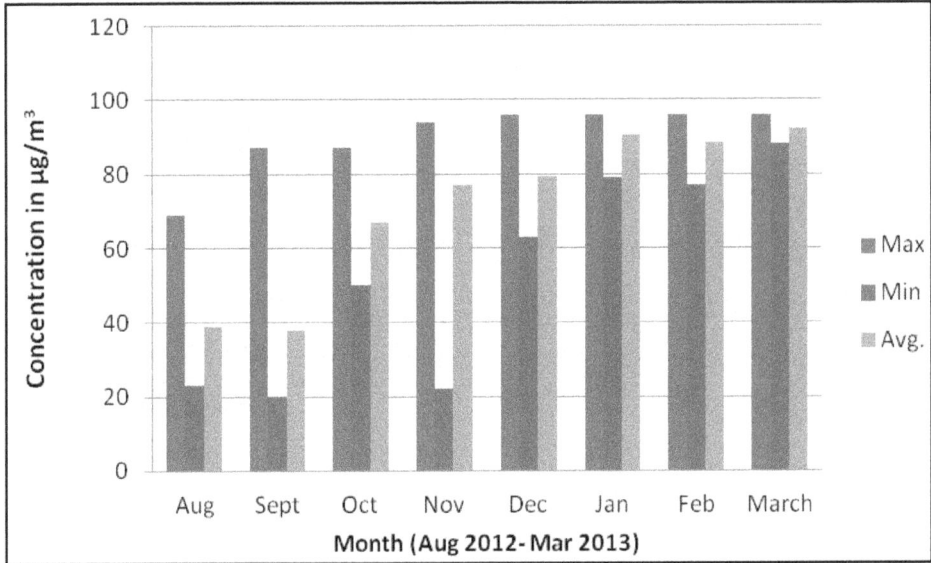

Figure 29.14: Concentration of PM$_{10}$ at Goalbandh.

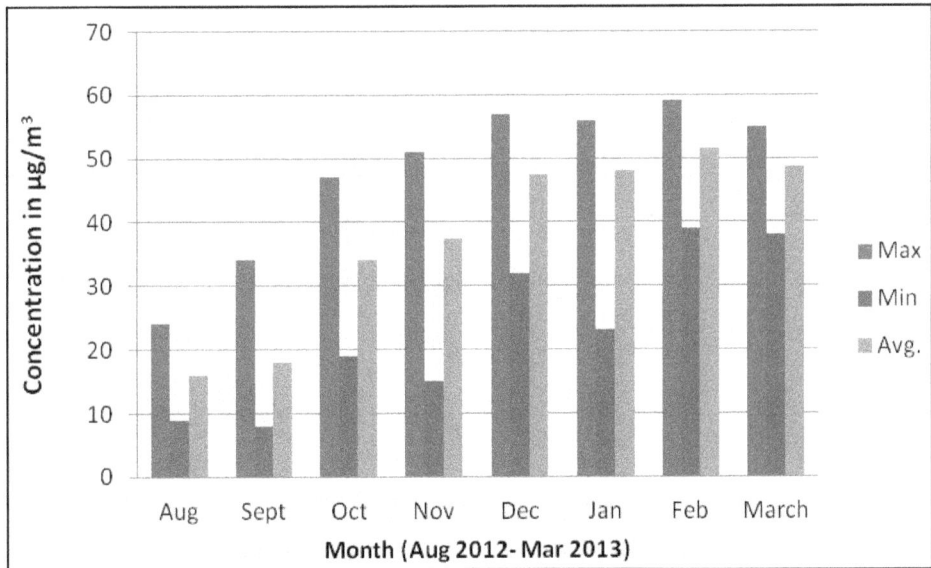

Figure 29.15: Concentration of PM$_{2.5}$ at Goalbandh.

Maximum concentration of NO$_x$ was found to be 51.6 µg/m^3, in the month January and the minimum value was in both August and September *i.e.* 4.5 µg/m^3. The average value was highest in February month *i.e.* 25 µg/m^3 whereas lowest in August and September (5 µg/m^3) (Figure 29.17).

Maximum PM$_{10}$ concentration in the study area was 99 µg/m^3, which was in the month January and minimum concentration in the month of August *i.e.* 11 µg/m^3.

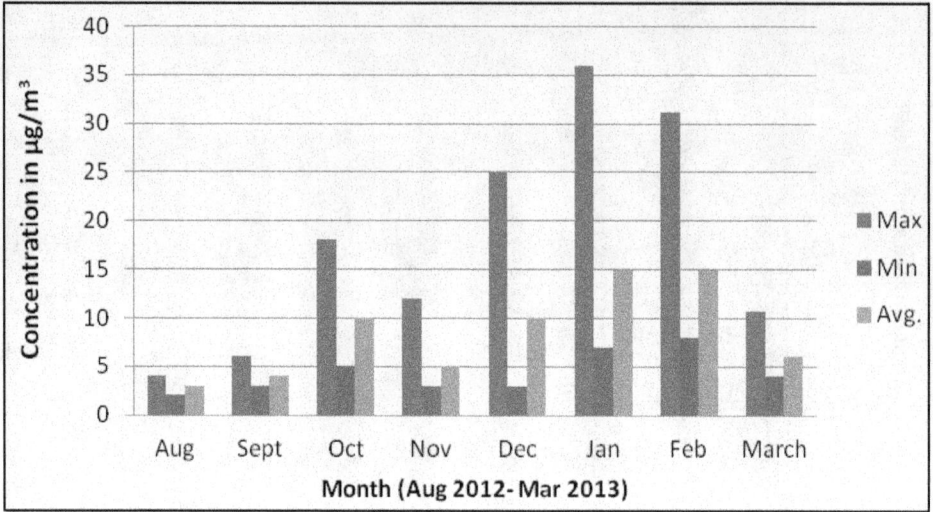

Figure 29.16: Concentration of SO_2 at Nisha.

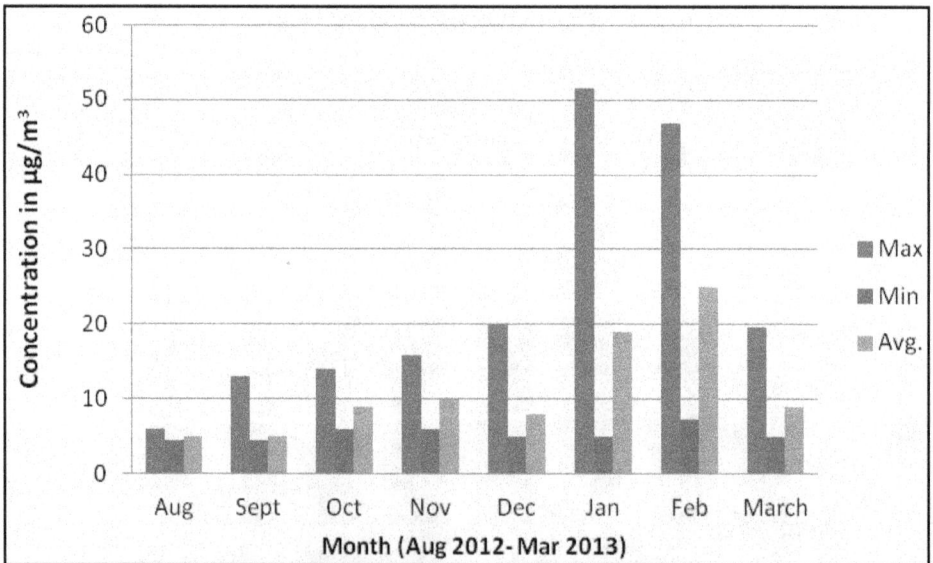

Figure 29.17: Concentration of NO_x at Nisha.

The average value was also high in January (87.55 µg/m³) whereas lowest (24 µg/m³) during August (Figure 29.18).

Here the maximum concentration of $PM_{2.5}$ was found to be 63 µg/m³, which was in the month of January and lowest in August *i.e.* 4 µg/m³. The average value was also highest in January *i.e.* 49.89 µg/m³ whereas lowest (13 µg/m³) in August (Figure 29.19).

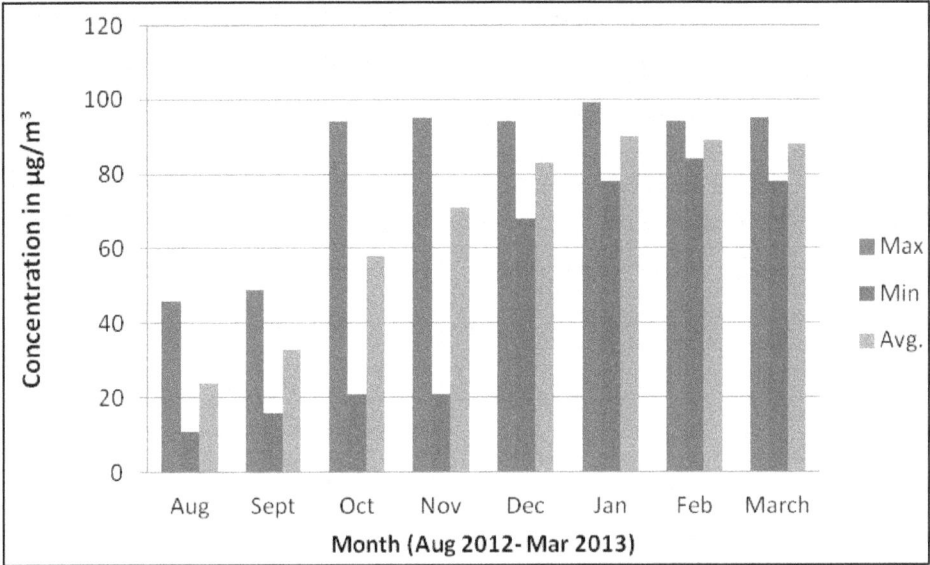

Figure 29.18: Concentration of PM$_{10}$ at Nisha.

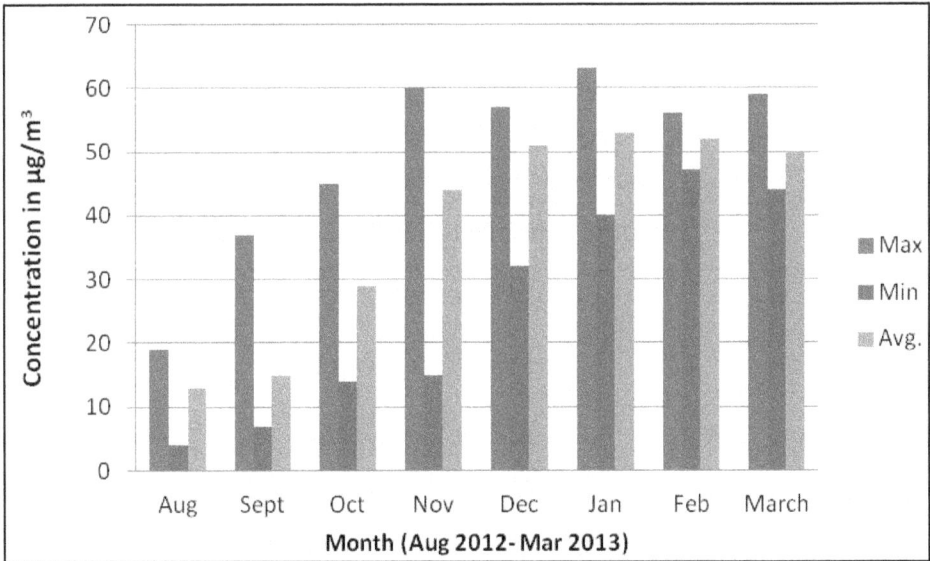

Figure 29.19: Concentration of PM$_{2.5}$ at Nisha.

Paranga

At Paranga monitoring station, the maximum concentration (69 µg/m^3) of SO$_2$ was found in the month of February and the minimum concentration *i.e.* 2 µg/m^3 during August. The average value is high in January month *i.e.*18 µg/m^3 whereas lowest (3 µg/m^3) in the month of August (Figure 29.20).

The maximum concentration of NO_x was found in the month of January *i.e.* 25 μg/m³ and the minimum concentration was during August 4.5 μg/m³. The average value was highest in the January month *i.e.* 13.5 μg/m³ whereas lowest in the month of August (5 μg/m³) (Figure 29.21).

Here, the maximum concentration of PM_{10} was found in the month of January and February which was 97 μg/m³ and the lowest concentration during the month of August which was 12 μg/m³. The average value was highest in both January and

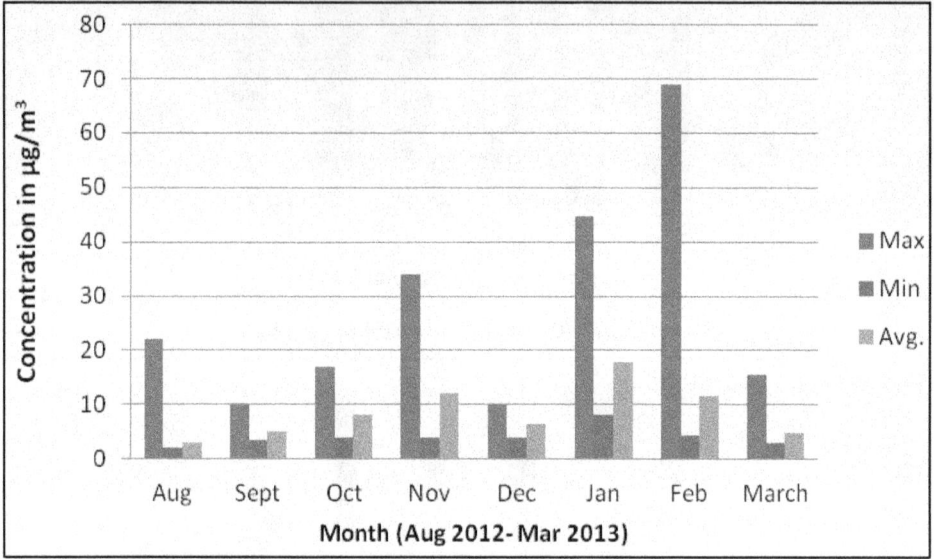

Figure 29.20: Concentration of SO_2 at Paranga.

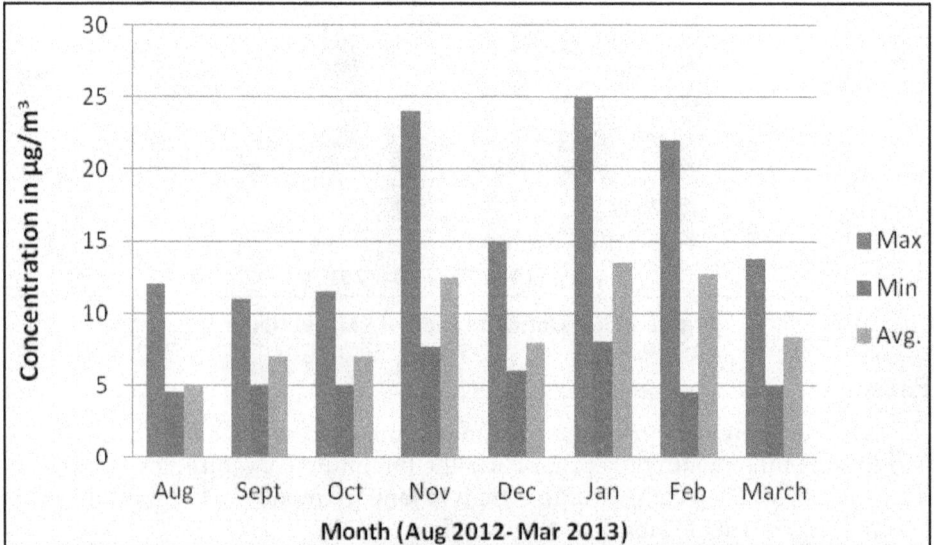

Figure 29.21: Concentration of NO_x at Paranga.

February months *i.e.* 89 µg/m³ whereas lowest (23 µg/m³) in the month of August (Figure 29.22).

Near Paranga, the maximum concentration of $PM_{2.5}$ was found in the month of January *i.e.* 59 µg/m³ and the lowest concentration for the month of August (5 µg/m³). The average highest $PM_{2.5}$ was recorded during January month *i.e.* 52 µg/m³ whereas lowest (11 µg/m³) in the month of August (Figure 29.23).

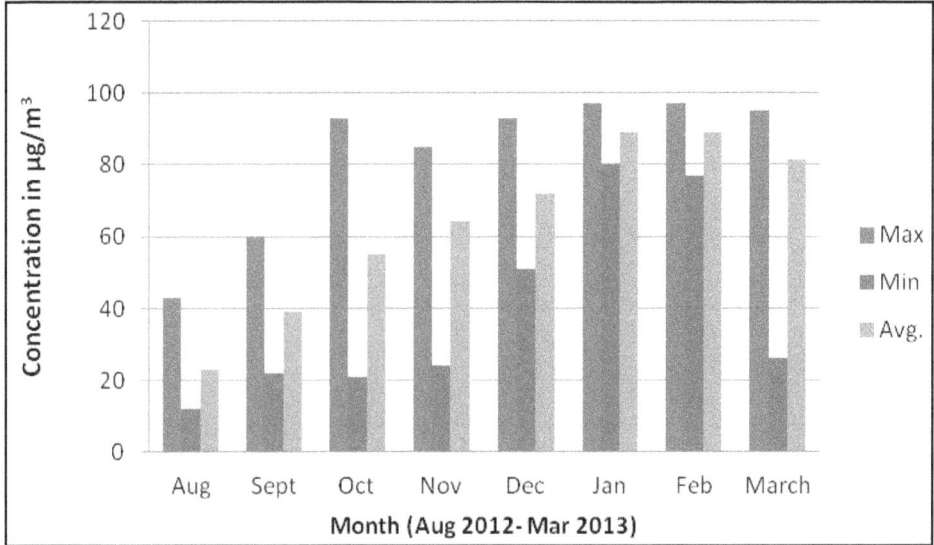

Figure 29.22: Concentration of PM₁₀ at Paranga.

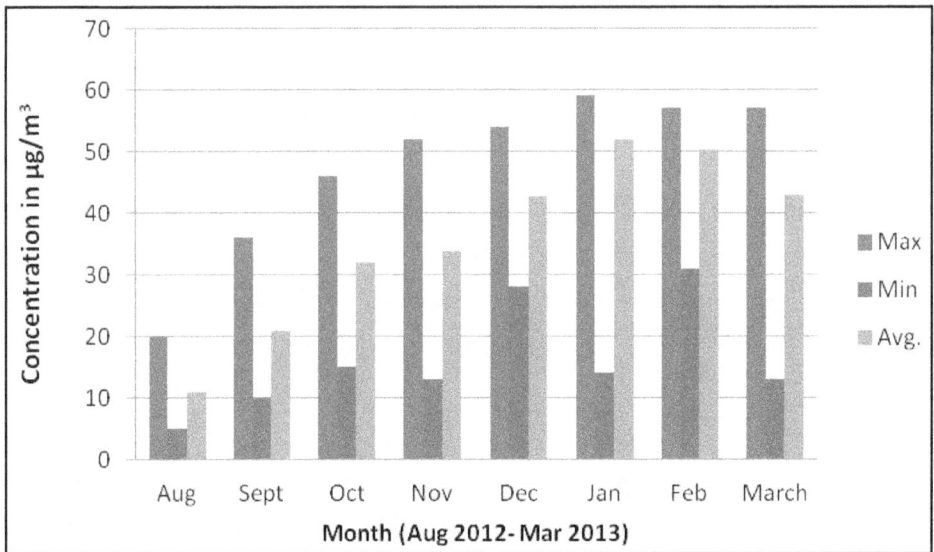

Figure 29.23: Concentration of PM₂.₅ at Paranga.

The ambient air quality standard as prescribed by the Central Pollution Control Board is given in table 29.2.

Table 29.2: AAQ Norms Prescribed by CPCB

Pollutants	Time Weighted Average	Industrial, Residential, Rural and Other Areas	Ecological Sensitive area (Notified by Central Govt.)
Sulphur Dioxide (SO$_2$),	Annual*	50 µg/m^3	20 µg/m^3
µg/m^3	24 hours**	80µg/m^3	80 µg/m^3
Nitrogen Dioxide (NO$_2$),	Annual*	40 µg/m^3	30 µg/m^3
µg/m^3	24 hours**	80 µg/m^3	80 µg/m^3
Particulate Matter Annual * 60 60–Gravimetric (Size less than 10µm) 24 Hours ** 100 100–TEOM	Annual*	60 µg/m^3	60 µg/m^3
or PM$_{10}$, µg/m^3	24 hours**	100 µg/m^3	100 µg/m^3
Particulate Matter Annual * 40 40 –Gravimetric (Size less than 2.5µm) 24 Hours ** 60 60–TEOM	Annual*	40 µg/m^3	40 µg/m^3
or PM$_{2.5}$, µg/m^3	24 hours**	60 µg/m^3	60 µg/m^3

*: Annual Arithmetic mean of minimum 104 measurements in a year at a particular site taken twice a week 24 hourly at uniform intervals.

**: 24 hourly or 8 hourly or 1 hourly monitored values, as applicable, shall be complied with 98 per cent of the time in a year. 2 per cent of the time, they may exceed the limits but not on two consecutive days of monitoring.

Air Quality Index (AQI) Calculation

The air pollution index for the four ambient air parameters were calculated as per the guideline of WHO and Central Pollution Control Board. The following computation was used to calculate the air quality index of the sites under consideration.

$$AQI = \tfrac{1}{4} \times (I_{PM2.5}/S_{PM2.5} + I_{PM10}/S_{PM10} + I_{SO2}/S_{SO2} + I_{NOX}/S_{NOX}) \times 100$$

where,

$I_{PM2.5}, I_{PM10}, I_{SO2}$ and I_{NOX} = Individual values of PM$_{2.5}$, PM$_{10}$, sulphur dioxide and oxides of nitrogen respectively.

$S_{PM2.5}, S_{PM10}, S_{SO2}$ and S_{NOX} = Standards of ambient air quality.

The AQI gives detailed about the cleanliness of ambient air. EPA has set National Ambient Air Quality Standard (NAAQS) to protect against harmful health effects. The higher the AQI value, greater is the level of air pollution and greater damage to Health.

The AQI scale is divided in to five categories. Each category describes the range of air quality and its associated potential health effects. The five level of AQI are described in the Table 29.3 and the AAQI of the study area during the study period is given in Table 29.4 and Figure 29.24.

Table 29.3: Index Value for Air Quality Index Calculation

Index Value	Remarks
0-25	Clean air
26-50	Light Air pollution
51-75	Moderate Air Pollution
76-100	Heavy Air pollution
Above 100	Severe air Pollution

Table 29.4: Ambient Air Quality Index of Monthly Average Value ($\mu g/m^3$) of the Study Area

Month	Location			
	Basudevpur	Golbandh	Nisha	Paranga
Aug	17	19	14	13
Sept	22	21	17	22
Oct	35	38	33	32
Nov	45	42	41	38
Dec	50	45	48	40
Jan	52	51	55	54
Feb	55	52	56	51
March	50	48	48	42

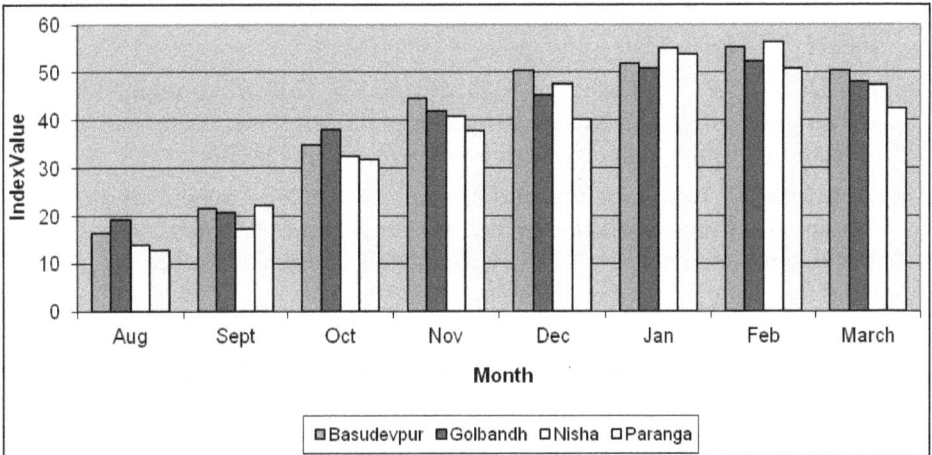

Figure 29.24: Ambient Air Quality Index ($\mu g/m^3$) of the Study Area.

The ambient air monitoring results of all four locations are within the stipulated standard of Central Pollution Control Board. However, the overall AQI result shows that, the study area is coming under the light pollution category. This might be due to the contribution of pollution load from the industrial and domestic activities and movement of vehicle in the nearby areas. In order to maintain the ambient air quality in the area, proper installation of pollution control devices in the industries, creation of green belt in the locality, conversion of earthen roads into concrete roads, proper maintenance of vehicles, regular sprinkling of water particularly in the industrial areas can be undertaken.

Conclusion

In India, outdoor air pollution is restricted mostly to urban areas because of automobiles and concentration of industries and thermal power plants. The ambient air quality monitoring network involves the measurement of a number of air pollutants at number of locations. From the above results it was found that, all the four monitoring stations are coming under light air pollution category. Measures for cleaning of flue gases and NO_x removal methods should be under taken in such a manner as to maintain the concentration of the gaseous pollutants like SO_2 and NO_x of the area. Regular sprinkling of water is required to control the fugitive emissions of the area. Development of green belts, wherever feasible around the pollution causing sources as well as the residential locations can result in air pollution control and management of the area.

References

Alias, M., Hamzah, Z. and Kenn, L. S. (2007): PM$_{10}$ and Total suspended particulates (TSP) measurements in various power stations. The Malayasian Journal of Analytical Sciences, 11(1): 255-261.

Balashanumugam, P., Ramanathan, A.R., Neheukumar, V. and Elango, E. (2012): Ambient air quality studies on Cuddalore. International journal of environmental science, 2(3): 1302- 1313.

Garg, A., Shukla, P.R., Bhattacharya, S., Dadhwal, V.K. (2001): Sub-region (district) and sector level SO$_2$ and NO$_x$ emissions for India: assessment of inventories and mitigation flexibility, Atmospheric Environment, 35: 703-713.

Giri, D., Murthi, K., Adhikary, P.R. and Khanal, S.N. (2006): Ambient air quality of Kathamandu valley as reflected by atmospheric particulacte matter concentrations (PM$_{10}$), International Journal of Environmental Science and Technology, 4: 413-410.

Goyal, P., Sidhartha, S. (2003): Present scenario of air quality in Delhi: a case study of CNG implementation. Atmospheric Environment, 37: 5423-5431.

Kaushik, K. S., Khare, M., Gupta, K. B. (2007): Suspended Particulate Matter Distribution in Rural- Industrial Satna and in Urban Industrial South Delhi, Environ Monitoring Assessment, 128: 431-445.

Lima, L. L., Hughesb, S. J., Hellawellb, E. E. (2005): Integrated decision support system for urban air quality assessment, Environmental Modeling and Software, 20: 947-954.

Sathe, Y. V. (2012): Air quality modeling in street canyons of Kollapur city, Maharashtra, India. Universal journal of Environmental research and Technology, 2(2): 97-105.

Sfetsos, A., Zoras, S., Bartzis, J.G., Triantafullou, A.G. (2006): Extreme value modeling of daily PM10 concentrations in an industrial area. Fresenius Environmental Bulletin, 15: 841-845.

Sharma, P., Chandra, A., Kaushik, S.C. and Jain, S. (2012): Predicting violations of national ambient air quality standards using extreme value theory for Delhi city. Atmospheric pollution research, 3: 170-179.

Silibello, C., Calori, G., Brusasca, G., Giudici, A., Angelino, E., Fossati, G., Peroni, E., Buganza, E. (2008): Modeling of PM_{10} concentrations over Milano urban area using two aerosol modules, Environmental Modeling and Software, 23: 333-343.

Simkhada, K., Murthy, K. and Khanal, S. N. (2005): Assessment of ambient air quality in Bishnumati corridor, Kathmandu metropolis. International Journal of Environmental Science and Technology, 2(3): 217-222.

Vedamadavan, V and Sarithabanbanuraman, S. (2012): Assessment of ambient air quality in Coimbatore city. Civil and environmental research, 2(1): 1-7.

Chapter 30

Assessment of Post Mining Water Quality of Pit Lakes in Sukinda Valley of Jajpur District, Odisha, India

Subhashree Pattanaik

*Department of Environmental Science and Engineering,
Indian School of Mines, Dhanbad – 826 004, Jharkhand, India*

ABSTRACT

The opencast chromite mining operation at Sukinda valley of Jajpur district, Odisha will cease by 2030 due to unavailability of ore within cost effective depth which would lead to the formation of number of residual pit lakes in the area. These pit lakes would form the majority of landscape which will affect the local climate, flora and fauna of the region. In order to know the potential use of the pit lake water and to implement the remediation if any required assessment of water quality of the pit lakes was taken up. Protection and conservation of groundwater through a regulatory mechanism is necessary against imposition of financial guarantee and other penal provision.

Keywords: Residual pit lake, Water quality, Remediation, Landscape.

Introduction

Sukinda valley in Jajpur District, Odisha, is well known for its chromite ore deposits producing nearly 98 per cent of chromite ore in India (Dhakate *et al.*, 2008). Sukinda ultramac complex is bounded in the north by Daitari Hill range and in the south by the Mahagiri Hill range, mostly composed of quartzites (Acharya *et al.*, 1998). Opencast Mining for chromium in Sukinda valley commenced since 1950.

Total 20 mining leases have been granted for chromite ore mining in Sukinda Ultramafic Belt (SUB) in Jajpur and Dhenkanal district, Odisha (www.orissaminerals.gov.in). Out of which 13 are operating (two are underground) and 7 are not in operation due to want of statutory clearances. The open cast mining is the most suitable method for mining confined to Band No. I to IV where weathered limonite is the host rock. Occurrence of groundwater at shallow depth and sub-vertical dipping chromite ore body (Banerjee, 1971) makes the open pit mining very difficult due to seepage of large quantities of groundwater and excavation of huge limonite overburden. This results in formation of large open pits (Pattanaik *et al.*, 2012) in which continuous pumping is carried out to facilitate mining operation. During conceptual period and after closure of the mines the seepage water as well as the runoff water will start accumulating at the quarry bottom leading to formation of large water bodies. The exploitation of the ore is carried out through open cast mining method since last few decades and is expected to continue till 2030 or so.

Pit lake will become an important part of post mining landscape. About 32 pit lakes exist in Sukinda chromite mining area in 12 miming leases. Few of them shall merge during conceptual period. Their water quality is mainly inuenced by the mobility of hexavalent chromium leachate. Lakes have particular ecological and socio-economic functions as habitat for future aquatic organisms, sites for shery and recreation, etc. Articial lakes and reservoirs should fulfill those functions as far as possible. The total surface area granted on mining lease basis in Jajpur district comes to 5305.537 ha. This is roughly the mineralized area of Sukinda valley. The possible surface areas already broken up and to be broken up shall more or less be confined to 30 open pit lakes after some of the quarries getting merged to cover an area of about 550 ha which is around 18 per cent of the effectively working lease area in the valley. This is likely to increase up to around 26 per cent during conceptual stage. The average depth of individual pit lakes shall not be less than 45mtr depth from ground level. Due to their total surface area of around 550 ha today, the sustainable use of the pit lakes is an important ecological and socio-economical factor.

Materials and Methods

Assessment of water quality has been done according to the APHA 1992, Standard methods for the examination of water and wastewater. Samples were collected from the pit lake. Few parameters were analyzed at the site and for analysis of other parameters the samples were carried to the laboratory with all precaution to avoid biased result. Basic information regarding the chromite mining lease holds area of all the working and temporarily discontinued leases were taken from www.orissaminerals.gov.in and the surface openings by open cast quarrying was calculated from the image available in Google earth. The total area covered by waste dump which are partly stabilized and partly under afforestation is computed from the image. The data collected from the grant order of Environment Clearance report of the respective mines displayed in the Govt. of Odisha portal is highly informative to estimate the total surface area of the prevailing pit lakes as well as future pit lakes proposed during conceptual period. Figure 30.1 shows the plan and section of the deepest pit lake in Sukinda Valley.

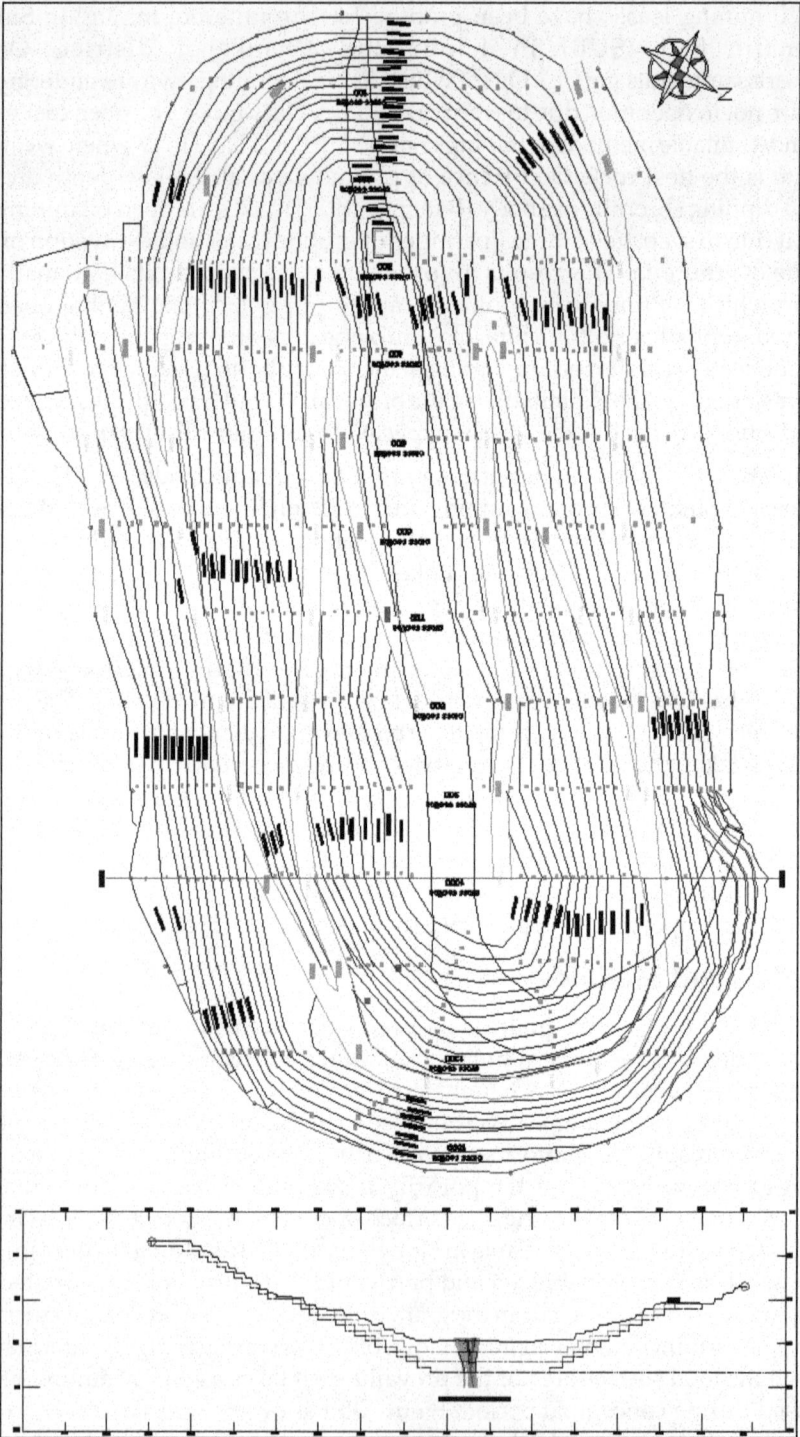

Figure 30.1: Plan and Section of the Deepest Pit Lake in Sukinda Valley.

Table 30.1: List of Mining Leases in Jajpur and Dhenkanal District of Odisha.

Name of the Lessee	Name of the Lease	Area in ha	Working/ Non-working	Area of Future Pit Lake (ha)	Percentage
M/s TISCO Ltd	Sukinda Chromite Mines	406.00	Working	550.00	18 per cent
M/s Misrilal Mines (P) Ltd	Saruabil Chromite Mines	246.858	Working		
M/s B.C.Mohanty and Sons (P) Ltd	Kamarda Chromite Mines	107.240	Working		
M/s O.M.C.Ltd.	Kaliapani Chromite Mines	971.245	Suspended		
M/s O.M.C Ltd.	Sukurangi Chromite Mines	382.709	Working		
M/s IMFA Ltd.	Chingudipal Chromite Mines	26.620	Discontinued		
M/s FACOR Ltd	Ostapal Chromite Mines	72.843	Working		
M/s IMFA Ltd.	Kaliapani Chromite Mines	116.760	Working		
M/s Balasore Alloys Ltd.	Kaliapani Chromite Mines	64.463	Working		
M/s Jindal Stainless Ltd.	Kaliapani Chromite Mines	89.00	Working		
M/s IDCOL	Talangi Chromite Mines	65.683	Working		
M/s I.C.C.L	Mahagiri Chromite Mines	73.777	Working		
M/s O.M.C Ltd.	S-Kaliapani Chromite Mines	552.457	Working		
M/s O.M.C.Ltd.	Kalarangi Chromite Mines	936.220	Discontinued		
M/s O.M.C.Ltd.	Balipada Chromite Mines	185.810	Discontinued		
M/s O.M.C.Ltd.	Kamarda Saruabil Chr Mines	23.243	Discontinued		
M/s FACOR Ltd.	Bhimtangar Chromite Mines	23.800	Working		
M/s FACOR Ltd	Kathpal Chromite Mines	113.312	Working		
M/s O.M.C Ltd.	Kathpal Chromite Mines	264.466	Discontinued		
M/s O.M.C Ltd.	Birasal Chromite Mines	583.021	Discontinued		
		5305.54			

Results and Discussions

Chromite is a strategic mineral and its extraction is inevitable irrespective of the fact that it pollutes the water environment of Sukinda valley. Sukinda contributes 97 per cent of the India's total requirement of chrome ore.

In order to conserve the groundwater which is as precious as chromite, it is the prime responsibility of all the stake holders to handle it in a sustainable manner so that the younger generation does not curse. The present economics of extraction of chromite should also cater the post mine closure effect of the region and hence the present concept of assessment of pit lake water existing and to happen in future for a balanced lake ecosystem.

The physico-chemical condition of hexavalent chromium generation and contamination of the water environment is a sensitive and very important issue due to its carcinogenetic effect beyond permissible limit. It is an attempt to increase awareness and induce regulation so that the pit lake is maintained as portable water for its use by the local inhabitants without any adverse impact on the biotic community of the region.

List of mining leases at Jajpur and Dhenkanal is given in Table 30.1. The water quality of a particular pit lake observed during four seasons in a year is given as below (Table 30.2).

Table 30.2: Water Quality of a Pit Lake in Sukinda Valley

Sl.No.	Parameter	Tolerance Limit	20° 57' 17' N - 85° 40' 22' E			
			Post Monsoon (Oct-Dec'10)	Winter (Jan-Mar'11)	Summer (Apr-Jun'11)	Monsoon (July-Sept'11)
1.	Colour	–	Colourless	Colourless	Colourless	Brown
2.	Odour	Unobjectionable	Unobjectionable	Unobjectionable	Unobjectionable	Unobjectionable
3.	Suspended solids, mg/l	100	18	13	10	75
4.	Particle size S.S.	Shall pass in 850 Micron	100% passed	100% passed	100% passed	100% passed
5.	Total dissolved Solids mg/l, Max.	2100	108	453	985	1120
6.	pH Value	5.5-9.0	7.8	8.0	7.5	7.8
7.	Oil and Grease mg/l	10	<0.01	<0.01	<0.01	<0.01
8.	Total residual chlorine (Cl), mg/l	1.0	Nil	Nil	Nil	Nil
9.	Total kjeldahl nitrogen (N), mg/l	100	3.2	3.4	3.6	3.7
10.	Ammoniacal nitrogen (N), mg/l	50	4.6	4.8	4.9	6.4
11.	Free ammonia (NH_3), mg/l	5.0	<0.05	<0.05	<0.05	<0.05
12.	B.O.D. (O_2), mg/l	30.0	9.36	10.3	12.7	15.5
13.	C.O.D. (O_2), mg/l	250	13.42	14.7	15.8	18.2

Contd...

Table 30.0–*Contd...*

Sl.No.	Parameter	Tolerance Limit	20° 57' 17" N - 85° 40' 22" E			
			Post Monsoon (Oct-Dec'10)	Winter (Jan-Mar'11)	Summer (Apr-Jun'11)	Monsoon (July-Sept'11)
14.	Arsenic (as As), mg/l, Max.	0.2	<0.001	<0.001	<0.001	<0.001
15.	Mercury (as Hg), mg/l, Max.	0.01	<0.001	<0.001	<0.001	<0.001
16.	Lead (as Pb), mg/l, Max.	0.1	<0.01	<0.01	<0.01	<0.01
17.	Cadmium (as Cd), mg/l, Max.	2.0	<0.005	<0.005	<0.005	<0.005
18.	Chromium (as Cr^{6+}), mg/l, Max.	0.1	0.638	0.84	0.735	0.84
19.	Total chromium (Cr),mg/l	2.0	0.874	1.02	0.98	1.85
20.	Copper (as Cu), mg/l, Max.	3.0	<0.01	<0.01	<0.01	<0.01
21.	Zinc (as Zn), mg/l, Max.	5.0	<0.01	<0.01	<0.01	<0.01
22.	Selenium (as Se), mg/l, Max.	0.05	<0.005	<0.005	<0.005	<0.005
23.	Nickel (Ni),mg/l	3.0	<0.1	<0.1	<0.1	<0.1
24.	Boron (as B) mg/l, max.	2.0	<0.1	<0.1	<0.1	<0.1
25.	Cyanide (as CN), mg/l, Max.	0.2	<0.01	<0.01	<0.01	<0.01
26.	Chloride (as Cl), mg/l, Max.	1000	8.0	6.7	6.7	7.8
27.	Fluoride (as F) mg/L, Max.	2.0	<0.01	<0.01	<0.01	<0.01
28.	Dissolved phosphate (P), mg/l	5.0	<0.1	<0.1	<0.1	<0.1
29.	Sulphate (SO_4), mg/l	1000	4.7	2.4	2.1	8.5
30.	Sulphide (S),mg/l	2.0	<0.01	<0.001	<0.001	<0.001
31.	Phenolic Compounds (as C_6H_5OH), mg/l Max.	1.0	<0.1	<0.01	<0.01	<0.01
32.	Manganese (as Mn), mg/l, Max.	2.0	<0.01	<0.01	<0.01	<0.01
33.	Vanadium (V), mg/l	0.2	<0.1	<0.1	<0.1	<0.1
34.	Nitrate (as NO_3), mg/l, Max.	10	1.63	1.83	1.85	2.20
35.	Iron (as Fe), mg/l, Max.	3.0	0.32	0.20	0.28	0.28

Abide by the regulations, huge overburden dumps generated during extraction of chromite ore are being stacked adjacent to the quarry due to paucity of space are susceptible for erosion by surface run off thereby carrying hexavalent chromium in to these pit lakes as original drainage channels ought to be disturbed by mining. This is due to the fact that the strike of the ore body is at a right angle to the 1st and 2nd order drainage system of the valley discharging to Damsal Nala ultimately flowing in south westerly direction meeting Brahmani River downstream. The waste dumps occupy almost 16 per cent of the total lease area. The waste dumps are placed almost parallel to the quarries for which the seasonal streams get vanished and re-appear as seepage water in the quarry. The leachate of the waste dump in the form of hexavalent chromium in the seepage water is therefore more in the quarries. Water samples

Figure 30.2: Waste Dump.

Figure 30.3: Pit Lake.

Figure 30.4: Disused Quarry.

collected from existing Pit Lake show high Cr^{+6} values. All the 1st to 3rd order seasonal drainage channels emanating from Mahagiri Hill will either discharge to these deep pits instead of Damsal Nala or get merged at the toe of the dump.

Conclusion

Water samples from some of the disused quarry indicate that the accumulated water is high in hexavalent chromium content. It is not suitable for human consumption. The author has tried to emphasize that such practice of abandonment of chromite quarries will lead to heavy environmental loss to the entire groundwater of the valley. Through the present paper it is indicated that during post mine closure utmost care is required to be taken to see that the water accumulation after mining in the pit is made free from hexavalent chromium. Just leaving the quarry as a lake does not end the mining process. The existing scenario in one or two pit lakes will replicate in future in the adjoining pit lakes contaminating the groundwater regime of the whole of Sukinda mining belt.

Stabilization of the waste dumps and channelizing the surface run off through suitable drain followed by suitable treatment method will minimize the problem. Amendment in the regulation of mine closure and reclamation of the open pits by backfilling the waste dump needs to be deliberated. The remediation by biological methods (Pattanaik *et al.*, 2012) is environmental friendly and can be tested in pilot studies. A well defined mechanism should be developed through a corporate social responsibility so that monitoring the quality of pit lake water during post mine closure is taken care of.

References

Acharya, S., Mahalik, N.K. and Mahapatra, S. (1998): Geology and mineral resources of Orissa, Society of Geo-scientists and allied Technology.

APHA (1992): Standard methods for the examination of water and wastewater. 18th American Public Health Association, American Water Works Association, Water Environmental Federation. Washington, D.C. 981.

Banerjee, P.K. (1971): The Sukinda Chromite field, Cuttack district, Orissa, Records of Geological Survey of India, 96:140-171.

Dhakate, R. and Singh, V.S. (2008): Heavy metal contamination in groundwater due to mining activities in Sukinda valley, Orissa-A case Study.

Pattanaik, S., Pattanaik, D.K., Das, M. and Panda, R.B. (2012): Environmental scenario of chromite ore mining at Sukinda valley beyond 2030, Discovery Sci, 1(2): 35-39.

Pattanaik, D.K., Pattanaik, S. and Panda R.B. (2012): Reduction of hexavalent chromium by *Spirogyra species*, JE (2012), 1(3): 100-104.

Index

www.ingramcontent.com/pod-product-compliance
Lightning Source LLC
Chambersburg PA
CBHW021438180326
41458CB00001B/327